The advent of semiconductor structures whose characteristic dimensions are smaller than the mean free path of carriers has led to the development of novel devices and given rise to many advances in our theoretical understanding of these mesoscopic systems or nanostructures. This book reviews the results of experimental research into mesoscopic devices and develops a detailed theoretical framework for understanding their behavior.

The authors begin by discussing the key observable phenomena in nanostructures, including phase interference and weak localization. They then describe quantum confined systems, transmission in nanostructures, quantum dots, and single electron phenomena. Separate chapters are devoted to interference in diffusive transport and temperature decay of fluctuations, and the book concludes with a chapter on non-equilibrium transport and nanodevices. Throughout, the authors interweave experimental results with the appropriate theoretical formalism.

The book will be of great interest to graduate students taking courses in mesoscopic physics or nanoelectronics, as well as to anyone working on semiconductor nanostructures or the development of new ultrasmall devices.

**Cambridge Studies in Semiconductor Physics
and Microelectronic Engineering: 6**

EDITED BY

Haroon Ahmed
Cavendish Laboratory, University of Cambridge

Michael Pepper
Cavendish Laboratory, University of Cambridge

Alec Broers
Department of Engineering, University of Cambridge

TRANSPORT IN NANOSTRUCTURES

TITLES IN THIS SERIES

TRANSPORT IN NANOSTRUCTURES

DAVID K. FERRY

Regents' Professor of Electrical Engineering
Arizona State University

and

STEPHEN M. GOODNICK

Professor and Chair of Electrical Engineering
Arizona State University

PUBLISHED BY THE PRESS SYNDICATE OF THE UNIVERSITY OF CAMBRIDGE
The Pitt Building, Trumpington Street, Cambridge CB2 1RP, United Kingdom

CAMBRIDGE UNIVERSITY PRESS
The Edinburgh Building, Cambridge CB2 2RU, United Kingdom
40 West 20th Street, New York, NY 10011-4211, USA
10 Stamford Road, Oakleigh, Melbourne 3166, Australia

First published 1997

Printed in the United States of America

Typeset in Times Roman

Library of Congress Cataloging-in-Publication Data

Ferry, David K.
Transport in nanostructures / David K. Ferry, Stephen Marshall Goodnick.
p.   cm. – (Cambridge studies in semiconductor physics and
microelectronic engineering ; 6)
ISBN 0-521-46141-3
1. Nanostructures.   2. Mesoscopic phenomena (Physics).   3. Solid
state electronics.   I. Goodnick, Stephen Marshall, 1955–   .
II.   Title   III.   Series.
QC176.8.N35F47
537.6′22 – dc21   1997   96-36825
                                                              CIP
A catalog record for this book is available from the British Library

ISBN 0 521 46141 3 hardback

Contents

Preface

This book has grown out of our somewhat disorganized attempts to teach the physics and electronics of mesoscopic devices over the past decade. Fortunately, these have evolved into a more consistent approach, and the book tries to balance experiments and theory in the current understanding of mesoscopic physics. Whenever possible, we attempt to first introduce the important experimental results in this field followed by the relevant theoretical approaches. The focus of the book is on electronic transport in nanostructure systems, and therefore by necessity we have omitted many important aspects of nanostructures such as their optical properties, or details of nanostructure fabrication. Due to length considerations, many germane topics related to transport itself have not received full coverage, or have been referred to by reference. Also, due to the enormity of the literature related to this field, we have not included an exhaustive bibliography of nanostructure transport. Rather, we have tried to refer the interested reader to comprehensive review articles and book chapters when possible.

The Introduction of Chapter 1 gives a general overview of the important effects that are observable in small systems that retain a degree of phase coherence. These are also compared to the needs that one forsees in future small electron devices. Chapter 2 provides a general introduction to quantum confined systems, and the nature of quasi-two-, quasi-one- and quasi-zero-dimensional systems including their dielectric response and behavior in the presence of an external magnetic field. It concludes with an overview of semi-classical transport in quantum wells and quantum wires including the relevant scattering mechanisms in quantum confined systems.

Chapter 3 begins with the general principle of quantum mechanical tunneling, and its historical evolution towards present day resonant tunneling diodes. The concepts of quantum mechanical flux, reflection, and transmission are introduced and applied towards understanding the characteristics of tunneling diodes. These concepts are then generalized to more complicated quantum wave guide systems, which introduces the tunneling/transmission connection upon which the Landauer formula is based. The concept of quantized conductance is introduced, and its connection to the experimentally observed conductance quantization in quantum point contacts. This is then followed by an elaboration of simulation techniques used for modeling wave guide structures and multi-terminal structures.

In Chapter 4, we focus exclusively on quantum dot structures beginning with their electronic structure, and then the experimental results and theoretical formalism related to single electron effects in such structures such as Coulomb Blockade. This is followed by discussion of more complicated systems of multiple quantum dots, and transport through such structures.

Chapter 5 begins to discuss the effects of weak localization and universal conduction fluctuations, which are direct results of inhomogeneities and phase coherence in small structures. First, the experiments and simple understanding are presented; then the more formal treatment by Green's functions provides the detailed understanding that is necessary. This chapter is closed by discussions of open quantum dot systems and the reinterpretation of "universality" that is necessary for the theory.

Chapter 6 extends the above treatments to real temperatures, and begins the discussion of how the phase breaking process, important for loss of coherence, occurs in these systems. The temperature, or Matsubara, Green's functions are introduced in order to handle the underlying physics for this process.

Finally, Chapter 7 discusses nonequilibrium transport in nanostructure under high bias potentials. A review of the important experimental observations under nonequilibrium conditions is given, followed by the introduction of the nonequilibrium, or real time, Green's functions, which provide the formal theoretical basis for treating transport in such systems.

Currently, we are teaching a two semester graduate sequence on the material contained in the book. In the first course, which is suitable for first-year graduate students, the experiments and simpler theory, such as that for tunneling, edge states, and the Landuer–Büttiker method, are introduced. This covers parts of each of the chapters, but does not delve into the topic of Green's functions. Rather, the much more difficult treatment of Green's functions is left to the second course, which is intended for more serious-minded doctoral students. Even here, the developments of the zero-temperature Green's functions in Chapter 6, followed by the Matsubara Green's functions in Chapter 7 and the nonequilibrium (real-time) Green's functions in Chapter 8, are all coupled closely to the experiments in mesoscopic devices.

In spite of the desire to consistently increase the level of difficulty and understanding as one moves through the book, there remain some anomalies. We have chosen, for example, to put the treatment of the lattice expansion and recursive Green's functions in the chapter with wave guide modal expansions, since these two quantities are closely coupled. Nevertheless, the reader would be well served to go through Chapter 5 and its introduction of the Green's functions prior to undertaking an in depth study of the recursive Green's function. This, of course, signals that topics have been grouped together in the chapters in a manner that lies on their connection to one another in physics, rather than in a manner that would be optimally chosen for a textbook. Nevertheless, we are convinced that one can use this book in graduate coursework.

Acknowledgements

The authors are indebted to a great number of people, who have read (and suggested changes) in all or parts of the manuscript during its presentation. In addition, there are the students who suffered through the courses mentioned above without an adequate textbook. In particular, we wish to thank (in alphabetical order) Richard Akis, John Barker, Jon Bird, Paolo Bordone, Kevin Connolly, Manfred Dür, Hal Grubin, Allen Gunther, Anu Krishnaswamy, David Pivin, Dragica Vasileska, and Christoph Wasshuber. In particular we would like to thank Andreas Ecker for substantial contributions to the material on recursive Green's functions in Chapter 3 as part of his Diplom Thesis at the University of Karlsruhe.

The authors are also indebted to several groups and institutes who supported the writing of this manuscript. These include Tom Zipperian at Sandia Laboratories, Peter Vogl at the Technical University of Munich, and Chihiro Hamaguchi at Osaka University. Finally, the authors would like to thank Larry Cooper, at the Office of Naval Research, for his support in the publication of this book.

1

Introduction

It is often said that *nanostructures* have become the system of choice for studying transport over the past few years. What does this simple statement mean?

First, consider transport in large, macroscopic systems. Quite simply, for the past fourscore years, emphasis in studies of transport has been on the Boltzmann transport equation and its application to devices of one sort or another. The assumptions that are usually made for studies are the following: (i) scattering processes are local and occur at a single point in space; (ii) the scattering is instantaneous (local) in time; (iii) the scattering is very weak and the fields are low, such that these two quantities form separate perturbations on the equilibrium system; (iv) the time scale is such that only events that are slow compared to the mean free time between collisions are of interest. In short, one is dealing with structures in which the potentials vary slowly on both the spatial scale of the electron thermal wavelength (to be defined below) and the temporal scale of the scattering processes.

In contrast to the above situation, it has become possible in the last decade or so to make structures (and devices) in which characteristic dimensions are actually smaller than the appropriate mean free paths of interest. In GaAs/GaAlAs semiconductor heterostructures, it is possible at low temperature to reach mobilities of 10^6 cm^2/Vs, which leads to a (mobility) mean free path on the order of 10 μm and an inelastic (or phase-breaking) mean free path even longer. (By "phase-breaking" we mean decay of the energy or phase of the "wave function" representing the carrier.) This means that transport in a regime in which the Boltzmann equation is clearly invalid becomes easily accessible. Each of the assumptions detailed above provides a factor that is neglected in the usual Boltzmann transport picture. Structures (and devices) can readily be built with dimensions that are much smaller than these dimensions, so new physical processes become important in the overall transport. These devices have come to be called *nanostructures*, *nanodevices*, or *mesoscopic devices*, depending upon the author. Perhaps the best description is that of a mesoscopic device, where the prefix "meso-" is used to indicate structures that are large compared to the microscopic (atomic) scale but small compared to the macroscopic scale upon which normal Boltzmann transport theory has come to be applied.

A simple consideration illustrates some of the problems. If the basic semiconductor material is doped to 10^{18} cm^{-3}, then the mean distance between impurity atoms is 10 nm, so that any discrete device size, say 0.1 μm, spans a countably small number of impurity atoms. That is, a cubic volume of 0.1 μm on a side contains only 1000 atoms. These atoms are not uniformly distributed in the material; instead they are randomly distributed with large fluctuations in the actual concentration on this size scale. Again, the variance in the actual number N in any volume (that is, the difference from one such volume to another) is roughly $\sqrt{N}$, which in this example is about 32 atoms (or 3.3% of the doping). Since these atoms often compose the main scattering center at low temperatures, the material is better described as a highly conducting but disordered material, since the material is certainly not uniform on the spatial scale of interest here. As the current lines distort to avoid locally high densities of impurities, the current density becomes non-uniform spatially within the material; this can be expected to lead to new effects. Since the dimensions can be smaller than characteristic scattering lengths, transport can be ballistic and highly sensitive to boundary conditions (contacts, surfaces, and interfaces). To complicate the problem, many new effects that can be observed depend upon the complicated many-body system itself, and simple one-electron theory no longer describes these new effects. Finally, the size can be small compared to the phase-breaking length, which nominally describes the distance over which the electron wave's phase is destroyed by some process. In this case, the phase of the particle becomes important, and many phase-interference effects begin to appear in the characteristic conductance of the material.

Our purpose in this book is twofold. First, we will attempt to review the observed experimental effects that are seen in mesoscopic devices. Second, we want to develop the theoretical understanding necessary to describe these experimentally observed phenomena. But in the remainder of this chapter, the goal is simply to give an introduction into the type of effects that are seen, and to discuss why these effects will be important to future technology, as well as for their interesting physics.

1.1 Nanostructures: The impact

1.1.1 Progressing technology

Since the introduction of the integrated circuit in the late 1950s, the number of individual transistors that can be placed upon a single integrated circuit chip (often just called the *chip*) has approximately quadrupled every three years. The fact that more functionality can be put on a chip when there are more transistors, coupled to the fact that the basic cost of the chip (in terms of \$/cm^2) changes very little from one generation to the next, leads to the conclusion that greater integration leads to a reduction in the basic cost per function for high-level computation as more functions are placed on the chip. It is this simple functionality argument that drives the chip progress. In 1980, Hewlett-Packard produced a single-chip microprocessor containing approximately 0.5 M devices in its 1 cm^2 area [1]. This chip was produced with transistors having a nominal 1.25 μm gate length and was considered a remarkable step forward. Today, the dynamic random-access memory chip (DRAM) is the technology driver; the 64 Mbit is currently in production, with the 256 Mbit expected in 1998. The former chip obviously contains on the order of 64 million transistors. With this progress, one can expect to see 10^9 devices on a single chip just after the turn of the century. In general, this rapid progress in chip density has followed a complicated scaling relationship [2]. The reduction in critical feature size, such as the gate length, is actually a factor of only 0.7 each generation, and this produces only a doubling of the device density. (Other factors are an increase in the actual chip area and changes in the circuit implementation, such as the introduction of trench capacitors.) Still, this leads to some remarkable projections. The 64 Mbit chip uses nominally 0.35-μm gate length transistors. Following the scaling relationships will lead to gate lengths of only 0.1 μm in just over a decade (for the 16 Gbit chip, which scaling suggests will arrive in full production in 2007).

From this discussion, one can reasonably ask just how far the size of an individual electron device can be reduced, and if we understand the physical principles that will govern the behavior of devices as we approach this limiting size. In 1972, Hoeneisen and Mead [3] discussed the minimal size expected for a simple MOS gate (as well as for bipolar devices). Effects such as oxide breakdown, source-drain punch-through, impact ionization in the channel, and so on were major candidates for processes to limit downscaling. Years later, Mead [4] reconsidered this limit in terms of the newer technologies that have appeared since the earlier work, concluding that one could easily downsize the transistor to a gate length of 30 nm *if macroscopic transport theory continued to hold.*

The above considerations tell us that the industry is pushing the critical dimensions downward at a very rapid rate. In contrast to this, research has led to the fabrication of really small individual transistors that operate (at room temperature) in a more-or-less normal fashion. For example, Schottky-gate FETs and high-electron-mobility FETs in GaAs have been made with gate lengths down to 20 nm [5]–[7], and MOSFETs in Si have been made with fabricated gate lengths down to 40 nm [8]. In the latter, the effective gate length was as short as 27 nm. While these devices appear to be normal, there is already evidence that the transport is changing, with tunneling through the gate depletion region becoming more important [9]. Perhaps the more important attribute is the variation that can be expected as one moves from device to device across a chip containing several million transistors. If there is a significant fluctuation in the number of impurities (and/or the number of electrons/holes), then the performance of the devices varies significantly across the chip. This is a major reliability problem, which translates into a dramatic reduction in the effective *noise margin* (the range over which a voltage level can vary without changing the state of a logic gate) of the devices in the chip. This in turn translates into reduced performance of the chip.

Granted that the technological momentum is pushing to ever smaller devices, and that the technology is there to prepare really small devices, it becomes obvious that we must now ask whether our physical understanding of devices and their operation can be extrapolated down to very small space and time scales without upsetting the basic macroscopic transport physics – or do the underlying quantum electronic principles prevent a down-scaling of the essential semi-classical concepts upon which this macroscopic understanding is based? Preliminary considerations of this question were presented more than a decade ago [10]. Suffice it to say, though, that experiment has progressed steadily as well, and ballistic (and therefore coherent and unscattered) transport has been seen in the base region of a GaAs/AlGaAs hot electron transistor [11]. From this, it is estimated that the inelastic mean free path for electrons in GaAs may be as much as 0.12 μm *at room temperature.* Moreover, there are simulations that suggest that it is less than a factor of two smaller in Si [12]. The inelastic mean free path is on the order of (and usually equal to) the energy relaxation length $l_e = v\tau_e$, where τ_e is the energy relaxation time and v is a characteristic velocity (which is often the Fermi velocity in a degenerate system). (There is some ambiguity here because the energy relaxation time is usually defined as the effective inverse decay rate for the mean electron energy, or temperature. The definition here talks about a mean-free path for energy relaxation, which is not quite the same thing. This is complicated by the fact that, in mesoscopic systems, one really talks about a phase-breaking time, which is meant to refer to the average time for relaxation of the coherent single particle phase of a charge carrier. Again, this is a slightly different definition. This ambiguity exists throughout the literature, and although we will probably succumb to it in later chapters, the reader should recognize these subtle differences.) This tells us that, even in Si, the electron inelastic mean free path may be 50–100 nm and comparable to gate lengths of the devices that are expected to be made in the not-too-distant future. Since the phase will likely remain coherent over these distances, it is quite natural to expect phase interference effects to appear in the transport, and to expect most of the assumptions inherent in the Boltzmann picture to be violated. The small device will then reflect the intimate details of the impurity distribution in the particular device, and macroscopic variations can then be expected from one device to another. These effects are, of course, well known in the world of mesoscopic devices. Thus, the study of mesoscopic devices, even at quite low temperatures, provides significant insight into effects that may well be expected to occur in future devices.

Consider, as an example, a simple MOSFET with a gate length of 50 nm and a gate width of 100 nm. If the number of carriers in the channel is 2×10^{12} cm^{-2}, there are only about 100 electrons on average in the open channel. If there is a fluctuation of a single impurity, the change in the conductance will not be 1%, but will be governed by the manner in which the phase interference of the carriers is affected by this fluctuation. This effect is traditionally taken to be of order e^2/h, which leads to a fluctuation in conductance of about 40 μS. If our device were to exhibit conductance of 1 S/mm (of gate width), the absolute conductance would only be 100 μS, so that the fluctuation is on the order of 40% of the actual conductance. This is a very significant fluctuation, arising from the lack of ensemble averaging in the limited number of carriers in the device. In fact, this may well be a limiting mechanism for the down-scaling of individual transistor sizes.

1.1.2 Some physical considerations

In macroscopic conductors, the resistance that is found to exist between two contacts is related to the bulk conductivity and to the dimensions of the conductor. In short, this relationship is expressed by

$$R = \frac{L}{\sigma A},$$ (1.1)

where σ is the conductivity and L and A are the length and cross-sectional area of the conductor, respectively. If the conductor is a two-dimensional conductor, such as a thin sheet of metal, then the conductivity is the conductance per square, and the cross-sectional area is just the width W. This changes the basic formula (1.1) only slightly, but the argument can be extended to any number of dimensions. Thus, for a d-dimensional conductor, the cross-sectional area has the dimension $A = L^{d-1}$, where here L must be interpreted as a "characteristic length." Then, we may rewrite (1.1) as

$$R = \frac{L^{2-d}}{\sigma_d}.$$ (1.2)

Here, σ_d is the d-dimensional conductivity. Whereas one normally thinks of the conductivity, in simple terms, as $\sigma = ne\mu$, the d-dimensional term depends upon the d-dimensional density that is used in this definition. Thus, in three dimensions, σ_3 is defined from the density per unit volume, while in two dimensions σ_2 is defined as the conductivity per unit square and the density is the sheet density of carriers. The conductivity (in any dimension) is not expected to vary much with the characteristic dimension, so we may take the logarithm of the last equation. Then, taking the derivative with respect to $\ln(L)$ leads to

$$\frac{\partial \ln(R)}{\partial \ln(L)} = 2 - d. \qquad (1.3)$$

This result is expected for macroscopic conducting systems, where resistance is related to the conductivity through Eq. (1.2). We may think of this limit as the *bulk* limit, in which any characteristic length is large compared to any characteristic transport length.

In mesoscopic conductors, the above is not necessarily the case, since we must begin to consider the effects of ballistic transport through the conductor. (For ballistic transport we generally adopt the view that the carrier moves through the structure with very little or no scattering, so that it follows normal phase space trajectories.) However, let us first consider a simpler situation. We have assumed that the conductivity is independent of the length, or that σ_d is a constant. However, if there is surface scattering, which can dominate the mean free path, then one could expect that the latter is $l \sim L$. Since $l = v_F \tau$, where v_F is the Fermi velocity in a degenerate semiconductor and τ is the mean free time, this leads to

$$\sigma_d = \frac{n_d e^2 \tau}{m} = \frac{n_d e^2 L}{m v_F}. \qquad (1.4)$$

Hence the dependence of the mean free time on the dimensions of the conductor changes the basic behavior of the macroscopic result (1.3). This is the simplest of the modifications. For more intense disorder or more intense scattering, the carriers are localized because the size of the conductor creates localized states whose energy difference is greater than the thermal excitation, and the conductance will be quite low. In fact, we may actually have the resistance only on the order of [13]

$$R = e^{\alpha L} - 1, \qquad (1.5)$$

where α is a small quantity. (The exponential factor arises from the presumption of tunneling between neighboring sites; the factor of -1 is required for the proper limit as $\alpha L \to 0$.) We think of the form of Eq. (1.5) as arising from the localized carriers tunneling from one site to another (hence the exponential dependence on the length), with the unity factor added to allow the proper limit for small L. Then, the above scaling relationship (1.3) becomes modified to

$$\frac{\partial \ln(R)}{\partial \ln(L)} \approx \alpha L. \qquad (1.6)$$

In this situation, unless the conductance is sufficiently high, the transport is localized and the carriers move by hopping. The necessary value has been termed the *minimum metallic conductivity* [13], but its value is not given by the present arguments. Here we just want to point out the difference in the scaling relationships between systems that are highly conducting (and bulk-like) and those that are largely localized due to the high disorder.

In a strongly disordered system, the wave functions decay exponentially away from the specific site at which the carrier is present. This means that there is no long-range wavelike behavior in the carrier's character. On the other hand, by bulk-like extended states we mean that the carrier is wavelike in nature and has a well-defined wave vector $\mathbf{k}$ and momentum $\hbar k$. Most mesoscopic systems have sufficient scattering that the carriers do not have fully wavelike behavior, but they are sufficiently ordered so that the carriers are not exponentially localized. Thus, when we talk about *diffusive* transport, we generally mean almost-wavelike states with very high scattering rates. Such states are neither free electron–like nor fully localized. We have to adopt concepts from both areas of research.

The rationale for such a view lies in the expectations of quantization in such small, mesoscopic conductors. We assume that the semiconductor sample is such that the electrons move in a potential that is uniform on a macroscopic scale but that varies on the mesoscopic scale, such that the states are disordered on the microscopic scale. Nevertheless, it is assumed that the entire conduction band is not localized, but that it retains a region in the center of this energy band that has extended states and a nonzero conductivity as the temperature is reduced to zero. For this material, the density of electronic states per unit energy per unit volume is given simply by the familiar dn/dE. Since the conductor has a finite volume, the electronic states are discrete levels determined by the size of this volume. These individual energy levels are sensitive to the boundary conditions applied to the ends of the sample (and to the "sides") and can be shifted by small amounts on the order of $\hbar/t$, where t is the time required for an electron to diffuse to the end of the sample. In essence, one is defining here a broadening of the levels that is due to the finite lifetime of the electrons in the sample, a lifetime determined not by scattering but by the carriers' exit from the sample. This, in turn, defines a maximum *coherence length* in terms of the sample length. This coherence length is defined here as the distance over which the electrons lose their phase memory, which we will take to be the sample length. The time required to diffuse to the end of the conductor (or from one end to the other) is L^2/D, where D is the diffusion constant for the electron (or hole, as the case may be) [14]. The conductivity of the material is related to the diffusion constant (we assume for the moment that $T = 0$) as

$$\sigma(E) = \frac{ne^2\tau}{m} = e^2 D \frac{dn}{dE}, \qquad (1.7)$$

where we have used the fact that $n = (2/d)(dn/dE)E$, where d is the dimensionality, and $D = v_F^2 \tau/d$. If L is now introduced as the effective length, and t is the time for diffusion, both from D, one finds that

$$\frac{\hbar}{t} = \frac{\hbar}{e^2} \frac{\sigma}{L^2} \frac{dE}{dn}. \qquad (1.8)$$

The quantity on the left side of Eq. (1.8) can be defined as the average broadening of the energy levels ΔE_a, and the dimensionless ratio of this width to the average spacing of the energy levels may be defined as

$$\frac{\Delta E_a}{dE/dn} = \frac{\hbar}{e^2} \sigma L^{-2}. \qquad (1.9)$$

Finally, we change to the total number of carriers $N = nL^d$, so that

$$\frac{\Delta E_a}{dE/dN} = \frac{\hbar}{e^2} \sigma L^{d-2}. \qquad (1.10)$$

This last equation is often seen with an additional factor of 2 to account for the double degeneracy of each level arising from the spin of the electron.

Another method of looking at Eq. (1.8) is to notice that a conductor connected to two metallic reservoirs will carry a current defined by the difference in the Fermi levels between the two ends, which may be taken to be eV. Now there are $eV(dn/dE)$ states contributing to the current, and each of these states carries a current e/t. Thus, the total conductance is $(e^2/t)(dn/dE)$, which leads directly to Eq. (1.8).

The quantity on the left side of Eq. (1.9) is of interest in setting the minimum metallic conductivity. In a disordered material, the ratio of the overlap energy between different sites and the disorder-induced broadening of the energy levels is important. The former quantity is related to the width of the energy bands. If this ratio is small, it is hard to match the width of the energy level on one site with that on a neighboring site, so that the allowed energies do not overlap and there is no appreciable conductivity through the sample. On the other hand, if the ratio is large, the energy levels easily overlap and we have bands of allowed energy, so that there are extended wave functions and a large conductance through the sample. The ratio (1.9) just expresses this quantity. The factor $e^2/\hbar$ is related to the fundamental unit of conductance and is just 2.43×10^{-4} siemens (the inverse is just 4.12 kΩ).

It is now possible to define a dimensionless conductance, called the *Thouless number* by Anderson and coworkers [15], in terms of the conductance as

$$g(L) = \frac{2\hbar}{e^2} G(L), \qquad (1.11)$$

where $G(L) = \sigma L^{d-2}$ is the actual conductance in the highly conducting system. These latter authors have given a scaling theory based upon renormalization group theory, which gives us the dependence on the scale length L and the dimensionality of the system. The details of such a theory are beyond the present work. However, we can obtain the limiting form of their results from the above arguments. The important factor is a critical exponent for the reduced conductance $g(L)$, which may be defined by

$$\beta_d \equiv \lim_{g \to \infty} \frac{d[\ln g(L)]}{d \ln L} \to d - 2, \qquad (1.12)$$

which is just Eq. (1.3) rewritten in terms of the conductance rather than the resistance. By the same token, one can rework Eq. (1.6) for the low conducting state to give

$$\beta_d \equiv \lim_{g \to 0} \frac{d[\ln g(L)]}{d \ln L} \to -\alpha L. \qquad (1.13)$$

What the full scaling theory provides is the connection between these two limits when the conductance is neither large nor small.

For three dimensions, the critical exponent changes from negative to positive as one moves from low conductivity to high conductivity, so that the concept of a *mobility edge* in disordered (and amorphous) conductors is really interpreted as the point where $\beta_3 = 0$. This can be expected to occur about where the reduced conductance is unity, or for a value of the total conductance of $e^2/\pi\hbar$ (the factor of $\pi/2$ arises from a more exact treatment). In two dimensions, there is no critical value of the exponent, as it is by and large always negative, approaching 0 asymptotically. Instead of a sharp mobility edge, there is a universal crossover from logarithmic localization at large conductance to exponential localization at small conductance. This same crossover appears in one dimension as well, except the logarithmic localization is much stronger. Hence, it may be expected that all states will be localized for $d \leq 2$ if there is any disorder at all in the conductor. This is the source of the size dependence that is observed in mesoscopic structures. In fact, we note that in the case of surface scattering (1.4), the additional factor of L in the conductivity leads immediately to variation in the conductance as L^{d-1}, which gives the value arising from (1.13) immediately for two dimensions.

1.2 Mesoscopic observables in nanostructures

1.2.1 Ballistic transport

In the simplest case of transport in small structures, one may assume that the particles move through the active region without scattering. This transport is termed *ballistic transport*. In this approach, we assume that the "device" region is characterized by the transmission and reflection of the incoming waves at both sides; essentially, this is a scattering matrix approach [16]. Particles flow through the active region without scattering (except for a possible reflection from a barrier) and move elastically. Thus, if the reservoirs are described by the Fermi levels E_{Fl} and E_{Fr}, which are separated by a small applied bias energy $e\delta V$, only those electrons flowing from filled states on the left (which is assumed to lie at higher energy) to the empty states on the right contribute to the current (see Fig. 1.1). This is easily seen for $T = 0$, where the states are completely filled up to the Fermi level and completely empty for energies above the Fermi level. Since the carriers move elastically, electrons leaving the left reservoir at the Fermi energy arrive at the right reservoir with an excess energy $e\delta V$, which must be dissipated in the latter reservoir (contact). If there is no barrier to this flow, the current is determined solely by the number of electrons which can leave the source contact, the left reservoir (actually the number which leave per unit time). When a barrier is present, a portion of the leaving electrons are reflected back into the source contact, and a fraction is transmitted through the barrier, based upon tunneling. We take the view of describing the properties of the barrier region in terms of the incoming currents from the contacts.

The process is described by the incoming particle currents j_i and j_i', from the left and the right sides, respectively. The Fermi level on the left side of the scatterer is raised slightly by an applied bias of $e\delta V$. This results in electrons appearing in the i channel, but those returning in the reflected r channel are also absorbed in the left reservoir described by its Fermi level. On the left side, there is an extra density of electrons in the levels near the Fermi surface (over the equilibrium values), and this density is given by

$$\delta n = \frac{dn}{dE} e\delta V. \tag{1.14}$$

At the same time, the extra density of electrons on the left side will be given by the sum of the magnitudes of the particle currents on the left minus those on the right, with each divided by the respective velocities,

$$\delta n = \frac{j_i + j_r}{v_l} - \frac{j_i' + j_o}{v_r} = \hbar R \frac{j_i - j_i'}{dE/dk_x}, \tag{1.15}$$

where j_o is the current transmitted from the left to the right, and we have assumed the velocity is the same on both sides of the barrier. Here, we have used the facts that $j_i - j_o =$

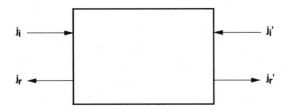

Fig. 1.1. The currents flowing into, and out of, a scattering barrier.

$(1 - T)j_i = Rj_i$ and $j'_i - j'_r = (1 - T)j'_i = Rj'_i$. The total current into the system is just $I = e(j_i - j_r) = e(j_o - j'_i) = eT(j_i - j'_i)$, so that

$$G = \frac{I}{\delta V} = \frac{e^2}{\hbar} \frac{T}{R} \frac{dn}{dE} \frac{dE}{dk_x}. \tag{1.16}$$

Now, in one dimension we can write the density of states as

$$dn = N(k) \left(\frac{dk}{dE} \right) dE = \frac{1}{\pi} \left(\frac{dk}{dE} \right) dE, \tag{1.17}$$

where a factor of 2 has been added for spin degeneracy. This may be combined with Eq. (1.16) to give

$$G = \frac{e^2}{\pi \hbar} \frac{T}{R}, \tag{1.18}$$

which is known as the *Landauer formula*. Actually, this latter form is a special one for four-terminal measurements, where the potentials are measured adjacent to the scattering site rather than in the reservoirs. In the case where the reservoirs are the source for the potential measurements, contact potentials would be possible, and the factor of R^{-1} is deleted. It should be noted that the transmission coefficient T is that for all possible energy states that can tunnel through the structure from one side to the other. When there are many such channels possible, the single factor of T is usually replaced with a summation over the individual T_i, as we shall see.

The Landauer formula is a generalization of a formula that is normally obtained in three dimensions for tunneling structures. Generally, one can describe the tunneling probability (the transmission coefficient) $T(E)$ for an arbitrary barrier structure. Then the current through this structure may be written semi-classically as

$$I = 2eA \int \frac{d^3\mathbf{k}}{(2\pi)^3} v_z(k_z) T(E_z)[f_l(E) - f_r(E + e\delta V)], \tag{1.19}$$

where the two distribution functions represent those on the left and right sides of the barrier. It will be shown in a later chapter that this can be rewritten as

$$I = \frac{e^2}{\pi \hbar} \bar{T}(E_F) N_\perp, \tag{1.20}$$

where we have introduced an average tunneling coefficient and where $N_\perp$ is the number of transverse modes, or

$$N_\perp = A \int \frac{d^2\mathbf{k}}{(2\pi)^2} = \frac{mA}{2\pi \hbar^2} \int dE_\perp. \tag{1.21}$$

This means that the simple Landauer formula is a result of a more complex three-dimensional theory when the latter is reduced to small structures with lateral quantization.

The significance of the Landauer formula, for a system such as a mesoscopic system of small size where phase coherence can be maintained over the tunneling distance, is that the conductance is a constant value that includes a fundamental value multiplied by the number of modes that are transmitting. This has been detected in simple but elegant sets of experiments by van Wees *et al.* [17] and by Wharam *et al.* [18]. In those of the former group, a very-high-mobility two-dimensional electron gas, formed at the interface between GaAs and doped AlGaAs, is used. A pair of gates is used to create a constriction between

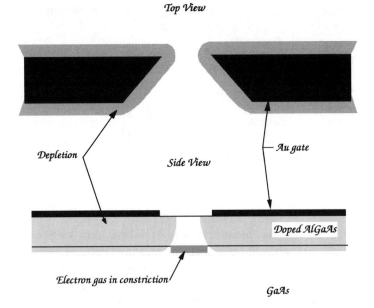

Fig. 1.2. Structure of the metal gates used to define a constriction to observe quantized conductance.

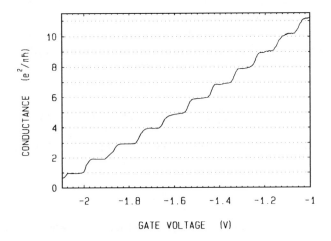

Fig. 1.3. The conductance obtained from a typical GaAs/AlGaAs structure such as that of Figure 1.2. [Reprinted with permission from van Wees *et al.* [17]. Copyright 1988 The American Physical Society.]

two parts of the electron gas, and the conductance between these two parts is measured. The structure of the gates is illustrated in Fig. 1.2, and the measurements are illustrated in Fig. 1.3. The gates form a short one-dimensional channel with several allowed modes of propagation between the two parts of the broad-area electron gas. Here, the transmission coefficient is either 0 or 1, depending upon whether the channel subband energy lies above or below the Fermi energy. The gates may be assumed to introduce a harmonic potential, so that the subbands are equally spaced. As the bias on the gates is varied, the number of channels below the Fermi level is changed as the saddle potential between the gates is raised or lowered. For a sufficiently large negative gate bias, all channels are pinched off and the conductance drops to zero. These results are a very dramatic verification of the Landauer formula and ballistic transport in mesoscopic systems discussed in detail in Chapter 3.

1.2.2 Phase interference

The relevant quantity for discussion of quantum interference effects is the phase of the carrier as it moves through the semiconductor. Interference between differing waves can occur over distances on the order of the coherence length of the carrier wave, and the latter distance is generally different from the inelastic mean free path for quasi-ballistic carriers (those with weak scattering). The latter is related to the energy relaxation length $l_e = v\tau_e$, where τ_e is the energy relaxation time and v is a characteristic velocity (typically the Fermi velocity in degenerate material and the thermal velocity in nondegenerate material). The inelastic mean free path can be quite long, on the order of several tens of microns for electrons at low temperatures in the inversion channel of a high-electron-mobility transistor in GaAs/AlGaAs. On the other hand, the coherence length is usually defined for weakly disordered systems by the diffusion constant (as we discuss below). Here, we are interested in the quasi-ballistic regime (nearly free, unscattered carriers) and so will use the inelastic mean free path as the critical length.

Consider two waves (or one single wave which is split into two parts which propagate over different paths) given by the general form $\psi_i = A_i e^{i\phi_i}$. Then, when the two waves are combined, the probability amplitude varies as

$$P = |\psi_1 + \psi_2|^2 = |A_1|^2 + |A_2|^2 + 4|A_1^* A_2| \cos(\phi_1 - \phi_2). \qquad (1.22)$$

The probability can therefore range from the sum of the two amplitudes to the differences of the two amplitudes, depending on how the phases of the two waves are related. In most cases, it is not important to retain any information about the phase in device problems because the coherence length is much smaller than any device length scale and because *ensemble averaging* averages over the phase interference factor so that it smooths completely away in macroscopic effects. This ensemble averaging requires that a large number of such small phase coherent regions are combined stochastically. In small structures this does not occur, and many observed quantum interference effects are direct results of the lack of ensemble averaging [19].

A particularly remarkable illustration of the importance of the quantum phase is the magnetic Aharonov-Bohm effect [20], as may be seen in quasi-two-dimensional semiconductor systems. The basic structure of the experiment is illustrated in Fig. 1.4. A quasi-one-dimensional conducting channel is fabricated on the surface of a semiconductor. This channel is usually produced in a high-electron-mobility heterostructure in which the channel is defined by reactive-ion etching [21]–[23], or by electrostatic confinement [24]. In either case, it is preferable to have the waveguide sufficiently small so that only one or a few electron modes are possible. The incident electrons, from the left of the ring in Fig. 1.4, have their wave split at the entrance to the ring. The waves propagate around the two halves of the ring to recombine (and interfere) at the exit port. The overall transmission through the structure, from the left electrodes to the right electrodes, depends upon the relative size of the ring circumference in comparison to the electron wavelength. If the size of the ring is small compared to the inelastic mean free path, the transmission depends on the phase of the two fractional paths. In the Aharonov-Bohm effect, a magnetic field is passed through

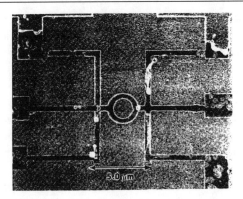

Fig. 1.4. Micrograph of the etched ring structure. [After Mankiewich *et al.* [23].]

the annulus of the ring, and this magnetic field will modulate the phase interference at the exit port.

The vector potential for a magnetic field passing through the annulus of the ring is azimuthal, so that electrons passing through either side of the ring will travel either parallel or antiparallel to the vector potential, and this difference produces the phase modulation. The vector potential will be considered to be directed counterclockwise around the ring. (We adopt cylindrical coordinates, with the magnetic field directed in the z-direction and the vector potential in the θ-direction.) The phase of the electron in the presence of the vector potential is given by the Peierl's substitution, in which the normal momentum vector $\mathbf{k}$ is replaced by $(\mathbf{p} + e\mathbf{A})/\hbar$,

$$\phi = \phi_0 + \frac{1}{\hbar}(\mathbf{p} + e\mathbf{A}) \cdot \mathbf{r}, \tag{1.23}$$

so that the exit phases for the upper and lower arms of the ring can be expressed as

$$\phi_{up} = \phi_0 + \int_\pi^0 \left(\mathbf{k} + \frac{e}{\hbar}\mathbf{A} \right) \cdot \mathbf{a}_\vartheta r d\vartheta,$$

$$\phi_{lo} = \phi_0 - \int_{-\pi}^0 \left(\mathbf{k} - \frac{e}{\hbar}\mathbf{A} \right) \cdot \mathbf{a}_\vartheta r d\vartheta, \tag{1.24}$$

and the phase difference is just

$$\delta\phi = \frac{e}{\hbar} \int_0^{2\pi} \mathbf{A} \cdot \mathbf{a}_\vartheta r d\vartheta = \frac{e}{\hbar} \int_{ring} \mathbf{B} \cdot \mathbf{n} dS = 2\pi \frac{\Phi}{\Phi_0}, \tag{1.25}$$

where $\Phi_0 = h/e$ is the quantum unit of flux and Φ is the magnetic flux coupled through the ring. The phase interference term in Eq. (1.22) goes through a complete oscillation each time the magnetic field is increased by one flux quantum unit. This produces a modulation in the conductance (resistance) that is periodic in the magnetic field, with a period h/eS, where S is the area of the ring. This periodic oscillation is the Aharonov-Bohm effect, and in Fig. 1.5 results are shown for such a semiconductor structure. While these oscillations are obvious in such a constructed ring, mesoscopic devices are described by a great many accidental rings that come and go as the electrochemical potential is varied.

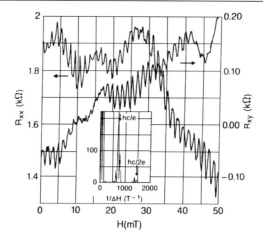

Fig. 1.5. Magnetoresistance and Hall resistance at 0.3 K for the ring of Fig. 1.4. The inset is the Fourier transform of the data showing the main peak at h/e. [After Mankiewich et al. [23].]

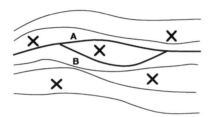

Fig. 1.6. Trajectories through the sample will depend on the specific impurity configuration and can be easily shifted.

1.2.3 Universal conductance fluctuations

The above discussion nicely indicates that there are certain conditions under which oscillations in the conductance that are periodic in h/e can be observed. However, this occurs when the sample is configured in the shape of a ring. In more realistic samples, which are usually long and thin, this will not be the normal observation. On the other hand, at low temperatures these samples are usually degenerate, and conduction (or any transport process) takes place mainly with the carriers at the Fermi energy. Impurities situated in the material create potential hills around which the current carriers must pass. As the Fermi energy is varied, either by some gate potential or by an imposed magnetic field (which depletes the conduction band), the *detailed* Fermi surface upon which the carriers travel can be shifted slightly. It is then possible for a carrier that moved on one side of an impurity to begin to travel on the other side of the impurity after the shift of the Fermi energy. This is shown in Fig. 1.6. For a given value of the Fermi energy, the carrier follows the bold trajectory marked "A." When the Fermi energy is shifted slightly, the carrier begins to follow the path marked "B." It may be noted that the change in the trajectory is exactly equivalent to coupling an Aharonov-Bohm loop composed of the A and B arms, and a change in the conductance of order $2e^2/h$ can be expected to occur by this process. Stone [25] has argued that this is a quite general phenomenon of quantum fluctuations in mesoscopic systems, such as those systems whose size scale is on the order of the phase-breaking length. These

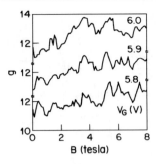

Fig. 1.7. Normalized conductance seen in a Si MOSFET at 4.2 K (the details are discussed in the text). [Reprinted with permission from Skocpol *et al.* [26]. Copyright 1986 The American Physical Society.]

fluctuations are *time independent*. That is, they occur with variations in the Fermi energy and are quite static in time. Since they will depend on the *exact* configuration of scattering centers within the sample, these fluctuations will be very sample specific and are in fact a sort of "fingerprint" for the sample. As long as the sample is maintained at low temperature, where the measurements are typically made, the latter will be perfectly repeatable. Warming the sample anneals the specific distribution of impurities and changes their configuration, hence changes the spectrum of fluctuations when the sample is again cooled.

Some of the earliest studies [26] in semiconductors occurred in silicon MOSFETs, in which the channel was made relatively long (0.5 μm) compared to the width (0.1 μm). The measured normalized conductance (in units of e^2/h) is shown as a function of the magnetic field for several values of the gate voltage in Fig. 1.7. Since the fluctuations are due to coupling of "rings" within the solid by the magnetic field (in this case), there should be a range of variation in the Fermi energy (whether induced by gate voltage or by magnetic field) for which the fluctuations are correlated. In later chapters we shall show that this is indeed the case. However, the correlation function defined by such a process has a half-width, either in gate voltage or in magnetic field, that characterizes the random interference phenomena of the trajectories in the samples. In the case of a magnetic field, this half-width, the correlation field, can be related to the flux coupled through a phase-coherent region, which is defined by the characteristic phase-breaking length of the sample. Thus, to see these effects, the sample must be comparable in size to the phase coherence length, and as the sample grows larger, the individual regions of this size ensemble average to smooth out these fluctuations.

1.2.4 Weak localization

In the previous paragraphs, the interference of two waves was considered. These waves could as well have been two groups of waves, say α and β, as shown in Fig. 1.8, where the two groups of waves propagate around a loop defined by three scattering centers. If these scattering events are elastic, such as due to ionized impurity scattering, they do not contribute to the phase-breaking process. We conceive of this process as arising from the circumstance that an electron has a certain probability of scattering from these three centers and *returning to its original site*. Now, the probability that the electron scatters around the loop clockwise is the same as that for its scattering around the loop counterclockwise. These

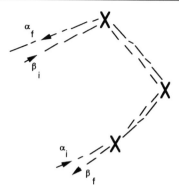

Fig. 1.8. Interference of two time-reversed paths a and b in the presence of three scattering centers.

two options are said to be paths which are *time-reversed* with respect to one another. Since the wave (the electron) returns to its original position, it can interfere with itself. More properly, it is said that the two time-reversed paths interfere continuously with one another. This is not unique to three scatterers, but it can occur for any number of scatterers. When the electron returns to its original position, it generally has a velocity that is negatively correlated to its original velocity, so that the interference creates an additional resistance through the interaction between the particle and itself, or between the two time-reversed paths. This additional resistance is called *weak localization*. The use of this term arises from a view that localization can be interpreted by total back-reflection of particles from potential barriers so that they become localized in a single potential well. Here, the back-reflection is weak and the particles are not truly localized; rather the weak interaction creates only an additional resistance, which is known as the weak localization contribution to the resistance.

If a magnetic field is now passed through the loop formed by the two time-reversed paths, the phases of the two paths are changed. For one direction, the phase is increased; for the other, it is decreased, just as in the Aharonov-Bohm effect. This has the result of modulating the interference of the two paths. In fact, the interference is reduced dramatically by the magnetic field. One might think that this effect would be periodic in the magnetic field, but it usually is not because the diffusion constant for motion around the loop is also reduced by the field, so that the probability of returning to the original site without phase breaking is greatly reduced by the magnetic field as well. The weak localization contribution to the resistance of a quantum wire is shown [27] in Fig. 1.9. The wire is defined by gate depletion in a GaAs/AlGaAs high-mobility layer and is 2 μm long and approximately 0.5 μm wide. The background resistance (the field-independent resistance upon which the data of Fig. 1.9 are superimposed) at zero magnetic field is about 590 ohms, and a quadratic contribution to the magnetoresistance has been removed. It may be seen that the excess resistance *decreases* with increasing magnetic field, which causes a negative magnetoresistance. But this is just the signature of the weak localization. This is an additional resistance at zero magnetic field due to the interference effects, and the magnetic field breaks up this resistance, which causes a net reduction in the total resistance of the wire. In this case, one can estimate (with formulas from later chapters) that the phase-breaking length is about 2.3 μm, which is slightly longer than the wire itself. Also apparent in the figure are significant fluctuations in the resistance. These d.c. fluctuations are the *universal conductance fluctuations* of the previous section.

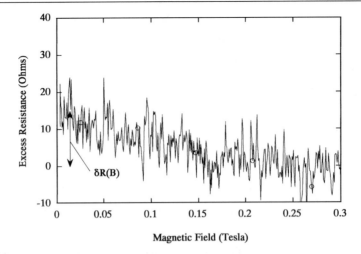

Fig. 1.9. The excess resistance for a quantum wire as a function of the magnetic field. The negative magnetoresistance is a characteristic of weak localization.

1.2.5 Carrier heating in nanostructures

One of the classic problems of semiconductors and semiconductor devices is carrier heating in the high electric fields present in the device. In general, low electric field transport is normally treated in the relaxation time approximation through a rigid shift of the distribution function in momentum space so that the carriers acquire a net velocity while maintaining a distribution function characterized by thermal equilibrium and the normal Fermi energy [28]. In the problem of carrier heating, however, this is no longer the case. The input of energy from the applied fields and potentials leads to an increase of the kinetic motion of the particles as well as a directed drift velocity. The distribution function is no longer the equilibrium one, and usually the major problem is finding this new distribution function. For this purpose one must rely upon a transport equation.

The Boltzmann transport equation (BTE) has been a cornerstone of semi-classical transport theory for many years. However, as we pointed out above, there are a great many assumptions built into this equation, and some of these will no longer be applicable in nanostructures. One of the strongest limitations on the BTE arises from the restrictions that the field and the scattering events are treated as noninteracting perturbations, or that each scattering event is treated as a separate entity which does not interact with other scattering events or processes. This result cannot be expected to hold either in high electric fields or in constrained geometries arising from nanostructures. This means that a new kinetic equation must be obtained to replace the BTE of semi-classical transport theory.

Virtually all quantum kinetic theories are based upon reductions of the Liouville–von Neumann equation for the density matrix ρ, with appropriate boundary conditions,

$$i\hbar \frac{\partial \rho}{\partial t} = [H_0, \rho] + [V, \rho] + [F, \rho], \tag{1.26}$$

where H_0 is the Hamiltonian for the carriers plus phonons plus impurities, V is the electron-scatterer interaction, and F is the Hamiltonian that couples the electrons (and scatterers in some cases) to external fields. In nanostructures, the latter can arise not only from applied fields but also from boundary and confining potentials such as surface roughness scattering and quantum confinement. The density matrix equation expressed here is one form of a quantum kinetic equation, but others exist. For example, the density matrix (and the Wigner distribution obtained from it) builds in correlation between different points, but it is a single-time function. In many cases, the correlation between different times is very important, and other approaches based upon nonequilibrium real-time Green's functions are required. One resultant problem in the case of quantum kinetic theories is that there are a variety of equations that result, as well as a diverse set of definitions of the so-called distribution function describing the one-electron probability in momentum and real space. It is not at all clear that quantum kinetic theories reduce to the semi-classical BTE, and there has been considerable work to try to establish this fact [29]. In most cases, however, additional assumptions, beyond those inherent in the BTE, must be made to connect a quantum kinetic equation to the BTE, even in the limiting case of low fields and macroscopic dimensions. Once spatial and temporal correlations are introduced into the quantum kinetic theory, these correlations must be addressed in the limiting process required to achieve any connection with the BTE, which is an equation for a *local* quantity. These correlations introduce retardation and memory effects into the equivalent one-electron quantum kinetic equation.

If the applied fields are time dependent, further retardation and memory effects become important. These are related first to the inertial response of the electron plus scatterer system, and second to the role the oscillating potentials can play in the extraction of energy in quanta, either from the field by the system or by the field from the system. The temporal nonlocality is important as well if the collision process cannot be completed within a single cycle of the oscillating potential. Although this normally is not a large problem, it can be important in systems with long phase memory, and is enhanced in those cases where the applied field can actually weaken the collisions (such as with impurity scattering).

Inertial effects are also anticipated as a consequence of the failure of the lattice to keep pace with the driven electron subsystem. Hence, spatial and temporal nonlocality can arise from the effect of a nonequilibrium phonon distribution. The extreme nonlinearity of the coupled electron and phonon systems at high driving fields makes the existence of a shock structure in the coupled distributions a distinct possibility. The self-consistent screening of the carriers can also lead to an indirect field-dependent effect, in which the high fields lead to a nonlinear distribution function that results in de-screening of the electron-phonon interaction. This variety of nonlinear effects just illustrates the difficulty of treating nonequilibrium transport in small structures.

Consequently, one of the central features of the nanostructure is that its transport cannot be treated in isolation. The quantum kinetic equations are strongly coupled to the boundaries, and the overall environment, through the boundary conditions on the equation. The basic Liouville equation and its causal boundary conditions are basically modified due to the strong nonlinear interaction between them. Moreover, the nonlinear system in the presence of the fields is strongly impacted by the boundaries, so that the resulting distribution function will largely be determined by these boundaries as well as the internal driving fields. While the incoherent coupling of the device to its environment and boundaries has traditionally been treated through weak scattering processes such as surface-roughness scattering, the effects of the boundaries on nanostructures can be much stronger and far more complicated to treat theoretically.

Table 1.1. *Some important parameters at low temperature.*

Parameter	GaAs	Si	Units
Density	4.0	4.0	10^{11} cm^{-2}
Mobility	10^5	10^4	cm^2/Vs
Scattering time	3.8	1.1	10^{-12} sec
Fermi wave vector	1.6	1.6	10^6 cm^{-1}
Fermi velocity	2.76	0.97	10^7 cm/s
elastic mfp	1.05	0.107	10^{-4} cm
inelastic mfp	5.0	0.5	10^{-4} cm
Phase-breaking time	1.8	0.57	10^{-11} s
Diffusion constant	1.45	0.52	10^3 cm^2/s
Phase coherence length	1.62	0.54	10^{-4} cm

1.3 Space and time scales

In the previous sections, we have discussed briefly the major new effects that cause the transport in nanostructures to differ significantly from that of macroscopic semiconductor devices. Effects such as phase interference, with the consequent universal conductance fluctuations and weak localization, and carrier heating all depend upon the relative sizes of important correlation lengths and the device dimensions. The key lengths of the above sections have been taken to be the inelastic and elastic mean free paths. It is worthwhile at this point to make a few estimates of some of the key parameters that will be utilized in the remainder of the book, and to assure ourselves that no fundamental limits of understanding are being violated. Some of these key parameters are listed in Table 1.1. In the case for which these numbers are tabulated, it is assumed that we have a quasi-two-dimensional electron gas localized at an interface. For the GaAs case, this assumption is predicated upon a modulation-doped AlGaAs/GaAs heterostructure with the dopant atoms in the AlGaAs and the free electrons forming an inversion layer on the GaAs side of the interface. For the Si case, it is assumed that electrons are introduced at an interface either between Si and an oxide or between Si and, for example, a strained SiGe layer utilizing modulation doping. However, we take the lower mobility of the former interface for the example. In each case, a modest inversion density (equal for the two examples) is chosen, and the characteristic lengths are worked out from these assumptions.

The density is the sheet density of carriers in the quasi-two-dimensional electron gas at the interface discussed above. The actual density can usually be varied by an order of magnitude on either side of this value; for example, 10^{10}–4×10^{12} cm^{-2} are possible in Si and somewhat lower at the upper end in GaAs. Again the mobility is a typical value, but in high-mobility structures, as much as 10^7 cm^2/Vs has been achieved in GaAs, and 2×10^5 cm^2/Vs has been achieved in Si (modulation doped with SiGe at the interface). Finding the scattering time from the mobility is straightforward, and masses of $0.067m_0$ and $0.19m_0$ are used for GaAs and Si, respectively. The Fermi wave vector is determined by the density through $k_F = (2\pi n_s)^{1/2}$. The Fermi velocity is then $v_F = \hbar k_F/m$. The elastic mean free path is then defined by the scattering time and the Fermi velocity (it is assumed that the quasi-two-dimensional electron gas is degenerate at low temperature) as $l_e = v_F \tau_{sc}$. The inelastic mean free path is estimated for the two materials based upon experiments that will be discussed in later chapters, but it also should be recognized that there will be a range of values for this parameter. The inelastic mean free time, or phase-breaking time, is found through the relationship $l_{in} = v_F \tau_\varphi$, and the inelastic, or phase-breaking, time τ_φ is found from the previous estimate of the inelastic mean free path. The diffusion constant is just $D = v_F^2 \tau_{sc}/d$ (in d dimensions). Finally, the phase coherence length is the normally defined Thouless length $l_\varphi = (D\tau_\varphi)^{1/2}$. It should be noted that the *inelastic mean free path* and the *phase coherence length* are two different quantities, with the latter being about a factor of three smaller than the former for the parameters used in the table, *even though these two quantities are often assumed to be interchangable and indistinguishable*. If this were the case, then the inelastic mean free time would be clearly defined to be one-half the scattering time from the definitions introduced here. Usually, in fact, the inelastic mean free time is larger than the scattering time, so that the inelastic mean free path is larger than the phase coherence length. Both of these quantities are important in phase interference, and we will try not to confuse the issue by using the two names inappropriately. But they do differ.

The inelastic mean free path describes the distance an electron travels *ballistically* in the phase-breaking (or energy relaxation) time τ_φ. Thus, this quantity is useful in describing those processes, such as tunneling, in which the dominant transport process is one in which the carriers move ballistically with little scattering through the active region of interest. On the other hand, the coherence length l_φ is defined with the *diffusion* constant D. This means that the carriers are in a region of extensive scattering, so that their transport is describable as a diffusion process in which the coherence length is the equivalent diffusion length defined with the phase-breaking time. In some small nanostructures, the transport is neither ballistic nor diffusive (consider the short quantum wire of Fig. 1.9 in which the wire length is less than the coherence length) but is somewhere between these two limits. For these structures, the effective phase relaxation length is neither the inelastic mean free path nor the coherence length. These structures are more difficult to understand, and they are usually still quite sensitive to the boundary conditions. Care must be taken to be sure that the actual length (and the descriptive terminology) is appropriate to the situation under study.

We will see in later chapters that there are two other lengths that are important. These are the *thermal length* $l_T = (\hbar D/k_B T)^{1/2}$ and the *magnetic length* $l_m = (\hbar/eB)^{1/2}$. The latter is important as it relates to the filling factor of the Landau levels and to the cyclotron radius at the Fermi energy $r_c = \hbar k_F/eB = k_F l_m^2$.

1.4 An introduction to the subsequent chapters

This book is of two general themes. The first is a systematic review of the experimental status of mesoscopic systems and devices, in particular the experiments that have been carried out to understand the relevant physics and the manner in which this physics affects the transport. We apologize up front that this review is not an exhaustive review. The second theme is the general development of quantum transport theory based primarily upon the Green's function (and quantum field theory) solutions to the Schrödinger equation in confined systems. To be sure, there are a great many review articles in the literature that cover mesoscopic systems and nanostructures [30]–[33]. Most of these review the experimental results for particular measurements or systems, but we do not feel that there is a systematic treatment of this at present. And, for sure, there is no systematic development of the theory to cover the range of measurements in mesoscopic systems, particularly in a manner suitable to use as a textbook. On the other hand, there are several good books on the use of quantum field theory and Green's functions in treating condensed matter systems [34]–[37]. It is not our aim to supplant these, but we will instead rely upon them as needed background on the general techniques. Here we want to develop a systematic treatment of the theory underlying mesoscopic processes in *semiconductor* nanostructures, which provide the basis for most of the interesting experiments.

In Chapter 2 we begin by reviewing important aspects of quantization and modes of quantum structures. We then examine the effect of a magnetic field. Finally, we briefly review homogeneous transport theory.

The general topics of Chapter 3 are tunneling and the extension of the Landauer formula to realistic, many-terminal mesoscopic devices. First, the general idea of *resonant* tunneling through a bound state and the experiments which probe this effect will be introduced. The extension of these experiments to many barriers and multiple wells will also be discussed. Then, the theoretical basis of wave transmission and reflection will be introduced so that the basis of the equations above can be established. We first discuss the experiments that describe the transition from a short gated region (the quantum point contact of Fig. 1.2) to a long waveguide in which the carriers move diffusively. We discuss the confined solutions to Schrödinger's equation through the mode picture of the quantum waveguide in analogy to a microwave waveguide. In general, the experiments tell us that the concept of coherent individual modes is rapidly eliminated in long guides, and the role of disorder in this process is discussed next. The role of contacts in modifying the basic forms of the Landauer equation leads us naturally into multi-terminal descriptions and the Büttiker scattering matrix approach with a magnetic field. This naturally leads into the quantum Hall effect and the gated quantum Hall effect in local and nonlocal geometries. A discussion of zero-temperature Green's functions for waveguides, and the development of Dyson's equation for their solution is given as well.

The general structure of a quantum *dot* is developed in Chapter 4 through the use of the two-dimensional harmonic oscillator. This introduces angular momentum states and the role of the magnetic field in producing a complex spectra. Experiments on magneto-tunneling have probed the delicate details of this spectra, and these are discussed and reviewed. We discuss the limit of almost nontransparent barriers for which single-electron tunneling becomes the norm. This single electron tunneling discussion begins with the experiments on metallic islands. We then consider the double junction, then turn to semiconductor dots in which quantized levels become important. The Russian orthodox theory is then discussed. Finally, the complexity of arrays of quantum dots is discussed through the idea of lateral surface superlattices.

Chapter 5 begins our discussion of higher-order transport properties through weak localization and the universal conductance fluctuations (UCF) and their experimental scaling with size and temperature. This leads us to develop the necessary higher-order perturbation theory necessary to incorporate the long-range correlations responsible for these effects. Here we will introduce the impurity ladder diagrams and the complex multi-particle interactions that lead to *diffusons* and *cooperons*.

In Chapter 6, the temperature Green's functions are introduced to investigate the temperature dependence of the UCF, weak localization, and other quantum interference. It is necessary here to give a deeper treatment of the electron-electron interaction to handle the cooperons and "screening". The latter is then treated through the random-phase approximation, which also leads us to discuss the lifetimes of single-particle states.

Finally, in Chapter 7, we return to the more complex picture of the nanodevice, which leads us into the real-time Green's functions. The more extensive family of these functions is necessitated by the fact that the system is now out of equilibrium so that the "distribution function" is no longer known in the system under bias. Unfortunately, considerably less is known about the application of these functions to meaningful discussions of the various effects in mesoscopic devices. However, there are experiments that illustrate the presence of carrier heating and the need for the more extensive description. Along the way, this approach allows us to connect to simulations that have been carried out in reduced descriptions with the density matrix and the Wigner distribution function (reduced in the sense that the latter are single-time functions which ignore temporal correlations). We end with an attempt to loop back to semi-classical transport concepts.

1.5 What is omitted

In the previous section, a short introduction to the material that is included in this book was presented. The reader will note that the list is quite selective, in spite of covering perhaps too much material. However, this book cannot be a self-contained book on either the theory or the experiments in nanostructures. The former would be much too long by the time we introduce the new concepts that have appeared since [34]–[37] (and comparable other books). (Nearly all the experiments described in this chapter have appeared in barely more than the last decade.) Equally, it is not our desire to exhaustively review the experimental literature, since there are many reviews which address this task. Rather, we have tried to select a systematic set of experiments that leads to an understanding of what must be described by theory, and then to develop the required theory. Consequently, much is omitted, among which is a comprehensive review of the literature for either theory or experiment.

Bibliography

[1] J. M. Mikkelson, L. A. Hall, L. A. Malhotra, S. D. Seccombe, and M. S. Wilson, IEEE J. Sol. State Circuits **16**, 542 (1981).

[2] G. Baccarani, M. R. Wordeman, and R. H. Dennard, IEEE Trans. Electron Dev. **31**, 452 (1984).

[3] B. Hoeneisen and C. A. Mead, Sol.-State Electron. **15**, 819, 891 (1972).

[4] C. A. Mead, J. VLSI Signal Processing **8**, 9 (1995).

[5] Y. Jin, D. Mailly, F. Carcenac, B. Etiene, and H. Launois, Microelectr. Engr. **6**, 195 (1987).

[6] G. Bernstein, D. K. Ferry, and P. Newman, Superlatt. Microstruc. **4**, 308 (1985); J. Han, D. K. Ferry, and P. Newman, IEEE Electron Dev. Lett. **11**, 209 (1990).

[7] A. Ishibashi, K. Funato, and Y. Mori, Jpn. J. Appl. Phys. **27**, L2382 (1988).

[8] M. Ono, M. Saito, T. Yoshitomi, C. Fiegna, T. Ohgura, and H. Iwai, in *Proc. 1993 Intern. Electron Dev. Mtg.* (IEEE Press, New York, 1993), pp. 6.2.1–6.2.4.

[9] D. K. Ferry, in *Granular Nanoelectronics*, eds. D. K. Ferry, J. R. Barker, and C. Jacoboni (Plenum Press, New York, 1990), p. 1.

[10] J. R. Barker and D. K. Ferry, Sol.-State Electron. **23**, 519, 531 (1980).

[11] These are reviewed by M. Heiblum, in *High Speed Electronics*, eds. B. Kallback and H. Beneking (Springer-Verlag, Berlin, 1986), and by J. R. Hayes, A. J. F. Levi, A. C. Gossard, and J. H. English, in *High Speed Electronics*, eds. B. Kallback and H. Beneking (Springer-Verlag, Berlin, 1986).

[12] T. Yamada, J.-R. Zhou, and D. K. Ferry, IEEE Trans. Electron. Dev. **41**, 1513 (1994).

[13] N. F. Mott, Philos. Mag. **22**, 7 (1970).

[14] D. J. Thouless, Phys. Rept. **13C**, 93 (1974); J. T. Edwards and D. J. Thouless, J. Phys. C **5**, 807 (1972); D. C. Licciardello and D. J. Thouless, J. Phys. C **8**, 4157 (1975); D. J. Thouless, Phys. Rev. Lett. **39**, 1167 (1977).

[15] E. Abrahams, P. W. Anderson, D. C. Licciardello, and T. V. Ramakrishnan, Phys. Rev. Lett. **42**, 673 (1979).

[16] R. Landauer, IBM J. Res. Develop. **1**, 223 (1957); Philos. Mag. **21**, 863 (1970).

[17] B. J. van Wees, H. van Houten, C. W. J. Beenakker, J. G. Williamson, L. P. Kouwenhoven, D. van der Marel, and C. T. Foxon, Phys. Rev. Lett. **60**, 848 (1988).

[18] D. A. Wharam, T. J. Thornton, R. Newbury, M. Pepper, H. Ahmed, J. E. F. Frost, D. G. Hasko, D. C. Peacock, D. A. Ritchie, and G. A. C. Jones, J. Phys. C **21**, L209 (1988).

[19] Y. Imry, in *Directions in Condensed Matter Physics*, eds. G. Grinstein and E. Mazenko (World Scientific Press, Singapore, 1986), pp. 103–163.

[20] Y. Aharonov and D. Bohm, Phys. Rev. **115**, 485 (1959).

[21] R. A. Webb, S. Washburn, C. P. Umbach, and R. B. Laibowitz, Phys. Rev. Lett. **54**, 2596 (1985).

[22] K. Ishibashi, Y. Takagaki, K. Gamo, S. Namba, S. Ishida, K. Murase, Y. Aoyagi, and M. Kawabe, Sol. State Commun. **64**, 573 (1987).

[23] P. M. Mankiewich, R. E. Behringer, R. E. Howard, A. M. Chang, T. Y. Chang, B. Chelluri, J. Cunningham, and G. Timp, J. Vac. Sci. Technol. **B6**, 131 (1988).

[24] C. J. B. Ford, T. J. Thornton, R. Newbury, M. Pepper, H. Ahmed, C. T. Foxon, J. J. Harris, and C. Roberts, J. Phys. C **21**, L325 (1988).

[25] A. D. Stone, Phys. Rev. Lett. **54**, 2692 (1985).

[26] W. J. Skocpol, P. M. Mankiewich, R. E. Howard, L. D. Jackel, D. M. Tennant, and A. D. Stone, Phys. Rev. Lett. **56**, 2865 (1986).

[27] The data for this figure were provided by K. Ishibashi of the Institute for Physical and Chemical Research (RIKEN) and Prof. Y. Ochiai of Chiba University.

[28] D. K. Ferry, *Semiconductors* (Macmillan, New York, 1991).

[29] D. K. Ferry and H. Grubin, in *Solid-State Physics*, eds. H. Ehrenreich and W. Turnbull (Academic Press, New York, in press).

[30] M. A. Reed, ed., *Nanostructured Systems* (Academic Press, New York, 1992).

[31] H. A. Cerdeira, F. Guinea López, and U. Weiss, eds., *Quantum Fluctuations in Mesoscopic and Macroscopic Systems* (World Scientific Press, Singapore, 1990).

[32] S. Namba, C. Hamaguchi, and T. Ando, eds., *Science and Technology of Mesoscopic Structures* (Springer-Verlag, Tokyo, 1992).

[33] B. L. Altshuler, P. A. Lee, and R. A. Webb, Eds., *Mesoscopic Phenomena in Solids* (North-Holland, Amsterdam, 1991).

[34] A. L. Fetter and J. D. Walecka, *Quantum Theory of Many-Particle Systems* (McGraw-Hill, New York, 1971).

[35] L. P. Kadanoff and G. Baym, *Quantum Statistical Mechanics* (Benjamin, New York, 1962).

[36] G. D. Mahan, *Many-Particle Physics* (Plenum Press, New York, 1981).

[37] A. A. Abrikosov, L. P. Gorkov, and I. Ye. Dzyaloshinskii, *Quantum Field Theoretical Methods in Statistical Physics*, 2nd ed. (Pergamon Press, Oxford, 1965).

2

Quantum confined systems

As discussed in the previous chapter, there are two issues that distinguish transport in nanostructure systems from that in bulk systems. One is the granular or discrete nature of electronic charge, which evidences itself in single-electron charging phenomena (see Chapter 4). The second involves the preservation of phase coherence of the electron wave over short dimensions. Artificially confined structures are now routinely realized through advanced epitaxial growth and lithography techniques in which the relevant dimensions are smaller than the phase coherence length of charge carriers. We can distinguish two principal effects on the electronic motion depending on whether the carrier energy is less than or greater than the confining potential energy due to the artificial structure. In the former case, the electrons are generally described as bound in the direction normal to the confining potentials, which gives rise to quantization of the particle momentum and energy as discussed in Section 2.2. For such states, the envelope function of the carriers (within the effective mass approximation) is localized within the space defined by the classical turning points, and then decays away. Such decaying states are referred to as *evanescent states* and play a role in tunneling as discussed in Chapter 3. The time-dependent solution of the Schrödinger equation corresponds to oscillatory motion within the domain of the confining potential.

The second type of motion we will be concerned with is that associated with propagating states of the system. Here the carrier energy is such that it lies above that of the confining potentials, or that the potentials are limited sufficiently in extent so that quantum mechanical tunneling through such barriers can occur. In this regime, where potential variations occur on length scales smaller than the phase coherence length, transport is more readily defined in terms of reflection and transmission of matter waves. This leads to definition of conduction in terms of the Landauer formula discussed in Chapter 3. In the present chapter, we will concern ourselves with quantum confined structures and the so-called reduced-dimensionality systems associated with such confinement of the particle motion.

The topic of electronic states in quantum confined systems, and the host of fundamental studies and device applications related to this subject, is quite extensive. Since the purpose of this book is to elucidate transport in nanostructures, we will focus only on those aspects of quantum confined systems that relate to electronic transport, and thus for example ignore the well-studied optical properties of these systems except as they pertain to the main theme. There are a number of excellent reviews on the topic such as that of Ando, Fowler, and Stern [1] on the pioneering research in this field during the 1970s related to quantization at the Si/SiO$_2$ interface in Si MOS (metal oxide semiconductor) devices. More recent books devoted to this topic in semiconductor heterojunction systems are now available [2], [3].

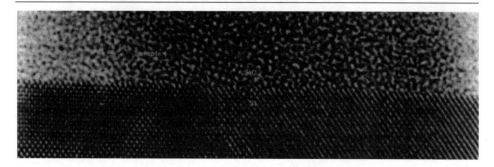

Fig. 2.1. High-resolution transmission electron microscope (HRTEM) image of the Si/SiO$_2$ interface on a Si(100) substrate [after S. M. Goodnick *et al.*, Phys. Rev. B **32**, 8171 (1985)].

2.1 Nanostructure materials

In the context of semiconductor materials and fabrication based on planar integrated circuit technology, quantum confinement may be realized in roughly two different ways: i) through the growth of inhomogeneous layer structures resulting in quantization perpendicular to the substrate surface, and ii) through lateral patterning using ultrafine lithography techniques. Historically, the development of quantum confined systems was realized in heterolayer structures grown on semiconducting substrates. Here, the discontinuities in the conduction and valence band edges between different materials behave as potential discontinuities within the framework of the effective mass approximation discussed in more detail later. The first demonstration of quantization of semiconductor states due to artificial confinement was in the inversion layers of Si MOS structures [4]. In this system, quantization of the carrier motion is due to the confining potential of the Si/SiO$_2$ interface barrier and the potential well in the other direction due to band bending. Part of the success of measurements made in the Si/SiO$_2$ system are due to the relatively ideal interface formed between Si and its oxide. Fig. 2.1 is a high-resolution transmission electron microscope (HRTEM) lattice image of a Si/SiO$_2$ interface on a Si(100) substrate. Here the lattice images in the Si are associated with the unresolved two-atom basis of the Si diamond lattice along the [110] direction. The apparent interface between the Si and the SiO$_2$ varies by only one or two atomic steps, which is sufficiently smooth to allow well-defined quantum states to exist at the interface, evidenced in optical and transport experiments. Later, with the development of precision epitaxial growth techniques such as molecular beam epitaxy (MBE) [5] and metal organic chemical vapor deposition (MOCVD), high-quality lattice-matched heterojunction systems could be realized. These systems exhibit quantum confinement effects far superior to those in the Si MOS system due to several factors, including the low surface state density at the interface of lattice-matched materials such as GaAs and Al$_x$Ga$_{1-x}$As, and the lower conduction band mass of III–V compound materials in general (compared to Si, which increases the energy spacing of the quantized levels). Figure 2.2 shows a scanning tunneling microscopy (STM) image of an InAs/GaSb *superlattice* grown by MBE. Since the lattice constants of InAs and GaSb are quite close, there is very little interface strain between the two *lattice-matched* binary systems. In the STM process, an ultrafine tungsten tip is scanned across the surface, the tunneling current used as a sensitive indicator of the feature height on the surface. As such, there is no averaging of the roughness due to projection through successive atomic layers as in the HRTEM measurement case. Although the system is lattice-matched, roughness on the order of one or two monolayers is readily apparent in this image.

Fig. 2.2. Scanning tunneling microscope (STM) image of an InAs/GaSb superlattice illustrating interface roughness in a lattice-matched system. [After R. M. Feenstra *et al.*, Phys. Rev. Lett. **72**, 2749 (1991), by permission.]

A further innovation which is an essential component of a large fraction of nanostructures studied today is the technique of modulation doping (MD) [6], in which the dopants responsible for free carriers in the heterostructure are spatially separated from the carriers themselves. This scheme usually involves growth of an undoped active layer with a smaller bandgap such as GaAs, onto which a doped wider bandgap material such as $Al_xGa_{1-x}As$ is grown with an undoped *spacer* layer. Figure 2.3 shows an illustration of a so-called high-electron mobility transistor (HEMT) based on the principle of modulation doping. In this structure, a layer of unintentionally doped GaAs (typically with carrier density between 10^{14}–10^{15}/cm^3) is grown on a semi-insulating GaAs substrate. Then an undoped spacer layer of $Al_xGa_{1-x}As$ is grown, followed by a thicker heavily doped layer of the same material. A cap layer of doped or undoped GaAs is usually grown on top to facilitate ohmic contact formation and to reduce oxidation of the AlGaAs layer. Since the free carriers forming a channel at the GaAs/AlGaAs interface are spatially remote from their parent donors, ionized impurity scattering is greatly reduced. Mobilities in excess of 1×10^7/cm^2/V-s are now obtainable in such structures, as discussed in Section 2.7.5 [7].

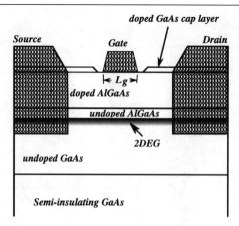

Fig. 2.3. Cross section of a high-electron moblity transistor (HEMT) based on modulation doping.

2.2 Quantization in heterojunction system

For simplicity, we will restrict our discussion of transport in nanostructures primarily to the conduction band states of the GaAs/AlGaAs system in which the vast majority of experiments have been performed. Here we consider the form of the electronic states in a layered heterojunction system introduced in the previous section. The electronic states in GaAs/AlGaAs systems are most easily described within the effective mass approximation (EMA). The criterion for the validity of this approach in heterojunctions has been discussed in detail [3]. As long as we associate ourselves with states of the same symmetry across the heterojunction, such as Γ valley electrons in both the GaAs and $Al_xGa_{1-x}As$ (x less than 0.44 for direct gap behavior), the motion of particles relative to the band edge is found by solving the envelope function equation

$$\left(\frac{\hbar^2}{2} \frac{\partial}{\partial z} \frac{1}{m(z)} \frac{\partial}{\partial z} + \frac{\hbar^2}{2m_\parallel} \nabla_\mathbf{r}^2 + V_{eff}(z) \right) \psi(\mathbf{r}, z) = E\psi(\mathbf{r}, z), \tag{2.1}$$

where $m(z)$ is the effective mass perpendicular to the heterointerface, $m_\parallel$ is the mass parallel to the interface, where $\mathbf{r}$ is the position vector parallel to the interface, and z is the direction perpendicular to the interface. The validity of the EMA requires that the envelope function $\psi(\mathbf{r}, z)$ be slowly varying over dimensions comparable to the unit cell of the crystal. The form for the kinetic energy operator for the motion perpendicular to the interface has been chosen to satisfy the requirement that the Hamiltonian be Hermitian (i.e., have real energy eigenvalues). $V_{eff}(z)$ in Eq. (2.1) is the effective potential energy normal to the interface,

$$V(z) = E_c(z) + V_D(z) + V_{ee}(z), \tag{2.2}$$

where $E_c(z)$ is the heterojunction conduction band discontinuities, $V_D(z)$ is the electrostatic potential due to ionized donors and accepters, and $V_{ee}(z)$ is the self-consistent Hartree and exchange potentials due to free carriers. Since for the heterojunction system the potential variation is only in the z-direction, the solution is separable as

$$\psi(\mathbf{r}, z) = \frac{1}{A^{1/2}} e^{i\mathbf{k}\cdot\mathbf{r}} \varphi_n(z), \tag{2.3}$$

corresponding to free electron motion in the plane parallel to the interface, where $\mathbf{k}$ is the wavevector in the plane parallel to the interface; A, the normalization factor, is the lateral area of the system; and n labels the eigenstates in the normal direction. The one-dimensional eigenfunctions $\varphi_n(z)$ satisfy

$$\left(\frac{\hbar^2}{2}\frac{\partial}{\partial z}\frac{1}{m(z)}\frac{\partial}{\partial z} + V_{eff}(z)\right)\varphi_n(z) = E_n\varphi_n(z). \tag{2.4}$$

The total energy relative to the band minima (maxima) is thus

$$E_{n,k} = \frac{\hbar^2 k^2}{2m_\parallel} + E_n, \tag{2.5}$$

assuming parabolic bands for simplicity. Continuity requires that the envelope function $\varphi_n(z)$ be continuous. For abrupt variations in the material, conservation of probability current requires that $m(z)^{-1}\partial\varphi_n(z)/\partial z$ be continuous, in analogy to the continuity of the electric field across an abrupt interface between two materials of differing dielectric constants.

2.2.1 Quantum wells and quasi-two-dimensional systems

The solution of Eq. (2.4) for $\varphi_n(z)$ allows both bound state and propagating solutions depending on the energy and the detailed form of the potential. As the simplest example of bound state behavior, consider the *quantum well* formed by a thin GaAs layer of thickness L sandwiched between two larger bandgap $Al_xGa_{1-x}As$ layers, as shown in Fig. 2.4. In this figure the bandgap difference is distributed between the valence and conduction bands in such a way that both electrons and holes are confined to the smaller bandgap GaAs layer. Such a heterojunction is referred to as a Type I system, to which we will primarily confine ourselves through the rest of the book.

The 300K bandgap of $Al_xGa_{1-x}As$ as a function of the Al mole fraction is given empirically [8] as $E_g^\Gamma(x) = 1.424 + 1.245x$ for $x < 0.45$. The actual fraction of the bandgap difference that appears across the conduction band, ΔE_c, compared to that in the valence band, ΔE_v, is a subject of continued investigation (see for example [9]). Most recent measurements based on photoluminesence studies of quantum well systems find that roughly 65% of the bandgap difference appears as the conduction band discontinuity [10]. Thus we see that for $x = 0.3$, for example, the bandgap discontinuity would be 243 meV according

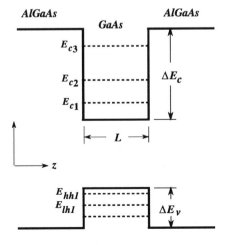

Fig. 2.4. Schematic of a Type I AlGaAs/GaAs/AlGaAs quantum well.

to the empirical relation above. If we assume a low free carrier density (less than $10^{10}/\text{cm}^2$) and low doping, then the terms $V_D(z)$ and $V_{ee}(z)$ in Eq. (2.2) are negligible, and the solution to Eq. (2.4) is that of a simple finite square well – a classic problem in elementary quantum mechanics.

As shown in Fig. 2.4, discrete levels form in the GaAs conduction and valence bands due to the confinement associated with the bandgap discontinuities. In the valence band, bound states associated with both light holes and heavy holes are possible. The quantum confinement further mixes the light and heavy hole bulk states so that the bound states in the valence band have the character of both types.

If we assume that the potential barriers are large compared to the bound state energies E_n, one can consider the even simpler solution corresponding to infinite potential barriers. Choosing the left interface as the origin, the boundary conditions in this case are that the envelope function vanishes at the points $z = 0, L$. The normalized solutions satisfying these conditions are simple sinusoids:

$$\varphi_n(z) = \sqrt{\frac{2}{L}} \sin\left(\frac{n\pi z}{L}\right) \quad n = 1, 2, \ldots. \tag{2.6}$$

The corresponding eigen-energies are given by

$$E_n = \frac{n^2 \hbar^2 \pi^2}{2 m_z L^2}. \tag{2.7}$$

E_n varies from several hundred meV for $L < 50$ nm to several meV as L approaches 100 nm. The motion in the plane parallel to the interface is free, whereas that normal to the epitaxial growth direction is confined. Such as system is referred to as a quasi-two-dimensional electron gas (2DEG) structure and plays a central role in the many nanostructure devices. For each solution n there exists a continuum of two-dimensional states called *subbands*, although later, in connection with quantum waveguide structures discussed in the next chapter, these solutions are sometimes referred to as *modes* in analogy to the standing wave solutions one would encounter for electromagnetic waves propagating between metallic plates.

To make a connection to the introductory remarks of the previous chapter, we note that the solutions (2.6) represent states that give rise to a coherent superposition of single-particle wave functions. Other possible solutions destructively interfere and thus decay with time. Obviously, the preservation of phase coherence is necessary to establish the quantization condition given by Eq. (2.7). As the length L becomes longer than the phase coherence length introduced in the last chapter, the particle loses its phase "memory" of the boundaries and thus behaves in a bulk-like fashion rather than as a quantized system.

It is instructive to consider the density of states (per unit area) of the 2DEG system above. The density of states may be calculated directly from

$$D(E) = \sum_{\alpha_s, \eta_v, n, \mathbf{k}} \delta(E - E_{n,k}), \tag{2.8}$$

where α_s and η_v are the spin and valley degeneracy (for a multi-valley minimum). Assuming a simple spherical conduction band minimum such that $E_{n,k}$ is given by Eq. (2.5), the sum over the quasi-continuous wavevectors $\mathbf{k}$ may be converted to an integration, and the

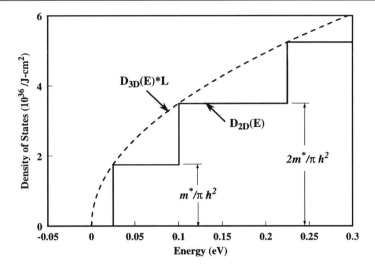

Fig. 2.5. Density of states versus energy for a 15-nm GaAs infinite quantum well (solid curve). The equivalent 3D density of states in the well are included here for comparison.

corresponding angular integration performed as

$$\sum_{\alpha_s, \eta_v, n, \mathbf{k}} \longrightarrow 2\eta_v \sum_n \frac{1}{2\pi} \int_0^\infty dk k \delta\left(E - E_n - \frac{\hbar^2 k^2}{2m_\parallel}\right), \qquad (2.9)$$

assuming two-fold spin degeneracy per state. The integration of k is easily performed to obtain

$$D(E) = \sum_n \frac{\eta_v m_\parallel}{\pi \hbar^2} \theta(E - E_n) = \sum_n D_o \theta(E - E_n), \qquad (2.10)$$

where θ is the unit step function. The density of states is thus a staircase in which the density of states for individual subbands is constant in energy characteristic of a two-dimensional system. Figure 2.5 shows the density of states for a 15 nm AlGaAs/GaAs quantum well versus energy. As shown by comparison to the parabolic 3D density of states in the well, the 2D density of states is a piecewise-continuous approximation to the 3D case.

The two-dimensional density of electrons in the well in equilibrium may be calculated by integrating the density of states (2.10) with the Fermi function for electrons to give

$$n_{2d} = \int_0^\infty dE D(E) f_o(E) = \sum_n N_n = k_B T D_o \sum_n \ln\left(1 + e^{(E_f - E_n)/kT}\right), \qquad (2.11)$$

where N_n is the 2D density of carriers in the nth subband, E_f is the Fermi energy of the system, and T is the temperature of the electron gas. As the well width decreases, the energy spacing of the levels increases, and the system behaves increasingly like a pure two-dimensional system. The *extreme quantum limit* corresponds to the case in which only the ground ($n = 1$) subband is occupied by electrons. This limit is realized at low temperatures where the electron density is sufficiently small such that E_F is below the second subband. Then Eq. (2.11) reduces to

$$n_{2d} = D_o E_F, \qquad (2.12)$$

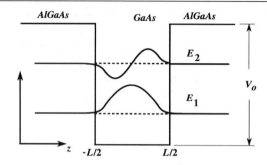

Fig. 2.6. Conduction band profile of a finite square well system showing the ground and the first excited subband.

and the system appears metallic with a well-defined Fermi surface (in this case, perimeter), with the Fermi wavevector given by

$$k_F = \sqrt{2\pi n_{2d}}. \tag{2.13}$$

For the problem of a finite-barrier-height quantum well, consider the conduction band profile shown in Fig. 2.6 with $V_o = \Delta E_c$. Generally, when considering such piecewise-continuous one-dimensional potentials, one would write the solution in each region in terms of the two independent solutions of the wave equation for a constant potential. States with energies below the potential barrier V_o are denoted *bound states*, and those above are known as *propagating* or *continuum states*. In characterizing the latter types of solutions, we would use the whole set of independent states from each region, using the matching conditions discussed earlier, and characterize the solutions in terms of their respective transmission and reflection characteristics (discussed in the next chapter). Of course, the bound states are important in connection with our discussion of quantum confined systems. First, we can simplify the solutions by recognizing the inversion symmetry around the center of the well and choosing the origin of our system there, as shown in Figure 2.6. The solutions are thus symmetric or antisymmetric with respect to the origin, which may be chosen as sine or cosine functions depending on the symmetry. The solutions in the classically forbidden regions are real exponential solutions (evanescent states), which must remain bounded away from the well. Therefore, only states that decay in space away from the well are allowed, and the solution in the three regions can be written as

$$\varphi_n(z) = \begin{cases} Be^{\kappa_z z} & \text{for } z < -\frac{L}{2} \\ A\cos(k_z z)\{\sin(k_z z)\} & \text{for } -\frac{L}{2} < z < \frac{L}{2}, \\ Be^{-\kappa_z z} & \text{for } z > \frac{L}{2}, \end{cases} \tag{2.14}$$

where $k_z = \sqrt{2m_I E/\hbar^2}$ and $\kappa_z = \sqrt{2m_{II}(V_o - E)/\hbar^2}$. Since we have used symmetry to equate the coefficient on the left and right evanescent states, we need only apply the matching conditions for $\varphi(z)$ and $m(z)^{-1}\partial\varphi_n(z)/\partial z$ at either boundary. The energy eigenvalues are then given for the even cosine solutions by the transcendental equations

$$(k_z/m_{II})\tan\left(\frac{k_z L}{2}\right) = \kappa_z/m_I \tag{2.15}$$

and similarly for the odd sine solutions as

$$(k_z/m_{II})\cot\left(\frac{k_z L}{2}\right) = -\kappa_z/m_I. \tag{2.16}$$

The corresponding energies must be solved graphically or numerically for all possible values lying in the well. (At least one bound state exists for any thickness and barrier height.) As is expected, the difference between the finite well solutions and the infinite well case are smallest for the ground ($n = 1$) solutions and become increasingly worse as one goes up in energy.

2.2.2 Coupled wells and superlattices

When we went from the case of the infinite well to a finite barrier well, one of the main differences is the existence in the latter case of decaying states in the barrier regions. If we have several quantum wells that are separated well beyond the characteristic decay length of a particular state, then they are expected to behave as isolated single wells. Such a system of uncoupled quantum wells is often referred to as *multiple quantum wells* (MQWs); the system would be characterized by N degenerate states for each level, where N is the number of wells.

If the barrier thickness between two wells is decreased, the overlap of the envelope function from each well into the other increases. If the overlap is small, one may treat this interaction using degenerate perturbation theory (see for example [11]). As is well known from this formalism, the degeneracy of the states of the two identical wells is split by the presence of the other, with the degree of splitting depending on the matrix element connecting the two wells which increases as the well separation is decreased. Figure 2.7 shows an exact numerical calculation of Eq. (2.4) for two 2.5 nm GaAs quantum wells separated by a 10 nm Al$_{.35}$Ga$_{.65}$As barrier for the two lowest states. As was discussed, the lowest two states are split by an amount ΔE. The two solutions for this symmetric well correspond to envelope functions that are symmetric and antisymmetric with respect to the two wells as shown. Figure 2.8 shows the calculated energy eigenvalues for two 10-nm quantum wells separated by various barrier widths. As expected, the splitting of the finite well states increases as the barrier separating the wells decreases. Significant splitting

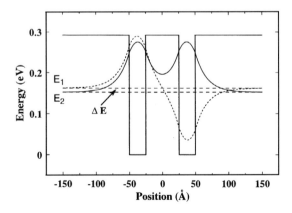

Fig. 2.7. Envelope functions of the lowest two conduction band states for two symmetric coupled Al$_{.35}$Ga$_{.65}$As/GaAs quantum wells.

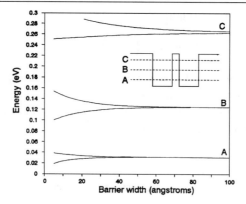

Fig. 2.8. Subband minima for two coupled 100-nm finite square wells, separated by an $Al_{.35}Ga_{.65}As$ barrier of variable width [after J. Lary, Ph.D. dissertation].

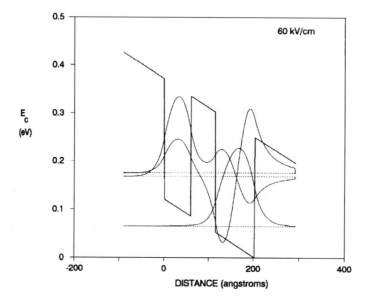

Fig. 2.9. Asymmetric coupled wells biased into resonance.

of the ground subband level does not occur until the barrier thickness is less than 4 nm, whereas for higher levels the splitting is more pronounced. This fact reflects the larger penetration [smaller κ_z in Eq. (2.14)] of the upper subband envelope functions into the AlGaAs due to the smaller effective barrier. The splitting of the levels of the symmetric coupled well may also occur in asymmetric wells when, for example, an external electric field is applied. Figure 2.9 shows a wide and narrow quantum well under the influence of a uniform electric field. Experimentally, this situation has been realized by imbedding the wells in the intrinsic region of a reverse-biased *p-i-n* diode [12]. As shown in Fig. 2.9, the two wells have been biased in such a way that the $n = 1$ state of the narrow well is in resonance with the $n = 2$ state of the wide well. The two levels split by an amount dependent on the overlap of the envelope functions, and again they have a symmetric/antisymmetric character in the two wells. Note that the $n = 1$ solution of the wide well is localized entirely in that well with little communication with the narrow well. The coupled well structure illustrates nicely the concept of tunneling, which we will come back to in the next chapter. Tunneling basically refers to the quantum mechanical phenomena of the possibility of a particle traversing a classically forbidden region such as the potential barrier shown in Fig. 2.9. For the asymmetric coupled well system, a particle existing in the $n = 1$ state of the wide well has very little probability of existing at some later time in the narrow well, as can be seen from the decay of the envelope function there. However, when two states coincide, a resonance condition exists in which there is a greatly enhanced probability of a particle tunneling from one well to another. Such a situation is referred to as *resonant tunneling*, to be discussed in detail in Chapter 3. Experimentally the resonant transfer of carriers from one well to another in coupled well systems has been observed both in optical studies [13] and from transport measurements [14].

The time-dependent solutions of such resonant coupled well systems have an interesting behavior. If one were able to prepare the system initially with an electron on one side or the other (i.e., by a mixed superposition of the two solutions), then the time-dependent solution of the Schrödinger equation predicts that the wavepacket will oscillate back and forth between the two wells with a characteristic frequency $\omega = \Delta E / \hbar$ until the phase coherence of the packet is destroyed through inelastic scattering, as discussed previously. Evidence for such coherent oscillations have in fact been reported in ultrafast pump and probe optical experiments in which the time resolution is sufficiently short to resolve the wavepacket motion [13]. Such charge oscillations have also been shown to generate terahertz frequency radiation characteristic of the oscillation frequencies predicted by the simple splitting of the levels, which may have practical applications as well [15].

We can extend the coupled well case to consider many identical wells. Figure 2.10 illustrates the behavior of the level splitting versus barrier width for 5 wells. Now the five-fold degeneracy of each level is split by the perturbation of the neighboring wells, giving rise to a band of discrete levels, with one state in each band arising from each well. Of course, this picture looks very similar to the broadening of the atomic levels in a solid to form energy bands within the tight binding or LCAO (linear combination of atomic orbitals) model discussed in solid state physics textbooks (see for example [16], [17]). If we increase the number of wells indefinitely, we form a semiconductor *superlattice*. One can impose periodic boundary conditions and express the solutions of Eq. (2.4) for the one-dimensional periodic potential in the form of Bloch functions. Again, if the doping and free carrier contributions to the potential are negligible, the solution for a Type I superlattice such as shown in Fig. 2.10. (extended infinitely) may be obtained from the textbook Kronig-Penney model (see for example [16]), generalized to the case of different effective masses in the barriers and wells. Thus the envelope function solutions to Eq. (2.4) can be expressed as

$$\varphi_n(z) = e^{ik_z z} u_{n,k_z}(z); \qquad u(z + L) = u(z), \qquad (2.17)$$

where L is the period of the superlattice, and k_z is the wavevector along the growth direction of the multilayer structure. Figure 2.11 shows a calculation based on the Kronig-Penney model for the allowed energies as a function of barrier and well width for a fixed barrier height of $V_o = 0.4$ eV [18]. The discrete bands of states shown in Fig. 2.10 now become a continuum of states (*minibands*), with *minigaps* separating them. Such minibands exist both above and below the energy barrier V_o. As the barrier width increases, the minibands shrink in width back to the discrete states of the individual wells becoming a system of multiple quantum wells rather than a superlattice.

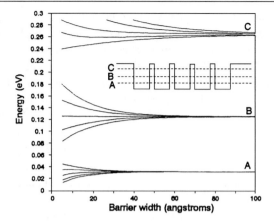

Fig. 2.10. Same calculation as Figure 2.8, but for 5 coupled wells [after J. Lary, Ph.D. dissertation].

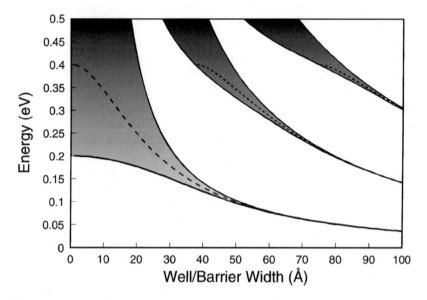

Fig. 2.11. Allowed energy bands versus well and barrier width for a superlattice with barrier height of $V_o = 0.4$ eV.

2.2.3 Doped heterojunction systems and self-consistent solutions

In the preceding discussion, we have ignored the effects of ionized donors and accepters, as well as the Coulomb interaction of free electrons in the solution of the bound states in hetero-junction systems. Undoped quantum well systems in which such effects are negligible are often used in optical studies and in optoelectronic devices. However, in electronic devices, free carriers are necessary for charge transport to occur. In particular, the modulation-doped structures discussed in Section 2.1 find widespread use in semiconductor mesoscopic devices due to the high mobilities, and hence long elastic mean free paths, one of the critical length scales discussed in the introduction. Such high-electron-mobility heterojunction material very often forms the basis for fabricating the quantum waveguide and quantum dot structures discussed in the subsequent chapters, in which a host of rich physical phenomena are observed.

In order to calculate the electronic states of doped heterolayer systems, one must include the additional terms in Eq. (2.2) neglected in the previous two sections, namely $V_D(z)$ and $V_{ee}(z)$. The many body contribution to the one-electron potential energy, $V_{ee}(z)$, may be approximated as

$$V_{ee}(z) = V_h(z) + V_{xc}(z). \tag{2.18}$$

$V_{xc}(z)$ is the exchange and correlation contributions to the one-electron potential energy, while $V_h(z)$ represents the *Hartree* contribution, the electrostatic potential energy arising from the average charge density of all the other electrons in the system. Since $V_h(z)$ is electrostatic in origin, we may combine it with the energy $V_D(z)$ obtained from the solution to Poisson's equation,

$$\nabla^2 \phi_e = \frac{q[\rho_I(z) - qn(z)]}{\epsilon_s}; \quad \phi_e = V_h + V_D, \tag{2.19}$$

where $\rho_I(z)$ is the charge density due to ionized donors and acceptors, $\epsilon_s = \kappa_s \epsilon_o$ is the permittivity of the semiconductor, q is the magnitude of the electric charge, and $n(z)$ is the 3D particle density of the 2DEG. This density is given by

$$n(z) = \sum_i N_i |\varphi_i(z)|^2, \tag{2.20}$$

where N_i is the 2D density of carriers in each subband given by Eq. (2.11). In writing (2.19), we have neglected the variation of the dielectric constant across the interface of a given heterojunction. For an abrupt interface between two dielectric materials, an additional image potential should be added to Eq. (2.18) to satisfy the boundary conditions of the normal electric field at a dielectric discontinuity. However, this leads to a noninterable singularity in the energy when the envelope function penetration into the barrier cannot be neglected. This singularity can be avoided by including a finite grading at the interface, which must exist at the atomic scale, as discussed by Stern [19], [20]. Fortunately, in the case of the AlGaAs/GaAs system, the difference in dielectric constants is relatively small, and thus the image potential is usually ignored.

The exchange-correlation contribution to the one-electron energy is usually included through the density-functional method using the Kohn-Sham equation with the local density-functional approximation [21], [22]. Various parameterizations for V_{xc} within the local-density-functional approximation give similar results. Das Sarma and Vinter [22] have used [23]

$$V_{xc}(z) = -[1 + 0.7734x \ln(1 + x^{-1})](2/\pi \alpha r_s)Ry^*$$

$$\alpha = (4/9\pi)^{1/3}, \quad x \equiv x(z) = r_s/21, \quad r_s \equiv r_s(z) = \left[\frac{4}{3}\pi a^{*3} n(z)\right]^{-1/3}, \tag{2.21}$$

where $a^* = \kappa_s \hbar^2/m^* q^2$ is the effective Bohr radius, $n(z)$ is given by Eq. (2.20), and the energy is measured in units of effective Rydbergs (approximately 5 meV for GaAs). The above equation assumes a uniform dielectric constant across the interface, which as mentioned earlier is not a bad approximation for the GaAs/AlGaAs system. The correct calculation of $V_{xc}(z)$ in the presence of a dielectric interface is still somewhat an open question [20].

The inclusion of many body effects due to V_h and V_{xc} couples the nonlinear envelope function equation (2.4) to Poisson's equation through $n(z)$. The simultaneous solutions of these two equations are referred to as *self-consistent* solutions, solved numerically in an iterative fashion [24]. The usual procedure is to begin the calculation with a trial solution for the 1D Schrödinger equation, $\varphi_n^0(z)$, such as the solution in the absence of the many-body potential or a solution derived from some model potential. The initial particle density $n^0(z)$ is calculated from Eq. (2.20), and Poisson's equation is solved numerically using for example finite differences. Once the electrostatic potential is calculated, together with the exchange-correlation potential (2.21) using $n^0(z)$, the Schrödinger equation is solved numerically using, for example, the Numerov method [25] or finite elements techniques [26] to obtain the new solutions, $\varphi_n^1(z)$ and E_n^1. These solutions are used as an approximation for the second iteration, and the procedure continues until the nth iteration converges (hopefully) to within some acceptable numerical error.

We return now to the problem of the modulation-doped single-heterojunction system shown in Fig. 2.3. Figure 2.12a shows the energy band diagram of a modulation-doped heterojunction between two semiinfinite regions of GaAs and $Al_xGa_{1-x}As$. A thin spacer layer of thickness d_s separates a region of uniform n-type doping of concentration N_d (Si for example) in the AlGaAs from the 2DEG at the interface. ΔE_c characterizes the effective potential due to the band offset of the heterojunction, which as discussed earlier is a function of the mole fraction x of Al in the $Al_xGa_{1-x}As$. The GaAs is assumed to be lightly p-doped with an acceptor concentration of N_a. The charge density of either side of the junction decays away into the bulk regions where charge neutrality exists. Electrons from ionized donors in the $Al_xGa_{1-x}As$ transfer to the lower-energy GaAs layer. Due to the confinement associated with the band-bending in the GaAs and the band offset, ΔE_c, the electrons are localized at the interface and form a quantum-confined system. To calculate the equilibrium charge density in the 2DEG, we can make the approximation that the free carrier concentration is zero inside of the depletion region of width W on the left, and of width z_d on the right (depletion approximation), except in the region close to the interface. Then the charge density due to ionized donors in the $Al_xGa_{1-x}As$, and to ionized acceptors in the GaAs is shown schematically in Fig. 2.12b. The depletion width in the GaAs may be given approximately by the inversion condition for the formation of the 2DEG at the interface:

$$z_d = \left(\frac{2\epsilon_s \psi_{s,inv}}{q N_a}\right)^{1/2}, \quad \psi_{s,inv} \approx 2(E_i - E_{fp})/q, \tag{2.22}$$

where $\psi_{s,inv}$ is the total electrostatic band-bending when electrons start to populate the inversion layer 2DEG, and $(E_i - E_{fp})$ is the difference between the intrinsic Fermi energy and the Fermi level in the GaAs bulk, calculated as discussed in the usual treatises on semiconductor physics (see for example Chapter 6, reference [17]). The depletion width W on the n-side and the sheet density of carriers in the 2DEG, n_{2d}, are coupled together through Gauss' law:

$$(W - d_s)N_d = n_{2d}(E_f) + n_{depl}, \tag{2.23}$$

where $n_{depl} = N_a z_d$ assuming overall charge neutrality. The total potential variation in the $Al_xGa_{1-x}As$, V_{bi}, is also related to the depletion width W and is found by integrating the

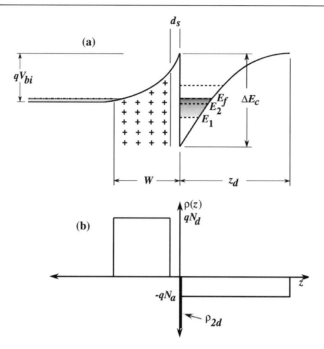

Fig. 2.12. (a) Conduction band profile through a modulation-doped heterojunction system. (b) Charge density versus distance due to ionized donors and acceptors.

charge distribution of Fig. 2.12b from $-\infty$ to 0 twice to obtain

$$V_{bi} = \frac{q N_d (W^2 - d_s^2)}{2\epsilon_s}. \tag{2.24}$$

From the band diagram, Fig. 2.12, we can deduce the following relation between the V_{bi} and the Fermi energy:

$$q\psi_n + \frac{q^2 N_d (W^2 - d_s^2)}{2\epsilon_s} = \Delta E_c - E_f. \tag{2.25}$$

Where $q\psi_n = E_c - E_{fn}$ is the difference between the conduction band edge and the Fermi energy in the bulk AlGaAs far from the interface. Further, n_{2d} is an explicit function of the Fermi energy through (2.11) once the subband energies E_n are given. Thus Eqs. (2.23) and (2.25) may be solved simultaneously to get W and E_f.

In order to obtain the correct subband energies and charge transfer in the well, the iterative self-consistent procedure described earlier is followed. However, now the total charge in the 2DEG, n_{2d}, must be updated according to the solution for E_f described above after every iteration step in which E_n is recalculated. Such an analysis can be further extended to the case in which the AlGaAs layer is finite in extent, terminated with a metal gate as in an actual HEMT. (For more details of charge control in HEMTs, the reader is referred to [27], Chapter 13.) Figure 2.13 shows the result of such a calculation for the probability density of the first three confined states superimposed on the conduction band profile. Here only the Hartree contribution has been included, and V_{xc} has been neglected. The parameters used in the calculation are shown in the figure key. There are several interesting features to notice about the solutions. The first three solutions are localized primarily in the GaAs with an exponential tail in the AlGaAs due to the finite barrier. However, if a lower conduction band offset had been chosen (for example $\Delta E_c = 0.125$ eV), the $n = 3$ state would become a bound state located in the AlGaAs barrier rather than the GaAs. Such solutions localized in the barrier are associated with an important effect in the transport properties of actual quantum confined systems, that of *real space transfer* [28]. If the carriers in the 2DEG of the channel are given sufficient energy, for example, via a d.c. electric field applied in the plane of the heterojunction, electrons may populate higher levels and effectively transfer into the AlGaAs layer. Charge transfer into the AlGaAs also gives rise to a problematic effect in HEMTs sometimes referred to as the *parasitic MESFET effect* [27]. As the gate voltage is made more positive, the Fermi energy in the 2DEG moves up in energy. Eventually, with sufficient forward bias, strong charge accumulation occurs in the AlGaAs. Since the transport properties in the heavily doped AlGaAs layer are far inferior to that of the 2DEG in the undoped GaAsAs, a noticeable saturation of channel current with gate bias occurs. A further point to note in the solutions shown in Fig. 2.13 is that the number of lobes in the probability density corresponds to the subband index n, as was the case of the square quantum well. However, in the present case, the effective width of the nth level,

$$\langle z_n \rangle = \int_{-\infty}^{\infty} dz z \varphi_n^2(z), \tag{2.26}$$

increases with increasing subband index.

Before leaving the topic of the modulation-doped, single-heterojunction system, we should mention one model solution that is often employed, the so-called triangular-well approximation. One can observe in Figures 2.12 and 2.13 that the potential variation on the GaAs side in the vicinity of the interface is roughly linear. Thus we can assume that the potential may be written

$$V(z) = q F_{eff} z \qquad z > 0$$

$$F_{eff} = \frac{q}{\epsilon_s} \left(n_{depl} + \frac{1}{2} n_{2d} \right), \tag{2.27}$$

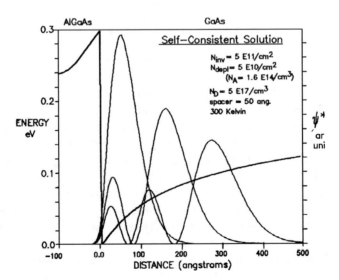

Fig. 2.13. Probability density versus distance for the first three levels using self-consistent solutions.

where F_{eff} is an effective surface field and the factor of 1/2 accounts for the average field in the channel due to the presence of the free carrier charge, n_{2d}. If the barrier height ΔE_c is sufficiently large, we can neglect the penetration of the envelope function in that region and assume that $\varphi_n(z) = 0$ for $z < 0$. Exact solutions exist for the linear or triangular potential in the form of Airy functions

$$\varphi_n(z) = C_n Ai \left\{ \left(\frac{2m_z}{\hbar^2 q^2 F_{eff}^2} \right)^{1/3} (q F_{eff} z - E_n) \right\}, \qquad (2.28)$$

where C_n is the normalization constant determined from the recursion relations for the Airy function, $Ai(z)$. The energy eigenvalues E_n are determined from the Dirichlet condition at $z = 0$, which to lowest order may be expressed analytically from the asymptotic expansion of the zeros of the Airy function (see [29]):

$$E_n \cong \left(\frac{\hbar^2}{2m_z} \right)^{1/3} \left(\frac{3\pi q F_{eff}}{2} \left[n - \frac{1}{4} \right] \right)^{2/3} \qquad n = 1, 2, \ldots. \qquad (2.29)$$

Table 2.1 shows a comparison of the first five energy eigenvalues, subband occupancy, and Fermi energy for the triangular-well approximation and the self-consistent calculation in the Hartree approximation for the parameters indicated. As expected, as the energy increases, the discrepancy between the two solutions increases due to the softening of the actual self-consistent potential with increasing energy.

2.3 Lateral confinement: Quantum wires and quantum dots

2.3.1 Nanolithography

In Section 2.2 we were concerned with both the quantization of energy states and the corresponding reduced dimensionality of the free carrier system due to quantum confinement in layered heterojunction systems. In such layered systems the motion of carriers in the plane perpendicular to the material growth direction is unconstrained. The next advance historically began with early attempts to confine the lateral motion in the plane of the

Table 2.1. *Comparison of the self-consistent (SC) and the triangular potential (TP) solutions for an AlGaAs/GaAs single heterojunction. Shown are the calculated energy eigenvalues and percent occupancy of each level. The calculation is at 300K for a 2D density of* 5×10^{11}/cm^2 *and a depletion density of* $n_{depl} = 5 \times 10^{10}$/cm^2. *The donor density in the AlGaAs is* 5×10^{17}/cm^3, *the spacer layer 5 nm, and the conduction band offset* $\Delta E_c = 0.3$ eV.

Level	Energy (SC) (meV)	Occupancy (SC) (%)	Energy (TP) (meV)	Occupancy (TP) (%)
1	55.3	60.2	50.0	70.4
2	87.9	19.8	87.3	20.0
3	107.0	9.9	118.0	6.4
4	120.0	6.0	145.0	2.3
5	130.0	4.1	170.0	0.9
Fermi	38.1		37.9	

2DEG using advanced *lithographic* techniques beginning in the late 1970s and early 1980s. *Lithography* refers to the transference of a desired pattern onto a substrate. The primary technique to accomplish this in modern IC technology involves a *resist*, which when selectively exposed to a particular radiation source, changes its chemical bonding in such a way that the irradiated portion dissolves when immersed in a developer, while the unexposed resist remains (positive resist), or vice versa (negative resist). To realize quantum confinement in the lateral directions, the dimensions must be such that the quantized energies are resolvable from thermal broadening and broadening due to the existing disorder in the system. As a rule of thumb, we want lithography tools capable of transferring patterns with critical dimensions less than 0.1 μm, although this is not absolutely necessary (discussed below). As one might expect, diffraction effects impose limitations on the wavelength of the radiation source employed to expose the resist, although clever schemes such as the use of phase-shift masks have extended conventional optical lithography down to dimensions less than 0.2 μm line width. Beyond this, however, extensive research and development is directed towards shorter-wavelength electromagnetic sources such as deep-ultraviolet and x-rays. The current challenges are not related to the sources themselves, but rather to the associated optics and mask materials that allow delineation of small features.

The most popular technique to date for realizing nanometer structures in the research laboratory setting has been the use of electron-beam (e-beam) lithography. The characteristic wavelength for electrons at the energies used (typically 30–50 keV) is on the order of interatomic distances. The actual limitation on the resolution is related to scattering due to the inelastic loss processes as the high-energy electrons lose energy to or are reflected by the substrate. Such factors limit the ultimate resolution to around 10–20 nm using current technology. Much of the current popularity of e-beam lithography can be ascribed to the fact that e-beam lithography can be performed with relatively simple modification to a conventional scanning electron microscope (SEM) through computer control of the x,y position of the beam. Thus the fabrication of individual device structures for transport experiments may be accomplished with relatively little capital investment. E-beam lithography is typically performed in a direct write fashion in which the electron beam is rastered over a resist material such as PMMA (poly-methyl-methacrylate), which is spin-cast as a thin coating over the substrate. The developed PMMA can be used as an etch mask for wet chemical or dry etching, or metal can be evaporated and a lift-off process employed to transfer metal lines to the semiconductor substrate. Interestingly enough, the idea of direct write e-beam lithography predated the use of optical lithography [30] at the time in which SEM technology was evolving. However, the limitations of the vacuum technology and material quality precluded effective use of this technology until the 1980s. A related but at the moment much more costly technique is focused ion beam (FIB) technology, in which charged ions replace electrons as the high-energy incident particles. FIB technology has the additional advantage of allowing direct in-situ etching with 10^{-2} μm resolution, as well as the possibility of implanting donor and acceptor species in the semiconductor, allowing for the definition of nanometer-scale doped regions.

The ultimate limit of nanolithography is the control of the placement of individual atoms on the surface of a material. Presently many groups are aggressively researching the techniques of scanning tunneling lithography and atomic force lithography, both of which in fact do allow such atomic control. Both techniques have evolved from the remarkable development in the 1980s of the scanning tunneling microscope (STM) for which researchers at IBM Zurich were awarded the Nobel prize in physics [31]. The basic principle of the

STM is that of the exponential relation between tunneling current and the thickness of the barrier separating the electrodes, as discussed earlier in connection with the double well problem, and as will be discussed in more detail in Chapter 3. In the case of the STM, one electrode is the semiconductor substrate while the other is a fine tungsten tip brought in close proximity to the semiconductor surface. As the tip is moved across the surface, the tunnel current provides a sensitive measure of the proximity of the tip to the charge density on the substrate. This technique has allowed atomic level mapping of the atomic structure on material surfaces. In extending the microscopy technique to lithography, the STM tip may be used to selectively drive off an adsorbed atomic or molecular species such as H on the surface, leaving behind a pattern determined by the tip movement. Another technique is to use the tip to drag or push a particular molecule or atom into position along the surface. The major obvious drawback of this technique is the scalability of such a technique to fabricate structures over an entire semiconductor wafer.

2.3.2 Quantum wire and quantum dot structures

With the use of nanoscale lithography techniques, an infinite variety of structures can be envisioned. Assuming we start with a 2DEG heterostructure, confinement of the 2DEG in another direction, say the y-direction, forms what is called a *quantum wire*, which corresponds to a quasi-one-dimensional system. Here translational invariance exists along one axis, while the electronic states are well discretized in the other two. Free electron motion is still possible along the axis of the wire, and a continuum of one-dimensional states exists. Additional confinement in the remaining direction completely confines the motion of charge carriers, giving rise to a discrete spectrum of bound states much like an isolated atom or molecule. Such systems are called *quantum dots*, *quantum boxes*, or sometimes *artificial atoms*, discussed in detail in Chapter 4.

Figure 2.14 shows two different methods of fabricating quantum wires by etching a 2DEG structure. In both of these processes, either a resist or some other material is patterned into thin lines above a 2DEG heterostructure. This top layer must serve as a mask to an etchant which removes the AlGaAs/GaAs structure. In Fig. 2.14a, the layers are etched below the 2DEG as shown. The dominant problem with this method (as well as all lithographic processes) for forming a quantum wire is that the quality of the sidewall/air interface is vastly inferior to that existing between the AlGaAs and GaAs layers. At best, roughness fluctuations on the order of 2–5 nm exist using present-day wet or dry etching techniques (compared with the 0.2–0.5 nm fluctuations of the heterojunction interface discussed earlier). In addition, the free surface of GaAs is invariably pinned upon exposure to the ambient

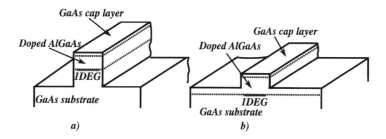

Fig. 2.14. Two different realizations of quantum wire structures by wet or dry etching.

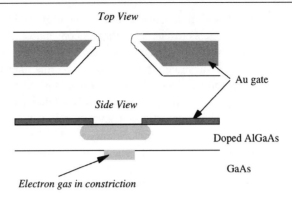

Fig. 2.15. Sketch of a split-gate Schottky contacts above a 2DEG structure. Negative potential applied to the gates depletes the 2DEG below the gates and laterally into the ungated regions.

due to the large density of surface states that form through defect formation [32]. Thus, instead of sharp confining potentials, the actual potential is that of lateral depletion layers, which looks more like a harmonic rather than "hard wall" potential. This lateral depletion, and the "soft" potential associated with it, limits the energy splitting between levels in the transverse direction. If the wire is made too thin, the space charge regions from either side overlap and deplete the free carriers from the wire, making them useless for transport measurements. Typically the lateral dimensions of such wires must be greater than 0.1–0.2 μm, realizable with clever optical techniques such as laser holography [33], [34] in which the interference lines between two beams can be used to expose a regular grating structure. Figure 2.14b shows a modification of the etch structure in which the AlGaAs layer – or even just the heavily doped GaAs cap layer – is partially etched. If a metal electrode is deposited covering the entire structure, the partially etched region is depleted first as the electrode bias is made increasingly negative, leaving behind a 1DEG in the wire region. Since the etching does not extend all the way to the active region of the wire, the detrimental effects associated with the poor sidewalls are less severe.

A popular technique in transport studies for realizing 1D channels is the so-called *split-gate technique*, shown in Fig. 2.15 [35]. Here metal Schottky gate electrodes are deposited on the surface of a 2DEG heterostructure between Ohmic source and drain contacts, as normally required in fabricating HEMT. However, the gate is split, leaving a narrow channel between the two electrodes. In contrast to etched wire structures, the metal electrodes (usually defined with e-beam lithography) are negatively biased to turn off or deplete the 2DEG under the gates, leaving a narrow region of undepleted channel between the two electrodes. The 2DEG is also depleted laterally from the gate electrodes as the gate bias is made more negative, in effect squeezing the 1D channel until eventually all the carriers are emptied from the region. In the intermediate regime, carriers in the channel feel a confining potential due to electrostatic potential of the reverse-biased Schottky contacts. A similar concept was in fact utilized some time earlier in studies of quantized electrons in Si/SiO$_2$ inversion layers using narrow channel MOSFETs [36]. P-n junctions were implanted adjacent to the channel of the MOSFET and reverse biased to laterally confine electrons in the channel in the direction perpendicular to current flow between the usual source and drain contacts of the device. The limitation of the split gate structure is the soft potential due to the lateral depletion, which

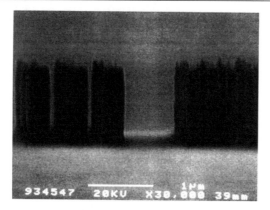

Fig. 2.16. SEM micrograph of quantum dot structures realized by etching resonant tunneling diode structures into one-dimensional columns [after Reed *et al.*, by permission].

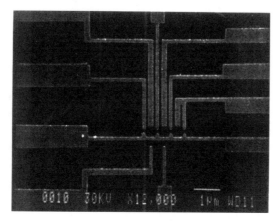

Fig. 2.17. SEM micrograph of quantum dot structures defined by electrostatic confinement via Schottky gates deposited on a 2DEG heterostructure.

limits the energy separation due to lateral confinement and limits the distance in which the electrodes may be spaced.

The fabrication techniques discussed above are similarly used to define more general geometries as will be discussed in later chapters. Quantum dot structures are realized by extending the lateral confinement of the wire structures to the remaining degree of freedom of the carriers. Figure 2.16 shows a micrograph of quantum dot structures realized by etching narrow columns into heterojunction material [37]. The heterojunction material itself consists of two AlGaAs barriers which form a resonant tunneling diode structure discussed in the next chapter. Electrons trapped in the narrow region between the two AlGaAs barriers experience additional confinement from the lateral sidewalls, resulting in a quantum dot. Figure 2.17 shows several quantum dot structures defined using electrostatic confinement as discussed in the split-gate quantum wires. The three dots in the center of this particular SEM micrograph are formed by the independently controlled electrodes shown, which are adjusted to completely isolate each dot while still maintaining carriers inside. Again, due to the lateral extent of the depletion regions under the gates, the dimensions of the dot structure have to be sufficiently large that total depletion does not occur before forming the dot.

2.4 Electronic states in quantum wires and quantum dots

In an ideal quantum wire system, the system is assumed sufficiently long along the axis of the wire that translational invariance holds. Within the EMA for a single band system, we can again write the envelope function as the product of free carrier solutions and confined solutions as

$$\psi(\mathbf{r}, z) = \varphi_{n,m}(\mathbf{r})e^{ik_x x}/L, \tag{2.30}$$

where $\mathbf{r}$ is the position vector in the yz-plane parallel to the wire cross section, and L is the normalization length of the wire. The function $\varphi_{n,m}(\mathbf{r})$ now satisfies the two-dimensional Schrödinger equation

$$\left(\frac{\hbar^2}{2m^*} \nabla_{\mathbf{r}}^2 + V_{eff}(\mathbf{r}) \right) \varphi_{n,m}(\mathbf{r}) = E_{n,m}\varphi_{n,m}(\mathbf{r}), \tag{2.31}$$

where again $V_{eff}(\mathbf{r})$ contains the potential of the band discontinuities, the electrostatic potential of the ionized donors and acceptors, the many body contribution due to free carriers, and the image potential (which is particularly important when considering freestanding wires, or wires with etched sidewalls). An isotropic effective mass, m*, has also been assumed for simplicity, corresponding to a spherically symmetric, parabolic energy dispersion relaltion for the band minima.

For simplicity, we can consider special cases as before. First consider a wire with a rectangular cross section and with infinite potential outside of the region defining the wire (i.e., hard walls). For the rectangular symmetry of this case, the solution of Eq. (2.31) is separable into y- and z-dependent solutions as

$$\varphi_{n,m}(y, z) = \left(\frac{4}{L_y L_z} \right)^{1/2} \sin(n\pi y/L_y) \sin(m\pi z/L_z) \quad n, m = 1, 2, 3, \ldots \tag{2.32}$$

where L_y and L_z are the dimensions in the y- and z-directions, and the origin of the coordinate system has been placed at one corner of the wire. The corresponding eigen-energies are given by

$$E_{k_x,n,m} = E_{k_x} + E_{n,m} = \frac{\hbar^2 k_x^2}{2m^*} + \frac{n^2 \hbar^2 \pi^2}{2m^* L_y^2} + \frac{m^2 \hbar^2 \pi^2}{2m^* L_z^2}, \tag{2.33}$$

where n and m are integers as before. In analogy to the 2DEG discussion earlier, we now have a continuum of one-dimensional (1D) states associated with each pair of integers n and m, called subbands, modes, or channels, depending on the context. In particular, the solutions of Eq. (2.33) are identical to the allowed solutions of eletromagnetic waves in a metallic rectangular waveguide for the transverse magnetic and electric solutions. In Chapter 3, we will connect different regions of such 1D waveguides together and look at the general propagation characteristics through the composite structure. Such systems are aptly named *quantum waveguides*.

The density of states for the quasi-1D system described above are again derivable using Eq. (2.8). Assuming the case of bands above, the density of states per unit length of the wire system is given by

$$D(E) = \eta_v \left(\frac{m^*}{\pi \hbar^2} \right) \left(\frac{\hbar^2}{2m^*} \right)^{\frac{1}{2}} \sum_{n,m} (E - E_{n,m})^{-1/2} \theta(E - E_{n,m}), \tag{2.34}$$

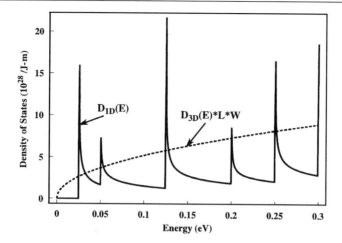

Fig. 2.18. Calculated density of states per unit length of a 15 nm–by–15 nm GaAs quantum wire (solid curve). The dashed line is the corresponding 3D density of states per unit length of wire.

where the two-fold degeneracy for spin has been included. Figure 2.18 shows the calculated density of states for a GaAs/Al$_x$Ga$_{1-x}$As rectangular quantum wire, illustrating the diverging density of states as the energy approaches that of one of the subband minima. Such divergences may lead to interesting effects in the transport properties and collective excitations of such systems. In reality, however, with the present limitations in lithography and processing discussed in the previous section, such divergences are usually broadened due to disorder, and strong effects have not been observed experimentally.

Numerical self-consistent solutions have been performed for electrons in Si quantum wires by Laux and Stern [38] and in GaAs quantum wires by Laux et al. [39] and Kojima et al. [40] Such calculations typically use a similar iterative procedure to treat the many body contributions within various approximation schemes. The calculated potential for a GaAs/Al$_x$Ga$_{1-x}$As wire formed in a split-gate structure is shown in Fig. 2.19. It is interesting to note that due to the depletion region confinement, the potential appears more parabolic than hard wall, particularly for increasing negative gate bias. Typical subband spacings are on the order of 5 to 10 meV.

A crude description of the states of a quantum dot can of course be obtained by extending the hard wall confinement in all three directions such that the energy is given by

$$E_{l,n,m} = \frac{l^2\,\hbar^2\pi^2}{2m^*L_x^2} + \frac{n^2\,\hbar^2\pi^2}{2m^*L_y^2} + \frac{m^2\,\hbar^2\pi^2}{2m^*L_y^2}, \tag{2.35}$$

where l, n, and m are positive, nonzero integers. The spectrum of energies and thus the density of states is completely discrete, with degeneracies due to spin and multiple valleys. In Section 4.1 we consider in more detail the states in a parabolic confining potential. However, as will be discussed in much more detail in Chapter 4, the states in real quantum dot systems are complicated by the granular nature of electronic charge. The change of one electron in a quantum dot system has a large influence on the electrostatic potential and many-body ground state of the system. Such effects are of great interest in themselves in the so-called *Coulomb blockade effect*, which is the basis of a number of interesting

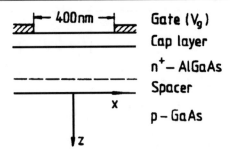

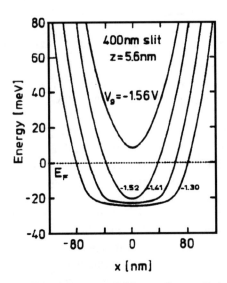

Fig. 2.19. Numerical self-consistent potential for a split-gate GaAs/AlGaAs structure [after Laux *et al.*, Surf. Sci. **196**, 101 (1988), by permission].

experiments and proposals for ultrasmall electronic devices, and thus we will defer full discussion of such systems until later.

2.5 Magnetic field effects in quantum confined systems

In nanostructure systems a static magnetic field may have a profound effect on their electronic and transport properties. External application of high magnetic fields to nanostructures is an invaluable degree of freedom available to the experimentalist in probing the system. In and of itself, magnetic fields give rise to new fundamental behavior not observed in bulk-like systems, such as the quantum hall effect (QHE). The fundamental quantity characterizing a magnetic field is the magnetic flux density, **B**, which in mks units is measured in Teslas. In the study of semiconductor nanostructures, low fields usually correspond to fields less than 1 T, which is the regime in which low-field magnetotransport experiments such as Hall effect measurements are usually performed. Magnetic intensities of 10–20 T are obtainable in the usual university and commercial research environments using superconducting alloy coils with high critical fields immersed in a liquid He dewer. Higher magnetic fields are obtainable only at a few large-scale facilities scattered around the world.

2.5.1 Magnetic field in a 2DEG

When we consider nanostructure systems such as quantum wells, wires, and dots, the effect of the magnetic field may be roughly separated into two cases. For the first, the magnetic field is parallel to one of the directions of free electron propagation; for the other, the field is perpendicular to the free electron motion of the system. Qualitatively, when free particles are subject to a magnetic field, they experience the Lorentz force

$$\mathbf{F} = q\mathbf{v} \times \mathbf{B}, \tag{2.36}$$

where $\mathbf{v}$ is the velocity of the carrier. Since the force is always perpendicular to the direction of travel of the particle, its motion in the absence of other forces is circular, with angular frequency given by the *cyclotron frequency*

$$\omega_c = \frac{eB}{m_c}, \tag{2.37}$$

where B is the magnitude of the magnetic flux density and m_c is the cyclotron mass of the particle, which is an average over the mass of a particle as it performs circular orbits in $\mathbf{k}$-space over a constant energy surface (since the particle does not gain or lose energy due to the angular acceleration of the magnetic field). For isotropic systems like those we have considered so far, the cyclotron mass becomes simply the electron effective mass, m^*. In the quantum mechanics picture, the circular orbits associated with the Lorentz force must be quantized in analogy to the orbital quantization occurring about a central potential, for example, about the nucleus of an atom. Since the particles execute time-harmonic motion similar to the motion in a harmonic oscillator potential, the energy associated with the motion in the plane perpendicular to the magnetic field is expected to be quantized. If we now consider the magnetic field applied perpendicular to the plane of a quasi-2D system, then the entire free electron–like motion in the plane parallel to the interface is quantized, and the energy spectrum becomes completely discrete.

In Hamiltonian mechanics, the equations of motion are formulated in terms of potentials rather than in terms of the electric and magnetic fields directly. The potential associated with a magnetic field is a vector rather than a scalar quantity, defined from the curl equation

$$\nabla \times \mathbf{A} = \mathbf{B}, \tag{2.38}$$

where $\mathbf{A}$ is the *vector potential*. The choice of $\mathbf{A}$ for a given $\mathbf{B}$ is not unique, for Eq. (2.38) allows for the *gauge transformation* $\mathbf{A} \longrightarrow \mathbf{A} + \nabla V(\mathbf{R})$ (where $V(\mathbf{R})$ is a scalar field) since $\nabla \times \nabla V \equiv 0$. The physical properties of the system, of course, cannot be affected by this choice of gauge. The Hamiltonian in the presence of a vector potential is obtained by making the Peierl's substitution $\mathbf{p} \rightarrow \mathbf{p} + q\mathbf{A}$, where $\mathbf{p}$ is the momentum operator. Consider now a quasi-two-dimensional system with a static magnetic field in the z-direction, $\mathbf{B} = (0, 0, B)$. One choice of a vector potential satisfying Eq. (2.38) is the so-called *Landau gauge*, $\mathbf{A} = (0, Bx, 0)$. The single-band effective mass Hamiltonian (2.1) may be generalized to include this vector potential as

$$\left(-\frac{\hbar^2}{2m_z} \frac{\partial^2}{\partial z^2} + \frac{1}{2m^*} \left(\frac{\hbar}{i} \nabla_{\mathbf{r}} + q\mathbf{A} \right)^2 + V_{eff}(z) \right) \psi(\mathbf{r}, z) = E\psi(\mathbf{r}, z), \tag{2.39}$$

where m^* is the effective mass in the plane of the 2DEG. Again the solution is separable as $\psi(\mathbf{r}, z) = \varphi(z)\chi(x, y)$, where $\varphi(z)$ is the solution of Eq. (2.4), and where $\chi(x, y)$

satisfies

$$\left(-\frac{\hbar^2}{2m^*}\frac{\partial^2}{\partial x^2} + \frac{m^*\omega_c^2}{2}(x - x_o)^2\right)\chi(x, y) = E_n\chi(x, y),$$

$$x_o = \frac{1}{eB}\frac{\hbar}{i}\frac{\partial}{\partial y}, \tag{2.40}$$

where E_n is the energy associated with the transverse motion, and electrons have been assumed. If we assume a solution for $\chi(x, y)$ of the form

$$\chi(x, y) = \chi(x)e^{ik_y y}, \tag{2.41}$$

then in Eq. (2.40) x_o may be written $x_o = \hbar k_y/eB$, which is referred to as the *center coordinate*. With this substitution, Eq. (2.40) becomes simply the harmonic oscillator equation for $\chi(x)$, with eigenfunctions given by

$$\chi_n(x) = (2^n n!\sqrt{\pi} l_m)^{-1/2}\exp\left(-\frac{(x - x_o)^2}{2l_m^2}\right)H_n\left(\frac{x - x_o}{l_m}\right) \quad n = 0, 1, 2, \ldots, \tag{2.42}$$

where $H_n(z)$ is the Hermite polynomial of order n, and $l_m = \sqrt{\hbar/eB}$ is the *magnetic length*, which is basically the cyclotron radius of the ground state. The expectation value of the position operator for the eigenfunctions (2.42) may be shown to be $\langle\chi_n|x|\chi_n\rangle = x_o$, and therefore the center coordinate represents the average position of the magnetic state. The corresponding energy eigenvalues are independent of the center coordinate (and thus the transverse momentum, k_y) and are given by the well-known expression for a harmonic oscillator,

$$E_n = \hbar\omega_c\left(n + \frac{1}{2}\right) \quad n = 0, 1, 2, \ldots. \tag{2.43}$$

The cyclotron radius corresponds to the classical orbit associated with the energy Eq. (2.43). Since the carrier velocity in a circular orbit, r_n, is given as $v = r_n\omega_c$, we can equate the kinetic energy $\frac{1}{2}m^*v^2$ with Eq. (2.43) to give

$$r_n = \sqrt{\frac{2\hbar(n + \frac{1}{2})}{eB}}. \tag{2.44}$$

For $n = 0$, we have $r_0 = l_m$, the magnetic length as asserted above.

As in the 3D case of magnetic quantization, each integer n corresponds to a different *Landau* level associated with a magnetic subband. In contrast to the bulk situation where a remaining degree of freedom exists due to motion parallel to the magnetic field, the total energy spectrum of the quasi-2D system is discrete, given by

$$E = E_i + \hbar\omega_c\left(n + \frac{1}{2}\right), \tag{2.45}$$

where E_i are the energy eigenvalues associated with the perpendicular confining potential. Even though the Hamiltonian contains the y-momentum, k_y, the total energy Eq. (2.45) is independent of k, and therefore of the group velocity

$$v = \frac{1}{\hbar}\frac{\partial E}{\partial k_y} = 0. \tag{2.46}$$

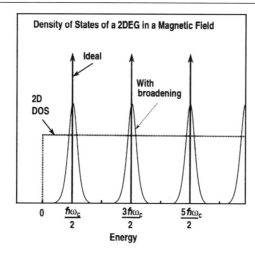

Density of States of a 2DEG in a Magnetic Field

Fig. 2.20. Density of states versus energy of a two-dimensional electron gas in the presence of a perpendicular magnetic field. The solid curve is the density of states in the presence of impurities and other homogeneities. For comparison, the dashed curve is the density of states for an ideal 2D system corresponding to $B = 0$.

In the above discussion, we neglected the effect of the magnetic field on the spin variables of the electrons. In fact, an extra term should be added to the Hamiltonian of the envelope Eq. (2.39),

$$\mathcal{H}_s = g^* \mu_B \sigma \cdot \mathbf{B}, \tag{2.47}$$

where σ is the electron spin operator (spinor), $\mu_B = \frac{e\hbar}{2m_0 c}$ is the Bohr magnetron, and g^* is the effective Landé g factor (which is equal to 2 in vacuum but differs in semiconductors). Electrons with spin up are raised in energy $\frac{1}{2}g^*\mu_B B$, and those with spin down are lowered by the same amount. The result is a splitting of the spin degeneracy, and thus there is a further splitting of the levels given by Eq. (2.45) by an amount $\pm\frac{1}{2}g^*\mu_B B$.

The density of states associated with a quasi-2D system in the presence of a perpendicular magnetic field is singular as shown in Fig. 2.20, where the density of states in the presence of a magnetic field is shown in comparison to that of an ideal, single subband 2D system. The lowest Landau level is shifted up $\hbar\omega_c/2$ with respect to the subband edge of the 2D system. The singular density of states corresponds to a zero-dimensional system, much like the states of the quantum dot systems discussed in Chapter 4. However, in contrast to the states of an isolated atomic-like quantum dot, each Landau level is highly degenerate, corresponding to all the 2D states with a range $\pm\hbar\omega_c/2$ that collapse into the level. The density of 2D states for a single subband system per unit area and energy was given by Eq. (2.10) as $D_o = \eta_v \eta_s m^*/(2\pi\hbar^2)$. Thus, the total number of states per unit area in each Landau level is given by this expression multiplied by $\hbar\omega_c$ yielding $D = \eta_v \eta_s e B/(2\pi\hbar)$. The degeneracy of each Landau level increases linearly as the magnetic field increases. As the density of carriers in the system is increased (for example, through gate bias or optical excitation), the Fermi level is pinned in the highest occupied magnetic subband until it is completely filled, then jumps discontinuously to the next Landau level. The highest occupied magnetic subband (at $T = 0K$) corresponds to the total density divided by the density per Landau level as

$$n_{\max} = Int\left(\frac{2\pi\hbar n_{2d}}{\eta_v\eta_s eB} + 1\right), \tag{2.48}$$

where $Int(\ )$ signifies taking the integer part. Correspondingly, if the carrier density is fixed but the magnetic field increases, the Fermi level moves with the highest occupied magnetic subband. However, as the magnetic field increases, the density of states in lower lying subbands increases, and at a certain critical field, the highest occupied Landau level becomes depopulated and the Fermi level jumps discontinuously to the next lower level. Qualitatively, this discontinuity in the Fermi energy with either density or magnetic field shows up in the magnetoresistance measured externally and results in oscillatory behavior known as *Shubnikov de-Haas oscillations*. Measurement of the period of the conductance or resistance as a function of magnetic field allows direct determination of the density of the occupied 2D subbands of the system, which is a powerful experimental technique used frequently to characterize quasi-2D systems (see Section 2.7).

If impurities or other nonidealities are present in the system, the delta-function density of states of the ideal system is broadened as shown by the dashed curve in Fig. 2.20. Using the so-called self-consistent Born approximation in perturbation theory, Ando [1] has shown that the density of states in the presence of short-range potential impurities has the form (for a review of the derivation, see page 536)

$$D(E) = \frac{1}{2\pi l_m^2} \sum_n \left[1 - \left(\frac{E - E_n}{\Gamma_n}\right)^2\right]^{1/2}, \tag{2.49}$$

where Γ_n is a broadening factor associated with the short-range impurities. If the magnetic length is such that $d < l_m/(2n + 1)^{1/2}$, where d is the range of the potential, then the broadening factor is simply given by

$$\Gamma^2 = \frac{2}{\pi}\hbar\omega_c\frac{\hbar}{\tau_f}, \tag{2.50}$$

where τ_f is the scattering time for the same impurities at the Fermi energy for $B = 0$ (as discussed in Section 2.7). If the scattering rate $(1/\tau_f)$ is sufficiently large, the Landau levels merge into one another, and the oscillatory behavior observed in Shubnikov de-Haas measurements is damped. Qualitatively, such oscillations will be observed only if distinct Landau levels exist, which implies that the broadening should be less than the Landau level spacing as $\Gamma < \hbar\omega_c$. Setting $\Gamma = \hbar\omega_c$ in Eq. (2.50), the criterion for observing clear magnetic quantization effects in a magnetotransport measurement is $\omega_c\tau_f > 2/\pi \approx 1$. Thus, as a rule, the poorer the mobility ($\mu = e\tau_f/m^*$) in a material, the larger the magnetic field necessary to observe oscillatory behavior.[1] Similarly, if the temperature is raised above absolute zero, carriers will be smeared out among the magnetic subbands, and the oscillatory motion of the Fermi energy will be damped. We will come back to this more quantitatively in Section 2.7 in the discussion of magnetotransport.

---

[1] Note, however, that τ_f appearing in the mobility is really the momentum relaxation time while τ_f in Eq. (2.50) is more related to the time between collisions, which differs substantially from the momentum relaxation time for anisotropic scattering mechanisms.

2.5.2 Magnetic field and 1D waveguides: Edge states

We now consider a more complicated problem, that of potential barriers such that the 2DEG is confined in one direction. As we saw in Section 2.4, such confinement leads to the formation of a quasi-one-dimensional system or quantum wire. For simplicity, assume for example a quantum wire formed by etching sidewalls in a 2DEG structure. Assume again that the direction perpendicular to the heterointerface is the z-direction, and that a magnetic field is applied in this direction. The confining potential due to the sidewall is assumed to be in the x-direction, and the axis of the wire is in the y-direction. The envelope function equation is again separable, and the z-dependent equation may be solved independently. The remaining equation in the xy-plane has the form as before:

$$\left(\frac{1}{2m^*}\left(\frac{\hbar}{i}\nabla_{\mathbf{r}} + q\mathbf{A}\right)^2 + V(x)\right)\chi(x, y) = E_n\chi(x, y), \qquad (2.51)$$

where $V(x)$ is the confining potential due to lateral confinement and spin splitting is ignored for simplicity. In the Landau gauge, where $A = (0, Bx, 0)$, Eq. (2.51) may be written

$$\left(-\frac{\hbar^2}{2m^*}\frac{\partial^2}{\partial x^2} + \frac{m^*\omega_c^2}{2}\left(x - \frac{1}{eB}\frac{\hbar}{i}\frac{\partial}{\partial y}\right)^2 + V(x)\right)\chi(x, y) = E_n\chi(x, y). \qquad (2.52)$$

If we again write the solution as in Eq. (2.41), then Eq. (2.52) becomes

$$\left(-\frac{\hbar^2}{2m^*}\frac{\partial^2}{\partial x^2} + \frac{m^*\omega_c^2}{2}(x - x_o)^2 + V(x)\right)\chi(x) = E_n\chi(x), \qquad (2.53)$$

with $x_o = \hbar k_y/eB$ as before.

A somewhat illustrative solution may be obtained by assuming that the lateral confining potential is parabolic of the form $V(x) = \frac{1}{2}m^*\omega_o^2$. For the etched quantum wire structure used as the model here, the assumption of a parabolic potential is not inappropriate, since the effect of surface states on either sidewall gives rise to such a quadratic potential due to band-bending. Expanding the quadratic term in the Hamiltonian operator appearing in Eq. (2.53) and combining with $V(x)$, we get

$$\mathcal{H} = \frac{p_x^2}{2m^*} + \frac{m^*\left(\omega_c^2 + \omega_o^2\right)}{2}x^2 - m^*\omega_c^2 x x_o + \frac{m^*\omega_c^2}{2}x_o^2. \qquad (2.54)$$

By defining a new center coordinate, $x_o' = x_o\omega_c^2/\omega^2$ with $\omega^2 = \omega_c^2 + \omega_o^2$, we can complete the square involving x and thus rewrite the Hamiltonian as

$$\mathcal{H} = \frac{p_x^2}{2m^*} + \frac{m^*\omega^2}{2}\left(x - x_o'\right)^2 + \frac{\hbar^2 k_y^2}{2M}, \qquad (2.55)$$

where

$$\frac{\hbar^2 k_y^2}{2M} = \frac{m^*\omega_c^2\omega_o^2}{2\omega^2}x_o^2; \qquad M = m^*\omega^2/\omega_o^2 = m^*\frac{\left(\omega_c^2 + \omega_o^2\right)}{\omega_o^2}. \qquad (2.56)$$

Comparing this to the case of an infinite 2D gas, there is an additional term on the right-hand side of Eq. (2.55) which looks like the free-electron dispersion of a one-dimensional particle with an effective mass M given by Eq. (2.56). This term may be combined with the energy E_n on the right side of Eq. (2.53) to obtain a harmonic oscillator–like equation in the form of Eq. (2.45). The energy becomes

$$E = E_{n,i}(k_y) = E_i + \hbar\omega_c\left(n + \frac{1}{2}\right) + \frac{\hbar^2 k_y^2}{2M}, \tag{2.57}$$

where E_i is again the quantized energy due to the perpendicular confinement. The Landau levels are no longer degenerate, being spread in energy by the momentum in the y-direction. Likewise, the density of states is no longer discrete as in the purely 2D case, but rather corresponds to a quasi-one-dimensional system as shown in Fig. 2.18. Also, the group velocity is no longer zero in the y-direction; instead, $v = \hbar k_y/M$. For $\omega_c \gg \omega_o$, the mass M goes to zero, giving the infinite 2D gas case. For the other extreme $\omega_c \ll \omega_o$, M goes to the effective mass, m^*, which is just the limit of a quantum wire with no magnetic field present, given by Eq. (2.33). It is interesting to note in this model that wavefunctions associated with the harmonic oscillator–type solutions in the transverse direction are localized on one side or the other of the wire, depending on the center coordinate, x_o' – which in turn depends on k_y, the momentum along the axis of the wire. For positive k_y, the envelope function is shifted to the right side of the wire; for negative k_y, the envelope function is shifted to the left side. Therefore, probability flux in one direction is localized on one side of the wire, while states with flux propagating in the opposite direction are localized on the opposite side.

The physical picture associated with the states in a quantum wire in a magnetic field is perhaps clearer if we consider a much less simple model, that of hard wall confinement (i.e., semiinfinite potential barriers) in the transverse x-direction. For this example, assume that the potential $V(x) = 0$ for $0 \le x \le W$ and is infinite otherwise outside in Eq. (2.53). The exact solution cannot be expressed analytically, and therefore the dispersion relation for energy and momentum has to be determined numerically. To calculate the energy numerically, it is convenient to expand the solution in the basis set of the infinite potential well as

$$\chi(x) = \sum_n a_n \sin\left(\frac{n\pi x}{W}\right), \tag{2.58}$$

which satisfies the boundary conditions at 0 and W. Substituting into Eq. (2.53), we get

$$\sum_n a_n\left(\frac{\hbar^2}{2m^*}\left(\frac{n\pi}{W}\right)^2 - E\right)\sin\left(\frac{n\pi x}{W}\right) + \frac{m^*\omega_c^2}{2}(x - x_o)^2 a_n \sin\left(\frac{n\pi x}{W}\right) = 0. \tag{2.59}$$

Multiplying this equation by $(2/W)\sin(m\pi x/W)$ and integrating over the width of the wire (using the orthogonality of the infinite well eigenfunctions), we obtain a coupled set of linear equations in the expansion coefficient a_m as

$$\frac{\hbar^2}{2m^*}\left(\frac{m\pi}{W}\right)^2 a_m + \sum_n F_{nm} a_n = E a_m, \tag{2.60}$$

where

$$F_{nm} = \frac{m^*\omega_c^2}{W}\int_0^W dx(x - x_o)^2 \sin\left(\frac{m\pi x}{W}\right)\sin\left(\frac{n\pi x}{W}\right). \tag{2.61}$$

Equation (2.61) can be rewritten as

$$F_{nm} = \frac{m^*\omega_c^2 W^2}{2\pi^3}\int_0^\pi dy(y - y_o)^2\{\cos([n-m]y) - \cos([n+m]y)\},$$

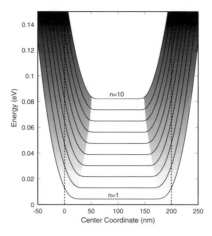

Fig. 2.21. Caculated energy versus center coordinate x_o for a 200-nm-wide wire and a magnetic field intensity of 5 T. The shaded regions correspond to skipping orbits associated with edge state behavior.

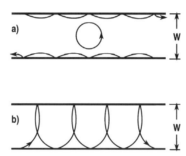

Fig. 2.22. Classical motion of a particle in a magnetic field for different energies and center coordinates.

with $y_o = x_o \pi / W$, which is easily evaluated as

$$F_{nm} = \frac{m^* \omega_c^2 W^2}{\pi^2} \left\{ \left(1 - \frac{x_o}{W}\right) \left(\frac{\cos \pi (n-m)}{(n-m)^2} - \frac{\cos \pi (n+m)}{(n+m)^2}\right) + \frac{x_o}{W} \frac{4nm}{(n^2 - m^2)^2} \right\} \quad (2.62)$$

for $n \neq m$, and

$$F_{mm} = \frac{m^* \omega_c^2 W^2}{\pi^2} \left\{ \frac{\pi^2 \left[\left(1 - \frac{x_o}{W}\right)^3 + \left(\frac{x_o}{W}\right)^3\right]}{6} - \frac{1}{4m^2} \right\} \quad (2.63)$$

for $n = m$. The eigenvalues of the determinant formed from the matrix equation (2.60) correspond to the allowed energies for each value of x_o, which are proportional to k_y. A plot of the dispersion relation obtained from this calculation is shown in Fig. 2.21, where we plot E versus x_o. Although x_o represents the momentum in reciprocal space, it is instructive to plot the energy versus the center coordinate itself, since this coordinate is a measure of the relative position of the envelope function associated with a state of momentum k_y. As can be seen from Fig. 2.21, the dispersion relation is almost flat in the middle of the waveguide and then shows strong dispersion close to either side. A simple physical interpretation of this behavior is possible from the classical motion of the particle in the waveguide, as shown in Fig. 2.22 (see for example [41]). The relevant parameters in interpreting this behavior are the cyclotron radius for a particular level (Eq. (2.44)), and the center coordinate, giving the center position of the circular cyclotron orbit of radius r_n. Three possible types of orbits in the confined structure are possible, based on the cyclotron radius and the center coordinate: pure cyclotron orbits, skipping orbits, and traversing orbits. In the case of pure cyclotron orbits, the distance from the center coordinate to one wall or the other is less than r_n, so that no scattering of the particle off the wall occurs, and pure cyclotron–type motion is possible, as shown by the circular orbits in Fig. 2.22. In this case, we expect the states to resemble those of simple magnetic quantization in a 2D gas, which is dispersionless. The flat regions of the dispersion curves shown in Fig. 2.21 correspond to these orbits. If the distance between the center coordinate to either wall is less than r_n, then boundary scattering occurs, giving rise to the skipping orbits shown in Fig. 2.22. If boundary scattering is assumed to be specular (rather than diffuse), the reflected momentum is such that the electron has a net momentum in the plus or minus y-direction, depending whether x_o (and thus k_y) is positive or negative. The states associated with these skipping orbits are referred to as *edge states*, and they correspond to the shaded region of Fig. 2.21 (the extent of which is determined by the classical overlap argument of the cyclotron radius about the center coordinate relative to the boundary). Because the edge states have a net velocity in the $\pm y$-direction, the dispersion relation is no longer flat, as observed in the shaded regions Fig. 2.21. The slope of the dispersion relation is related to the group velocity, and as we can see from the figure, edge states on the right side have positive velocity while those on the left have negative velocity. Finally, for large energy, the cyclotron orbit is sufficiently large that the electrons perform traversing orbits in which the electron interacts with both boundaries. As one may qualitatively observe, specular scattering from either boundary results in a net momentum along the axis of the waveguide, which corresponds to a transition from dispersionless behavior to more 1D waveguide behavior for the higher magnetic subbands.

This concept of edge states that propagate in opposite directions on opposite sides of a 1D waveguide is sometimes compactly shown by the representation given in Fig. 2.23. The flow lines represent the direction and relative center coordinate of the edge state with respect to the waveguide boundaries. We will come back to this description later in Chapter 3 when we discuss mode matching and the general behavior of edge states in inhomogeneous structures.

2.6 Screening and collective excitations in low-dimensional systems

So far we have discussed the electronic properties of quantum confined systems in terms of a single-particle or one-electron picture. The other electrons in the system were either ignored or treated as an average static potential, as in the case of self-consistent solutions discussed in Section 2.2.3. However, in transport and other nonequilibrium phenomena, one has to consider the dynamic response of the system of electrons to the presence of an

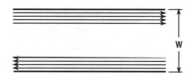

Fig. 2.23. Schematic representation of edge states in a quantum waveguide.

external perturbation. The most familiar example is the dielectric response of a material to an external, time-varying electric field. Due to the presence of the electric field in the solid, charges rearrange themselves in response to the external force, resulting in an additional induced charge in the solid. This induced charge gives rise to a polarization field in addition to the external field, and the net effect is described in terms of the frequency- (and wavevector-) dependent dielectric function of the material.

In a bulk semiconductor, we generally can separate various frequency-dependent contributions to the dielectric response due to the lattice (in polar semiconductors), the valence electrons (which determines the electronic polarizability of the material and thus the "high-frequency" dielectric constant), and finally the free mobile carriers in the conduction band and unoccupied states in the valence band (holes). In a bulk system, the dielectric response of the system may be represented in the Fourier domain with respect to time and space by the *dielectric function, $\epsilon(\mathbf{Q}, \omega)$*. For an external potential $V^{ext}(\mathbf{R}, t)$, the effective, or total, potential in the system is the sum of the external potential and the potential due to the induced charge by the external potential, which in the Fourier domain is given by

$$V^{tot}(\mathbf{Q}, \omega) = \frac{V^{ext}(\mathbf{Q}, \omega)}{\epsilon(\mathbf{Q}, \omega)}. \tag{2.64}$$

In a bulk system, the total dielectric function may be expressed as (see for example [17])

$$\epsilon(\mathbf{q}, \omega) = \epsilon(\infty) + \epsilon_l(\omega) + \frac{e^2}{\epsilon_o q^2} \sum_{\mathbf{k}} \frac{f(\mathbf{k}) - f(\mathbf{k} + \mathbf{q})}{E(\mathbf{k} + \mathbf{q}) - E(\mathbf{k}) - \hbar\omega - i\alpha\hbar}, \tag{2.65}$$

where ϵ_o is the permittivity of free space, $f(\mathbf{k})$ is the one-particle distribution function for a particle of wavevector $\mathbf{k}$, $E(\mathbf{k})$ is the energy of state $\mathbf{k}$, and α is a small convergence parameter. The first term on the right is the relative permittivity due to core and valence electrons, the second term is the contribution due to the lattice in polar materials, and the third term is the so-called *Lindhard dielectric function* for the free carriers (electrons and holes in general). The latter contribution due to free carriers is responsible for effects such as free carrier absorption of photons, screening of external potentials such as those responsible for scattering in the system, and collective excitations of the free electron gas, which in the parlance of many body theory are referred to as *plasmons*.

In this section, we are interested in the dielectric response of free carriers residing in quantum confined systems such as quantum wells and quantum wires. In particular, we are interested in how the reduced dimensionality of the system influences the collective behavior of the electron gas as compared to a bulk system, and also the consequences this has for transport in such systems. The main difference of the quantum confined systems discussed in this chapter compared to bulk systems is the quantization of the perpendicular motion resulting in a series of quantum subbands, each with free electron behavior in the parallel direction. While the dimensionality of the quantum confined system is reduced due to the confining potential, perturbations of the system generally are three-dimensional. An example is the 3D potential due to an ionized impurity located in the vicinity of a quantum well or wire. The quantum confined system itself responds (polarizes) to the presence of this potential, both in the free direction of the system (intrasubband polarization) and in the confined directions (intersubband polarization). The consequence of this is that instead of being able to define a single dielectric function of the system, as in the bulk case, the dielectric response in a quantum confined system must be defined in terms of a dielectric matrix [41] rather than a simple (although already complicated) scalar function. Here we derive the form of the dielectric matrix for quasi-2D and quasi-1D systems using a simple time-dependent perturbation theory approach used to derive the Lindhard dielectric function for bulk 3D systems (see for example [42]). In Chapter 7, we will revisit more formally the derivation of the dielectric function for low-dimensional systems using Green's function methods.

2.6.1 Dielectric function in quasi-2D systems

We begin with the case of a quasi-2D system as discussed earlier. In the spirit of perturbation theory, we assume that the ideal system with no external perturbation is described by an unperturbed Hamiltonian whose properly normalized eigenstates are given by

$$\psi_{\mathbf{k},n}(\mathbf{r}, z) = \varphi_n(z)e^{i\mathbf{k}\cdot\mathbf{r}} \Rightarrow |\mathbf{k}, n\rangle, \tag{2.66}$$

as given in Eq. (2.3) with the normalization factor set equal to unity for convenience. To the ideal system we add an external, generally time-varying potential, $V^{ext}(\mathbf{r}, z)e^{-i\omega t}e^{\alpha t}$, where α is a small parameter corresponding to the adiabatic "turning on" of the perturbation. Due to this external potential, a particle initially in a state $\mathbf{k}$ and with n given by Eq. (2.66) finds itself with a nonzero probability of being in a mixed state [43]

$$\psi(t) = |\mathbf{k}, n\rangle + b_{\mathbf{k}+\mathbf{q},n'}(t)|\mathbf{k} + \mathbf{q}, n'\rangle, \tag{2.67}$$

where from time-dependent perturbation theory the coefficient $b_{\mathbf{k}+\mathbf{q},n'}$ is given by

$$b_{\mathbf{k}+\mathbf{q},n'}(t) = \frac{V_{nn'}^{ext}(-\mathbf{q})e^{-i\omega t}e^{\alpha t}}{E_{n'}(\mathbf{k} + \mathbf{q}) - E_n(\mathbf{k}) - \hbar\omega - i\alpha\hbar} = b_{\mathbf{k}+\mathbf{q},n'}e^{-i\omega t}e^{\alpha t}. \tag{2.68}$$

$E_n(\mathbf{k})$ is the energy associated with the eigenstate (2.66), and

$$V_{nn'}^{ext}(-\mathbf{q}) = \langle \mathbf{k} + \mathbf{q}, n'|V^{ext}(\mathbf{r}, z)|\mathbf{k}, n\rangle = \int d\mathbf{r} \int_{-\infty}^{\infty} dz \varphi_{n'}^*(z)\varphi_n(z)V^{ext}(\mathbf{r}, z)e^{-i\mathbf{q}\cdot\mathbf{r}}. \tag{2.69}$$

The actual state vector of the system of course would actually be a superposition of all states $\mathbf{q}$ in Eq. (2.67). However, within the so-called *random phase approximation* (RPA) it is assumed that we can calculate the response of the system associated with one Fourier component (in the unconstrained direction) independent of all the other components and superimpose the results at the end, thus ignoring cross correlations, which are assumed to cancel on the average.

In understanding the response of the many body system to the external perturbation, what we wish to find is the induced charge density, or change in charge density, caused by the external potential. The induced charge density will contribute an additional potential that combines with the external potential to give the net additional potential in the system. In a multisubband quasi-2D system, the induced charge density may be generalized from the bulk case as

$$\rho_{ind}(\mathbf{r}, z, t) = -2e \sum_{\mathbf{k},n,n'} f_n(\mathbf{k})\{|\psi(t)|^2 - |\psi_{\mathbf{k},n}(\mathbf{r}, z)|^2\}, \tag{2.70}$$

where a two-fold spin degeneracy is assumed, and a single-valley band extremum is assumed for simplicity. The distribution function $f_n(\mathbf{k})$ is defined such that

$$2 \sum_{\mathbf{k},n} f_n(\mathbf{k}) = n_{2d}, \tag{2.71}$$

where the factor of 2 accounts for the spin degeneracy of each state $\mathbf{k}$. We further assume that the external perturbation is sufficiently weak that the response is linear, that is, that we may neglect terms on the order of $b^2_{\mathbf{k}+\mathbf{q},n'}$, yielding

$$\rho_{ind}(\mathbf{r}, z, t) = -2e \sum_{\mathbf{k},n,n'} f_n(\mathbf{k})$$

$$\times \left\{ b_{\mathbf{k}+\mathbf{q},n'} \varphi_n^*(z)\varphi_{n'}(z)e^{i\mathbf{q}\cdot\mathbf{r}}e^{-i\omega t}e^{\alpha t} + b^*_{\mathbf{k}-\mathbf{q},n'}\varphi_n(z)\varphi_{n'}^*(z)e^{-i\mathbf{q}\cdot\mathbf{r}}e^{i\omega t}e^{\alpha t} \right\}. \tag{2.72}$$

This includes both an in-phase and a 180-degree out-of-phase component of the induced density with respect the perturbation. Since the external potential is a real quantity, we should add the complex conjugate of the potential, $V_{nn'}^{ext}(\mathbf{q})e^{i\omega t}e^{\alpha t}$, to the induced charge density so that Eq. (2.72) becomes

$$\rho_{ind}(\mathbf{r}, z, t) = -2e \sum_{\mathbf{k},n,n'} f_n(\mathbf{k}) \times \left\{ \frac{\varphi_n^*(z)\varphi_{n'}(z)V_{nn'}^{ext}(\mathbf{q})}{E_{n'}(\mathbf{k}+\mathbf{q}) - E_n(\mathbf{k}) - \hbar\omega - i\alpha\hbar} \right.$$

$$\left. + \frac{\varphi_n(z)\varphi_{n'}^*(z)V_{nn'}^{ext}(-\mathbf{q})}{E_{n'}(\mathbf{k}-\mathbf{q}) - E_n(\mathbf{k}) + \hbar\omega + i\alpha\hbar} \right\} e^{i\mathbf{q}\cdot\mathbf{r}}e^{i\omega t}e^{\alpha t} + c.c. \tag{2.73}$$

Since $V^{ext}(\mathbf{r}, z)$ is real, $V_{nn'}^{ext}(\mathbf{q}) = V_{nn'}^{ext}(-\mathbf{q})$, and the second term can be combined with the first term by letting $\mathbf{k} - \mathbf{q} = \mathbf{k}$. Switching n and n' in the summation over the second term yields

$$\rho_{ind}(\mathbf{r}, z, \omega) = -2e \sum_{\mathbf{k},n,n'} \varphi_n^*(z)\varphi_{n'}(z)V_{nn'}^{ext}(\mathbf{q}) \frac{f_n(\mathbf{k}) - f_n(\mathbf{k}+\mathbf{q})}{E_{n'}(\mathbf{k}+\mathbf{q}) - E_n(\mathbf{k}) - \hbar\omega - i\alpha\hbar} e^{i\mathbf{q}\cdot\mathbf{r}}$$

$$= -e \sum_{n,n'} \varphi_n^*(z)\varphi_{n'}(z)V_{nn'}^{ext}(\mathbf{q})L_{nn'}(\mathbf{q}, \omega)e^{i\mathbf{q}\cdot\mathbf{r}}, \tag{2.74}$$

where the sum over $\mathbf{k}$ (including the factor of 2 for spin) is included in the term $L_{nn'}(\mathbf{q}, \omega)$, and $\rho_{ind}(\mathbf{r}, z, \omega)$ is assumed to have the same harmonic time dependence as the potential. The induced potential energy resulting from the induced charge is given from electrostatics assuming an isotropic media (i.e., neglecting image potential effects) as

$$V^{ind}(\mathbf{r}, z) = \frac{-e}{4\pi\epsilon_s} \int d\mathbf{r}' \int_{-\infty}^{\infty} dz' \frac{\rho_{ind}(\mathbf{r}', z', \omega)}{\{|\mathbf{r} - \mathbf{r}'|^2 + (z - z')^2\}^{1/2}}, \tag{2.75}$$

where ϵ_s, the permittivity of the semiconductor, can be taken as the first two terms of Eq. (2.65) multiplied by ϵ_o. Introducing the 2D Fourier transform of the Coulomb potential,

$$\frac{1}{4\pi\epsilon_s\{|\mathbf{r} - \mathbf{r}'|^2 + (z - z')^2\}^{1/2}} = \int d\mathbf{q}' \frac{e^{-q'(z-z')}e^{i\mathbf{q}'\cdot(\mathbf{r}-\mathbf{r}')}}{2q'\epsilon_s}, \tag{2.76}$$

the induced potential (2.75) may be written using Eq. (2.74) as

$$V^{ind}(\mathbf{r}, z) = \frac{e^2}{2\epsilon_s q} \sum_{n,n'} \int_{-\infty}^{\infty} dz' \varphi_n^*(z')\varphi_{n'}(z')e^{-q(z-z')}V_{nn'}^{ext}(\mathbf{q})L_{nn'}(\mathbf{q}, \omega)e^{i\mathbf{q}\cdot\mathbf{r}}, \tag{2.77}$$

where the integral over $\mathbf{r}'$ yields $\delta(\mathbf{q} - \mathbf{q}')$, which then collapses the integration over $\mathbf{q}'$.

We look now at the matrix element of the induced potential over initial and final basis states $\mathbf{k}$ and $\mathbf{k}'$ in subbands m and m', respectively. Such matrix elements occur in scattering problems in transport (as discussed in Section 2.7) related to calculating the transition rate between states due to presence of an external potential (such as that due to an ionized impurity). However, as we have seen in the present section, the polarization of the electron gas gives rise to an additional induced potential. Thus what is really desired for scattering problems is the matrix element of the total potential, that is, the external plus the induced potential. The matrix element of the induced potential follows from Eq. (2.77) as

$$V_{mm'}^{ind}(\mathbf{k} - \mathbf{k}') = \frac{e^2}{2\epsilon_s q} \sum_{n,n'} \int_{-\infty}^{\infty} dz \int_{-\infty}^{\infty} dz' \varphi_{m'}^*(z)\varphi_m(z)\varphi_n^*(z')\varphi_{n'}(z')e^{-q(z-z')}$$

$$\times V_{nn'}^{ext}(\mathbf{q})L_{nn'}(\mathbf{q}, \omega) \int d\mathbf{r} e^{i(\mathbf{q}+\mathbf{k}-\mathbf{k}')\cdot\mathbf{r}}. \tag{2.78}$$

The latter integration simply gives a delta function. This may be written more concisely in terms of the form factor, $F_{nn'}^{mm'}(\mathbf{q})$, as

$$V_{mm'}^{ind}(\mathbf{q}) = \frac{e^2}{2\epsilon_s q} \sum_{n,n'} F_{nn'}^{mm'}(\mathbf{q})L_{nn'}(\mathbf{q}, \omega)V_{nn'}^{ext}(\mathbf{q}). \tag{2.79}$$

At this point, Eq. (2.79) gives the relation of the induced potential to the external potential. However, the net potential is now reduced, which causes the induced potential to be reduced, and so on. In order to account for this effect self-consistently, we replace $V_{nn'}^{ext}(\mathbf{q})$ on the right side with the total potential

$$V_{nn'}^{tot}(\mathbf{q}) = V_{nn'}^{ext}(\mathbf{q}) + V_{nn'}^{ind}(\mathbf{q}).$$

The external potential may thus be related to the total potential using Eq. (2.79) (with $V_{nn'}^{ext}(\mathbf{q})$ replaced with $V_{nn'}^{tot}(\mathbf{q})$) as

$$V_{mm'}^{ext}(\mathbf{q}) = V_{mm'}^{tot}(\mathbf{q}) - V_{mm'}^{ind}(\mathbf{q}) = \sum_{n,n'} \epsilon_{nn'mm'}(\mathbf{q}, \omega)V_{nn'}^{tot}(\mathbf{q})$$

$$= \sum_{n,n'} \left[\delta_{nm}\delta_{n'm'} - \frac{e^2}{2\epsilon_s q} F_{nn'}^{mm'}(\mathbf{q})L_{nn'}(\mathbf{q}, \omega) \right] V_{nn'}^{tot}(\mathbf{q}), \tag{2.80}$$

where $\epsilon_{nn'mm'}(\mathbf{q}, \omega)$ defines the four-dimensional dielectric matrix. Thus, the effective potential residing in the 2DEG associated with an external perturbation may be written as the inverse of (2.80), or

$$V_{mm'}^{tot}(\mathbf{q}) = \sum_{n,n'} \epsilon_{nn'mm'}^{-1}(\mathbf{q}, \omega)V_{nn'}^{ext}(\mathbf{q}). \tag{2.81}$$

The relation (2.81) determines the screening of an external potential by the 2DEG. Diagonal contributions contribute to the intrasubband polarization of the system, while off-diagonal elements are related to intersubband polarizations. Some clue to the relative contributions of the intrasubband versus intersubband contributions are given by the form factor,

$$F_{nn'}^{mm'}(\mathbf{q}) = \int_{-\infty}^{\infty} dz \int_{-\infty}^{\infty} dz' \varphi_{m'}^*(z)\varphi_m(z)\varphi_n^*(z')\varphi_{n'}(z')e^{-q(z-z')}. \tag{2.82}$$

Since the envelope functions are orthonormal functions with respect to the subband indices,

$$\lim_{\mathbf{q}\to0} F_{nn'}^{mm'}(\mathbf{q}) = \delta_{nn'}\delta_{mm'}. \tag{2.83}$$

The long-wavelength limit, $\mathbf{q} \to 0$, is reached when the wavelength $(2\pi/q)$ is large compared to the spatial extent of the subband envelope functions. Thus, in the limit of a purely

2D system of zero width, Eq. (2.83) holds for all $\mathbf{q}$. As a consequence of (2.83), intersubband transitions are completely unscreened in the long-wavelength limit

$$V_{mm'}^{tot}(\mathbf{q}) = V_{mm'}^{ext}(\mathbf{q}), \tag{2.84}$$

whereas for intrasubband transitions Eq. (2.80) reduces to

$$V_{mm}^{ext}(\mathbf{q}) = V_{mm}^{tot}(\mathbf{q}) + \sum_n \frac{e^2}{2\epsilon_s q} L_{nn}(\mathbf{q}, \omega) V_{nn}^{tot}(\mathbf{q})$$

$$= V_{mm}^{tot}(\mathbf{q}) + \sum_n \chi_{nn}(\mathbf{q}, \omega) V_{nn}^{tot}(\mathbf{q}), \tag{2.85}$$

where $\chi_{nn}(\mathbf{q}, \omega)$ represents the intrasubband polarizability of subband n. The matrix equation (2.85) may be cast in the form of a matrix equation

$$\mathbf{V}^{ext} = \bar{\epsilon} \cdot \mathbf{V}^{tot}; \quad \bar{\epsilon} = \bar{\mathbf{I}} + \bar{\mathbf{u}} \cdot \bar{\mathbf{v}}, \tag{2.86}$$

where the dielectric matrix may be factored into the identity matrix plus the product of an identity column vector and a polarizability vector

$$\bar{\mathbf{v}} = [\, \chi_{11} \ \chi_{22} \ \chi_{33} \ \cdots \,]. \tag{2.87}$$

A matrix in this symmetric form may be inverted analytically [44], and thus (2.85) is inverted to give

$$V_{mm}^{tot}(\mathbf{q}) = \frac{V_{mm}^{ext}(\mathbf{q}) + \sum_{n \neq m} \chi_{nn}(\mathbf{q}, \omega)\left(V_{mm}^{ext}(\mathbf{q}) - V_{nn}^{ext}(\mathbf{q})\right)}{\epsilon_D(\mathbf{q}, \omega)}, \tag{2.88}$$

where the denominator is the determinant of $\bar{\epsilon}$, which is the scalar dielectric function associated with the independent polarizabilities in each subband,

$$\epsilon_D(\mathbf{q}, \omega) = 1 + \sum_n \chi_{nn}(\mathbf{q}, \omega). \tag{2.89}$$

The second term in (2.88) is expected to be small, and thus under the conditions of neglecting the intersubband polarizabilities and the correction terms due to the intrasubband polarizabilities of the other subbands, the multisubband dielectric matrix reduces to a scalar relation as in the bulk case

$$V_{mm}^{tot}(\mathbf{q}) = \frac{V_{mm}^{ext}(\mathbf{q})}{\epsilon_D(\mathbf{q}, \omega)}. \tag{2.90}$$

Close to equilibrium, the intraband polarizability is a function of the distribution function $f_n(E)$, which is a function of only the energy. In this case, Maldague [45] has shown that the polarizability at finite temperature may be calculated as an integral over the zero temperature function as

$$\chi_{nn}(\mathbf{q}, \omega; T, E_f) = \int_0^\infty dE'_f \frac{\chi_{nn}(\mathbf{q}, \omega; 0, E'_f)}{4k_B T \cosh^2\left(\frac{E_f - E'_f}{2k_B T}\right)}, \tag{2.91}$$

where E_f is the Fermi energy, which appears as a parameter in the distribution function.

Static screening

If the potential is static (i.e., $\omega = 0$), such as the potential due to an impurity atom in the system, some further simplification is possible. Taking the long-wavelength limit again, the polarization function $L_{nn}(\mathbf{q}, 0)$, defined in (2.74), reduces to

$$\lim_{q \to 0} L_{nn}(\mathbf{q}, 0) = \lim_{q \to 0} 2 \sum_{\mathbf{k}} \frac{f_n(\mathbf{k}) - f_n(\mathbf{k} + \mathbf{q})}{E_n(\mathbf{k} + \mathbf{q}) - E_n(\mathbf{k})}$$

$$= \frac{-2}{4\pi^2} \int_0^{2\pi} d\theta \int_0^{\infty} dk\, k \frac{\partial f_n(\mathbf{k})}{\partial E_n(\mathbf{k})}. \qquad (2.92)$$

If we assume a simple parabolic dispersion relation for each subband, then the integral over k may be converted to one over energy to yield

$$\lim_{q \to 0} L_{nn}(\mathbf{q}, 0) = \frac{m_n^*}{\pi \hbar^2} f_n(0) = D_n f_n(0), \qquad (2.93)$$

where D_n is the 2D density of states from Eq. (2.10) (generalizing to different masses in different subbands), and $f_n(0)$ is the occupancy of the bottom of subband n. From Eq. (2.93), the scalar dielectric function Eq. (2.89) may be written

$$\lim_{q \to 0} \epsilon_D(\mathbf{q}, 0) = 1 + \frac{e^2}{2\epsilon_s q} \sum_n D_n f_n(0) = 1 + \frac{q_s}{q}, \qquad (2.94)$$

where q_s is the 2D Thomas-Fermi screening constant. At low temperature in the extreme quantum limit, only the $n = 1$ state is occupied, and the occupancy $f_n(0) = 1$, so that $q_s = e^2 D_1/(2\epsilon_s)$. For a one-subband system at high temperature, Eq. (2.11) may be inverted to solve for the Fermi energy. This is then substituted in the high-temperature distribution function, $f_n(0) = e^{(E_f - E_0)/kT}$, and assuming $n_{2d}/(kT D_1) \ll 1$, the screening constant is $q_s = e^2 n_{2d}/(2\epsilon_s kT)$. Typically the screening length, q_s^{-1}, representing the fall off of the potential with distance, is shorter than the corresponding quantity in a bulk 3D system. In 3D, however, the real-space decay of the screened Coulomb potential found by inverse transforming the 3D equivalent of Eq. (2.90) is exponential, whereas the corresponding 2D decay goes as r^{-3}, which is weaker. Thus one cannot say quantitatively whether screening is stronger or weaker in low-dimensional systems.

Collective excitations

In a bulk system, the dielectric function may vanish for certain $\mathbf{q}$ and ω. This implies that a potential of wavevector $\mathbf{q}$ and frequency ω gives rise to an infinite response in the system. Thus for certain frequencies, potential oscillations are sustained in the absence of an external potential. Such self-oscillations are referred to as *plasma oscillations* and correspond to the collective modes of vibration of the entire electron gas about its positive background. In a bulk system, the zeros of the Lindhard dielectric function, Eq. (2.65), give the dispersion relation for the long-wavelength plasma oscillations as

$$\omega_p = \left(\frac{n_{3d} e^2}{\epsilon_s m^*} \right)^{1/2}. \qquad (2.95)$$

For long-wavelength oscillations, the plasma frequency is independent of wavevector.

For a multiband system, the plasma frequency is complicated by the matrix nature of the dielectric function as given, for example, by Eq. (2.80). The zeros of the dielectric matrix correspond to the zeros of the determinant of the matrix. Solving the determinant equation is of course quite complicated, and may sustain both intraband plasma oscillations as well as intersubband oscillations. However, if we again limit ourselves to $\mathbf{q} \to 0$, the dielectric matrix is two-dimensional, and the determinant may be written from Eq. (2.89)

$$\det |\epsilon| = 1 + \chi_{11}(\mathbf{q}, \omega) + \chi_{22}(\mathbf{q}, \omega) + \chi_{33}(\mathbf{q}, \omega) + \cdots$$

and thus the plasma frequencies are given by the zeros of the above relation. Let us now consider the long-wavelength plasma frequency ($\mathbf{q} \to 0$) by looking at

$$0 = 1 + \sum_i \chi_{ii}(\mathbf{q}, \omega_p) = 1 + \frac{e^2}{2\epsilon_s q} \sum_i L_{ii}(\mathbf{q}, \omega_p)$$

$$= 1 + 2\frac{e^2}{2\epsilon_s q} \sum_{\mathbf{k},i} \frac{f_i(\mathbf{k}) - f_i(\mathbf{k} + \mathbf{q})}{E_i(\mathbf{k} + \mathbf{q}) - E_i(\mathbf{k}) - \hbar\omega_p - i\alpha\hbar}, \tag{2.96}$$

where ω_p is the plasma frequency. Unfolding the summation and combining terms, Eq. (2.96) becomes

$$0 = 1 + 2\frac{e^2}{2\epsilon_s q} \sum_{\mathbf{k},i} \left\{ \frac{f_i(\mathbf{k})}{E_i(\mathbf{k} + \mathbf{q}) - E_n(\mathbf{k}) - \hbar\omega_p - i\alpha\hbar} \right.$$

$$\left. - \frac{f_i(\mathbf{k})}{E_n(\mathbf{k}) - E_n(\mathbf{k} - \mathbf{q}) - \hbar\omega_p - i\alpha\hbar} \right\} \tag{2.97}$$

$$= 1 + \frac{e^2}{\epsilon_s q} \sum_{\mathbf{k},i} \frac{f_i(\mathbf{k})[2E_i(\mathbf{k}) - E_i(\mathbf{k} + \mathbf{q}) - E_i(\mathbf{k} - \mathbf{q})]}{(E_i(\mathbf{k} + \mathbf{q}) - E_i(\mathbf{k}) - \hbar\omega_p - i\alpha\hbar)(E_i(\mathbf{k}) - E_i(\mathbf{k} - \mathbf{q}) - \hbar\omega_p - i\alpha\hbar)}. \tag{2.98}$$

In the limit that $\mathbf{q}$ becomes small, the denominator goes to $(\hbar\omega)^2$ and the term in brackets in the numerator goes to $-\hbar^2 q^2/m_i^*$. The sum over $\mathbf{k}$ over the subband distribution function gives n_{2d}^i, the 2D sheet density in subband i. Thus the long-wavelength plasma dispersion relation becomes

$$0 = 1 + \frac{e^2 q}{\epsilon_s \omega_p} \sum_i \frac{n_{2d}^i}{m_i^*}.$$

If the subband masses are assumed to be the same, the 2D plasma frequency becomes

$$\omega_p^{2d} = \left(\frac{n_{2d} e^2 q}{\epsilon_s m^*} \right)^{1/2}, \tag{2.99}$$

where n_{2d} is the total 2D density. In contrast to the dispersionless 3D relationship, ω_p^{2d} goes as $q^{1/2}$ and thus the plasma frequency goes to zero as q approaches zero. As q becomes larger, the q dependence of the form factor (2.82) can no longer be neglected, and a more complicated dependence on q arises.

As in the bulk case, the plasma-like oscillations of the 2D electron gas may be described using collective coordinates for a boson-like system of quasi-particles called *plasmons* as described by the dispersion relation above (which is analogous to the collective representation of the vibrational modes of the lattice in terms of quasi-particles called phonons). The plasmons may interact with the free electrons in the system or with incident electromagnetic energy, emitting and absorbing energy in quanta corresponding to Eq. (2.99). In transport, plasmons may evidence themselves as an additional energy loss mechanism for electrons; this effect should be considered along with the other scattering mechanisms in the system.

Evidence for 2D plasmons was observed early on in the Si/SiO_2 system in far-infrared transmission experiments [46], [47] and in inversion layers excited by an electric current [48].

2.6.2 Dielectric function for quasi-1D systems

The derivation of the matrix dielectric function for a quasi-1D system follows the basic one described above for quasi-2D systems. The differences are that the wavevector for propagation in the parallel direction is now restricted to one dimension, and that the envelope function describing the confined states is a function of two perpendicular directions, as given by Eq. (2.30). To simplify the index notation, we express the transverse modes of the quantum wire in terms of a single subband index n and write the envelope function (setting L to unity for convenience) as

$$\psi(\mathbf{r}, x) = \varphi_n(\mathbf{r}) e^{ik_x x}, \qquad (2.100)$$

where x is the direction parallel to the axis of the quantum wire, and $\mathbf{r}$ is the y, z confinement direction perpendicular to the axis of the wire. Without loss of generality, the induced charge due to the same time-dependent perturbation considered in the 2D case above gives

$$\rho_{ind}(\mathbf{r}, x, \omega) = -2e \sum_{k_x, n, n'} \varphi_n^*(\mathbf{r}) \varphi_{n'}(\mathbf{r}) V_{nn'}^{ext}(q_x) \frac{f_n(k_x) - f_n(k_x + q_x)}{E_{n'}(k_x + q_x) - E_n(k_x) - \hbar\omega - i\alpha\hbar} e^{iq_x \cdot x}$$

$$= -e \sum_{n, n'} \varphi_n^*(\mathbf{r}) \varphi_{n'}(\mathbf{r}) V_{nn'}^{ext}(q_x) L_{nn'}(q_x, \omega) e^{iq_x \cdot x}. \qquad (2.101)$$

The 1D Fourier transform of the Coulomb potential differs from Eq. (2.76) as

$$\frac{1}{4\pi\epsilon_s \{|\mathbf{r} - \mathbf{r}'|^2 + (x - x')^2\}^{1/2}} = \frac{1}{4\pi^2\epsilon_s} \int_{-\infty}^{\infty} dq_x' K_0(|q_x'||\mathbf{r} - \mathbf{r}'|) e^{iq_x' \cdot (x - x')} \qquad (2.102)$$

where K_0 is the zeroth-order modified Bessel function of the second kind. Thus, following the same steps as in the 2D case, the matrix element of the 1D induced potential energy is given by

$$V_{mm'}^{ind}(q_x) = \frac{e^2}{2\pi^2\epsilon_s} \sum_{n, n'} F_{nn'}^{mm'}(q_x) L_{nn'}(q_x, \omega) V_{nn'}^{ext}(q_x), \qquad (2.103)$$

where the form factor in the quasi-1D case is given by

$$F_{nn'}^{mm'}(q_x) = \int d\mathbf{r} \int d\mathbf{r}' \varphi_n^*(\mathbf{r}) \varphi_{n'}(\mathbf{r}) \varphi_{m'}^*(\mathbf{r}) \varphi_m(\mathbf{r}) K_0(|q_x||\mathbf{r} - \mathbf{r}'|). \qquad (2.104)$$

Thus, as in the 2D case, we may define the matrix dielectric function for a quasi-1D quantum wire as

$$V_{mm'}^{ext}(q_x) = \sum_{n, n'} \left[\delta_{nm} \delta_{n'm'} - \frac{e^2}{2\pi^2\epsilon_s} F_{nn'}^{mm'}(q_x) L_{nn'}(q_x, \omega) \right] V_{nn'}^{tot}(q_x)$$

$$= \sum_{n, n'} \epsilon_{nn'mm'}(q_x, \omega) V_{nn'}^{tot}(q_x). \qquad (2.105)$$

The form factor (2.104) is complicated by the Bessel function, which diverges logarithmically as $K_0(x) \sim -\ln(x)$ as $x \to 0$. However, as is discussed for the 2D case, it is expected that the intersubband contributions to the polarizability should be small. For the intrasubband case, Hu and Das Sarma [49] have shown that if one assumes that the width of the wire is infinitesimally narrow in one transverse direction (for example, in the growth direction), and assumes a square quantum well of width a bounded by infinite potential barriers in the other transverse direction, the intrasubband form factor for the ground subband reduces to

$$F_{11}^{11}(q_x) = K_0(|q_x|a) + 1.9726917\ldots \quad \text{for } |q_x|a \to 0. \tag{2.106}$$

In the long-wavelength, static limit, $L_{nn}(q_x, 0)$ in the 1D case may be written

$$\lim_{q_x \to 0} L_{nn}(q_x, 0) = \frac{-2}{2\pi} \int_{-\infty}^{\infty} dk_x \frac{\partial f_n(k_x)}{\partial E_n(k_x)}. \tag{2.107}$$

If we consider the quantum limit at low temperature, the derivative of the Fermi-Dirac function becomes a delta function, $-\delta(E - E_f)$, and therefore (2.107) is simply the definition of the 1D density of states at the Fermi energy given by Eq. (2.34),

$$L_{nn}(q_x, 0) = D_{1d}(E_f) = \frac{2m^*}{\pi^2 \hbar^2} E_f^{-1/2}, \tag{2.108}$$

where the Fermi energy is measured with respect to the lowest subband minima, and a single-valley system is again assumed ($\eta_v = 1$). For long-wavelength, static screening in the quantum limit, the dielectric function in 1D becomes

$$\epsilon_{1d}(q_x) = 1 - \frac{e^2}{2\pi^2 \epsilon_s} \ln(|q_x|a) D_{1d}(E_f), \tag{2.109}$$

where a may be taken as the average width of the 1D gas.

The long-wavelength limit of the of the plasmon dispersion relation is again derived in an analogous fashion to the 2D case given by for Eq. (2.99). Using the same unfolding $L_{nn}(q_x, \omega)$ used in Eq. (2.98), the zeros of the 1D dielectric function for an infinite square well become

$$\lim_{q_x \to 0} \epsilon_{1d}(q_x, \omega_p) = 0 = 1 + \frac{e^2 n_{1d} q^2}{\pi^2 \epsilon_s m^* \omega_p^2} \ln(|q_x|a), \tag{2.110}$$

and thus the plasmon dispersion goes as

$$\omega_p^{1d} = \left(\frac{e^2 n_{1d}}{\pi^2 \epsilon_s m^*} \right)^{1/2} q_x \sqrt{-\ln(|q_x|a)} \qquad \ldots |q_x|a \ll 1. \tag{2.111}$$

2.7 Homogeneous transport in low-dimensional systems

In the quasi-two-dimensional system discussed in Section 2.2.1, one degree of freedom of the electrons is restricted by the confinement potential associated with an interface(s) leading to quantization of the momentum in that direction, while motion in the remaining two directions is unconstrained. Likewise, in a quasi-one-dimensional system such as the quantum wire discussed above, the motion along the axis of the wire is free electron–like. If we restrict ourselves to carrier transport parallel to confining potentials, then we can discuss the homogeneous transport properties of such reduced-dimensionality systems in the same context as homogeneous transport in bulk systems, that is, in terms of macroscopic phenomenological parameters such as mobility, conductivity, thermopower, and so on, which may be measured in an appropriately designed experiment. Transport in this context is quite different from the case of transport perpendicular to the confining barriers discussed in Chapter 3 in which quantum mechanical reflection and transmission from the confining barriers themselves play a central role.

The same framework of theoretical techniques in nonequilibrium statistical mechanics used to explicate the bulk transport properties of materials may be applied to the homogeneous properties of the quasi-2D and quasi-1D systems. Indeed, an enormous volume of experimental and theoretical literature has been published on the parallel transport properties of quasi-2D systems, beginning in the late 1960s with studies of the Si/SiO_2 system and continuing with the heterojunction semiconductor systems in the mid- to late 70s until today. Perhaps the most striking manifestation of reduced dimensionality is in the transport properties of the 2DEG subject to a perpendicular magnetic field. Measurement of the Hall coefficient versus magnetic field or Fermi energy of a 2DEG under strong magnetic fields at low temperature reveals that the transverse resistivity, ρ_{xy}, is quantized extremely accurately in integer multiples of $h/e^2(1/n)$, where n is an integer. The discovery of the integer or "normal" *quantum Hall effect* (NQHE) [50] led to the Nobel prize in physics for Klaus von Klitzing in 1985. While elegant arguments based on gauge invariance have related this phenomenon to the bulk properties of the 2DEG [51], a simple alternative model associated with lateral confinement of the 2DEG and the formation of so-called *edge states* has been given by Büttiker [52], which we will come back to in Chapter 3. At even lower temperatures and in very-high-mobility samples, new plateaus appear at fractional values of h/e^2, in which the fraction is a ratio of two integers [53]. This *fractional quantum Hall effect* (FQHE) is theorized to arise from the condensation of the interacting electron system into a new many-body ground state characteristic of an incompressible fluid [54]. Currently, theoretical and experimental investigation of the FQHE is an active area of research in condensed matter physics.

The transport properties of quantum wire systems is an emerging body of research, which at the time of this writing is still evolving as fabrication methods continually improve. The transport properties of 1D systems are of particular interest, as many interesting predictions of unusual many body behavior have been suggested, such as the formation of a Luttinger liquid [55] or the presence of a charge-density-wave ground state due to the lattice Peierls distortion [56]. However, evidence for such behavior in present quantum wire systems is lacking due to the energy broadening effects of disorder in present technology today, which leads to normal Fermi liquid–like behavior [49]. Thus, in the following section we will discuss parallel transport in quantum wires on the same footing as parallel transport in quantum wells in the context of semiclassical transport.

2.7.1 Semiclassical transport

In discussing semiclassical transport in low-dimensional systems, we first begin by assuming that the description of the system in terms of the solutions of the effective mass equation is still a "good" solution to the problem in the presence of disorder. In low-dimensional systems, we consider the perfect or unperturbed system to be defined by the static lattice, with perfectly smooth boundaries defining the system, free of impurities or other random inhomogeneities. A "good" solution basically implies that the broadening of the energy levels due to disorder (i.e., the real part of the self-energy) is small, so that crystal momentum conservation is approximately preserved. In this weak coupling limit, we can then construct a kinetic equation in which the particle density function (or distribution function) evolves in time under the streaming motion of external forces and spatial gradients, and the randomizing influence of nearly point-like (in space-time) scattering events. Let $f_i(\mathbf{r}, \mathbf{k})$ represent the one-particle distribution function in a $2n$-dimensional phase space, where n is the dimensionality of the system and i labels the subband index. Here $\mathbf{r}$ and $\mathbf{k}$ refer to the position and wavevector in the propagating direction(s) of the system (i.e., parallel to the quantum wire axis or interface in a quantum well). By describing the system in terms of $f_i(\mathbf{r}, \mathbf{k})$, which gives the probability of finding a particle in a infinitesimal volume $d\mathbf{r}d\mathbf{k}$ around $\mathbf{r}$ and $\mathbf{k}$, we implicitly neglect the fact that $\mathbf{r}$ and $\mathbf{k}$ are quantum-mechanically noncommuting variables. This assumption does not pose a serious problem in considering homogeneous transport in a semiinfinite medium, as discussed in the present section, further assuming that the phase information carried by the carrier is lost between subsequent collisions. As we will see in Chapters 5 and 6, conditions exist where the phase is preserved between elastic scattering events giving rise to experimentally observable phenomena such as negative magnetoresistance.

Under the influence of a driving force such as an electric field applied in the plane parallel to the interface for 2D systems (or parallel to the longitudinal axis in the 1D case), the in-plane crystal momentum of carriers in the unperturbed system evolves according to the acceleration theorem

$$\frac{d(\hbar\mathbf{k})}{dt} = \mathbf{F} = e(\mathbf{E} + \mathbf{v} \times \mathbf{B}), \tag{2.112}$$

where $\mathbf{F}$ is the force acting on the particle, $\mathbf{E}$ is the electric field, $\mathbf{B}$ is the magnetic flux density, and $\mathbf{v}$ is the particle velocity given by the group velocity

$$\mathbf{v} = \frac{1}{\hbar}\nabla_{\mathbf{k}}E(\mathbf{k}), \tag{2.113}$$

where E is the energy associated with state $\mathbf{k}$. The *Boltzmann transport equation* (BTE) may be derived by writing the continuity equation in the $2n$-dimensional phase space in terms of the particle flux through a small hypervolume of this space centered around $\mathbf{r}$ and $\mathbf{k}$,

$$\frac{\partial f_i}{\partial t} = -\frac{1}{\hbar}\nabla_{\mathbf{k}}E(\mathbf{k}) \cdot \nabla_{\mathbf{r}}f_i - \frac{1}{\hbar}\mathbf{F} \cdot \nabla_{\mathbf{k}}f_i + \left.\frac{\partial f_i}{\partial t}\right|_{collisions}, \tag{2.114}$$

where the last term represents the rate of change of the distribution function due to scattering. Assume instantaneous (again in both space and time) phase-randomizing collisions, in the classical sense as well as the quantum mechanical, meaning that the particle loses any correlation it has with other particles, allowing one to decouple higher-order two-, three-, etc., particle distribution functions from (2.114). Under this assumption, the last term may be written as a detailed balance of in-scattering and out-scattering events as the so-called *collision integral*

$$\left.\frac{\partial f_i}{\partial t}\right|_{collisions} = \sum_{j,\mathbf{k}'} S_{j,i}(\mathbf{k}', \mathbf{k})[f_j(\mathbf{k}')(1 - f_i(\mathbf{k}))] - S_{i,j}(\mathbf{k}, \mathbf{k}')[f_i(\mathbf{k})(1 - f_j(\mathbf{k}'))],$$

$$\tag{2.115}$$

where $S_{j,i}(\mathbf{k}', \mathbf{k})$ represents the scattering rate from a state in subband j of wavevector $\mathbf{k}'$ to a state in subband i with wavevector $\mathbf{k}$. The case of $i = j$ refers to *intrasubband* scattering whereas $i \neq j$ is denoted *intersubband* scattering. For a quasi-1DEG, $\mathbf{k}$ is no longer a vector quantity, but rather a scalar. The only possible states after scattering are either in the forward or the backward direction, thus the sum over $\mathbf{k}'$ reduces to just two terms.

The assumption of instantaneous collisions in space and time allows us to write the detailed balance above as affecting only the **k**-dependent part of f_i in the infinitesimal volume in phase space that we consider in the general continuity equation (2.114). However, when we look at the problem of transport through generalized structures with dimensions that vary on the order of the de Broglie wavelength of carriers (i.e., nanostructures), the nonlocality of the particle invalidates this simplifying approximation. The collision integral in general couples the distribution function from subband j into the BTE for subband i through the *intersubband* scattering terms involving $S_{i,j}, i \neq j$, thus leading to the necessity of solving a set of coupled partial differential equations for $f_i(\mathbf{r}, \mathbf{k})$. In reality, one has the same complication in a bulk 3D system if one considers transport including multiple bands (e.g., light-hole/heavy-hole transport), or multiple valleys such as the problem of intervalley transfer in high-field transport in the conduction bands. However, in the case of quasi-2D or quasi-1D systems, the spacing of the subbands is usually tenths of an electron volt, and thus multiband transport is not negligible except in the extreme quantum limit in which only one subband is occupied.

2.7.2 Relaxation time approximations

To determine phenomenological transport parameters such as the carrier mobility and diffusion coefficient in low-dimensional systems, the relaxation time approximation may be employed as in bulk systems. (For a derivation of the relaxation time approximation in bulk systems, see for example [17], Section 8.2.1.) First consider the system to be stationary, so that the time derivative in Eq. (2.114) is zero, and bring the streaming terms involving gradients in phase space to the left side of the BTE. If the system is not driven far from equilibrium, as in the case of measuring the low-field mobility, the collision integral (2.115) may be written in terms of a relaxation time for the ith subband as

$$\left.\frac{\partial f_i}{\partial t}\right|_{collisions} = -\frac{f_i(\mathbf{k}) - f^0(E)}{\tau_i}, \qquad (2.116)$$

where $f^0(E)$ is the equilibrium Fermi-Dirac function, and τ_i is the relaxation time. The energy E is understood to be the total energy for a particle in a given subband, $E = E_i + E_\mathbf{k}$, where $\mathbf{k}$ is the crystal momentum in subband i. Again for the case of a 1DEG, $\mathbf{k}$ is no longer a vector quantity but rather a scalar along the propagation direction of the system. In the linear response regime, only the lowest-order solution is kept on the LHS of the BTE.

$$LHS = \frac{\mathbf{F}}{\hbar} \cdot \nabla_\mathbf{k} f^0(E) = \frac{\mathbf{F}}{\hbar} \cdot \nabla_\mathbf{k} E_\mathbf{k} \frac{\partial f^0}{\partial E} = \mathbf{F} \cdot \mathbf{v}_\mathbf{k} \frac{\partial f^0}{\partial E}. \qquad (2.117)$$

The nonequilibrium distribution function may thus be written

$$f_i(\mathbf{k}) = f^0(E) - \tau_i \mathbf{F} \cdot \mathbf{v}_\mathbf{k} \frac{\partial f^0}{\partial E} = f^0(E) + f_i^1(\mathbf{k}), \qquad (2.118)$$

where $f_i^1(\mathbf{k})$ represents the perturbation to the equilibrium distribution due to the applied field. To evaluate τ_i, we further restrict ourselves to elastic scattering, so that $S_{i,j}(\mathbf{k}, \mathbf{k}') = S_{j,i}(\mathbf{k}', \mathbf{k})$. Substituting (2.118) into (2.115) and observing that in equilibrium the collision integral must vanish, (2.115) becomes

$$\left.\frac{\partial f_i}{\partial t}\right|_{collisions} = -\sum_{j,\mathbf{k}'} S_{i,j}(\mathbf{k}, \mathbf{k}')\big(f_i^1(\mathbf{k}) - f_j^1(\mathbf{k}')\big). \qquad (2.119)$$

Using (2.118), (2.119) may be written

$$\left.\frac{\partial f_i}{\partial t}\right|_{collisions} = -f_i^1(\mathbf{k}) \sum_{j,\mathbf{k}'} S_{i,j}(\mathbf{k},\mathbf{k}')\left(1 - \frac{\tau_j(E)}{\tau_i(E)}\frac{\mathbf{v_k}\cdot\mathbf{v_{k'}}}{v^2}\right). \qquad (2.120)$$

If we assume for simplicity that the states are derived from a spherically symmetric constant energy minimum, then the velocities may be written as $\mathbf{v_k} = \hbar\mathbf{k}/m^*$. Then comparing (2.120) to (2.118) and (2.116), the relaxation time τ_i is given by

$$\frac{1}{\tau_i(E)} = \sum_{j=i,\mathbf{k}'} S_{i,j}(\mathbf{k},\mathbf{k}')\left(1 - \frac{\tau_j(E)}{\tau_i(E)}\cos\theta\right), \qquad (2.121)$$

where θ is the angle between $\mathbf{k}$ and $\mathbf{k}'$. Again, for a quasi-1D system, the only possible choice of scattering is either forwards or backwards, so that $\cos\theta$ is either 1 or -1. As can be seen from Eq. (2.121), the solution for τ_i is coupled to the relaxation times of all the other participating subbands, and thus (2.121) represents a set of simultaneous equations which must be solved for each energy, E. In the case that the intersubband scattering rate is negligible (i.e., $S_{i,j} \approx 0$ for $i \neq j$), or when only one subband is occupied (the extreme quantum limit), a simple momentum relaxation time in closed form may be defined:

$$\frac{1}{\tau_i(E)} = \sum_{\mathbf{k}'} S_{i,i}(\mathbf{k},\mathbf{k}')(1 - \cos\theta). \qquad (2.122)$$

To calculate the mobility, we assume that the system is driven by a small external electric field. Thus for electrons, $\mathbf{F} = -q\mathbf{E}$, where $\mathbf{E}$ is the electric field. The current carried by the ith subband in steady state is determined only by the antisymmetric part of the distribution function, $f_i^1(\mathbf{k})$, given by (2.118). For electrons, the current density is an average over the velocity and particle density,

$$\mathbf{J}_i = -q \int d^d\mathbf{k} N(\mathbf{k})\mathbf{v_k} f_i^1(\mathbf{k}), \qquad (2.123)$$

where d is the dimensionality of the system and $N(\mathbf{k})$ is the density of states per unit volume in $\mathbf{k}$ space,

$$N(\mathbf{k}) = \left(\frac{1}{2\pi}\right)^d. \qquad (2.124)$$

The integration over $\mathbf{k}$ in Eq. (2.123) may be converted to an integration over energy as

$$\mathbf{J}_i = -q^2 \int_0^\infty dE\, D_i(E)\tau_i(E)\mathbf{v_k}(\mathbf{E}\cdot\mathbf{v_k})\left(\frac{\partial f^0}{\partial E}\right), \qquad (2.125)$$

where $D_i(E)$ is the density of states per unit energy of subband i, and the energy $E = E_\mathbf{k}$ is the kinetic energy of the particle in subband i relative to the subband minima. Factoring out the electric field and the density per subband, n_i, Eq. (2.125) becomes

$$\mathbf{J}_i = -q^2 n_i \mathbf{E}\frac{\int_0^\infty dE\, D_i(E)\tau_i(E)v_E^2(\partial f^0/\partial E)}{\int_0^\infty dE\, D_i(E)f^0(E)}, \qquad (2.126)$$

where v_E is the magnitude of the velocity in the direction of the field. By arguing that the drift velocity is much smaller than the thermal velocity for the linear response regime considered here, we can replace component v_E^2 with the total velocity as $v_E^2 = v^2/d = 2E/(m^*d)$, where again d is the dimensionality of the system. Further, since $D_i(E) \propto E^{d/2-1}$, we may write

$$\mathbf{J}_i = \frac{q^2 n_i \mathbf{E}}{m^*} \frac{\int_0^\infty dE\, E^{d/2} \tau_i(E)(\partial f^0/\partial E)}{\int_0^\infty dE\, E^{d/2}(\partial f^0/\partial E)} = q n_i \left(\frac{q \langle \tau_i \rangle}{m^*} \right) \mathbf{E} = q n_i \mu_i \mathbf{E}, \qquad (2.127)$$

where integration by parts was used to write the denominator of (2.126) on the same footing as the numerator (which also contributes a factor of $2/d$), and μ_i is the subband mobility related to the averaged relaxation time, given by

$$\langle \tau_i \rangle = \frac{\int_0^\infty dE\, E^{d/2} \tau_i(E)(\partial f^0/\partial E)}{\int_0^\infty dE\, E^{d/2}(\partial f^0/\partial E)}. \qquad (2.128)$$

The total current is given by the sum over the current of the individual subbands

$$\mathbf{J} = q \sum_i n_i \mu_i \mathbf{E} = q n_T \bar{\mu} \mathbf{E}, \qquad (2.129)$$

where n_T is the total density, and the average mobility $\bar{\mu}$ is defined by

$$\bar{\mu} = \left(\sum_i n_i \mu_i \right) \Big/ n_T. \qquad (2.130)$$

2.7.3 Elastic scattering mechanisms

In the weak coupling limit discussed above, in which the time between collisions is relatively long, the transition or scattering rate may be calculated from Fermi's golden rule. The rate is derived from first order using time-dependent perturbation theory (see for example [11]) as

$$S_{i,j}(\mathbf{k}, \mathbf{k}') = \frac{2\pi}{\hbar} |\langle \mathbf{k}', j | V_s(\mathbf{r}) | \mathbf{k}, i \rangle|^2 \delta(E_{\mathbf{k}'} + E_j - E_{\mathbf{k}} - E_i \mp \hbar\omega_{\mathbf{q}}), \qquad (2.131)$$

where $V_s(\mathbf{r})$ is the potential associated with a particular scattering mechanism, and $\omega_{\mathbf{q}}$ is the frequency associated with a harmonic time-dependent perturbation such as the normal modes of the crystal lattice or a time-varying electromagnetic field, with the upper sign referring to absorption (and the lower sign, emission) of a quasi-particle excitation of the field. As is well known from the standard derivation of Fermi's rule, the delta function in (2.131) is only approximately true as the time between collisions approaching infinity, with an uncertainty in the final energy after the collision that decreases as $\hbar/\tau_c$, where τ_c is the time after the collision.

For many scattering mechanisms, the scattering rate in a low-dimensional system differs from that in 3D only in the different initial and final states, which become increasingly restrictive as the dimensionality is reduced. In particular, in a quasi-1D system there are only two possible types of scattering, forward and backward with respect to the wire axis. The density of states tends to be reflected in the scattering rates. As the dimensionality is reduced, the density of states at the subband edge increases. In 1D, the scattering rates can be divergent at low energies because the singularity in the 1D density of states of the ideal system at the subband minimum, as seen in Section 2.4. New mechanisms that are not present per se in bulk systems, such as surface or boundary roughness scattering, interface states, interface phonons, remote impurities, etc., also must be included. In addition, the nature of the allowed phonon modes themselves are modified by the inhomogeneous structure associated with low-dimensional systems. Folding of the acoustic branches is possible, giving rise to nonzero frequencies for long-wavelength excitations. The optical branches may be confined as well, giving rise to interface modes and "guided" modes, in analogy to electromagnetic waveguide modes. In the following Sections, we briefly summarize the important scattering mechanisms in quasi-2D and quasi-1D systems, and later connect this to the phenomenological transport parameters of the homogeneous systems.

Coulomb scattering

Coulomb scattering is generally associated with ionized impurities in bulk systems, and it is dominant at low temperatures. Additional Coulomb scatterers exist due to the presence of the interface in the form of surface states and fixed charges at or near the interface. In modulation-doped heterojunction systems, the distribution of impurities is inhomogeneous, with high doping occurring in a region spatially separated from the 2DEG. In general, then, we have to consider a variety of distributions of impurities when discussing Coulomb scattering in low-dimensional systems.

The general scattering rate in a 2DEG due to point charges is discussed in detail in [1] with regards to transport in the Si/SiO_2 system, which is applicable to the heterojunction case as well. The unscreened potential energy due to a charge located at $\mathbf{r}_i$ and z_i is given as

$$V_i^o(\mathbf{r}, z) = \frac{Z_i e^2}{4\pi \varepsilon_s \{(\mathbf{r} - \mathbf{r}_i)^2 + (z - z_i)^2\}^{1/2}} + V_{image}, \quad (2.132)$$

where ε_s is the permittivity of the semiconductor and Z_i is the charge state of the impurity. As discussed earlier, the image potential is necessary to satisfy the usual electrostatic boundary conditions at the interfaces between layers of different permittivity. Such conditions may be satisfied by adding additional image charges outside the domain of solution which, combined with the potential of the source charge, satisfy the required boundary conditions. If the system consists of a single interface at $z = 0$ between two materials of differing permittivity, and if the impurity lies in the region 1 corresponding to $z_i \leq 0$ and the scattering particle is in region 2 corresponding to $z > 0$, V_{image} may be accounted for by replacing ε_s with the average dielectric constant, $\bar{\varepsilon} = \frac{\varepsilon_1 + \varepsilon_2}{2}$. For an impurity charge in the same layer as the scattering electron (for a single interface) in region 2, the image potential corresponds to a charge located at $-z_i$ as

$$V_{image}(\mathbf{r}, z) = \frac{(\varepsilon_1 - \varepsilon_2)Z_i e^2}{4\pi \varepsilon_2 (\varepsilon_1 + \varepsilon_1)\{(\mathbf{r} - \mathbf{r}_i)^2 + (z + z_i)^2\}^{1/2}}. \quad (2.133)$$

For the GaAs/AlGaAs system, the difference in permittivity is a small, and we can neglect V_{image} in Eq. (2.132). For simplicity we will assume such a system in the following. The matrix element for scattering introduces the 2D Fourier transform of this potential (neglecting V_{image}) as given by Eq. (2.76):

$$\langle \mathbf{k}', n | V_i^o(\mathbf{r}, z) | \mathbf{k}, m \rangle = \frac{Z_i e^2}{2\varepsilon_s q} e^{i\mathbf{q}\cdot\mathbf{r}_i} \left(e^{qz_i} \int_{-\infty}^{\infty} dz e^{-qz} \rho_{mn}(z) \right) = V_{i,nm}^o(q, z_i) e^{i\mathbf{q}\cdot\mathbf{r}_i}$$

$$\rho_{mn}(z) = \varphi_n(z)\varphi_m(z), \quad (2.134)$$

where $\varphi_n(z)$ is the envelope function solution where m and n refer to the initial and final subbands, respectively, and $\mathbf{q} = \mathbf{k} - \mathbf{k}'$ is the scattered wavevector with $q = |\mathbf{k} - \mathbf{k}'|$. The matrix element above relates only to the bare potential of the impurity, and does not account for the additional potential due to the polarization of the 2DEG, which is included through the wavevector dependent dielectric function. As discussed in Section 2.6, the screened potential is a complicated matrix function due to the multisubband nature of the low-dimensional system. Under the simplifying assumptions discussed in Section 2.6.1, the screened potential for intrasubband scattering ($n = m$) may be obtained by dividing (2.134) by the scalar static dielectric function given by (2.94). For intersubband scattering, the screening function is unity to first order.

Equation (2.134) represents the scattering matrix element due to a single impurity. The potential due to all the impurities in the system may be written as

$$V_{nm}^{ii}(q) = \sum_i e^{i\mathbf{q}\cdot\mathbf{r}_i} \frac{V_{i,nm}^o(q, z_i)}{\epsilon_D(q)}. \tag{2.135}$$

To calculate the scattering rate from Fermi's rule (2.131), the square of the matrix element is needed. As a first approximation, we assume that in any given plane parallel to interface that the position of the impurities is completely uncorrelated. The cross terms arising from squaring (2.135) then cancel on the average (similar to the random phase approximation), and using (2.131) and (2.135), we can write the elastic scattering rate for impurity scattering as

$$S_{nm}^{ii}(\mathbf{k}, \mathbf{k}') = \frac{2\pi}{\hbar} \left| V_{nm}^{ii}(q) \right|^2 \delta(E_{\mathbf{k}'} - E_{\mathbf{k}} + E_n - E_m)$$

$$= \int_{-\infty}^{\infty} dz_i N_i(z_i) \left| \frac{V_{i,nm}^o(q, z_i)}{\epsilon_D(q)} \right|^2 \delta(E_{\mathbf{k}'} - E_{\mathbf{k}} + E_n - E_m), \tag{2.136}$$

where $N_i(z_i)$ is the density of impurities per unit volume as a function of z_i. If intersubband scattering is weak, the momentum relaxation time for each subband is given by Eq. (2.122). Writing the matrix element explicitly using (2.134), we get

$$\frac{1}{\tau_n^{ii}(E)} = \frac{\pi Z_i^2 e^4}{2\hbar\varepsilon_s^2} \sum_{\mathbf{k}'} \int_{-\infty}^{\infty} dz_i N_i(z_i) \frac{e^{2qz_i}}{q^2\epsilon_D^2(q)} \mathcal{F}_{nn}^2(q)(1 - \cos\theta)\delta(E_{\mathbf{k}'} - E_{\mathbf{k}}), \tag{2.137}$$

where θ is the angle between $\mathbf{k}$ and $\mathbf{k}'$, and

$$\mathcal{F}_{nn}(q) = \int_{-\infty}^{\infty} dz e^{-qz} \rho_{nn}(z), \tag{2.138}$$

with $\rho_{nn}(z)$ defined as in (2.134). The sum over $\mathbf{k}'$ may be converted to an integral; assuming parabolic bands,

$$\sum_{\mathbf{k}'} \rightarrow \frac{1}{4\pi^2} \int_0^{2\pi} d\theta \int_0^{\infty} dk'k' = \frac{2m^*}{\hbar^2} \int_0^{2\pi} d\theta \int_0^{\infty} dE. \tag{2.139}$$

The delta function over energy can be used to reduce the latter integral, giving $|\mathbf{k}| = |\mathbf{k}'|$, and therefore Eq. (2.137) may be written

$$\frac{1}{\tau_n^{ii}(E)} = \frac{Z_i^2 e^4 m^*}{4\pi\hbar^3\varepsilon_s^2} \int_0^{2\pi} d\theta \int_{-\infty}^{\infty} dz_i N_i(z_i) \frac{e^{2qz_i}}{q^2\epsilon_D^2(q)} \mathcal{F}_{nn}^2(q)(1 - \cos\theta), \tag{2.140}$$

where

$$q = |\mathbf{k} - \mathbf{k}'| = \sqrt{k^2 + k'^2 - 2kk'\cos\theta} = \sqrt{2k^2(1 - \cos\theta)} = 2k\sin\frac{\theta}{2}. \tag{2.141}$$

The momentum relaxation time (2.140) is written assuming one type of impurity. For multiple types of impurity distributions for example from interface, remote impurities, and bulk impurities, the momentum relaxation time may be written as

$$\frac{1}{\tau_n^{ii}(E)} = \frac{1}{\tau_{1,n}^{ii}(E)} + \frac{1}{\tau_{2,n}^{ii}(E)} + \cdots, \tag{2.142}$$

where (2.140) is used to calculate the relaxation time for each individual impurity distribution.

Further simplification of (2.140) is possible if we consider impurity scattering due to a 2D sheet of impurities, $N_i(z_i) = N_{ss}\delta(z_i)$ at the interface $z_i = 0$, which may for example exist due to interface states at the hetero- or oxide-semiconductor interface. If the 2DEG is assumed to be extremely narrow, and we neglect screening altogether ($\epsilon_D(q) = 1$), then (2.140) reduces to

$$\frac{1}{\tau^{ii}(E)} = \frac{Z_i^2 e^4 N_{ss}}{8\,\hbar\varepsilon_s^2 E}. \tag{2.143}$$

The mobility limited by interface impurity scattering under the above approximations is calculated using the average (2.128). At low temperatures, $E \to E_f$; at high temperatures, the average (2.128) results in the substitution $E \to k_B T$ which gives the mobility

$$\mu_{ii} = \frac{8\,\hbar\varepsilon_s^2 k_b T}{Z_i^2 e^3 m^* N_{ss}}. \tag{2.144}$$

This equation shows that the impurity limited mobility decreases as $1/N_{ss}$ and increases with increasing temperature. In reality, the inclusion of screening through the temperature-dependent dielectric function (2.91) gives a more complicated temperature dependence, since screening tends to decrease with increasing temperature, resulting in an increased scattering rate (from consideration of screening alone).

If we wish to consider scattering in a quantum wire due to an impurity, the Coulomb matrix element for scattering is calculated between the quasi-one-dimensional electronic states labeled $\varphi_n(\mathbf{r})e^{ik_x x}$, where x is the longitudinal direction of the wire. For an impurity located at point x_i, y_i, z_i, the unscreened matrix element for scattering from an electron initially with wavevector k_x in subband m is written

$$
\begin{aligned}
H(q_x) &= \langle k_x', n | V_i^o(\mathbf{r}, x) | k_x, m \rangle \\
&= \frac{e^2}{4\pi\varepsilon_s} \int dz \int dy \varphi_n^*(y, z)\varphi_m(y, z)\frac{1}{L}\int dx \frac{e^{-i(k_x - k_x')x}}{\sqrt{(x - x_i)^2 + (y - y_i)^2(z - z_i)^2}} \\
&= \frac{e^2 e^{-iq_x x_i}}{2\pi\varepsilon_s L} \int dz \int dy \varphi_n^*(y, z)\varphi_m(y, z)K_o\big(|q_x|\sqrt{(y - y_i)^2 + (z - z_i)^2}\big), \tag{2.145}
\end{aligned}
$$

where $q_x = k_x - k_x'$, L is the normalization length of the 1D wire, and K_o is the zeroth order modified Bessel function. The total scattering rate is calculated from a sum over all final states

$$S_m^{ii}(k_x) = \frac{2\pi}{\hbar} \sum_n \sum_{k_x'} |H(q_x)|^2 \delta\left(\frac{\hbar^2 k_x^2}{2m_m} + E_m - \frac{\hbar^2 k_x'^2}{2m_n} - E_n,\right), \tag{2.146}$$

where $m_{m,n}$ refer to masses in the individual subbands m and n, and $E_{m,n}$ are the respective 1D subband minima. In 1D, the delta-function only allows two possible solutions for the final wavevector and subband

$$k'_x = \pm\sqrt{k_x^2 + \frac{2m_m}{\hbar^2}(E_m - E_n)}. \tag{2.147}$$

Assuming we have n_i impurities per unit length uniformly distributed along the wire, the total scattering rate may therefore be written [57]

$$S_m^{ii}(k_x) = \frac{e^4 n_i}{4\pi^2 \varepsilon_s^2 \hbar^3} \sum_n \left[\frac{m_n}{|k'_x|}\left(|H(k_x - k'_x)|^2 + |H(k_x + k'_x)|^2\right) \right]. \tag{2.148}$$

In the case that the initial and final subbands are the same, only one scattering process is possible that changes the momentum of the particle, that is complete backscattering from k_x to $-k_x$ (the second term above). If we want to include screening, than the we need to replace $H(q_x) \to H(q_x)/\varepsilon(q_x)$ where ε is the one-dimensional dielectric function defined in Section 2.6.2.

Surface roughness

Surface roughness is a term generically applied to the random fluctuations of the boundaries that nominally form the confining potential to low-dimensional systems. On a microscopic level, roughness appears as atomic layer steps in the interface between two differing materials, even those which are lattice matched. As evidenced by the high-resolution images obtained by HRTEM shown in Fig. 2.1 and by STM shown in Fig. 2.2, the extent of the roughness fluctuations is only one or two monolayers at the growth interface between two materials. In contrast, the roughness evident in lateral patterning using present-day lithographic techniques is at least an order of magnitude larger due to the "Neolithic" nature of the pattern transfer process which usually involves etching of some sort. As a consequence, while roughness scattering may be relatively weak in heterojunction 2DEG systems, in 1D wires it is still a dominant effect.

To date, a truly microscopic model of interface roughness scattering in semiconductor nanostructures has not been attempted. Instead, most treatments of roughness scattering assume that the fluctuations of the interface from its ideal flat boundary are described by a two-dimensional roughness function, $\Delta(\mathbf{r})$, where as before, $\mathbf{r}$ is the two-dimensional position vector in the plane of the interface, illustrated in Fig. 2.24. The potential associated with the roughness $\Delta(\mathbf{r})$ can be viewed as a combination of a boundary perturbation which causes the envelope functions to be displaced from their unperturbed values, and electrostatic contributions due to the imposed fluctuation of the electric fields and charge density at the rough interface. The combined effect on a two-dimensional system has been considered in detail elsewhere [58].

For simplicity, assume that in a 2D system the potential may be expanded as

$$V^{sr}(\mathbf{r}, z) = V_{eff}(z + \Delta(\mathbf{r})) - V_{eff}(z) \simeq \Delta(\mathbf{r})\frac{\partial V_{eff}(z)}{\partial z}, \tag{2.149}$$

where V_{eff} is the one-dimensional potential including many-body and image potential corrections. The scattering matrix element of the roughness potential (2.149) is

$$V_{nm}^{sr}(\mathbf{k} - \mathbf{k}') = \int d\mathbf{r}e^{i(\mathbf{k}-\mathbf{k}')\cdot\mathbf{r}}\Delta(\mathbf{r})\int_{-\infty}^{\infty} dz\varphi_m^*(z)\frac{\partial V_{eff}(z)}{\partial z}\varphi_n(z) = e\Delta(\mathbf{q})F_{avg}^{nm}, \tag{2.150}$$

where F_{avg}^{nm} is the average surface field defined from the z integration above, and the 2D Fourier transform of the roughness function has been introduced

$$\Delta(\mathbf{q}) = \int d\mathbf{r} \Delta(\mathbf{r}) e^{i\mathbf{q}\cdot\mathbf{r}}, \qquad (2.151)$$

where $\mathbf{q} = \mathbf{k} - \mathbf{k}'$. The intraband momentum relaxation time or scattering rate is again calculated from Fermi's rule,

$$\frac{1}{\tau_{sr}^{nn}(E)} = \frac{2\pi}{\hbar} |F_{avg}^{nn}|^2 \sum_{\mathbf{k}'} S(\mathbf{q})(1 - \cos\theta) \left| \frac{\Gamma(q)}{\epsilon_D(q)} \right|^2 \delta(E_{\mathbf{k}'} - E_{\mathbf{k}}), \qquad (2.152)$$

where the scalar static screening function has been introduced, and where $\Gamma(\mathbf{q})$ represents corrections for image potential and electric field modification at the deformed interface [58] which are on the order of unity. The function $S(\mathbf{q}) = |\Delta(\mathbf{q})|^2$ is the power spectrum of the roughness fluctuations, which depends only on the magnitude of the fluctuations, and thus the phase information can be neglected. As before, the sum over final states can be converted to an integral, and the delta function can be used to reduce the integration of the magnitude of $\mathbf{k}'$, again assuming parabolic bands, to give

$$\frac{1}{\tau_{sr}^{nn}(E)} = \frac{e^2 |F_{avg}^{nn}|^2 m^*}{2\pi\hbar^3} \int_0^{2\pi} d\theta\, S(\mathbf{q})(1 - \cos\theta) \left| \frac{\Gamma(q)}{\epsilon_D(q)} \right|^2, \quad q = 2k\sin(\theta/2). \quad (2.153)$$

The statistical properties of the roughness function $\Delta(\mathbf{r})$ are contained in the autocovariance function

$$C(\mathbf{r}) = \langle \Delta(\mathbf{r}')\Delta(\mathbf{r}' - \mathbf{r}) \rangle, \qquad (2.154)$$

where the brackets denote the ensemble average of the random variable $\Delta(\mathbf{r})$. The autocovariance function measures the probability that given a certain value of the roughness function at $\mathbf{r}'$, the function has the same value at $\mathbf{r}' - \mathbf{r}$. It is apparent that this probability should decay in some fashion as the distance $\mathbf{r}' - \mathbf{r}$ increases due to the random nature of $\Delta(\mathbf{r})$. In early works, it was usual to choose a Gaussian decay for the autocovariance function (2.154) as

$$C(\mathbf{r}) \approx \Delta^2 e^{-r^2/L^2}, \qquad (2.155)$$

where Δ is the rms value of the roughness fluctuations and L is the autocovariance length which roughly (no pun intended) may be interpreted as the mean distance between "bumps" along the surface as shown in Fig. 2.24. The main justification for the use of a Gaussian autocovariance function is the rather simple mathematical connection to the power spectrum, $S(\mathbf{q})$. By the Wiener-Kitchine theorem, the Fourier transform of the autocovariance function of a random variable is the power spectrum, $S(\mathbf{q})$. The Gaussian function has the desirable property that the Fourier transform of a Gaussian is again a Gaussian, which in 2D gives

$$S(\mathbf{q}) = \pi \Delta^2 L^2 e^{-q^2 L^2/4} \qquad (2.156)$$

which may be substituted into (2.153) to facilitate evaluation of the scattering rate. However, as was shown by Goodnick *et al.* [59], the actual autocovariance function is well fit by an exponential rather than Gaussian autocovariance function. Figure 2.25 shows the estimated power spectrum associated with the HRTEM interface of the Si/SiO$_2$ interface shown in Fig. 2.21. The power spectrum has been fit assuming a Gaussian autocovariance function and an exponential autocovariance function, where the latter is clearly a much

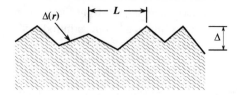

Fig. 2.24. Schematic representation of the surface roughness function, including the rms height Δ and autocovariance length L.

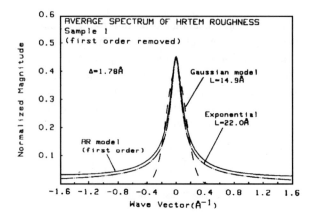

Fig. 2.25. Power spectrum corresponding to the 1D roughness function of the Si/SiO$_2$ interface from HRTEM measurements (after Goodnick et al. [59]).

better representation of the power spectrum. The roughness measured in HRTEM or STM is always a one-dimensional roughness rather than the true 2D roughness function, and thus consideration must be given to the actual form of the 2D roughness estimated from measurement of the 1D function. For an exponential model, the autocovariance function is given by

$$C(\mathbf{r}) \approx \Delta^2 e^{-\sqrt{2}r/L}, \tag{2.157}$$

where the parameters Δ and L have the same meaning as the Gaussian model. The fact that an exponential model provides a better fit to the autocovariance function is perhaps no surprise as an exponential autocovariance is characteristic of a Markov process, which in the analysis of stochastic processes is the lowest-order model for a random process. The 1D transform corresponding to the fit in Fig. 2.25 is a Lorentzian function, whereas if an isotropic roughness in 2D is assumed, the power spectrum in 2D is given by the 2D transform of (2.157) as

$$S(\mathbf{q}) = \frac{\pi \Delta^2 L^2}{(1 + (q^2 L^2/2))^{3/2}}. \tag{2.158}$$

We will revisit the topic of surface roughness scattering in Section 7.3 in connection with nonequilibrium transport in inversion layers, in which the influence of the different roughness power spectrum models is compared in a more formal manner.

Surface roughness scattering in 1D wires has also been treated in a similar fashion in terms of a parameterized roughness power spectrum associated with the different boundaries [60].

In that work a Gaussian autocovariance was employed. As mentioned earlier, the roughness in lateral boundaries defined by present day lithographic techniques typically results in rms heights Δ which are an order of magnitude larger than those found at the oxide-semiconductor or heterointerface. The effect on the calculated mobility in 1D can therefore be severe and limit this quantity even at room temperature.

2.7.4 Lattice scattering

Phonons themselves are associated with the propagating coupled vibrational modes of the individual atoms of the crystal lattice, which when Fourier transformed to so-called "normal coordinates," form an independent collection of harmonic oscillators, one for each mode $\mathbf{q}$ and branch μ. For the semiconductors of interest in this book, the two-atom basis of the diamond or zincblende crystal lattice results in three acoustic branches and three optical branches. The three branches roughly correspond to one longitudinal and two transverse modes relative to the direction of propagation of the lattice wave. Acoustic phonons are associated with the low-lying vibrational modes of the lattice whose energy $\hbar\omega_{\mathbf{q}}$ goes to zero as the wavevector $\mathbf{q}$ goes to zero. The dispersion relation is linear close to $\mathbf{q} = 0$, with group velocity v_s, which is the sound velocity in the crystal. On the other hand, the optical modes go to a nonzero value, ω_{lo} or ω_{to}, as $\mathbf{q} = 0$ associated with the longitudinal optical or transverse optical modes. The optical modes are almost dispersionless which means zero group velocity. Thus it is convenient to consider optical phonons as interacting with electrons and holes at one frequency, ω_{lo} or ω_{to}, exchanging fixed quanta of energy $\hbar\omega_o$.

As mentioned above, the modes of vibrations in normal coordinates correspond to independent harmonic oscillators, each with energy

$$E_{\mathbf{q},\mu} = \hbar\omega_{\mathbf{q},\mu}\left(n_{\mathbf{q},\mu} + \frac{1}{2}\right) \cdots n_{\mathbf{q},\mu} = 0, 1, 2, \ldots. \qquad (2.159)$$

The excitation state index, $n_{\mathbf{q},\mu}$, is given the physical picture of representing the number of quasi-particles called phonons in each mode $\mathbf{q}$ and μ. Thus, the total energy of a particular mode is the zero-point energy plus $\hbar\omega_{\mathbf{q},\mu}n_{\mathbf{q},\mu}$, the number of phonons multiplied by the energy per phonon. $n_{\mathbf{q},\mu}$ can be greater than 1, as the phonons are a system of bosons, obeying Bose-Einstein statistics in equilibrium

$$n_{\mathbf{q},\mu} = \frac{1}{e^{\hbar\omega_{\mathbf{q},\mu}/k_BT} - 1}. \qquad (2.160)$$

It is convenient to represent the lattice vibrational states using the so-called *second quantized* representation (see for example [17]). The collection of phonons is represented by a state vector $|n_{\mathbf{q}_1}, n_{\mathbf{q}_2}, n_{\mathbf{q}_3}, \ldots\rangle$, which is determined by the occupation factors for each mode $\mathbf{q}$ and μ (the branch index has been dropped for simplicity). Ladder operators may be defined for the simple harmonic oscillator model, which are referred to as *creation* and *annihilation* operators, as

$$a_{\mathbf{q}_i}^{\dagger}|n_{\mathbf{q}_1}, n_{\mathbf{q}_2}, n_{\mathbf{q}_3}, \ldots, n_{\mathbf{q}_j}, \ldots\rangle = \sqrt{n_{\mathbf{q}_j} + 1}|n_{\mathbf{q}_1}, n_{\mathbf{q}_2}, n_{\mathbf{q}_3}, \ldots, n_{\mathbf{q}_j} + 1, \ldots\rangle$$

$$a_{\mathbf{q}_i}|n_{\mathbf{q}_1}, n_{\mathbf{q}_2}, n_{\mathbf{q}_3}, \ldots, n_{\mathbf{q}_j}, \ldots\rangle = \sqrt{n_{\mathbf{q}_j}}|n_{\mathbf{q}_1}, n_{\mathbf{q}_2}, n_{\mathbf{q}_3}, \ldots, n_{\mathbf{q}_j} - 1, \ldots\rangle, \qquad (2.161)$$

where $a_{\mathbf{q}_i}^{\dagger}$ represents the creation operator since mode $\mathbf{q}_i$ is increased by one quanta after the application of this operator, while $a_{\mathbf{q}_i}$ represents the annihilation operator, which destroys one quanta in mode $\mathbf{q}_i$. Using (2.161), we see from the operator relations defined above, that $|n_{\mathbf{q}_1}, n_{\mathbf{q}_2}, n_{\mathbf{q}_3}, \ldots\rangle$ is an eigenstate of the number operator

$$a_{\mathbf{q}_i}^\dagger a_{\mathbf{q}_i} |n_{\mathbf{q}_1}, n_{\mathbf{q}_2}, n_{\mathbf{q}_3}, \ldots, n_{\mathbf{q}_j}, \ldots\rangle = n_{\mathbf{q}_j} |n_{\mathbf{q}_1}, n_{\mathbf{q}_2}, n_{\mathbf{q}_3}, \ldots, n_{\mathbf{q}_j}, \ldots\rangle. \tag{2.162}$$

Various operators may be constructed from the creation and annihilation operators defined in this fashion, such as, for example, the Hamiltonian operator for the phonon system,

$$H = \sum_{\mathbf{q}, \mu} \hbar\omega_{\mathbf{q}, \mu} \left(a_{\mathbf{q}, \mu}^\dagger a_{\mathbf{q}, \mu} + \frac{1}{2} \right), \tag{2.163}$$

which when acting on the second quantized state $|n_{\mathbf{q}_1}, n_{\mathbf{q}_2}, n_{\mathbf{q}_3}, \ldots\rangle$ gives the total energy as a sum over the mode energies (2.159). In connection with transport, an important operator associated with the phonon coordinates is the *displacement operator*

$$\mathbf{u}(\mathbf{R}, t) = -i \sum_{\mathbf{q}} \left(\frac{\hbar}{2\omega_{\mathbf{q}} \rho V} \right)^{1/2} \frac{\mathbf{q}}{q} \left\{ a_{\mathbf{q}} e^{-i(\omega_{\mathbf{q}} t - \mathbf{q} \cdot \mathbf{R})} + a_{\mathbf{q}}^\dagger e^{i(\omega_{\mathbf{q}} t - \mathbf{q} \cdot \mathbf{R})} \right\}, \tag{2.164}$$

where $\mathbf{R}$ represents the 3D position vector, ρ is the mass density and V is the volume of the system. This operator appears quite frequently in the discussion of the electron-phonon interaction, since qualitatively one expects the perturbed potential due to lattice vibrations to be proportional to the displacement of the lattice from its equilibrium site, as we shall see below.

When we discuss phonons in nanostructures, we have to pay attention to the fact that the modes themselves are modified by the geometry. In particular, lattice vibrational modes may be localized in the various layers existing in the system, as well as forming surface or interface modes at the boundaries between different regions. In general, microscopic calculations are necessary to truly account for the behavior of lattice vibrations in general nanoscale structures, although such methods are usually expensive computationally. Some features are obvious, however, from experimental studies as well as simple analytical models. One effect is folding or confinement of the acoustic modes due to breaking of the translational invariance in a nanostructure. Due to this folding, modes with frequency not equal to zero at $\mathbf{q} = 0$ are possible. Evidence for zone-folded acoustic phonons is readily apparent from inelastic light scattering experiments [61]. For optical phonons in polar materials, it is possible to calculate the possible modes in a quantum-confined structure from electrostatic considerations using the dielectric continuum model (see [62] and references therein) by solving Laplace's equation in different regions and applying the appropriate electrostatic boundary conditions at the interfaces. This model has been found to compare well to first principles calculation of the lattice displacements in quantum well structures [63]. The main picture that emerges from this model, at least for a simple quantum well system, is that two types of modes appear. One type of mode corresponds to guided wave modes existing in the layer of quantum confined structures, sometimes referred to as *slab modes*. A second type of solution is associated with surface or interface modes, which decay spatially away from an interface. The interaction of electrons with all these possible modes is rather detailed, and significant effects due to the confinement of the phonon modes do not occur until very small dimensions are reached. Therefore, for simplicity, we assume in the following that electrons interact with bulk-like modes unless otherwise stated.

Acoustic phonons

The interaction of acoustic modes with electrons and holes in semiconductors is usually treated using the deformation potential Ansatz. Here, the local deformation of the crystal lattice due to an acoustic vibration is associated with a shift of the band edges, which is the acoustic deformation potential. For a spherical constant energy surface, the square of the matrix element for scattering in a quasi-2DEG system may be written [64, 65]

$$|\langle \mathbf{k}', n|V^{ac}|\mathbf{k}, m\rangle|^2 = \sum_{q_z} \frac{\hbar D_{ac}^2 q^2}{2\omega_\mathbf{q}\rho V}\left(n_q + \frac{1}{2} \mp \frac{1}{2}\right)\delta_{\mathbf{q}_t,\mathbf{k}-\mathbf{k}'}|G_{nm}(q_z)|^2,\qquad(2.165)$$

where D_{ac} is the acoustic deformation potential, the upper sign is for absorption and the lower sign emission of a phonon, and $q^2 = q_t^2 + q_z^2$ where z is the direction perpendicular to the interface. The form factor is given by

$$G_{nm}(q_z) = \int_{-\infty}^\infty dz\varphi_n^*(z)\varphi_m(z)e^{\mp iq_z z} = \int_{-\infty}^\infty dz\rho_{nm}(z)e^{\mp iq_z z}.\qquad(2.166)$$

For longitudinal acoustic modes in an isotropic system, $\omega_\mathbf{q} = u_l q$, where u_l is the longitudinal sound velocity. If we assume high temperatures, the equipartition limit of (2.160) may be used:

$$n_q \approx n_q + 1 \approx \frac{k_B T}{\hbar\omega_q}.\qquad(2.167)$$

With these simplifications, the matrix element squared may be written as

$$|\langle \mathbf{k}', n|V^{ac}|\mathbf{k}, m\rangle|^2 = \sum_{q_z} \frac{D_{ac}^2 k_B T}{u_l^2\rho V}\int_{-\infty}^\infty dz\rho_{nm}(z)\int_{-\infty}^\infty dz'\rho_{nm}(z')e^{iq_z(z-z')}.\qquad(2.168)$$

The sum over q_z may be converted to an integral

$$\sum_{q_z} \to \frac{L}{2\pi}\int_{-\infty}^\infty dq_z e^{iq_z(z-z')} = L\delta(z-z'),\qquad(2.169)$$

where L is the length of the system in the z-direction, which reduces (2.168) to

$$|\langle \mathbf{k}', n|V^{ac}|\mathbf{k}, m\rangle|^2 = \frac{D_{ac}^2 k_B T I_{ac}^{nm}}{u_l^2\rho A}.\qquad(2.170)$$

The overlap factor is given by

$$I_{ac}^{nm} = \int_{-\infty}^\infty dz\rho_{nm}^2(z),\qquad(2.171)$$

which for an infinite square well of width W is easily integrated to give

$$\begin{aligned}I_{ac}^{nm} &= \frac{3}{2W} \quad && n = m\\ &= \frac{1}{W} \quad && n \neq m.\end{aligned}\qquad(2.172)$$

Thus, the form factor for intrasubband scattering for the acoustic phonons is only 1.5 times that of the intersubband rate. To complete the derivation of the intrasubband relaxation time, use Fermi's rule and sum over final states as before:

$$\frac{1}{\tau_{ac}^{mm}} = \frac{2\pi}{\hbar}\frac{D_{ac}^2 k_B T I_{ac}^{mm}}{u_l^2\rho A}\sum_{\mathbf{k}'}\delta(E_{\mathbf{k}'} - E_\mathbf{k})(1 - \cos\theta).\qquad(2.173)$$

In the energy-conserving delta function, the energy of the acoustic phonon has been assumed small compared to the initial and final energies since the energy of the phonon goes to zero for long-wavelength phonons. In this approximation then, acoustic phonon scattering is quasi-elastic. The scattering rate is independent of angle; hence the factor $1 - \cos\theta$ may be removed, and therefore the momentum relaxation time is equal to the total inverse scattering rate. The sum over the delta function just gives the density of states of the final subband times the area of the system, and thus (2.173) becomes

$$\frac{1}{\tau_{ac}^{mm}} = \frac{D_{ac}^2 k_B T m^* I_{ac}^{mm}}{u_l^2 \rho \hbar^3}. \tag{2.174}$$

Since (2.174) is independent of energy, the average relaxation time is the same as the above expression. The scattering rate increases with increasing lattice temperature due to the phonon occupancy factors which gave rise to the factor $k_B T$ above. The overlap integral decreases as the width of the system increases, decreasing the scattering rate. Of course, as the width of the well increases, the number of available subbands within a given energy increases, increasing the total scattering rate due to intra- and intersubband scattering, which approaches the bulk acoustic phonon scattering rate as the width becomes infinite.

Another form of interaction with acoustic phonons in polar materials is through the piezoelectric interaciton. This mechanism is often found to be important at intermediate tempartures in bulk materials before the onset of strong scattering due to polar optical phonons at higher temperatures as discussed in the next section. Piezoelectric scattering in 2DEG structures has been considered for example by Price [66].

Polar optical phonons

For III-V compound materials such as GaAs and AlGaAs, the dominant energy relaxation mechanism for electrons in the central valley below the $\Gamma - L$ threshold is that of polar-optical-phonon (POP) emission. This mechanism is usually treated within the continuum model described by the Fröhlich interaction relating the polarization field associated with the charge separation between cation and anion to the scattering potential due to the optical-mode displacement. The correct model in quantum confined system for the coupling with confined and surface modes as discussed above is still somewhat of an open research area at present, particularly in quantum wire systems. For bulk-like polar optical phonons, the scattering rate for an electron initially in subband n to a final subband m may be written [67], [68], [69]

$$S_{nm}^{pop}(k) = \frac{eE_o}{2\hbar}\left[\left(n_{\omega_{lo}} + \frac{1}{2} \mp \frac{1}{2}\right)\int_0^\infty d\theta \frac{H_{nm}(q_\pm)}{q_\pm}\right], \tag{2.175}$$

where $n_{\omega_{lo}}$ is the longitudinal optical phonon occupancy, $q_\pm$ is the scattered wavevector in the plane parallel to the interface, and the upper and lower signs correspond to phonon absorption and emission, respectively. The effective field eE_o is given by

$$eE_o = \frac{m^* e^2 \hbar \omega_{lo}}{4\pi \hbar^2}\left(\frac{1}{\varepsilon_\infty} - \frac{1}{\varepsilon_0}\right), \tag{2.176}$$

where ε_0 and ε_∞ are the low- and high-frequency permittivities of the material. The function $H_{nm}(q_\pm)$ is given by

$$H_{nm}(q_\pm) = \int_{-\infty}^\infty dz \rho_{nm}(z) \int_{-\infty}^\infty dz' \rho_{nm}^*(z') e^{-q_\pm|z-z'|}, \tag{2.177}$$

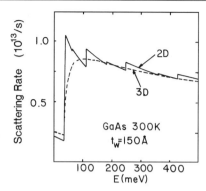

Fig. 2.26. Total scattering rate due to LO phonon scattering at 300 K for a 15 nm GaAs quantum well for electrons in the first subband (solid line). The dashed line corresponds to the 3D scattering rate. [After Goodnick and Lugli, Phys. Rev. B **37**, 2578 (1988).]

where the scattered wavevector is fixed by energy conservation to be

$$q_{\pm} = |\mathbf{k} - \mathbf{k}'| = \left[2k^2 \pm \frac{2\omega_{nm}^* m^*}{\hbar} - 2k \left(k^2 \pm \frac{2\omega_{nm}^* m^*}{\hbar} \right)^{1/2} \cos\theta \right]^{1/2},$$

$$\hbar\omega_{nm}^* = \hbar\omega_{lo} \pm (E_n - E_m), \tag{2.178}$$

where $\hbar\omega_{nm}^*$ represents an effective phonon energy, which may be zero or negative. For intersubband scattering, the rate given by (2.175) is a maximum when $q_{\pm}$ in the denominator is a minimum, which occurs when $\hbar\omega_{lo} = E_n - E_m$, resulting in a resonance which may be observed experimentally.

Figure 2.26 shows the calculated LO phonon scattering rate versus energy in a 15 nm GaAs quantum well. In comparison to the 3D rate, the 2D rate shows a much sharper emission threshold, and sharp discontinuities corresponding to the onset of emission and absorption for higher subbands. The sharpness of the onsets is related to the discontinuous density of states of a 2D versus 3D system. For quantum wires, these discontinuities are even more pronounced due to the divergent density of states.

Figure 2.27 shows the calculated intersubband scattering rates from the second subband to the first subband due to LO phonons, acoustic phonons, and impurity scattering in a 27 nm wide GaAs well. For low temperatures, the acoustic rate cannot be calculated using the equipartition approximation. The rate shown in Fig. 2.27 is calculated numerically directly from the matrix element (2.161). As can be seen, the LO phonon emission rate is much larger than that due to acoustic phonons. However, since the subband spacing in this case (14 meV) is less than the optical phonon energy, electrons below this threshold cannot emit to the ground subband, thus leading to a bottleneck in the intersubband transition rate governed by the long-time acoustic phonon rather than LO phonon rate [70]. With impurities present, however, another channel is present for electrons to scatter from the higher to lower subband.

Optical deformation potential and intervalley scattering

In Si, the nonpolar optical deformation potential is the predominant interaction of carriers with optical phonons. This may occur through intravalley scattering, or via intervalley scattering between equivalent minima. For the case of GaAs, the intravalley scattering

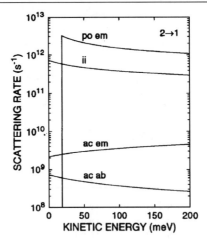

Fig. 2.27. The total scattering rates for electrons from the second to the first subband due to polar optical and acoustic phonon modes (both emission and absorption), and ionized impurities ($n_i = 1.5 \times 10^{11}$ cm^{-1}) for a 27-nm-wide GaAs well at 10 K.

contribution is forbidden by symmetry for zero-order scattering. However, as carriers are heated, or in cases of large carrier confinement so that the satellite valleys are accessible, the intervalley scattering mechanism is important.

For 2D electrons, if we assume that the states are quantized for all valleys, we may write the scattering rate for nonpolar optical deformation scattering as [69]

$$S_{nm}^o = \sum_{v_f} m_f^* \frac{E_o^2 \left(n_{\omega_o} + \frac{1}{2} \mp \frac{1}{2}\right) I_{nm}^o}{2\rho\omega_o\,\hbar^2},\qquad(2.179)$$

where the sum is over the final valleys if intervalley scattering is considered, m_f^* is the mass of electrons in the final valley, and ω_o is the relevant phonon frequency. For simple nonpolar optical scattering, ω_o is either the longitudinal optical or transverse optical frequency. For intervalley scattering, this frequency corresponds to the wavevector of the phonon that couples one valley to another, and may correspond to several different values. E_o is the optical deformation potential in the case of intravalley scattering, whereas it is the intervalley deformation for intervalley scattering. The overlap integral in (2.179) is given by

$$I_{nm}^o = \int_{-\infty}^{\infty} dz \left|\varphi_{v_i,n}(z)\right|^2 \left|\varphi_{v_f,m}(z)\right|^2,\qquad(2.180)$$

where v_i and v_f refer to the initial and final valley envelope functions.

Lattice scattering in 1D systems

Various authors have calculated the scattering rates for the different lattice scattering mechanisms above for quasi-one-dimensional systems. Calculation of the phonon scattering rates in GaAs quantum wires was performed by Jovanovic and Leburton [71]. The rates are essentially calculated from the standard models for the interaction potentials described above, with the matrix elements taken between the initial and final quasi-one-dimensional envelope functions. The phonons in these studies are taken as bulk-like while the electronic states are confined. The calculated rates for acoustic, piezoelectric, and polar optical phonons as

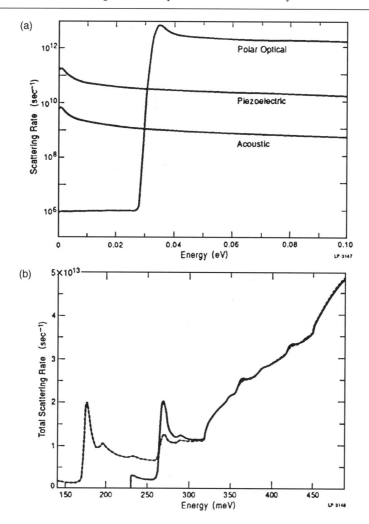

Fig. 2.28. Scattering rates for polar optical, acoustic, and unscreened piezoelectric phonon scattering. (a) Individual scattering rates in the first subband for different mechanisms at 30 K relative to the subband minimum. (b) Combined scattering rates for the first two subbands at 300K [after Jovanovic and Leburton, from *Monte Carlo Device Simulation: Full Band and Beyond,* Ed. K. Hess (Kluwer Academic Publishers, Boston, 1991), by permission].

a function of energy are shown in Fig. 2.28. As expected, once the threshold for optical phonon emission is exceeded, the polar optical rate is largest. The 1D density of states leads to the peaked behavior in the scattering rate when the energy corresponds to one of the subband minima of the multisubband structure. These rates were used in a Monte Carlo simulation of transport in quasi-one-dimensional systems.

2.7.5 Experimental mobility in 2DEG heterostructures

As mentioned in Section 2.1, the inclusion of a spacer layer separates the 2DEG from the heavily doped AlGaAs region. This gives rise to an exponential attenuation of the scattering rate for a given q, as seen in (2.140). Figure 2.29 shows the measured mobility

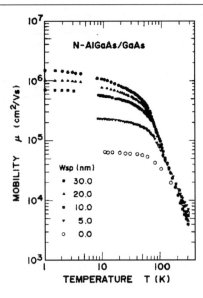

Fig. 2.29. Hall mobility versus temperature of n-modulation doped AlGaAs/GaAs heterostructures of varying spacer layer thickness W_{sp} [after Solomon *et al.*, IEEE Elec. Dev. Lett. EDL-**5**, 379 (1984), by permission].

versus temperature for a number of different modulation-doped samples for different spacer layer thicknesses. As expected, the low-temperature mobility increases with increasing spacer layer thickness compared to the bulk case, which is dominated by ionized impurity scattering.

Another limiting scattering mechanism at low temperature is the background impurity concentration. Unintentional impurities are incorporated into the epitaxial layers during growth due to the residual background concentration of impurities such as carbon and oxygen in the growth chamber itself. As the quality of epitaxial growth has improved over the years, so has the low-temperature mobility in modulation-doped structures. Figure 2.30 illustrates the mobility versus temperature reported by various groups which is quite similar in trend to the data shown in Fig. 2.29. However, here the increase in low-temperature mobility is not due to increasing spacer layer thickness, but rather to decreasing background impurity concentration as the quality of epitaxial growth systems improves [7]. For the best data shown in Fig. 2.30, electron mobilities in excess of $1 \times 10^7 \mathrm{cm}^2/\mathrm{V}$-s are evident at low temperature, which correspond with an estimated background ionized impurity concentration in the channel of $2 \times 10^{13}/\mathrm{cm}^3$.

For the data shown in Figs. 2.29 and 2.30, the 2DEG mobility in heterolayers shows a strong temperature dependence at high temperature that eventually plateaus with very little temperature dependence at low temperature. Figure 2.31 shows a comparison of the experimental mobility as a function of temperature for one sample reported by Lin *et al.* [72]. The dashed lines in this figure compare the calculated contributions due to the remote impurities with the various lattice scattering mechanisms discussed in Section 2.7.4 treated within the relaxation time approximation. Although it is not proper to treat inelastic mechanisms such as LO phonon scattering within this approximation, the results are still illustrative of the dominant contributions within different temperature ranges. As the temperature approaches room temperature, the contribution due to polar optical (LO) phonon scattering is dominant,

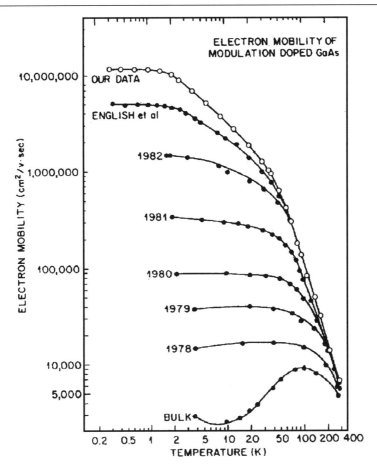

Fig. 2.30. Hall mobility versus temperature of modulation-doped AlGaAs/GaAs structures by various groups, showing the continued improvement in epitaxial material quality with time [after Pfeiffer *et al.*, Appl. Phys. Lett. **55**, 1888 (1989), by permission].

as in bulk materials. Close to the knee of the mobility curve around 90K, the mobility is more dominated by the combined contributions of acoustic and piezoelectric scattering. Finally, at low temperature, the mobility is entirely dominated by impurities, here assumed to be all in the form of remote impurities. Comparing Fig. 2.31 to the more recent data shown in Fig. 2.30, it is clear that the contributions due to acoustic and piezoelectric scattering have been overestimated in order to fit the experimental data (since the mobilities limited by these mechanisms are predicted to be less than 1×10^7). Part of the problem is choosing the exact deformation potential to use in describing the acoustic interaction.

Interest in the transport properties of quantum wires was spurred by predictions of enhanced mobility due to the reduction in scattering associated with reduced number of possible final states after scattering [73]. However, the actual mobilities measured in long wires is typically less than that of the 2DEG structure from which they are fabricated, as shown in several early studies (see for example [74], [75], [76]). As mentioned earlier, this reduction in mobility in 1D compared to 2D systems is primarily associated with the roughness and impurity induced scattering induced during the lateral patterning process necessary to realize quantum wire structures. As discussed further in Chapters 5 and 6, quantum interference effects also play a role in the transport properties of quantum wires, which lead to effects such as universal conductance fluctuations and negative magnetoresistance, which go beyond the scope of the present chapter. Improvements in lateral fabrication technology are expected to eventually reduce these inhomogeneities and allow improved mobilities in quantum wires, as indicated by recent evidence of mobility enhancement in such systems [77].

2.7.6 Magnetotransport in quantum confined systems

The transport properties of low dimensional systems under an applied magnetic field provides valuable insight into the electronic properties of such systems as discussed in Section 2.5. The most well studied system is that of the 2DEG with an applied d.c. magnetic field perpendicular to the plane of the gas. As mentioned in the beginning of this chapter, the first clear evidence of purely two-dimensional behavior in the Si/SiO_2 inversion layer system was reported by Fowler et al. [4] who looked at the oscillations in the channel conductance with magnetic field in this system. Such *magnetoresistance* oscillations in bulk materials are often referred to as Shubnikov de-Haas (SdH) oscillations, and provide an important tool for investigating the Fermi surface of conductors (see for example [16]). In contrast to the bulk SdH effect, the observed oscillations in the conductance or resistance depended only on the component of the magnetic field perpendicular to the 2DEG, and not on the in-plane component, demonstrating the reduced dimensionality of the electronic states. Figure 2.32 shows an experimental curve of similar SdH oscillations in the longitudinal resistance (left axis) of a high mobility GaAs/AlGaAs heterolayer at cryogenic temperature. Due to the higher mobility and hence longer scattering times in modulation-doped heterojunction systems compared to the Si/SiO_2 system, features arising from the magnetic quantization are better resolved in the former system due to reduced broadening of the underlying Landau levels as discussed earlier in Section 2.5. As seen in this figure, the longitudinal resistance increases slightly with increasing magnetic field before entering a regime of sinusoidal modulation around 0.5 T. As discussed in Section 2.5, when the separation of the Landau levels exceeds the broadening of the levels due to scattering ($\omega_c \tau \approx 1$), the discrete nature of the density of states becomes observable in the transport properties, where ω_c is the cyclotron frequency and τ is the scattering time. However, the apparent onset of the SdH oscillations at 0.5 T corresponds to a scattering time τ which is significantly shorter than that associated with the mobility, $\mu = e\tau_m/m^*$, which for this sample is close to $10^6 cm^2$/V-s. As discussed in Section 2.5, τ_m is the momentum relaxation time discussed previously in the present section whereas τ is the total scattering time which may be much less in the case of anisotropic scattering mechanisms such as impurity scattering [78].

The oscillations shown in Fig. 2.32 are periodic in $1/B_z$ and may be related to the electron density of the 2DEG in a simple way. In Eq. (2.45), the energy of the electrons was given as $E = E_i + \hbar\omega_c(n + \frac{1}{2})$, where n is an integer, and E_i is the 2D subband energy. Since electrons follow circular paths in coordinate space due to the magnetic field, their motion in k-space is orbital as well for a spherical conduction band minimum. The area in k-space traversed by such an orbit in the $k_x - k_y$ plane becomes

$$A_k = (2\pi B_z/\hbar)\left(n + \frac{1}{2}\right). \tag{2.181}$$

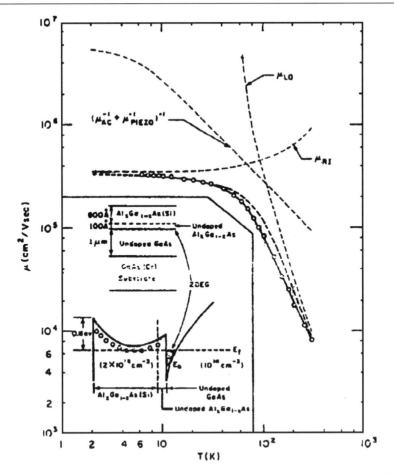

Fig. 2.31. Temperature dependence of the 2DEG mobility in a sample with $n_{2D} = 5.35 \times 10^{11}$ cm^{-1}. The open circles are experimental data, and the dashed lines are the various theoretical contributions to the mobility [after Lin *et al.*, Appl. Phys. Lett. **45**, 695 (1984), by permission].

The Fermi radius in k-space for a given subband follows from the parabolic dispersion relation and the density of states in 2D

$$k_f = \left(\frac{4\pi n_{2d}^i}{n_s n_v}\right)^{1/2},$$ (2.182)

where η_s and η_v are the spin and valley degeneracies, respectively, and n_{2d}^i the two-dimensional density of the ith subband. The area in k-space enclosed by the Fermi radius is therefore

$$A_F = \pi k_f^2 = \frac{4\pi^2 n_{2d}^i}{n_s n_v}.$$ (2.183)

A peak in the conductivity occurs when a given Landau level passes through the Fermi perimeter, which occurs when 2.181 and 2.183 are equal:

$$(2\pi e B_z/\hbar)\left(n + \frac{1}{2}\right) = \frac{4\pi^2 n_{2d}^i}{n_s n_v}.$$ (2.184)

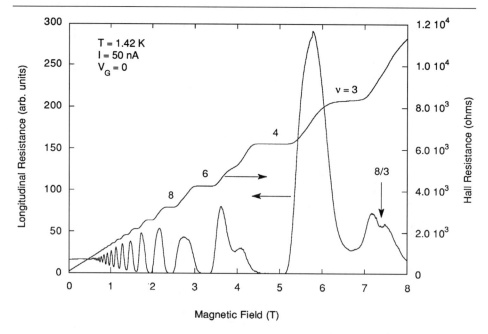

Fig. 2.32. Experimental plot of the longitudinal resistance (left axis) and Hall resistance (right axis) as a function of magnetic field for an AlGaAs/GaAs 2DEG structure.

The spacing between successive Landau levels is thus periodic in $1/B_z$, similar to the three-dimensional SdH oscillations

$$\Delta 1/B_z = \frac{n_s n_v}{2\pi \hbar n_{2d}^i}. \tag{2.185}$$

In contrast to the three-dimensional SdH effect, however, Eq. (2.185) has no mass dependence, rather it depends inversely on the two-dimensional density of a given subband. If multiple subbands are occupied, different frequency oscillations are observed corresponding to the unequal densities in each subband. If the magnetic field is held constant, but the two-dimensional sheet density is changed by an external gate potential, oscillations will be seen as is evident from Eq. (2.184).

The conductivity of a 2DEG in the presence of a perpendicular magnetic field was derived by Ando using the Kubo formula in linear response theory assuming a system of short range scatterers (see [1] for a details). For weak magnetic fields ($\omega_c \tau \leq 1$) in the extreme quantum limit, the lowest order corrections to the longitudinal conductivity are written

$$\sigma_{xx} = \sigma_o \frac{1}{1 + \omega_c^2 \tau^2} \left[1 - \frac{2\omega_c^2 \tau^2}{1 + \omega_c^2 \tau^2} \frac{2\pi^2 k_B T_e}{\hbar \omega_c} \right.$$

$$\left. \operatorname{csch}\left(\frac{2\pi^2 k_B T_e}{\hbar \omega_c} \right) \cos\left(\frac{2\pi^2 E_f}{\hbar \omega_c} \right) \exp\left(-\frac{\pi}{\omega_c \tau} \right) + \ldots \right], \tag{2.186}$$

where τ is the scattering time at the Fermi energy, E_f, T_e is the electron temperature (which in general can be different from that of the lattice due to carrier heating), and $\sigma_o = n_{2d} e^2 \tau / m^*$ is the zero field conductivity. The cosine term gives the oscillatory behavior with a period dependent on either $1/B_z$ or E_f (which is directly proportional to density) as discussed above. The hyperbolic cosecant term leads to a damping behavior with temperature which depends explicitly on the effective mass through the cyclotron frequency. This temperature dependence of the oscillation amplitude is sometimes used as a means of ascertaining the effective cyclotron mass in 2D systems by fitting Eq. (2.186) to experiment. As mentioned earlier, we have to be careful regarding the interpretation of τ and its strict relation to the actual momentum relaxation time in the presence of long range scatterers like ionized impurities.

As the magnetic field becomes increasingly larger in Fig. 2.32, the sinusoidal characteristic becomes distorted, and the longitudinal resistance goes to zero between the peaks. The peaks themselves split due to lifting of the spin degeneracy (as well as the valley degeneracy in multivalley semiconductors), as discussed in Section 2.5. Even more remarkable is the behavior of the Hall resistance ($R_H = V_H/I$) which plateaus precisely in the regime where the longitudinal resistance vanishes. The plateaus correspond almost exactly (to less than one part in 10^{-8}) to

$$R_H = \left(\frac{h}{e^2}\right)\frac{1}{n}, \quad n = 1, 2, 3, \ldots, \tag{2.187}$$

where n corresponds to the number of occupied single spin degenerate Landau levels. The corresponding numbers are indicated in Fig. 2.32 where for low magnetic fields, the splitting of the spin degeneracy cannot be resolved. This effect is the quantum Hall effect [50] discussed earlier in the introduction to the present section. The effect is quite robust, and observed in a variety of different systems, in relatively impure samples.

To understand this behavior somewhat, we need to relate the resistivity to the conductivity, both of which are described as tensor properties with off-diagonal elements due to the magnetic field. The longitudinal and transverse components of the resistivity tensor are given by inversion of the conductivity tensor as

$$\rho_{xx} = \frac{\sigma_{xx}}{\sigma_{xy}^2 + \sigma_{xx}^2}; \quad \rho_{xy} = \frac{\sigma_{xy}}{\sigma_{xy}^2 + \sigma_{xx}^2}. \tag{2.188}$$

If one considers the collisionless motion of electrons in crossed electric (in the plane) and magnetic fields (perpendicular to the plane) in a 2D system, the electronic orbits form cycloids, which have zero net motion in the direction of the electric field, and a constant net motion perpendicular to the electric field with drift velocity E/B. The longitudinal conductivity σ_{xx} vanishes, and the transverse conductivity is just $\sigma_{xy} = en_{2d}/B$. According to Eq. (2.188), the longitudinal resistivity vanishes, and the transverse resistivity is given by $\rho_{xy} = B/en_{2d}$. The classical transverse resistivity increases linearly with magnetic field, which in fact corresponds to the increasing background of the quantum Hall plateaus of Fig. 2.32. In the quantum mechanical picture, the electrons form Landau levels, which at high magnetic fields have a single spin and valley degeneracy due to splitting of the levels by the field. The density of states (with broadening) was illustrated in Fig. 2.20. As the magnetic field is increased, the $n + 1$ level depopulates until the nth level is completely full, the $n + 1$ level is empty and the Fermi energy lies in between. Under such conditions, electrons cannot scatter without exciting across the gap separating the two levels, and the longitudinal conductivity vanishes just as in any completely full band at zero temperature. However, in the presence of the magnetic field, motion perpendicular to the electric field is still possible, and in this collision free situation, we can argue that the classical transverse conductivity gives the correct value. As discussed in Section 2.5, the density of occupied states per Landau level (assuming $n_s n_v = 1$) is $eB/(2\pi\hbar)$, and n of these levels are occupied.

Substituting this expression for n_{2d} in the classical transverse conductivity formula above gives $\rho_{xy} = h/ne$, where n is an integer. This is precisely the experimental result Eq. (2.187) above.

The difficulty in this model is in explaining the finite plateau width observed experimentally. In the ideal picture described in Section 2.5, the Fermi energy does not smoothly move from one Landau level to another, but jumps discontinuously. In order to explain the finite plateau width, one must invoke a picture in which *localized* states exist in the tail regions of the broadened Landau levels. In the transition from one Landau level to the next, the Fermi energy is pinned by these localized states such that for a nonzero range of magnetic fields, the transverse resistance is only determined by the occupied *extended* states below. Localized states occur naturally due to disorder in the system, and are characterized by an activated conductivity which vanishes at low temperature [1]. Models which invoke disorder induced localization to explain the normal quantum Hall effect are sometimes referred to as "bulk" theories as the effect is a property of the 2DEG itself, and not the geometry. Extensive review of the normal quantum Hall effect and its explanation within this context is given in [79]. In Section 3.10 we will come back to a different explanation of the NQHE in the context of the Landauer-Büttiker equation and transport through electro-magneto edge states which also explains the observed quantization of the transverse resistivity.

Bibliography

[1] T. Ando, A. Fowler, and F. Stern, Rev. Mod. Phys. **54**, 437 (1982).
[2] C. Weisbuch and B. Vinter, *Quantum Semiconductor Structures* (Academic Press Inc., Boston, 1991).
[3] G. Bastard, J. A. Brum, and R. Ferreira, Sol. State Phys. **44**, 229 (1991).
[4] A. B. Fowler, F. F. Fang, W. E. Howard, and P. J. Stiles, Phys. Rev. Lett. **16**, 901 (1966).
[5] A. Y. Cho and J. R. Arthur, Prog. Solid State Chem. **10**, 157 (1975).
[6] R. Dingle, H. L. Störmer, A. C. Gossard, and W. Wiegmann, Appl. Phys. Lett. **33**, 665 (1978).
[7] L. Pfeiffer, K. W. West, H. L. Störmer, and K. W. Baldwin, Appl. Phys. Lett. **55**, 1888 (1989).
[8] S. Adachi, J. Appl. Phys. **58**, 62 (1985).
[9] F. Capasso and G. Margaritondo, *Heterojunction Band Discontinuities: Physics and Device Applications* (Elsevier Science Publishing Company, Amsterdam, 1987).
[10] D. J. Wolford, T. F. Kuech, J. A. Bradley, M. A. Gell, M. A. Ninno, and M. Jaros, J. Vac. Sci. Technol. B **4**, 1043 (1986).
[11] L. I. Schiff, *Quantum Mechanics* (McGraw-Hill Inc., New York), 1955.
[12] D. Y. Oberli, J. Shah, T. C. Damen, T. Y. Chang, C. W. Tu, D. A. B. Miller, J. E. Henry, R. F. Kopf, N. Sauer, and A. E. DiGiovanni, Phys. Rev. B **40**, 3028 (1989).
[13] K. Leo, J. Shah, E. O. Göbel, T. C. Damen, S. Schmitt-Rink, W. Schäfer, and K. Köhler, Phys. Rev. Lett. **66**, 201 (1991).
[14] J. P. Eisenstein, L. N. Pfeiffer, and K. W. West, Phys. Rev. Lett. **26**, 3804 (1992).
[15] N. M. Froberg, B. B. Hu, X.-C. Zhang, and D. H. Auston, IEEE J. Quan. Elec. 28, 2291 (1992).
[16] N. W. Ashcroft and N. D. Mermin, *Solid State Physics* (Holt, Rinehart and Winston, New York, 1976).
[17] D. K. Ferry, *Semiconductors* (Macmillan, New York, 1991).
[18] L. Esaki, in *Recent Topics in Semiconductor Physics*, eds. H. Kaminmura and Y. Toyozawa, 1983.
[19] F. Stern, Phys. Rev. B **17**, 5009 (1978).
[20] F. Stern and S. Das Sarma, Phys. Rev. B **30**, 840 (1984).
[21] T. Ando, Phys. Rev. B **13**, 3468 (1976).
[22] S. Das Sarma and B. Vinter, Phys. Rev. B **23**, 6832 (1981).

[23] L. Hedin and B. I. Lundqvist, J. Phys. C **4**, 2064 (1971).

[24] F. Stern, J. Comput. Phys. **6**, 56 (1970).

[25] P. C. Chow, AJP **40**, 730 (1972).

[26] S. Bhobe, W. Porod, S. Bandyopadhyay, and D. J. Kirkner, Surface and Interface Analysis **14**, 590 (1990); S. M. Bhobe, *Analysis and Simulations of the Velocity Modulation Transistor*, M.S. Thesis, University of Notre Dame, 1989.

[27] M. Shur, *GaAs Devices and Circuits*, (Plenum Press, New York 1987).

[28] K. Hess, Physica B **117B–118B**, 723 (1983).

[29] M. Abramowitz and I. A. Stegun, *Handbook of Mathematical Functions* (National Bureau of Standards Applied Mathematics Series, No. 55), U.S. Government Printing Office, Washington, 1964.

[30] R. F. W. Pease, J. Vac. Sci. Technol. B, 10, 278 (1992).

[31] G. Binnig, H. Rohrer, Ch. Gerber, and E. Weibel, Phys. Rev. Lett. **49**, 57 (1982); G. Binnig and H. Rohrer, Rev. Mod. Phys. **59**, 615 (1987); J. A. Stroscio and W. J. Kaiser, eds., *Scanning Tunneling Microscopy* in Methods of Experimental Physics, Vol. **27** (Academic Press, Boston), 1993.

[32] W. E. Spicer, Z. L. Liliental-Weber, E. Weber, N. Newman, T. Kendelewicz, R. Cao, C. McCants, P. Mahowald, K. Miyano, and I. Lindau, J. Vac. Sci. Technol. B**6**, 1245 (1988).

[33] W. Hansen, M. Horst, J. P. Kotthaus, U. Merkt, Ch. Sikorski, and K. Ploog, Phys. Rev. Lett. **58**, 2586 (1987).

[34] F. Brinkop, W. Hansen, J. P. Kotthaus, and K. Ploog, Phys. Rev. B **37**, 6547 (1988).

[35] T. J. Thornton, M. Pepper, H. Ahmed, D. Andrews, and G. J. Davies, Phys. Rev. Lett. **56**, 1198 (1986).

[36] A. B. Fowler, A. Hartstein, and R. A. Webb, Phys. Rev. Lett. **48**, 196 (1982).

[37] M. Reed, J. N. Randall, R. J. Aggarwal, R. J. Matyi, T. M. Moore, and A. E. Wetsel, Phys. Rev. Lett. **60**, 535 (1988).

[38] S. E. Laux and F. Stern, Appl. Phys. Lett. **49**, 91 (1986).

[39] S. E. Laux, D. J. Frank, and F. Stern, Surf. Sci. **196**, 101 (1988).

[40] K. Kojima, K. Mitsunaga, and K. Kyuma, Appl. Phys. Lett. **55**, 862 (1989).

[41] C. W. J. Beenakker, H. van Houten, and B. J. van Wees, Superlattices and Microstructures **5**, 127 (1989); C. W. J. Beenakker and H. van Houten, *Quantum Transport in Semiconductor Nanostructures*, in Solid State Physics **44** (H. Ehrenreich and D. Turnbull, eds., Academic Press, Boston), pp. 1–228, 1991.

[42] E. D. Siggia and P. C. Kwok, Phys. Rev. B **2**, 1024 (1970).

[43] J. M. Ziman, *Principles of the Theory of Solids* (Cambridge University Press, Cambridge, 1972).

[44] C. R. C. Handbook of Mathematical Sciences, 6th ed. (CRC Press, West Palm Beach, 1987).

[45] P. F. Maldague, Surf. Sci. **78**, 296 (1978).

[46] S. J. Allen, Jr., D. C. Tsui, and R. A. Logan, Phys. Rev. Lett. **38**, 980 (1977).

[47] T. N. Theis, J. P. Kotthaus, and J. P. Stiles, Solid State Comm. **26**, 603 (1978).

[48] D. C. Tsui, E. Gornik, and R. A. Logan, Solid State Comm. **35**, 875 (1980).

[49] B. Yu-Kuang Hu and S. Das Sarma, Phys. Rev. B **48**, 5469 (1993).

[50] K. von Klitzing, G. Dorda, and M. Pepper, Phys. Rev. Lett. **45**, 494 (1980); K. von Klitzing and G. Ebert, Springer Ser. Solid State Sci. **59**, 242 (1984).

[51] R. B. Laughlin, Phys. Rev. B **25**, 5632 (1981); R. B. Laughlin, Springer Ser. Solid State Sci. **59**, 272, 288 (1984).

[52] M. Büttiker, Phys. Rev. B **38**, 9375 (1988).

[53] D. C. Tsui, H. L. Störmer, and A. C. Gossard, Phys. Rev. Lett. **48**, 1559 (1982).

[54] R. B. Laughlin, Phys. Rev. Lett. **50**, 1395 (1983).

[55] J. M. Luttinger, Phys. Rev. **121**, 924 (1961).

[56] R. E. Peierls, *Quantum Theory of Solids* (Clarendon, Oxford, 1955).

[57] L. Rota and S. M. Goodnick, in the *Proceedings of the Third Int. Workshop on Computational Electronics*, (S. M. Goodnick, ed., Oregon States University Press, Corvallis OR) pp. 231–234 (1994).

[58] T. Ando, J. Phys. Soc. Japan **43**, 1616 (1977).

[59] S. M. Goodnick, D. K. Ferry, C. W. Wilmsen, Z. Liliental, D. Fathy, and O. L. Krivanek, Phys. Rev. B**32**, 8171 (1985).

[60] H. Akera and T. Ando, Phys. Rev. B **41**, 11967 (1990); Phys. Rev. B **43**, 11676 (1991).

[61] M. Cardona, Superlattices and Microstructures, 7, 183 (1990)

[62] K. W. Kim and M. A. Stroscio, J. Appl. Phys. **68**, 6289 (1990).

[63] H. Rücker, E. Molinari, and P. Lugli, Phys. Rev. B **45**, 6747 (1992).

[64] S. Kawajii, J. Phys. Soc. Japan **27**, 906 (1969).

[65] H. Ezawa, T. Kuroda, and K. Nakamura, Surf. Sci. **27**, 218 (1971).

[66] P. J. Price, Surf. Sci. **143**, 145 (1984).

[67] P. J. Price, Ann. Phys. **133**, 217 (1981).

[68] F. A. Riddoch and b. K. Ridley, J. Phys. C **16**, 6971 (1983); Physica **134B**, 342 (1985).

[69] S. M. Goodnick and P. Lugli, Phys. Rev. B **37**, 2578 (1988).

[70] M. Duer, S. M. Goodnick, and P. Lugli, Phys. Rev. B **54**, 17794 (1996).

[71] D. Jovanovic and J. P. Leburton, "Monte Carlo Simulation of Quasi-One-Dimensional Systems," from *Monte Carlo Device Simulation: Full Band and Beyond*, ed. by K. Hess (Kluwer Acadmic Publishers, Boston, 1991).

[72] B. J. F. Lin, D. C. Tsui, M. A. Paalanen, and A. C. Gossard, Appl. Phys. Lett. **45**, 695 (1984).

[73] H. Sakaki, Japan J. Appl. Phys. **19**, L735 (1980).

[74] T. J. Thornton, M. L. Roukes, A. Scherer, and B. P. Van de Gaag, Phys. Rev. Lett. **63**, 2128 (1989).

[75] K. Ismail, D. A. Antoniadis, and H. I. Smith, Appl. Phys. Lett. **54**, 1130 (1989).

[76] N. Sawaki, R. Sugimoto, and T. Hori, Semicond. Sci. Technol. **9**, 946 (1994).

[77] C. Wirner, H. Momose, and C. Hamaguchi, Physica B **227**, 34 (1996).

[78] J. P. Harrang, R. J. Higgins, R. K. Goodall, P. R. Jay, M. Laviron, and P. Delescluse, Phys. Rev. B **31**, (1985).

[79] *The Quantum Hall Effect*, 2nd Ed., eds. R. E. Prange and S. M. Girvin, (Springer-Verlag, Heidelberg, 1990).

3

Transmission in nanostructures

In Chapter 2, we introduced the idea of low-dimensional systems arising from quantum confinement. Such confinement may be due to a heterojunction, an oxide-semiconductor interface, or simply a semiconductor-air interface (for example, in an etched quantum wire structure). When we look at transport *parallel* to such barriers, such as along the channel of a HEMT or MOSFET, or along the axis of a quantum wire, to a large extent we can employ the usual kinetic equation formalisms for transport and ignore the phase information of the particles. Quantum effects enter only through the description of the basis states arising from the confinement, and the quantum mechanical transition rates between these states are due to the scattering potential. This is not to say that quantum interference effects do not play a role in parallel transport. As we will see in Chapters 5 and 6, several effects manifest themselves in parallel transport studies such as *weak localization* and *universal conductance fluctuations*, which at their origin have effects due to the coherent interaction of electrons.

In contrast to transport parallel to barriers, when particles traverse regions in which the medium is changing on length scales comparable to the phase coherence length of the particles, quantum interference is expected to be important. By "quantum interference" we mean the superposition of incident and reflected waves, which, in analogy to the electromagnetic case, leads to constructive and destructive interference. Such a coherent superposition of states is of course what leads to the quantization of momentum and energy in the formation of low-dimensional systems discussed in the previous chapter. However, we now want to look at the general case of transport in *open systems* when carriers are incident on such barriers, and how the reflected and transmitted amplitudes of the particle waves through such structures determine transport. Prior to the early 1980s, most artificially engineered structures for studying such quantum transport were in the form of planar barrier structures, in which, for example, one thin layer of material is grown on top of another, such as the simple AlGaAs/GaAs heterostructure discussed in the previous chapter. The change in potential is therefore in only one dimension (the vertical growth direction), with the other two degrees of freedom unconstrained. We will begin our discussion of transmission in nanostructures by first discussing transport in these structures, with particular emphasis on the so-called *resonant tunneling diode* discussed in the Sections 3.1–3.3. During the mid-1980s until the present, researchers became increasingly engaged in the fabrication of fully three-dimensional quantum interference structures through the rapidly evolving technologies related to lateral patterning and ultra-small device fabrication discussed in Chapter 2. The generalization of quantum transmission concepts to these more complicated 3D structures will be pursued in the remaining sections of this chapter.

3.1 Tunneling in planar barrier structures

Perhaps the most well studied transport phenomenon associated with quantum transmission is that of tunneling. In general, the term "tunneling" refers to particle transport through a classically forbidden region, where we mean a region in which the total energy of a classical point particle is less than its potential energy. This is illustrated in Fig. 3.1, where a particle of energy E is incident on an arbitrary-shaped potential energy barrier of height $V_o > E$. In classical mechanics, a particle is completely reflected at the so-called *classical turning points* labeled A and B in Fig. 3.1, that is, points where the total energy equals the potential energy. Quantum mechanically, the underlying equation of motion is the Schrödinger equation, in which the role of the potential is analogous in electromagnetics to that of a spatially varying permittivity. In electromagnetics, the solution of the wave equation must satisfy certain boundary conditions at the abrupt interface between two dielectrics of different permittivity, which leads to a certain portion of an incident wave being transmitted and a certain portion reflected. Likewise, in quantum mechanics the wavefunction and its normal derivative must be continuous across a boundary of two regions of different potential energy, which similarly leads to reflection and transmission of probability waves at the boundary. As shown in Fig. 3.1, the wavefunction associated with a particle incident from the left on the potential barrier has nonzero solutions inside the barrier and on the right side. Because the square of the wavefunction represents the probability density for finding a particle in a given region of space, it follows that quantum mechanically a particle incident on a potential barrier has a finite probability of *tunneling* through the barrier and appearing on the other side.

Historically, the phenomenon of tunneling was recognized soon after the foundations of quantum theory had been established in connection with field ionization of atoms and nuclear decay of alpha particles. Shortly thereafter, tunneling in solids was studied by Fowler and Nordheim [1] in the field emission of electrons from metals, that is, the electric field–aided thermionic emission of electrons from metal into vacuum. Later, interest developed in tunneling through thin insulating layers (such as thermally grown oxides) between metals (MIM), and between semiconductors and metals (MIS). After the development of the band theory of solids, Zener [2] proposed the concept of *interband* tunneling, in which electrons tunnel from one band to another through the forbidden energy gap of the solid. The time period of the late 1940s and early 1950s saw tremendous breakthroughs in the development of semiconductor device technology, and conditions favorable to the experimental observation of Zener tunneling in p-n junction diodes were realized. In the late 1950s, Esaki [3] proposed the so-called Esaki diode, in which negative differential resistance (NDR) is observed in the I-V characteristics of heavily doped p-n diodes due to interband Zener tunneling between the valence and conduction bands. The Esaki diode continues to be important technologically and finds many applications in microwave technology. In the 1960s, a flourish of activity developed related to measurements of tunneling between superconductors and normal metals [4] and between superconductors themselves separated by thin insulating layers [5], which

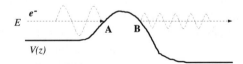

Fig. 3.1. Quantum mechanical tunneling through a potential barrier. The points A and B correspond to the classical turning points.

reveal striking evidence of the superconducting density of states and the associated super-conducting gap. Theoretically, this led to development of perturbative many-body theories of tunneling embodied by the *tunneling* or *transfer Hamiltonian method* [6]. Recent theories associated with *single electron charging* in quantum dots are descendants of the transfer Hamiltonian method and will be revisited in Chapter 4 in connection with single electron phenomenon. The transfer Hamiltonian model was also utilized quite extensively in the study of independent particle tunneling [7, 8, 9], which became the basis for interpreting a host of experimental tunneling studies in normal metals and semiconductors. A thorough review of the status of experimental and theoretical tunneling related research prior to the 1970s is given by Duke [10].

During the 1970s, advances in epitaxial growth techniques such as MBE increasingly allowed the growth of well-controlled heterostructure layers with atomic precision and low background impurity densities. In their pioneering work in this field, Tsu and Esaki [11, 12] at IBM predicted that when bias is applied across the structure, the current-voltage (I-V) characteristics of GaAs/Al_xGa_{1-x}As double and multiple barrier structures should show NDR similar to that in Esaki diodes. However, NDR in this case occurs due to resonant tunneling through the barriers within the same band. Resonant tunneling refers to tunneling in which the electron transmission coefficient through a structure is sharply peaked about certain energies, analogous to the sharp transmission peaks as a function of wavelength evident through optical filters, such as a Fabry-Perot étalon consisting of two parallel dielectric interfaces. A schematic representation of the resonant tunneling process and the associated NDR in the I-V characteristics of such a *resonant tunneling diode* (RTD) is shown for the structure reported by Sollner *et al.* in Fig. 3.2 [13]. This particular structure consists of two Al_xGa_{1-x}As barriers ($x \approx 0.25$–0.30) separated by a thin GaAs quantum well. For the Al mole fraction of this structure, the estimated barrier height due to the conduction band offset, ΔE_c, is approximately 0.23 eV. The thickness of the barriers (here 50 Å) is sufficiently thin that tunneling through the barriers is significant. The energy E_1 corresponds to the lowest resonant energy, which as discussed above is the energy where the transmission coefficient is very peaked, and in fact may approach unity in some cases. This energy may qualitatively be thought of as the bound state associated with the quantum well formed between the two confining barriers as discussed in Chapter 2. However, since the electron may tunnel out of this bound state in either direction, there is a finite lifetime τ associated with this state, and the width of the resonance in energy (i.e., the energy range in which the transmission coefficient is sizable) is inversely proportional to this lifetime, approximately as $\hbar/\tau$. Depending on the well width and barrier heights, there may exist several such *quasi-bound states* in the system. The double barrier structure is surrounded by heavily doped GaAs layers, as shown in Fig. 3.2, which provide low-resistance emitter and collector contacts to the tunneling region, forming the RTD structure. With a positive bias applied to the right contact relative to the left, the Fermi energy on the left is pulled through the resonant level E_1. As the Fermi energy passes through the resonant state, a large current flows due to the increased transmission from left to right. At the same time, the back flow of carriers from right to left is suppressed as electrons at the Fermi energy on the right see only a large potential barrier, as shown in the figure. Further bias pulls the bottom of the conduction band on the left side through the resonant energy, which cuts off the supply of electrons available at the resonant energy for tunneling. The result is a marked decrease of the current with increasing voltage, giving rise to a region of NDR as shown schematically by the I-V characteristics in Fig. 3.2.

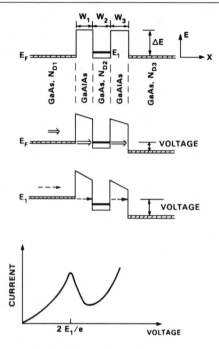

Fig. 3.2. Energy band diagram and I-V characteristics of a GaAs/AlGaAs resonant tunneling diode. The upper part shows the electron energy as a function of position in the quantum well structure. The parameters are $N_{D1} = N_{D3} = 10^{18}\,\mathrm{cm}^{-3}$, $N_{D2} = 10^{17}\,\mathrm{cm}^{-3}$, and $W_1 = W_2 = W_3 = 50\,\text{Å}$. The aluminum mole fraction in the barriers is $x \approx 25\%-30\%$ [after Sollner *et al.*, Appl. Phys. Lett. **43**, 588 (1983), by permission].

    The first experimental evidence for resonant tunneling in double barrier structures was reported by the IBM group in MBE grown structures [12] where weak NDR was observed in the I-V characteristics. The difficulty at the time in realizing pronounced NDR as observed in interband Esaki diodes was due to the difficulty in achieving low background impurity concentrations in epitaxial layers at the time as discussed in Chapter 2. The effect of such inhomogeneities is to broaden the expected resonances, in effect washing them out. It was not until the 1980s that epitaxial material quality had improved sufficiently to observe marked resonant tunneling behavior in III-V epitaxial layer structures. Sollner *et al.* at MIT reported pronounced NDR in the low temperature I-V characteristics of the double barrier structure [13] shown in Fig. 3.2. Figure 3.3 shows the I-V characteristics for both forward and reverse bias at three different temperatures. The I-V curves are approximately symmetric about the origin although it is clear that the peak current is slightly lower in the negative bias direction. Such asymmetry in RTDs is typical in many structures and may arise from several sources such as out-diffusion of impurities during MBE growth, or due to differences in interface roughness between AlGaAs grown on GaAs versus GaAs grown on AlGaAs, where the latter is believed to be rougher. For many applications, NDR devices should have a large peak current and a small valley current, where the latter is the minimum current following the peak current as the magnitude of the voltage increases. Therefore an important figure of merit for an NDR device such as the RTD is the *peak to valley ratio* (PVR). For the data in Fig. 3.3, the PVR is 6:1 at low temperature (<50 K), while at increasingly higher temperatures the PVR decreases until the NDR effect completely vanishes. The reduction

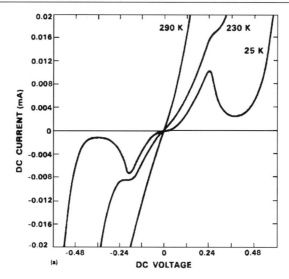

Fig. 3.3. Current-voltage characteristics of the RTD structure shown in Fig. 3.2 at three different temperatures [after Sollner *et al.*, Appl. Phys. Lett. **43**, 588 (1983), by permission].

of the PVR as temperature increases is simply related to the increase in off-resonant current over the barriers due to thermionic emission, as well as the spreading of the distribution function around the resonant energy, which decreases the peak current as discussed in more detail below. The valley current is also increased at higher temperatures due to inelastic phonon-assisted tunneling that further degrades the PVR of the diode.

A considerable volume of research into RTDs has evolved since the MIT group's successful demonstration of NDR effects. Interest stems not only from the fundamental physics aspects of this deceptively simple structure, but also from the potential practical applications in high-speed microwave systems and novel digital logic circuits. One advantage of the RTD for electronic applications is that the fundamental time associated with the intrinsic tunneling process itself may be quite short, often taken as the lifetime of the quasi-bound state (i.e., the inverse of the resonance width). In their original work, Sollner *et al.* demonstrated NDR up to frequencies of 2.5 THz, which qualitatively implies charge transport on the order of $\tau = 6 \times 10^{-14}$ s. In reality, there are several time constants that come into play in the frequency response of an RTD, including the transit time across the nontunneling regions of the device and the RC time constant associated with the capacitance of the structure. With proper design, the various time constants can be minimized, and high-frequency performance may be obtained in analog applications such as oscillators operating up to 420 GHz in GaAs/AlAs double barrier structures [14].

Improvements in material growth as well as an understanding of the physics and device design of RTDs has led to ever improving performance. Room-temperature NDR was soon obtained in similar structures to Fig. 3.2 [15]. There have been systematic studies of the dependence of the I-V characteristics upon structural parameters of $Al_xGa_{1-x}As$/GaAs RTDs such as the GaAs well width [16], the $Al_xGa_{1-x}As$ barrier height [17], and thickness [18], [19]. To minimize impurity scattering due to dopant interdiffusion into the barrier regions, thin spacer layers of undoped or reduced-doping material are usually included on either side of the tunneling barriers (see for example [16]). Studies of the dependence of RTD performance on the thickness of such spacers were reported by Yoo *et al.* [20]. For good devices with thin AlAs barriers, PVRs close to 4:1 and peak current densities in excess of $1 \times 10^5 A/cm^2$ [21] may be obtained at 300 K, although not in the same structures, since there is usually a trade-off between these two parameters in terms of device design.

Alternative material systems to the AlGaAs/GaAs system have yielded greatly improved performance benchmarks. In$_{.53}$Ga$_{.47}$As lattice matched to InP has been employed as the narrow gap material in double barrier structures. PVRs of 14:1 at room temperature using an InGaAs well have been reported [22]. Even better PVRs of 30:1 and 63:1 at 300 K and 77 K were reported using a strained InAs well [23] between AlAs barriers. More recently, attention has focused on the use of Type II staggered bandgap systems such as the InAs/AlSb system for resonant tunneling applications [24]. In contrast to a Type I heterojunction system [eq. GaAs/AlGaAs], the valence band edge of the wider bandgap AlSb lies higher in energy than that of InAs, so that electrons are confined in the InAs layer, but holes are not [see Fig. 3.4]. Söderström *et al.* reported room-temperature PVRs of 3.2 and peak current densities of 3.7×10^5 A/m^2 in double barrier structures [25] consisting of InAs wells and cladding layers and AlSb barriers (grown on GaAs substrates). High-frequency oscillations were measured up to 712 GHz [26], the highest frequency reported to that date of any solid state oscillator. Part of the improvement in performance in this material system is associated with the favorable transport and Ohmic contact properties of InAs compared to GaAs, reducing parasitic contributions to the total delay time across the device.

One final and interesting innovation in resonant tunneling diodes has been the successful development of resonant interband tunneling diodes (RITs) in which both electrons and holes participate in transport [27]. Figure 3.4 shows the band diagram (for the Γ point) of an RIT showing the staggered alignment of the Type-II system. Of particular importance is the fact that between InAs and GaSb, the band offset is so large that the top of the valence band of GaSb is coincident with the bottom of the conduction band of InAs, here separated by thin AlSb barriers. When bias is applied, the Fermi energy in the InAs emitter aligns with the quasi-bound state associated with the quantized hole state in the GaSb well, and electrons may tunnel through the valence band of the GaSb well into the InAs collector on the opposite side. As further bias is applied, this resonant energy drops below the CB edge of the emitter, and current flow is quickly suppressed. In contrast to the conventional intra-band tunneling diodes discussed so far, in which the barrier is lowered with increasing bias, the barrier to electrons tunneling from the InAs actually becomes larger as the electrons have to tunnel not only through the AlSb barriers but also through the bandgap of the GaSb. The advantage is that the valley current is greatly suppressed compared to a conventional RTD. In fact, the device operation may be thought qualitatively as a combination of an Esaki diode (where tunneling is from valence to conduction band across the bandgap) and an RTD. This fact is reflected in the large PVRs measured in these and similar structures (see, e.g. [28], [29]), which are on the order of 20:1 at room temperature and 150:1 at low temperature.

3.2 Wavefunction treatment of tunneling

In the present section, we will begin an elementary analysis of tunneling in planar barrier structures such as the RTD, developing a model based on the independent electron approach. As alluded to earlier, transport in nanoscale structures such as the RTD depends strongly on the quantum mechanical flux throughout the structure. In wave mechanics, we can identify

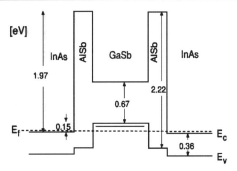

Fig. 3.4. Band-edge diagram (Γ point) for a resonant interband diode (RIT) at room temperature. The valence-band edge of the GaSb and the quantized hole state of the well are both above the InAs conduction band edge at zero bias [after Söderström *et al.*, Appl. Phys. Lett. **55**, 1094 (1989), by permission].

the probability density from the quantity

$$\rho(\mathbf{r}, t) = \psi(\mathbf{r}, t)\psi^*(\mathbf{r}, t), \tag{3.1}$$

where $\psi(\mathbf{r}, t)$ is the time-dependent solution of the Schrödinger equation. If we multiply the time-dependent Schrödinger equation by $\psi^*(\mathbf{r}, t)$ and rearrange terms, it may be written in the form of a continuity equation for the probability density, $\rho(\mathbf{r}, t)$, as

$$\frac{\partial \rho(\mathbf{r}, t)}{\partial t} = \frac{-\hbar}{2m_e i}\nabla \cdot [\psi^*(\mathbf{r}, t)\nabla\psi(\mathbf{r}, t) - \psi(\mathbf{r}, t)\nabla\psi^*(\mathbf{r}, t)], \tag{3.2}$$

where m_e is the mass of the particle. The probability flux density, or "current," can therefore be identified with the quantity

$$\mathbf{f} = \frac{\hbar}{2im_e}[\psi^*(\mathbf{r}, t)\nabla\psi(\mathbf{r}, t) - \psi(\mathbf{r}, t)\nabla\psi^*(\mathbf{r}, t)]. \tag{3.3}$$

When we have a situation such as the RTD structure of Fig. 3.2, where particles are incident on a barrier, we can define reflection and transmission coefficients in terms of the ratio of incident fluxes to reflected fluxes, or incident fluxes to transmitted fluxes, for a given energy with the fluxes defined in terms of Eq. (3.3). Later introduction of the statistical mechanical distribution functions for the various states allows a formalism for calculating the current through the structure, as will be developed further in Section 3.3. In the present section we look in more detail at the definition and calculation of the transmission and reflection coefficients for planar structures.

3.2.1 Single rectangular barrier

Symmetric barrier

To begin, we consider the simple example of a single symmetric planar barrier consisting of a layer of $Al_xGa_{1-x}As$ of width $W = 2a$ imbedded in GaAs, as shown in Fig. 3.5. For simplicity, we consider an electron in the conduction band of the GaAs incident on the barrier from the left. Here the barrier height V_o is simply the conduction band offset, ΔE_c, associated with the Γ point in the two materials, as long as the incident electron energies are lower than the X-point in both materials. We can then consider the single-band effective mass model as long as nonparabolicity effects are ignored, so that the dispersion relation is parabolic. The solution is again separable into parallel and perpendicular parts relative to the barrier, and we can solve the envelope equation (2.4) for the stationary solutions in the z-direction (perpendicular to the barriers),

$$\left(\frac{-\hbar^2}{2}\frac{\partial}{\partial z}\frac{1}{m^*(z)}\frac{\partial}{\partial z} + V_{eff}(z)\right)\varphi(z) = E\varphi(z). \tag{3.4}$$

If we assume that space charge effects are negligible, the effective potential is simply that due to the conduction band offset, ΔE_c. Then the solution of this piecewise-continuous function may be written in each region as

$$\varphi(z) = \begin{cases} Ae^{ikz} + Be^{-ikz} & z < -a \\ Ce^{\gamma z} + De^{-\gamma z} & -a < z < a \\ Ee^{ikz} + Fe^{-ikz} & z > a \end{cases} \tag{3.5}$$

where

$$k = \frac{\sqrt{2m^*E}}{\hbar},$$
$$\gamma = \frac{\sqrt{2m^*(V_o - E)}}{\hbar}. \tag{3.6}$$

The coefficients A and B are associated respectively with incoming and outgoing waves on the left side relative to the barrier. Likewise, the coefficients E and F are respectively outgoing and incoming waves on the right. For energies lying below the top of the barrier, $V_o > E$, the solutions in the barrier are exponentially increasing and decaying functions associated with *evanescent states*. For energies above the barrier, the solutions in the barrier are complex exponential solutions as are the solutions on either side of the barrier. The coefficients A through F are related to one another through the boundary conditions requiring the continuity of the envelope function and its derivative at the interfaces $z = -a$ and $z = a$. Therefore, at $z = -a$, we have that

$$\varphi_I(-a_-) = \varphi_{II}(-a_+)$$

$$\frac{1}{m_-^*}\frac{\partial \varphi_I}{\partial z}\bigg|_{-a_-} = \frac{1}{m_+^*}\frac{\partial \varphi_{II}}{\partial z}\bigg|_{-a_+}, \tag{3.7}$$

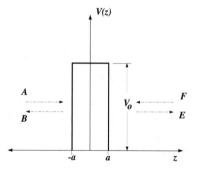

Fig. 3.5. A simple rectangular tunneling barrier.

where m_-^* and m_+^* are the masses on the respective sides of the barrier. If we further assume that the masses are the same for these two materials, then application of Eq. (3.7) to the envelope function solution (3.5) gives

$$Ae^{-ika} + Be^{ika} = Ce^{-\gamma a} + De^{\gamma a},$$

$$ik[Ae^{-ika} - Be^{ika}] = \gamma[Ce^{-\gamma a} - De^{\gamma a}].$$

$$(3.8)$$

Rewriting the two equations, the coefficients on the left side, A and B, may be related to C and D through the matrix relation

$$\begin{bmatrix} A \\ B \end{bmatrix} = \begin{bmatrix} \left(\frac{ik+\gamma}{2ik}\right)e^{(ik-\gamma)a} & \left(\frac{ik-\gamma}{2ik}\right)e^{(ik+\gamma)a} \\ \left(\frac{ik-\gamma}{2ik}\right)e^{-(ik+\gamma)a} & \left(\frac{ik+\gamma}{2ik}\right)e^{-(ik-\gamma)a} \end{bmatrix} \begin{bmatrix} C \\ D \end{bmatrix}. \tag{3.9}$$

Similarly, the boundary conditions at $z = a$ give two more equations

$$Ce^{\gamma a} + De^{-\gamma a} = Ee^{ika} + Fe^{-ika},$$

$$\gamma[Ce^{\gamma a} - De^{-\gamma a}] = ik[Ee^{ika} - Fe^{-ika}].$$

$$(3.10)$$

Similarly, we can relate C and D to E and F through the matrix relation

$$\begin{bmatrix} C \\ D \end{bmatrix} = \begin{bmatrix} \left(\frac{ik+\gamma}{2\gamma}\right)e^{(ik-\gamma)a} & -\left(\frac{ik-\gamma}{2\gamma}\right)e^{-(ik+\gamma)a} \\ -\left(\frac{ik-\gamma}{2\gamma}\right)e^{(ik+\gamma)a} & \left(\frac{ik+\gamma}{2\gamma}\right)e^{-(ik-\gamma)a} \end{bmatrix} \begin{bmatrix} E \\ F \end{bmatrix}. \tag{3.11}$$

We can now replace the column vector containing C and D on the right side of Eq. (3.9) with the right side of (3.11) to give

$$\begin{bmatrix} A \\ B \end{bmatrix} = \begin{bmatrix} M_{11} & M_{12} \\ M_{21} & M_{22} \end{bmatrix} \begin{bmatrix} E \\ F \end{bmatrix}, \tag{3.12}$$

where the elements of the 2×2 array arise from the matrix multiplication of the 2×2 arrays appearing in (3.11) and (3.9). The elements are given by

$$M_{11} = \left(\frac{ik+\gamma}{2ik}\right)\left(\frac{ik+\gamma}{2\gamma}\right)e^{2(ik-\gamma)a} - \left(\frac{ik-\gamma}{2ik}\right)\left(\frac{ik-\gamma}{2\gamma}\right)e^{2(ik+\gamma)a}$$

$$= \left[\cosh(2\gamma a) - \frac{i}{2}\left(\frac{k^2 - \gamma^2}{k\gamma}\right)\sinh(2\gamma a)\right]e^{2ika}, \tag{3.13}$$

$$M_{21} = \left(\frac{ik-\gamma}{2ik}\right)\left(\frac{ik+\gamma}{2\gamma}\right)e^{-2\gamma a} - \left(\frac{ik+\gamma}{2ik}\right)\left(\frac{ik-\gamma}{2\gamma}\right)e^{2\gamma a}$$

$$= -\frac{i}{2}\left(\frac{k^2 + \gamma^2}{k\gamma}\right)\sinh(2\gamma a), \tag{3.14}$$

$$M_{22} = M_{11}^*, \quad M_{12} = M_{21}^*. \tag{3.15}$$

It is a simple exercise to show that the determinant of the matrix $\overline{\mathbf{M}}$ is unity. However, it is not a unitary matrix, because the diagonal elements are complex.

As postulated above, the transmission and reflection coefficients are associated with the ratio of fluxes. We want to look at the system excited from one side only; that is, consider that a wave is incident from the left of the barrier, while assuming that there is no incoming wave from the right side. Thus we assume that $F = 0$ in Eq. (3.12). Superposition will allow us to superimpose the separate results for incoming waves from both sides, as will be necessary later when we consider the total current through such a structure. Assuming an incoming wave from the left with amplitude A, as in Eq. (3.5), the probability flux incident on the barrier is given by (3.3) for this solution as

$$f_{inc} = |A|^2 \frac{\hbar k}{m^*} = v|A|^2, \tag{3.16}$$

which is basically the group velocity v of a particle of wavevector k multiplied by the magnitude squared of the probability amplitude. For this symmetric barrier problem, the outgoing or *transmitted* wave associated with the amplitude E has the same group velocity,

$$f_{tran} = |E|^2 \frac{\hbar k}{m^*} = v|E|^2. \tag{3.17}$$

The transmission coefficient is defined as the ratio of the transmitted to incident probability flux density as

$$T(E) = \frac{f_{tran}}{f_{inc}} = \frac{|E|^2}{|A|^2} = \frac{1}{|M_{11}|^2}, \tag{3.18}$$

since for $F = 0$, $A = M_{11}E$ in Eq. (3.12). The energy E is related to k and γ as in Eq. (3.6). From (3.13) we may write $T(E)$ as

$$T(E) = \left[\cosh^2(2\gamma a) + \left(\frac{k^2 - \gamma^2}{2k\gamma} \right)^2 \sinh^2(2\gamma a) \right]^{-1}$$

$$= \frac{1}{1 + \left(\frac{k^2 + \gamma^2}{2k\gamma} \right)^2 \sinh^2(2\gamma a)}. \tag{3.19}$$

We can consider two limits to the transmission coefficient above. If $2\gamma a \ll 1$, corresponding to strong transmission through a thin barrier, Eq. (3.19) becomes

$$T(E) \to \frac{1}{1 + k^2 a^2}. \tag{3.20}$$

In the opposite limit, $2\gamma a \gg 1$, the positive exponential in the sinh function dominates giving

$$T(E) \to \left(\frac{4k\gamma}{k^2 + \gamma^2} \right)^2 e^{-4\gamma a} \propto e^{-2W\sqrt{2m^*(V_o - E)}/\hbar}, \tag{3.21}$$

which gives the commonly cited tunneling dependence, for example, through a thin oxide barrier in which the tunneling probability decreases exponentially with the width of the barrier and the square root of the barrier height.

The reflection coefficient is similarly defined as the ratio of the incident to reflected flux using Eq. (3.14):

$$R(E) = \frac{|B|^2}{|A|^2} = \frac{|M_{21}|^2}{|M_{11}|^2} = |M_{21}|^2 T(E) = \frac{\left(\frac{k^2 + \gamma^2}{2k\gamma} \right)^2 \sinh^2(2\gamma a)}{1 + \left(\frac{k^2 + \gamma^2}{2k\gamma} \right)^2 \sinh^2(2\gamma a)}. \tag{3.22}$$

It is clear that the sum of the transmission and reflection coefficients is $R + T = 1$.

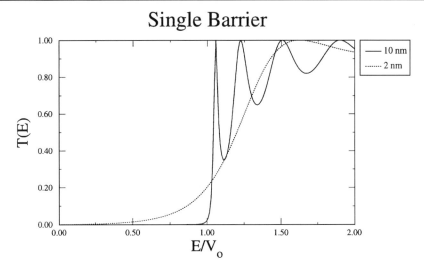

Fig. 3.6. Transmission coefficient versus energy for a simple symmetric barrier.

The above discussion basically assumed that the energy of the particle was less than the barrier height V_o. The same analysis holds for the case of energies greater than the barrier height, with γ complex. Writing $\gamma = -ik'$, the corresponding transmission coefficient is then oscillatory, with Eq. (3.19) becoming

$$T(E > V_o) = \frac{1}{1 + \left(\frac{k^2 - k'^2}{2kk'}\right)^2 \sin^2(2k'a)}. \tag{3.23}$$

Combining the two results, for $E < V_o$ and $\dot{E} > V_o$, the transmission probability versus density is plotted in Fig. 3.6 for a typical structure. As is apparent, the transmission probability decays exponentially as the energy difference below the barrier increases. For energies above the barrier height, the transmission coefficient is oscillatory approaching unity as E becomes large, as predicted by (3.23).

Asymmetric rectangular barrier

Before continuing to the case of interest of the double barrier structure, it is instructive to consider the case of an asymmetric single barrier as shown in Fig. 3.7. The asymmetric barrier could be thought of as a crude approximation to the case of a single barrier with an applied voltage qV_1 to the right side relative to the left side of the structure (of course, the effective barrier V_o would have to be reduced as well, but for simplicity we will ignore this effect). The wavefunction is slightly more complicated with

$$\varphi(z) = \begin{cases} Ae^{ikz} + Be^{-ikz} & z < -a \\ Ce^{\gamma z} + De^{-\gamma z} & -a < z < a \\ Ee^{ik_1 z} + Fe^{-ik_1 z} & z > a \end{cases} \tag{3.24}$$

where

$$k_1 = \frac{\sqrt{2m^*(E + V_1)}}{\hbar}. \tag{3.25}$$

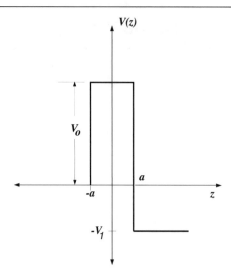

Fig. 3.7. Asymmetric single tunneling barrier.

The matching condition between C, D, E, and F is slightly modified as

$$Ce^{\gamma a} + De^{-\gamma a} = Ee^{ik_1 a} + Fe^{-ik_1 a},$$
$$\gamma[Ce^{\gamma a} - De^{-\gamma a}] = ik_1[Ee^{ik_1 a} - Fe^{-ik_1 a}], \tag{3.26}$$

where again the same mass has been assumed. The connecting matrix may be formed once more, and the result cascaded with the matrix of the opposite interface to form the composite transmission matrix for the entire barrier. The matrix elements of the matrix relation (3.12) are given by

$$M_{11} = \left(\frac{ik + \gamma}{2ik}\right)\left(\frac{ik_1 + \gamma}{2\gamma}\right)e^{(ik+ik_1-2\gamma)a} - \left(\frac{ik - \gamma}{2ik}\right)\left(\frac{ik_1 - \gamma}{2\gamma}\right)e^{(ik+ik_1+2\gamma)a}$$
$$= \left[\frac{1}{2}\left(1 + \frac{k_1}{k}\right)\cosh(2\gamma a) - \frac{i}{2}\left(\frac{kk_1 - \gamma^2}{k\gamma}\right)\sinh(2\gamma a)\right]e^{i(k+k_1)a}, \tag{3.27}$$

$$M_{21} = \left(\frac{ik - \gamma}{2ik}\right)\left(\frac{ik_1 + \gamma}{2\gamma}\right)e^{-2\gamma a - i(k-k_1)a} - \left(\frac{ik + \gamma}{2ik}\right)\left(\frac{ik_1 - \gamma}{2\gamma}\right)e^{2\gamma a - i(k-k_1)a}$$
$$= -\left[\frac{i}{2}\left(\frac{kk_1 + \gamma^2}{k\gamma}\right)\sinh(2\gamma a) + \frac{1}{2}\left(\frac{k_1}{k} - 1\right)\cosh(2\gamma a)\right]e^{-i(k-k_1)a}, \tag{3.28}$$

and the complex conjugate relationships (3.15) are still valid. The determinant of the matrix is no longer unity but rather the ratio k_1/k. As in the case of the symmetric barrier, the transmission and reflection coefficients from left to right are given by the ratio of the fluxes of the incident over the transmitted and over the reflected waves respectively, which are now slightly more complicated due to the difference in group velocities in the two regions

$$R(E) = \frac{f_{refl}}{f_{inc}} = \frac{|B|^2}{|A|^2} = \frac{|M_{21}|^2}{|M_{11}|^2}, \tag{3.29}$$

$$T(E) = \frac{f_{tran}}{f_{inc}} = \frac{v_1|E|^2}{v|A|^2} = \frac{v_1}{v}\frac{1}{|M_{11}|^2}$$

$$= \frac{\frac{4k_1 k}{(k_1+k)^2}}{1 + \left(\frac{(k^2+\gamma^2)(k_1^2+\gamma^2)}{\gamma^2(k_1+k)^2} \right) \sinh^2(2\gamma a)}, \tag{3.30}$$

which reduces to Eq. (3.19) when $k_1 = k$. If we consider the transmission coefficient from the opposite direction at the same energy, then we set $A = 0$ and look at the ratio of the incident flux from right to left

$$T_{rl}(E) = \frac{k}{k_1} \frac{|B|^2}{|F|^2}. \tag{3.31}$$

Rather than inverting Eq. (3.12), we can solve for B in terms of F directly from the two equations with $A = 0$ as

$$B = \left(M_{22} - \frac{M_{21}M_{12}}{M_{11}} \right) F = \left(\frac{|M_{11}|^2 - |M_{12}|^2}{M_{11}} \right) F = M_{11}^*(1 - R_{lr})F,$$

$$\tag{3.32}$$

$$T_{rl}(E) = \frac{k}{k_1} \frac{|B|^2}{|F|^2} = \frac{k|M_{11}|^2}{k_1} T_{lr}^2 = T_{lr}(E),$$

where R_{lr} is the reflection coefficient from left to right. Equation 3.32 illustrates an important symmetry property of the transmission coefficient, which is independent of the direction of the incident wave upon the barrier. This reciprocity relation is a general result of the time-reversal symmetry that reflects the microscopic reversibility of quantum mechanics itself. In contrast, when we consider current flow associated with the quantum mechanical fluxes above, the system itself is irreversible. Macroscopic irreversibility in the system is not imposed until we consider contacts placed on the system, which destroy the phase relationship between the incident and transmitted particle when the transmitted particle is absorbed in the contact (see Section 3.3).

Scattering matrix

We formulated the matrix relation between the coefficients on the left and right of the barrier in terms of the matrix $\overline{\mathbf{M}}$ in Eq. (3.12). Obviously there is nothing unique about this representation, and we could easily have defined a different matrix relation in terms of outgoing fluxes, B and E, and incoming fluxes, A and F, on either side of the barrier as

$$\begin{bmatrix} B \\ E \end{bmatrix} = \begin{bmatrix} S_{11} & S_{12} \\ S_{21} & S_{22} \end{bmatrix} \begin{bmatrix} A \\ F \end{bmatrix}, \tag{3.33}$$

where the matrix $\overline{\mathbf{S}}$ defines the so-called scattering matrix or S-matrix. The transmission and reflection coefficients are again obtained by letting $F = 0$ above, which gives

$$T = |S_{21}|^2 v_1/v,$$

$$R = |S_{11}|^2. \tag{3.34}$$

If we had considered waves incident from the right rather than left, the transmission and reflection coefficients would be determined by letting $A = 0$ to give

$$T = \frac{v|B|^2}{v_1|F|^2} = |S_{12}|^2 v/v_1, \tag{3.35}$$

which is equal to the transmission coefficient from left to right by reciprocity.

It is apparent that the S-matrix is a more natural representation for scattering problems in that the diagonal elements are simply related to reflection coefficients and the off-diagonal elements to transmission coefficients. Due to time reversal symmetry, the product of the S-matrix with its complex conjugate in the general asymmetric case is the unit matrix

$$\overline{\mathbf{S}}^*\overline{\mathbf{S}} = \mathbf{1}. \tag{3.36}$$

This relation gives, for example, that

$$|S_{11}|^2 + S_{12}^* S_{21} = 1, \tag{3.37}$$

which according to Eqs. (3.34) and (3.35) is just a statement of flux conservation, $R + T = 1$.

In the symmetric case this matrix also has the property of being a unitary matrix

$$\overline{\mathbf{S}}^\dagger\overline{\mathbf{S}} = \mathbf{1}, \tag{3.38}$$

where $\mathbf{1}$ is the identity matrix and $\overline{\mathbf{S}}^\dagger$ denotes the Hermitian conjugate. This relation implies that

$$|S_{11}|^2 + S_{21}^* S_{21} = 1. \tag{3.39}$$

For the asymmetric case, this relation is clearly in violation of the current conservation $T + R = 1$ due to the difference in group velocities on either side. If unitarity is desired, however, the coefficients of the S-matrix may be normalized by the flux ratio $S_{nm}\sqrt{v_n/v_m}$ (where the indices indicate the respective sides of the structure) such that the overall matrix takes the form

$$\overline{\mathbf{S}} = \begin{bmatrix} r & t' \\ t & r' \end{bmatrix}, \tag{3.40}$$

where the elements r, t, r', and t' represent the reflection and transmission amplitudes from left to right and right to left, respectively, with $T = |t|^2$ and $R = |r|^2$.

For one-dimensional problems such as the planar barrier considered here, there is no apparent advantage for using either the $\mathbf{M}$ or the $\mathbf{S}$ matrices. However, for multidimensional problems, the latter is more frequently encountered. We will come back to this representation later in connection with quantum waveguide problems in Section 3.5.

3.2.2 The double barrier case

We now turn to the case of a double barrier structure in connection with the technologically important resonant tunneling diode discussed in Section 3.1. As shown in Fig. 3.2, the RTD structure may be thought of as two single-barrier tunneling structures such as those considered in the previous section, separated by a quantum well. As discussed qualitatively in Section 3.1, the NDR arises due to the presence of one or more resonant levels associated with the quasi-bound states of the well. Such states are not true bound states as in the quantum well structures discussed in Chapter 2 because an electron localized in such a state has a finite probability of tunneling out. We now want to generalize the discussion of the previous section and consider the resonant transmission properties of this double barrier case before discussing the current-voltage characteristics in Section 3.3. In the schematic behavior of the RTD in Fig. 3.2, the barriers are symmetric (for that particular experimental study) with no bias applied. However, when the device is biased into resonance (i.e., when the Fermi energy on one side is aligned with the quasi-bound state of the well), it is obvious that this symmetry is broken, and in fact the transmission properties are a strong function of the voltage applied across the RTD. In the following we will discuss first the more ideal symmetric case and then the more general asymmetric case such as that which occurs with bias, and calculate the transmission coefficient as a function of energy.

Rather than going through the entire procedure of matching all the incoming and outgoing wave coefficients through boundary conditions at the various interfaces, we can make use of the results established so far for single barriers to construct the transmission properties of the composite system. One needs to recognize that the transmission matrix derived for the single barrier is independent of the location of the barriers in space, which is somewhat intuitive. Figure 3.8 shows the representation of the two barrier problem to be considered here. The transmission matrices connecting A and B with E and F, as well as A' and B' with E' and F', may be obtained from the single barrier result. The trick is to connect the coefficients E, F with A', B'. Since these coefficients correspond to different points in a region of uniform potential, the difference between the coefficients is simply a phase constant for the rightward wave

$$A' = E e^{ikb}, \tag{3.41}$$

where b is the thickness of the well in Fig. 3.8 and k is the propagation constant in the well region. Similarly, for the leftward wave

$$B' = F e^{-ikb}. \tag{3.42}$$

These two relations allow us to define a propagation or *transfer* matrix, $\overline{\mathbf{M}}_W$, in connecting the coefficients of the two barriers

$$\begin{bmatrix} E \\ F \end{bmatrix} = \begin{bmatrix} e^{-ikb} & 0 \\ 0 & e^{ikb} \end{bmatrix} \begin{bmatrix} A' \\ B' \end{bmatrix}. \tag{3.43}$$

Given the connection between these coefficients with their respective coefficients on the other side of the left and right barriers, we can write the composite matrix representing the coefficients on the left and right as

$$\begin{bmatrix} A \\ B \end{bmatrix} = \overline{\mathbf{M}}_L \cdot \overline{\mathbf{M}}_W \cdot \overline{\mathbf{M}}_R \begin{bmatrix} E' \\ F' \end{bmatrix} = \overline{\mathbf{M}}_T \begin{bmatrix} E' \\ F' \end{bmatrix}, \tag{3.44}$$

where $\overline{\mathbf{M}}_L$ and $\overline{\mathbf{M}}_R$ are the transmission matrices of the left and right barriers, respectively. The transmission coefficient of the double barrier structure again will be related to the inverse square of the coefficient M_{T11} of the composite matrix, which using (3.43) is simply

$$M_{T11} = M_{L11} M_{R11} e^{-ikb} + M_{L12} M_{R21} e^{ikb}. \tag{3.45}$$

Resonance behavior may occur in this expression through the phase factors, which may lead to cancellation of terms that minimize M_{T11}, giving rise to peaked behavior in the transmission coefficient. To see this explicitly, we now consider the simple case of symmetric barriers followed by the more general asymmetric case.

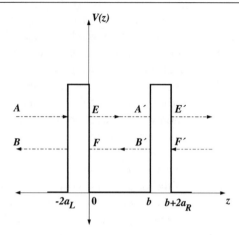

Fig. 3.8. Two generic barriers combined to form a double barrier structure.

Identical barriers

For the case of symmetric barriers, the propagation constant k is the same in the well, left and right regions. Further, both the barriers are the same width, $a_L = a_R$, and same height with the same attenuation constant, γ. We may thus use the results from Section 3.2.1 for a symmetric single barrier given by Eqs. (3.12)–(3.15) for the elements of $\overline{\mathbf{M}}_L$ and $\overline{\mathbf{M}}_R$ appearing in (3.45). To simplify the notation, write the matrix element M_{11} of the single barrier in polar coordinates as

$$M_{11} = m_{11}e^{i\theta}, \tag{3.46}$$

where from Eq. (3.13)

$$m_{11} = \sqrt{\cosh^2(2\gamma a) + \left(\frac{k^2 - \gamma^2}{2k\gamma}\right)^2 \sinh^2(2\gamma a)} \tag{3.47}$$

is the magnitude and

$$\theta = -\arctan\left[\left(\frac{k^2 - \gamma^2}{2k\gamma}\right)\tanh(2\gamma a)\right] + 2ka \tag{3.48}$$

is the phase of M_{11}. Substituting into (3.45) and squaring we may write (using the fact that $M_{12} = M_{21}^*$)

$$|M_{T11}|^2 = |m_{11}|^4 + |M_{21}|^4 + 2|m_{11}|^2|M_{21}|^2 \cos[2(kb - \theta)]$$

$$= \left(|m_{11}|^2 - |M_{21}|^2\right)^2 + 4|m_{11}|^2|M_{21}|^2 \cos^2(kb - \theta). \tag{3.49}$$

The first term in brackets on the right side is the determinant of the matrix $\overline{\mathbf{M}}$, which for the case of a symmetric rectangular barrier was unity as discussed earlier. Thus the overall transmission coefficient is given by

$$T_{tot}(E) = \frac{1}{|M_{T11}|^2} = \frac{1}{1 + 4|m_{11}|^2|M_{21}|^2 \cos^2(kb - \theta)} = \frac{T_1^2}{T_1^2 + 4R_1 \cos^2(kb - \theta)}, \tag{3.50}$$

where T_1 and R_1 are the transmission and reflection coefficients of the single symmetric barrier defined by (3.19) and (3.22).

Off resonance, the cosine function in (3.50) is nonzero. The minimum value of the total transmission coefficient occurs when $kb - \theta = n\pi$, giving

$$T_{tot}^{min} = \frac{T_1^2}{T_1^2 + 4R_1} \sim \frac{T_1^2}{4}, \tag{3.51}$$

where in the last expression the transmission coefficient of the single barrier is assumed to be much smaller than unity (typically the case in real structures) while $R_1 = 1 - T_1$ is on the order of unity. Thus the off-resonance transmission appears to be the cascaded transmission of two successive identical barriers independent of the well in between, apart from the factor of 4.

Resonance occurs when the cosine vanishes. In the symmetric barrier case, the transmission coefficient approaches unity

$$T_{tot}^{max} = 1, \qquad kb - \theta = (2n + 1)\frac{\pi}{2}, \qquad n = 0, 1, \ldots. \tag{3.52}$$

Thus the transmission coefficient may change over orders of magnitude to approach unity over a narrow range of energies. Note that the wavevectors satisfying the resonance condition above correspond to the quantized values of the wavevector in a finite square well associated with the bound states there as discussed in Chapter 2. In the limit of an infinite well, the factor θ just approaches the value $-\pi/2$. Thus, as asserted previously, the resonant levels may be thought of as the bound state levels of the finite well formed between the two barriers, and pronounced transmission of the electron wave occurs when the energy of the electron is aligned with one of these quasi-bound states. On resonance, an electron wave is reflected back and forth between the two barriers in a way that adds coherently. The incoming wave excites the occupancy of the resonance level until a steady state is reached in which the incoming wave is balanced by an outgoing wave, and the overall transmission is unity. This unity transmission due to the perfect match between barriers is broken by any asymmetry as we see below.

To illustrate the above discussion for a resonant tunneling diode, Fig. 3.9 shows the calculated transmission coefficient for three different well widths for a symmetric GaAs/AlAs double barrier. The potential diagram corresponds to the Γ-valley conduction band minimum even though for AlAs, the X-valley is the lowest conduction band minimum, and tunneling may occur via this band as well. The complication of multiband tunneling is neglected in this calculation for simplicity. The transmission coefficient is unity on resonance, and then it falls off rapidly with energy on either side of the resonance. As one would expect from the discussion above, there are several resonance peaks satisfying the condition (3.52). The spacing of these resonances is increasingly closer as the well width increases, which follows from the identification of these resonances with the quasi-bound states of the finite well formed in the RTD structure. Note in Fig. 3.9 that the widths of the resonances decrease as the well width increases, and that higher-energy resonances for the same well width are invariably broader than the lowest one. As we qualitatively asserted earlier, the width of the resonance is inversely proportional to the quasi-bound state lifetime as $\tau \approx \hbar/\Gamma_r$, where Γ_r is the half-width of the resonance. Electrons occupying a higher lying resonance state see an effectively smaller barrier to tunnel through, and thus the "time" to tunnel out is correspondingly shorter, giving a broader resonance. Likewise, if the barriers became thicker, the tunneling probability for a single barrier would decrease exponentially, and the resonance width would narrow considerably. The trend of decreasing resonance width with increasing well width can be understood in terms of the decrease in bound state energy as well as the transit time across the well (over which the electron makes several reflections before tunneling out). The latter increases as the well width increases, first because of the increased traversal distance, and second because of the reduced group velocity for a given resonance state, which lowers in energy as the width increases.

Asymmetric barriers

The band diagram of an asymmetric double barrier is shown in Fig. 3.10. Although the analysis that follows is for a general structure, the barriers shown in Fig. 3.10 could represent an approximation to the symmetric double barrier case with a bias applied between the left and right regions, as was assumed by Tsu and Esaki in their original analysis of the structure [11]. Now we must consider individual wavevectors for the regions to the left and right of the composite barrier, and the well regions labeled k, k_2, and k_1, respectively. In addition, the attenuation or decay constants of the two barriers differ, so that the thickness and height of the two barriers may also be different. Nevertheless, the result (3.45) is still valid and will serve as the basic equation to which we seek solutions.

We can now use the more general results derived for the asymmetric single barrier case in Eqs. (3.27) and (3.28) for the matrix elements of the left and right barriers, which are now connected by the propagation matrix of the well region. Again introducing polar coordinates,

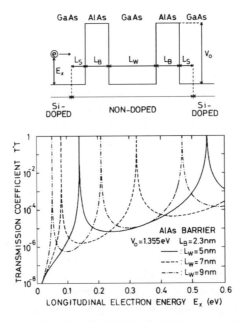

Fig. 3.9. Schematic illustration of an AlAs/GaAs/AlAs double barrier structure and the transmission coefficient calculated as a functions of the electron energy E_x. The full widths at half-maximum of the lowest resonances are 180, 54, and 22 μeV going from the narrowest to widest well [after Tsuchiya and Sakaki, Appl. Phys. Lett. **49**, 88 (1986), by permission].

the matrix elements of interest for (3.45) are written as

$$m_{L11} = \sqrt{\frac{1}{4}\left(1 + \frac{k_1}{k}\right)^2 \cosh^2(2\gamma a_L) + \frac{1}{4}\left(\frac{kk_1 - \gamma^2}{k\gamma}\right)^2 \sinh^2(2\gamma a_L)}, \quad (3.53)$$

$$m_{L12} = \sqrt{\frac{1}{4}\left(1 - \frac{k_1}{k}\right)^2 \cosh^2(2\gamma a_L) + \frac{1}{4}\left(\frac{kk_1 + \gamma^2}{k\gamma}\right)^2 \sinh^2(2\gamma a_L)}, \quad (3.54)$$

$$m_{R11} = \sqrt{\frac{1}{4}\left(1 + \frac{k_2}{k_1}\right)^2 \cosh^2(2\gamma_1 a_R) + \frac{1}{4}\left(\frac{k_1 k_2 - \gamma_1^2}{k_1 \gamma_1}\right)^2 \sinh^2(2\gamma_1 a_R)}, \quad (3.55)$$

$$m_{R21} = \sqrt{\frac{1}{4}\left(1 - \frac{k_2}{k_1}\right)^2 \cosh^2(2\gamma_1 a_R) + \frac{1}{4}\left(\frac{k_1 k_2 + \gamma_1^2}{k_1 \gamma_1}\right)^2 \sinh^2(2\gamma_1 a_R)}, \quad (3.56)$$

for the magnitudes, and for the phases

$$\theta_{L11} = -\arctan\left[\frac{kk_1 - \gamma^2}{(k + k_1)\gamma}\tanh(2\gamma a_L)\right] + (k + k_1)a_L, \quad (3.57)$$

$$\theta_{L12} = -\arctan\left[\frac{kk_1 + \gamma^2}{(k_1 - k)\gamma}\tanh(2\gamma a_L)\right] + \pi + (k - k_1)a_L, \quad (3.58)$$

$$\theta_{R11} = -\arctan\left[\frac{k_1 k_2 - \gamma_1^2}{(k_1 + k_2)\gamma_1}\tanh(2\gamma_1 a_R)\right] - (k_1 + k_2)a_R, \quad (3.59)$$

$$\theta_{R21} = \arctan\left[\frac{k_1 k_2 + \gamma_1^2}{(k_2 - k_1)\gamma_1}\tanh(2\gamma_1 a_R)\right] + \pi + (k_1 - k_2)a_R. \quad (3.60)$$

These results for the magnitudes and phases of the individual matrix elements can now be substituted into Eq. (3.45) to yield the composite transmission matrix element squared

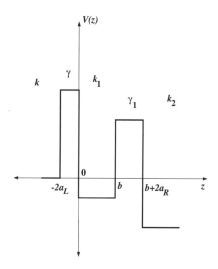

Fig. 3.10. The potential for a general double barrier.

as before

$$|M_{T11}|^2 = (m_{L11}m_{R11} - m_{L12}m_{R21})^2$$
$$+ 4m_{L11}m_{R11}m_{L12}m_{R21} \cos^2\left(k_1 b + \frac{\theta_{L12} + \theta_{R21} - \theta_{L11} - \theta_{R11}}{2}\right).$$

(3.61)

In contrast to the symmetric result, the first term on the right side is greater than unity unless the two barriers are equal. The condition for resonance still corresponds to energies in which the argument of the cosine function is $\pi/2$. However, now the transmission is no longer unity on resonance due to the mismatch of the effective "impedances" of the two barriers, which always gives some reflection.

The transmission coefficient of the composite structure is again given by (3.29) with (3.61). Equation 3.61 may be factored as

$$|M_{T11}|^2 = m_{L11}^2 m_{R11}^2 \left[\left(1 - \frac{m_{L12}m_{R21}}{m_{L11}m_{R11}}\right)^2 + 4\frac{m_{L12}m_{R21}}{m_{L11}m_{R11}} \cos^2(\Phi)\right],$$

(3.62)

where Φ represents the sum of the various phase factors. We can then write the matrix elements appearing in (3.62) in terms of the reflection and transmission coefficients of the single barriers on the right and left using (3.29) as

$$T(E) = \frac{k_2}{k}\frac{1}{|M_{T11}|^2} = \frac{T_L T_R}{(1 - \sqrt{R_L R_R})^2 + 4\sqrt{R_L R_R}\cos^2(\Phi)},$$

(3.63)

where T_L, T_R, R_L, and R_R are the transmission and reflection coefficients of the left and right barriers.

At resonance, $\Phi = (2n + 1)\pi/2$, and the right term of the denominator vanishes. If the transmission coefficients, T_L and T_R are small (as they generally are), the reflection coefficients in the denominator may be expanded as

$$(1 - \sqrt{R_L R_R}) \approx \left(1 - \left(1 - \frac{T_L}{2}\right)\left(1 - \frac{T_R}{2}\right)\right) = \left(\frac{T_L + T_R}{2}\right).$$

(3.64)

The total transmission coefficient is thus

$$T_{res} = \frac{4T_L T_R}{(T_L + T_R)^2}.$$

(3.65)

Although we assumed that $T \ll 1$ in deriving (3.65), it still approaches the correct asymptotic value of $T_{res} = 1$ when $T_L = T_R$. If the transmission coefficients for the left and right barriers are significantly different, the result is that

$$T_{res} \approx \frac{4T_{\min}}{T_{\max}}, \qquad T_{\min} \ll T_{\max},$$

(3.66)

where $T_{\min}$ and $T_{\max}$ refer to the minimum and maximum of the two transmission coefficients T_L and T_R. Thus the transmission on resonance is given by the ratio of the minimum of the two individual barrier transmissions to the maximum of these two. Since the peak current is proportional to the peak transmission coefficient as we will see below in Section 3.3, it is desirable to have the transmission of the two barriers equal under bias conditions that cause the resonance to appear.

The minimum transmission off resonance occurs when the cosine function in Eq. (3.61) is unity. The right side of the denominator now dominates since it is on the order of unity, compared for small transmission factors, $T_{L,R} \ll 1$. Thus the off-resonant current is given by

$$T_{off} = \frac{T_L T_R}{(1 - \sqrt{R_L R_R})^2 + 4\sqrt{R_L R_R}} \approx \frac{T_L T_R}{4}. \tag{3.67}$$

This result indicates that for energies off resonance, the well does not play an essential role; that is, the double barrier structure behaves as two independent barriers. Only at those energies where the wavefunction may bounce back and forth in the well region coherently is the amplitude able to build up such that transmission is enhanced. In the off-resonance case, the wavefunction decays exponentially in the classically forbidden barrier regions from the incident region to the transmitted region, which is consistent with the product of transmission coefficients given by (3.67). At one of the resonance energies corresponding to one of the quasi-bound states of the well, the particle is strongly localized in the well region, since the bound state solutions are themselves localized in the well. This case corresponds to solutions in the barrier that exponentially grow from the left and right regions, which allows the transmission to be large.

3.2.3 Tunneling time

A seemingly simple question we might ask is what the time associated with the tunneling process is. As it turns out, this deceptively simple question does not lend itself to easy answers. In fact, despite a great deal of published literature on the topic, no general consensus has yet emerged. Besides the academic interest in a question that goes to the heart of the interpretation of quantum mechanics, the issue of tunneling time also relates to whether tunneling in RTDs and other planar barrier structures is a coherent or incoherent process, and ultimately relates to the maximum speed of operation of such structures. It is far beyond the scope of the present book to attempt a cohesive presentation of this subject. The interested reader is referred to the recent reviews connected with nanostructure transport by Jonson [30] and Jauho [31].

One measure of the time involved in tunneling through a resonant structure like the RTD is the decay time associated with an electron initially localized in the well. Close to resonance, we can expand the cosine squared term appearing in the denominators of (3.50) and (3.63) around one of the resonance energies, E_n. The zero- and first-order terms vanish, and so to lowest order (assuming that the other factors are relatively slow functions of energy) the transmission coefficient (3.50) for the symmetric double barrier structure has a *Lorentzian* form close to a resonant energy

$$T(E) \approx \frac{\Gamma_n^2/4}{\Gamma_n^2/4 + (E - E_n)^2}, \tag{3.68}$$

where Γ_n, the full width at half maximum of the resonance associated with E_n, is given from Eq. (3.50) as

$$\Gamma_n = \left(\frac{2\hbar^2 E_n T_1^2}{m^* b^2 R_1} \right)^{1/2}. \tag{3.69}$$

Without writing out the terms explicitly, the asymmetric double barrier transmission may be similarly expanded around resonance as

$$T(E) \approx \frac{\Gamma_n^L \Gamma_n^R}{\Gamma_n^2/4 + (E - E_n)^2} = T_{res} \frac{\Gamma_n^2/4}{\Gamma_n^2/4 + (E - E_n)^2}, \quad (3.70)$$

where $\Gamma_n = \Gamma_n^L, +\Gamma_n^R$ where Γ_n^L and Γ_n^R are the partial "widths" associated with the left barrier and the right barrier, and T_{res} is the on-resonance transmission (3.65) [32]. The fit of the Lorentzian lineshape to the full transmission coefficient is actually accurate over several orders of magnitude [33]. The Lorentzian nature is certainly evident in the numerical calculations of $T(E)$ shown in Fig. 3.9.

Equations (3.68) and (3.70) have the form of a *Breit–Wigner formula*, which first appeared in relation to the decay of resonant states in nuclear problems [34], [35]. For open systems such as those encountered in scattering problems, solutions of the Schrödinger equation may be constructed of outgoing waves from a localized scattering center as discussed in various quantum textbooks (see, e.g., [36]). A discussion in connection with double barrier RTD structures is given in a review by García-Calderón [37]. Writing the solution in terms of outgoing states only is incomplete with regard to the usual prescriptions of quantum mechanics because the Hamiltonian is not Hermitian. The energy eigenvalues are therefore complex with $\overline{E}_n = E_n - i\Gamma_n/2$. The transmission coefficient may then be shown to be given by the form of Eqs. (3.68) and (3.70). The real part of the complex energy corresponds to the position of the quasi-bound states of the well given by the resonance condition (3.52), while the imaginary part corresponds to the decay of the quantum-well probability density $|\varphi(z, t)|^2 \sim e^{-\Gamma_n t}$. Thus, we can associate the inverse width of the resonance peak with the decay time out of the well, which is qualitatively associated with the intrinsic delay associated with the tunneling process. The sharper the resonance, the longer the decay time from the well.

Another definition more closely coupled with the concept of the traversal time across the tunneling region is that associated with the time delay of a wavepacket [38], referred to as the *phase time*. If, for example, a Gaussian wavepacket is constructed initially localized on the left side of a barrier, the transmitted wavepacket picks up an additional phase associated with the complex transmission amplitude

$$\mathcal{T} = M_{T11}^{-1} = m_{T11}^{-1} e^{i\alpha z},$$

$$T = \frac{v_1}{v} \mathcal{T}\mathcal{T}^*, \quad (3.71)$$

where $\mathcal{T}$ is the transmission amplitude, M_{T11} is the composite transfer matrix coefficient (3.45), and α is the phase of $\mathcal{T}$. If the wavepacket is a peaked function in momentum space about a wavevector k_i, the expectation value of the position after a time t of the packet is given by [31]

$$\langle x(t) \rangle - \langle x(0) \rangle \approx \frac{\hbar k_i t}{m^*} - \left. \frac{d\alpha(k)}{dk} \right|_{k=k_i}, \quad (3.72)$$

where $\langle x(0) \rangle$ is the initial average position of the particle, and use has been made of the peaked nature of the wavepacket to expand the asymptotic solution. The first term is just the semiclassical delay associated with the group velocity, whereas the second term represents a delay associated with multiple reflections in the tunneling region. For off-resonant transport, this second term is small and so there is little delay compared to the classical trajectory due to tunneling. On resonance, the wavepacket may bounce back and forth numerous times, giving rise to a sizable delay. If, for example, we define the traversal distance in our symmetric double barrier problem as the length $L = 2a + b$, the phase time in (3.72) becomes (setting $t = \tau_\phi$)

$$\tau_\phi(E_i) = \left(\frac{L}{v} + \hbar \frac{d\alpha}{dE} \right) \Bigg|_{E=E_i}, \qquad (3.73)$$

where $v = \hbar k / m^*$ is the group velocity of the packet assuming parabolic bands. In reality, if we consider a system like the double barrier system where the resonances are very narrow in energy, then the expansion leading to the form (3.73) is no longer valid since $T(E)$ is not slowly varying compared to the energy variation of the wavepacket. However, Hauge *et al.* have shown that the more general form reduces to (3.73) on resonance [39]. Therefore, close to resonance where the Lorentzian form holds, it is apparent comparing Eqs. (3.71) and (3.68) that the complex transmission amplitude may be written in the form

$$\mathcal{T} = \frac{\Gamma_n/2}{(E - E_n) + i\Gamma_n/2}; \qquad \alpha = \tan^{-1} \frac{-\Gamma_n/2}{(E - E_n)}. \qquad (3.74)$$

Differentiating the phase above, the phase time on resonance $E = E_n$ is simply

$$\tau_\phi(E_n) = \left(\frac{L}{v(E)} + \frac{\hbar \Gamma_n/2}{(E - E_n)^2 + \Gamma_n^2/4} \right) \Bigg|_{E=E_n} = \left(\frac{L}{v_n} + \frac{2\hbar}{\Gamma_n} \right), \qquad (3.75)$$

where the second term is the additional delay due to resonant tunneling. Not surprisingly, the delay is no more than the inverse of the resonance width as before. One can imagine the time-dependent propagation through the barrier as a trapping process where the particle is captured by the well before leaking out again. The above definition of phase time is found to be consistent with numerical studies of wavepacket motion through barriers [40], [41]. For the model barrier structures shown in Fig. 3.9, the phase times associated with the full widths at half maximum for the lowest resonance state (listed in the figure caption) vary from 7.3 to 60 ps. These times are in fact large compared to the characteristic scattering times in typical semiconductor materials, particularly at room temperature. As a consequence, the actual tunneling process may be incoherent rather than coherent, as discussed in Section 3.3.2.

The whole concept of studying the dynamic behavior of wavepackets in order to extract information that can be related to experiment in a meaningful way is of course open to criticism. Alternate schemes have been proposed for calculating the tunneling time involving quantum clocks, such as a time-modulated barrier scheme which results in the so-called *Büttiker-Landauer time* [42], and the Larmor time based on the precession of the particle spin in a strong magnetic field as a clock for the traversal time [43]. Other schemes involving quantum trajectories other than wavepackets have also been studied. Regardless of these differing interpretations of the quantum tunneling time, we may at least qualitatively associate a tunneling delay time associated with the inverse resonance width, which is apparent from the frequency response of RTDs fabricated with varying barrier widths [44].

3.3 Current in resonant tunneling diodes

In the previous section we analyzed several one-dimensional potentials in terms of their transmission and reflection properties. Although we considered only the single and double barrier problems, it is apparent from the cascading process used to determine the composite transmission matrix that the same process applies for more complicated structures. In real RTD structures, the actual potential contains contributions from the ionized dopants, free carriers, and the applied potential itself, in addition to the band offsets of the heterojunctions. We previously considered such effects in Chapter 2 in connection with quantum confinement effects. Using the results of the previous section, we may cascade a large number of piecewise-continuous potentials together to approximate the actual potential in a real RTD structure to an arbitrary accuracy as the basis for a numerical solution of the transmission and reflection coefficients. Given that we can in general calculate the transmission and reflection properties through such a structure, we now want to turn to the problem of using these results to calculate the current flow through a three-dimensional device such as an RTD.

3.3.1 Coherent tunneling

When we looked at the tunneling process in Section 3.2, we considered a single one-dimensional traveling wave that propagates through the barrier region and out the other side. As such, we viewed tunneling as an elastic process involving no loss of energy of the particle. In a real structure such as an RTD, the ideal problem consists of a plane wave incident on a barrier potential that is semiinfinite in extent in the two transverse directions and that varies only in the third direction. For almost all practical planar barrier devices, this variation in potential is in the growth direction due to the bandgap discontinuities of the heterojunction interfaces, and the space charge due to doping and the applied bias. The plane wave has some component of its wavevector (and hence momentum) in the transverse direction parallel to the barrier. Along with our assumption that tunneling is an energy conserving process, we will further assume that the transverse momentum is conserved, that is, that it remains the same before and after tunneling. This latter assumption is violated in real structures if random inhomogeneities exist in the lateral direction, such as interface roughness and ionized impurities. The main effect is to broaden the effective transmission resonance, reducing the PVR in measured structures compared to the ideal model.

To connect the quantum mechanical fluxes to charge current, we need to introduce the statistical mechanical distribution function to tell us the occupancy of current-carrying states incident and transmitted on the barriers. Exactly what distribution function to use is perhaps one of the central issues of describing nonstationary transport in a phase-coherent system such as the nanostructures discussed in this book. The starting model we will use assumes that we have contacts or reservoirs on the left and right sides of a barrier structure that are essentially in equilibrium and are described by the single-particle distribution function such as the Fermi-Dirac distribution characterized by a Fermi energy. However, since current is flowing, the distribution function cannot truly be characterized by the equilibrium distribution function. We will consider the consequences of this fact later.

The general problem is shown in Fig. 3.11 for a generic tunneling barrier. The applied bias separates the Fermi energies on the left and right by an amount eV. The Hamiltonian on either side of the barrier is assumed separable into perpendicular (z-direction) and transverse components. If we choose the zero-reference of the potential energy in the system to be the conduction band minimum on the left, $E_{c,l} = 0$, the energy of a particle before and after tunneling may be written as

$$E = E_z + E_t = \frac{\hbar k_{z,l}^2}{2m^*} + \frac{\hbar k_{t,l}^2}{2m^*} \qquad (3.76)$$

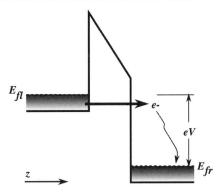

Fig. 3.11. Band diagram for a tunnel barrier under bias, illustrating charge flow.

on the left side, and

$$E = \frac{\hbar k_{z,r}^2}{2m^*} + \frac{\hbar k_{t,r}^2}{2m^*} + E_{c,r} \tag{3.77}$$

on the right side, where $E_{c,r}$ is the conduction band minimum on the right side and k_z and k_t are the perpendicular and transverse components of the wavevector relative to the barrier. A single, parabolic, isotropic conduction band minimum has been assumed for simplicity. Since the transverse momentum is assumed to be conserved during the tunneling process, then $k_{t,l} = k_{t,r}$, and the transverse energy $E_{t,l} = E_{t,r}$ is the same on both sides for the tunneling electron. Therefore, the z-component of the energy is

$$E_z = \frac{\hbar k_{z,l}^2}{2m^*} = \frac{\hbar k_{z,r}^2}{2m^*} + E_{c,r} \tag{3.78}$$

on the left and right sides of the barrier.

As a further approximation, necessary to introduce irreversibility into the formalism, we assume that the contacts are perfectly absorbing. This means that when a particle injected from one side reaches the contact region of the other side, its phase coherence and excess energy are lost through inelastic collisions with the Fermi sea of electrons in the contact. Thus we assume that an electron injected from one contact at a certain energy E has a certain probability of being transmitted through the boundary determined by $T(E)$, exits the barrier with the same energy and transverse momentum, and finally is absorbed in the opposite contact, where it loses the energy and memory of its previous state. Current flow in this picture is essentially the net difference between the number of particles per unit time transmitted to the right and collected versus those transmitted to the left. This view of tunneling is referred to as *coherent* since the particles maintain their phase coherence across the whole structure before losing energy in the contacts.

To proceed with this picture, consider the current density perpendicular to the barrier in the z-direction for a given energy E with corresponding z-component E_z. The incident current density on the barrier from the *left* due to particles in a infinitesimal volume of momentum space $d\mathbf{k}_l$ around $\mathbf{k}_l$ may be written

$$j_i = -e D(\mathbf{k}_l) f_l(\mathbf{k}_l) v_z(\mathbf{k}_l) d\mathbf{k}_l, \qquad D(\mathbf{k}) = \frac{2}{(2\pi)^3}, \tag{3.79}$$

where f_l is the distribution function on the left side of the barrier, $D(\mathbf{k})$ is the density of states in **k**-space, and the velocity perpendicular to the barrier from the left is

$$v_z(\mathbf{k}_l) = \frac{1}{\hbar} \frac{\partial E(\mathbf{k}_l)}{\partial k_{z,l}} = \frac{\hbar k_{z,l}}{m^*} \tag{3.80}$$

using the parabolic relation (3.78). Here we neglect the possibility that the energy states on the left or right side of the barriers may be quantized due to, for example, band bending, and we therefore treat the states as three-dimensional. The transmitted current density from the left to right is simply Eq. (3.79) weighted by the transmission coefficient

$$j_l = \frac{-2e\hbar}{(2\pi)^3 m^*} T(k_{z,l}) f_l(\mathbf{k}_t, k_{z,l}) k_{z,l} \, dk_{z,l} \, d\mathbf{k}_t, \tag{3.81}$$

where $T(k_z)$ is the transmission coefficient which for the ideal case is only a function of the perpendicular momentum and energy. Similarly, the transmitted current from right to left may be written for the same energy E and E_z

$$j_r = \frac{-2e\hbar}{(2\pi)^3 m^*} T(k_{z,r}) f_r(\mathbf{k}_t, k_{z,r}) k_{z,r} \, dk_{z,r} \, d\mathbf{k}_t. \tag{3.82}$$

At a given perpendicular energy E_z, the transmission coefficient is symmetric so that $T(E_{z,l}) = T(E_{z,r})$. Further, $k_{z,l} \, dk_{z,l} = k_{z,r} \, dk_{z,r} = m^* dE_z/\hbar^2$ if we differentiate both sides of (3.78). Therefore, the net current in the direction of the voltage drop is the difference between the left and right currents integrated over all **k**, or

$$J_T = \frac{2e}{(2\pi)^3 \hbar} \int_0^\infty dE_z \int_0^\infty dk_t k_t \int_0^{2\pi} d\theta T(E_z)[f_l(E_z, k_t) - f_r(E_z, k_t)], \tag{3.83}$$

where the integration over E_z is from zero to infinity because tunneling from right to left below $E_z = 0$ is forbidden.

At this point, no further reduction to Eq. (3.83) can be made unless we make assumptions concerning the nature of the distribution functions on the left and right sides. The lowest-order approximation is to assume that these distributions are given by the equilibrium Fermi-Dirac functions determined by the bulk Fermi levels on the respective sides of the barrier,

$$f_{l,r}(E_z, E_t) = \frac{1}{1 + e^{\left(E_z + E_t - E_F^{l,r}\right)/k_B T_L}}, \tag{3.84}$$

where T_L is the lattice temperature and $E_F^{l,r}$ is the Fermi energy on the left and right side, respectively. The difference between the two is just the applied bias, $E_f^l = E_f^r + eV$. Since the Fermi function is isotropic, the angular integration gives 2π. Likewise, the integration over the transverse wavevector may be converted to an integral over energy. Assuming parabolic bands, Eq. (3.83) becomes

$$J_T = \frac{4\pi e m^*}{(2\pi)^3 \hbar^3} \int_0^\infty dE_z T(E_z) \int_0^\infty dE_t [f_l(E_z, E_t) - f_r(E_z, E_t)]. \tag{3.85}$$

For the Fermi function (3.84), the integration over energy is easily evaluated to give

$$J_T = \frac{e m^* k_B T_L}{2\pi^2 \hbar^3} \int_0^\infty dE_z T(E_z) \ln\left(\frac{1 + e^{(E_F^l - E_z)/k_B T_L}}{1 + e^{(E_F^l - eV - E_z)/k_B T_L}}\right), \tag{3.86}$$

sometimes referred to as the *Tsu-Esaki formula*, where the particular form was popularized [11] in connection to resonant tunneling diodes. (Similar such equations appear much earlier in single-particle tunneling; see for example [10].) The logarithmic term is sometimes called the *supply function* [45], since it more or less determines the relative weight of available carriers at a given perpendicular energy. At low temperature, the supply function becomes step-like, and Eq. (3.86) becomes simply

$$J_T = \frac{em^*}{2\pi^2\hbar^3}\left[\int_0^{E_F} dE_z T(E_z)(E_F - E_z) - \int_0^{E_F - eV} dE_z T(E_z)(E_F - eV - E_z)\right]. \quad (3.87)$$

If we now consider current through a resonant structure such as an RTD, the current density is dominated by the resonant portion of the transmission coefficient. For example, if the transmission coefficient is assumed to be very peaked around $E_z = E_n$ using the Lorentzian form (3.70), we can approximate it as a delta function so that at low temperature, Eq. (3.87) may be integrated to give

$$J_T = \frac{em^* T_{res}\Gamma_n}{4\pi\hbar^3}(E_F - E_n) \qquad 0 < E_n < E_F, \quad (3.88)$$

where the asymptotic approximation for the delta function has been employed:

$$\delta(E_z - E_n) = \frac{1}{\pi}\lim_{\Gamma_n \to 0}\frac{\Gamma_n/2}{\Gamma_n^2/4 + (E_z - E_n)^2}. \quad (3.89)$$

The voltage dependence enters through E_n, the bound state energy. Figure 3.10 showed the band diagram of a general asymmetric double barrier. Assume that the voltage drop is equally divided between the two barriers. Thus the well is lowered in potential energy by an amount $eV/2$ with respect to the emitter (the left side), and is higher in energy than the collector by the same amount. E_n may therefore be replaced in (3.88) by $E_o - eV/2$, where E_o is the quasi-bound state energy relative to the well bottom. This gives a sudden turn-on of current when $eV = 2(E_o - E_F)$, and cuts off when $eV = 2E_o$ giving rise to NDR. The peak occurs when $E_n = 0$, giving a peak current density

$$J_P = \frac{em^* T_{res}\Gamma_n E_F}{4\pi\hbar^3}. \quad (3.90)$$

As can be seen, the peak current depends physically on the Fermi energy in the emitter, and hence the doping there, as well as on the product of the peak transmission probability and resonance width. Since both the resonance width and the resonant transmission amplitude increase as the barrier thickness decreases, thin barriers are essential for high peak current densities.

As an example of the numerically calculated current in an RTD, consider the simple bias model above. The general transmission coefficient is determined by Eq. (3.65) using the transfer matrix element $|M_{T11}|^2$ in (3.61). In order to calculate the current-voltage characteristics using Eq. (3.86), the transmission coefficient versus energy is tabulated for each bias point (since the barrier shape continuously changes as a function of bias), and the integration is performed numerically for the current. Figure 3.12 shows the result of such a calculation at low temperature for the structure of Sollner *et al.* shown in Fig. 3.2. The I-V characteristics indeed show a pronounced NDR region. However, in comparison to the measured results for the same structure shown in Fig. 3.3, the calculated peak current and the peak-to-valley ratio are much larger than experiment, while the voltage for the occurrence for NDR does not exactly correspond. Part of the latter problem may be explained by including a series resistance, R_s, which reduces the voltage across the diode by an amount $V_d = V - IR_s$ [46]. However, this does not rectify the other discrepancies, particularly the magnitude of the valley current which is much larger than ideal model.

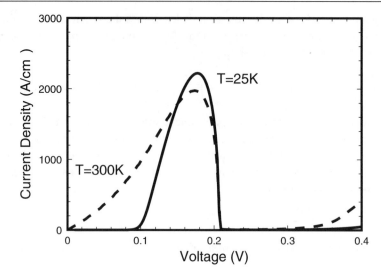

Fig. 3.12. Calculated current-voltage characteristics for the RTD structure of Fig. 3.2 at low temperature.

The extra valley current represents contributions due to scattering, both elastic and inelastic. Both types of scattering allow a relaxation of the parallel-momentum conservation rule and thus increase the amount of current which may flow off-resonance. We may nominally associate elastic scattering in RTDs with interface roughness at the heterojunction interfaces, unintentional doping in the tunneling region, and alloy disorder when barrier materials like $Al_xGa_{1-x}As$ are employed. The first two depend strongly on the quality of epitaxial material growth, which helps explain the dramatic improvements in PVRs in the GaAs/AlGaAs system over time, similar to the improvements in the 2DEG mobility discussed in Chapter 2. Calculations that incorporate roughness effects have been reported by [47] and [48], for example. Another factor is the limitation of the single-band model used here. As bias is increased, carriers are injected at higher and higher energies into the AlGaAs barriers and GaAs well. Ultimately, high conduction band minima are accessible to these carriers, particularly the X-valley minimum in AlGaAs, which provide additional channels for current to flow [49] and necessitates a multiband calculation of the tunnel current [50].

Inelastic scattering via phonons and collective excitations not only break the assumption of transverse momentum conservation but also lead to loss of phase coherence, as mentioned in the introductory chapter. As a consequence, if such interactions are strong, the tunneling process is no longer characterized as coherent, since the phase relationship leading to the buildup of the resonant amplitude is only partially preserved. We will defer a rigorous discussion of this phenomenon until Chapter 7, when inelastic effects are included formally into the transport model. Here we basically consider inelastic scattering from a phenomenological point of view in regards to planar barrier tunneling, and we discuss some of the consequences of this interaction on the tunneling process in RTDs.

3.3.2 Incoherent or sequential tunneling

Inelastic or phonon-assisted tunneling has long been recognized as an important process in tunneling, as discussed in the review of pre-1970 tunneling by Duke [10]. Luryi [51]

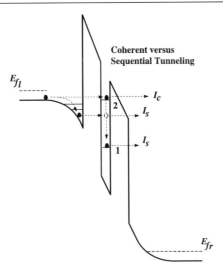

Fig. 3.13. Illustration of incoherent (sequential) tunneling in an RTD structure. I_c corresponds to the coherent component of the current, while I_s represent incoherent components.

pointed out in connection with double barrier RTD structures that NDR could be explained with a *sequential* tunneling process in which phase coherence is lost. In this model, shown schematically in Fig. 3.13, carriers tunnel through the first barrier into the quasi-2DEG residing in the well and subsequently lose their phase memory through a scattering process there. The phase-randomized carriers tunnel out of the second barrier through a second un-correlated tunneling process. Luryi argued that in order to explain NDR in an RTD structure, it is sufficient to consider the tunneling of a 3D electron from the emitter into a 2DEG under the assumption of transverse momentum conservation. Assuming that the excess kinetic energy in the quantum confined well is associated only with the transverse motion, we may equate the total energy of a tunneling electron from the emitter into the well as

$$E = E_c + \frac{\hbar^2 k_z^2}{2m^*} + \frac{\hbar^2 k_{t,e}^2}{2m^*} = E_n + \frac{\hbar^2 k_{t,w}^2}{2m^*}, \qquad (3.91)$$

where $k_{t,e}$ and $k_{t,w}$ denote the transverse wavevector in the emitter and well, respectively. If tunneling itself is energy-conserving, and the transverse moment is also conserved, the condition for tunneling to occur becomes

$$E_c + \frac{\hbar^2 k_z^2}{2m^*} = E_n. \qquad (3.92)$$

Since the perpendicular wavevector k_z is real, no electrons can tunnel and conserve trans-verse momentum when $E_c > E_n$, that is, when the conduction band edge in the emitter rises above the bound state energy in the well. At low temperature, no electrons tunnel until the Fermi energy crosses the bound state energy. Then the supply of electrons available for tunneling rises as $E_f - \dot{E}_n$ until the criterion (3.92) is violated and the current drops abruptly, giving rise to NDR. The qualitative shape of the I-V is the same as that predicted in the coherent model of the previous section, although at first glance the current predicted by the two should be quite different. However, as was subsequently shown [52], [53], [54], the peak current density of the two models is in fact the same and cannot be used to distinguish between coherent and incoherent tunneling.

To illustrate this, we use the simple model given by Stone and Lee [55] to include inelastic scattering in the resonant tunneling amplitude. There it is assumed within the context of the Breit-Wigner formula that part of the incident flux is absorbed by inelastic processes that can be described by adding an additional imaginary part [36], $i\Gamma_i/2$, to the complex energy eigenvalue discussed in Section 3.2.3. This imaginary part may be qualitatively associated with the scattering time, $\tau_s = \hbar/\Gamma_i$, due to inelastic processes. The transmission coefficient in the general asymmetric barrier case then becomes

$$T_c(E) = T_{res} \frac{\Gamma_n^2/4}{\Gamma_T^2/4 + (E - E_n)^2} = T_{res} \left(\frac{\Gamma_n}{\Gamma_T}\right)^2 \frac{\Gamma_T^2/4}{\Gamma_T^2/4 + (E - E_n)^2}, \tag{3.93}$$

where $\Gamma_T = \Gamma_n + \Gamma_i$. Here we use T_c to denote the coherent transmission since the damping term Γ_i represents just the loss of flux from that transmitted coherently. Because of this flux loss, probability current is no longer conserved, i.e., $T_c + R_c < 1$. However, since charge is not created or destroyed, this lost flux must appear somewhere else as transmitted flux through another channel in order to conserve the total flux. This extra flux may be thought as that lost due to scattering in the well to lower energy (as envisioned through the sequential model) and subsequently transmitted in a separate process at a different energy. This inelastic fraction is therefore that lost from the coherent part

$$T_i + R_i = 1 - T_c - R_c = T_{res} \frac{\Gamma_i \Gamma_n/2}{\Gamma_T^2/4 + (E - E_n)^2}, \tag{3.94}$$

where T_i and R_i are the inelastic forward and backward scattering probabilities, and the expression for the coherent reflection coefficient has been similarly expanded about the resonance energy. If these are assumed to be the same, the total transmission coefficient including both elastic and inelastic scattering is therefore

$$T(E) = T_i + T_c = T_{res} \left(\frac{\Gamma_n}{\Gamma_T}\right) \frac{\Gamma_T^2/4}{\Gamma_T^2/4 + (E - E_n)^2}. \tag{3.95}$$

This differs from the form of (3.70) in that the maximum transmission is reduced due to inelastic scattering by the ratio $\Gamma_n/(\Gamma_n + \Gamma_i)$ and the resonance is broadened compared to the purely coherent case. A similar form was obtained by Jonson and Grincwajg [54] using ray-tracing arguments and by Büttiker [56], [32] using the multichannel Landauer-Büttiker formula introduced in Section 3.5.

If we go back to our derivation of the peak current density leading to Eq. (3.90), if the same assumptions are made at low temperature and for peaked transmission coefficients, it is apparent that in the more general case

$$J_P \propto T_{res} \left(\frac{\Gamma_n}{\Gamma_T}\right) \Gamma_T = T_{res} \Gamma_n, \tag{3.96}$$

which is independent of the inelastic scattering time. Therefore, as long as the resonance is not broadened too much by inelastic scattering, the peak currents are expected to be the same in the coherent and incoherent cases.

As mentioned earlier, the characteristic time for tunneling associated with the coherent width Γ_n may be quite long compared with that due to inelastic scattering. Thus, pure coherent transport in RTDs probably does not occur except in very thin barrier structures such as those characteristic of state of the art devices. The role of such inelastic scattering on the frequency response of RTDs is still unclear. For more discussion in this regard, the reader is referred to the review by Liu and Sollner [57].

3.3.3 Space charge effects and self-consistent solutions

The ideal model we developed so far neglected the electrostatic potential arising from ionized donors and acceptors as well as the free carrier charge distribution. In reality, space charge layers form in the device, leading to additional potential drops inside the device other than that across the barriers themselves. Further, free carriers may be stored in the well, particularly on resonance, which modifies the potential distribution across the RTD structure. Evidence for charge storage is observed in magnetotunneling experiments in which a magnetic field is applied perpendicular to the plane of the tunnel barrier [58], and Shubnikov–de Haas (SdH)–like oscillations in the conductance are observed related to magnetic quantization of the quasi-2D states in the well [59]. Space charge build-up has also been probed optically [60] where luminescent transitions between the quasi-bound electron and hole states in the well of the RTD are observable. The storage of free carriers in the well as the device is biased through resonance has been suggested as a source of *intrinsic bistability*, giving rise to the experimentally observed hysteresis in the I-V characteristics of RTDs in the NDR regime [61]. Practically, however, there is great difficulty in separating bistability due to physical charge storage in the quantum well from external measurement circuit effects when probing the unstable NDR regime of an RTD [62], [62a] unless asymmetric RTD structures are prepared which accentuate charge storage effects [63].

In Chapter 2 we discussed the necessity for self-consistent solutions in which the coupled Schrödinger-Poisson equations were solved with appropriate approximations for the many-body potential for modulation-doped structures. For an RTD structure, similar self-consistent solution methods must be employed to account for the space charge layer that forms in the diode under bias, and to account for the pile-up of free charge in the well and in the accumulation layer adjacent to the double barrier structure on the emitter side. However, in contrast to the quantum well case, an RTD represents an *open* structure in which the wavefunction does not vanish on the boundaries (at least when we conceptually separate the microscopic system from the macroscopic external environment). In the open system, particles move in and out of the boundary of the microscopic system into the macroscopic external circuit, which we have modeled as ideal absorbing and emitting contacts. Therefore, care must be taken in specifying the boundary conditions in such a system and the statistical weighting associated with normalization of the single-particle wavefunctions in order to conserve charge, and to give the appropriate charge density (when summed over all states) in the contact regions. In this sense, for an open system one cannot formulate the problem purely in a reversible quantum mechanical framework. Rather, nonequilibrium statistical mechanics also plays a central role in the system description.

In order to move beyond the simple rectangular barrier approximations used above, we can still assume that flat band conditions are reached at some point far away from the tunneling region on the right and left sides. These somewhat arbitrary points serve as the boundaries of the system across which the interior solution is matched to asymptotic plane-wave solutions in the contact regions. In the interior, we have not only the potential due to the band discontinuities, $E_c(z)$, but also the potential due to ionized donors and free carriers,

$$\frac{d}{dz}\left(\epsilon(z)\frac{d\phi}{dz}\right) = -e\big[N_D^+(z) - N_A^-(z) - n(z)\big], \qquad (3.97)$$

where $\phi(z)$ is the electrostatic potential, $N_D^+(z)$ and $N_A^-(z)$ are the ionized donor and acceptor concentrations respectively, and $n(z)$ is the free carrier concentration. In the n-type asymptotic contact regions, charge neutrality is assumed so that $n = N_D - N_A$ independent of position (assuming complete ionization).

A semi-classical approximation for the free carrier density is to assume the Thomas-Fermi screening model in which the density is determined by the position of the Fermi energy relative to the band edge

$$n(z) = N_c \mathcal{F}_{1/2} \{ [E_f - (q\phi + E_c)] \}, \tag{3.98}$$

where $\mathcal{F}_{1/2}$ is the Fermi-Dirac integral, and N_c is the effective density of states in the conduction band. When substituted into Eq. (3.97), the resulting nonlinear Poisson equation must be solved numerically. The Thomas-Fermi model is strictly valid only in equilibrium, but it is typically extended to nonequilibrium cases by defining quasi-Fermi energies governing the local density. In an RTD structure, this is usually accomplished by partitioning the device into regions governed either by the Fermi energy of the left contact or that of the right contact. Of course, this is arbitrary to a large extent, and so accordingly are the solutions of the problem one obtains.

Within the Hartree approximation (see Chapter 2), we need to combine the one-electron envelope functions with the appropriate distribution function. Due to the flux of particles from the left, the density on the left of the barrier may be written

$$n_l(z) = \frac{2}{(2\pi)^3} \int d\mathbf{k}_t \int_0^\infty dk_z |\varphi_l(k_z)|^2 f_l(k_t, k_z), \tag{3.99}$$

where the prefactor is the density of states in $\mathbf{k}$-space with a factor 2 for spin degeneracy, and where $f_l(k_t, k_z)$ is the one-particle distribution function on the left side. The integration over k_z is only over positive values because the envelope function, $\varphi_l(k_z)$, includes both the incident and reflected amplitudes in the form of Eq. (3.5). In various treatments of self-consistent solutions in RTD structures [64]–[67], the distribution function is assumed to be a Fermi-Dirac function in the contact region characterized by an effective temperature, T_L. Then the easily performed integration over the transverse motion gives

$$n_l(z) = \frac{m^* k_B T_L}{2\pi^2 \hbar^2} \int_0^\infty dk_z |\varphi_l(k_z)|^2 \ln\left(1 + e^{(E_{f,l} - E_z)/k_B T_L}\right) = \int_0^\infty dk_z |\varphi_l(k_z)|^2 f_l(E_z), \tag{3.100}$$

where $E_{f,l}$ is the Fermi energy on the left and $f_l(E_z)$ is the Fermi function integrated over $\mathbf{k}_t$. A similar equation may be written for the density due to carriers flowing from the right, with the appropriate change of $l \rightarrow r$ above. The total density is the sum of the two.

However, the assumption of equilibrium distribution functions in the contacts under nonequilibrium conditions when current is flowing leads to an unphysical pile-up or depletion of charge in the contact region unless an artificial readjustment of charge is made [68]. This fact is apparent if one looks at the asymptotic density on the left side; for example,

$$n_l = \int_0^\infty dk_z \{ [2 - T(E_z)] f_l(E_z) + T(E_z) f_r(E_z) \}, \tag{3.101}$$

where f_r is the averaged distribution function in the right contact defined by (3.100), with $E_{f,r} = E_{f,l} - eV$ with V the applied voltage. The first term in the integral is the incident and reflected density due to carriers injected from the left contact, $1 + R = 2 - T$ (invoking current conservation). The second term is the transmitted density from the right (which properly should be written over negative wavevectors, but assuming f_r is symmetric can be combined above). Here the incident waves from left and right have been normalized to unity. In equilibrium ($V = 0$), the terms involving $T(E_z)$ cancel, and the bulk carrier density is recovered. Under forward bias, the right term vanishes, and the density reduces below its equilibrium value, causing a depletion in the contact. Correspondingly, there is a pile-up of charge in the right contact above the equilibrium value under forward bias given by

$$n_r = \int_0^\infty dk_z \{[2 - T(E_z)] f_r(E_z) + T(E_z) f_l(E_z)\}. \tag{3.102}$$

This evident rearrangement of free charge due to the barrier region, and the inconsistency of the simple use of the equilibrium contact distribution function, is initmately related to the derivation of the *Landauer formula* (see Section 3.4). In the present context of self-consistent solutions in RTDs, Pötz [68] used a drifted Fermi-Dirac distribution for the contacts in order to conserve charge which when integrated over the transverse momentum gives:

$$f_i(E_z) = \frac{m^* k_B T_i}{2\pi^2 \hbar^2} \ln\left(1 + e^{(E_{f,i} - \hbar^2 [k_{z,i} - k_{o,i}]^2 / 2m^*)/k_B T_i}\right), \tag{3.103}$$

where $i = l, r$ refers to the left or right side, T_i is the electron temperature in the corresponding region, and $\hbar k_{o,i}/m^*$ is the drift velocity in region i. The drift wavevectors and/or electron temperatures in the respective contact regions are used as parameters to establish charge neutrality in the contacts.

In order to calculate the current self-consistently, we still need the transmission coefficient through the tunneling region. For quasi-1D problems like planar barrier tunneling, a transfer matrix technique is frequently employed [69], [70]. Here, a one-dimensional grid is defined, and the transfer matrix from one grid point to the next calculated. The simplest approximation is just to assume constant potential in each grid, so that the transfer matrix is just that due to a finite step at every grid point. A better approximation is to assume that the potential varies linearly between grid points, in which the solution in each segment is given by Airy functions [64], and to define the transfer matrices accordingly. The total potential used to solve the transfer matrix problem is the sum of the electrostatic potential and that due to the band offsets in the simple one-band model. Since the electrostatic potential also depends on the solution of the Schrödinger equation through the density (3.100), the total solution must be found iteratively [66] in a similar fashion to that discussed in Chapter 2.

Figure 3.14 shows an example of the calculated band profile of an RTD structure at several bias voltages. The RTD structure has 5-nm spacer layers adjacent to the barrier on either side which give rise to band bending even for zero applied bias. For increasing bias, it is apparent that a substantial fraction of the bias drops across the region outside of the double barrier structure in contrast with the simple model introduced in the previous sections. One consequence is that the voltage corresponding to the peak current is pushed to higher voltages than that predicted by the ideal model, which in some sense resembles a series resistance effect. Another interesting feature is the presence of a potential notch on the emitter side of the double barrier structure. This notch resembles the quantum confining potential on a single-heterojunction system discussed in Chapter 2 and may trap charge there in quasi-bound states (quasi-bound as they still have a finite lifetime) forming an accumulation layer adjacent to the tunnel barrier. In fact, the existence of such a layer is evident in magnetotunneling experiments of the type discussed earlier in this Section [71] where SdH-like oscillations are observed associated with quantum confined electrons there. The contribution to the current due to tunneling out of these states is quantitatively

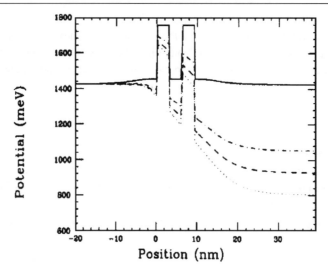

Fig. 3.14. Effective potential for a double barrier structure with 5-nm undoped spacer layers and $N_D = 2 \times 10^{18}\ \text{cm}^{-3}$ for various applied bias. Solid line, 0 V; dot-dashed line, 0.375 V (close to resonance); dashed line, 0.500 V (above resonance); and the dotted line, 0.625 V [after W. Pötz, J. Appl. Phys. **66**, 2458 (1989), by permission].

different than that due to electrons in the continuum above [72] and not accounted for in the heretofore described flux formalism. In fact, in our ballistic paradigm developed so far, there is no possibility of charge becoming trapped there because this requires some sort of inelastic process to relax carriers into the accumulation layer. This paradox really points to the limitations of a description in which dissipation occurs only in the contacts. In reality, the distribution function for electrons cannot be described by a local semi-classical function across the device. This has led to consideration of quantum distribution functions such as the Wigner distribution [73] for RTD modeling [74], [75]. We will come back to issues regarding dissipative nonequilibrium transport in more detail in Chapter 7.

3.4 Landauer formula

In the previous sections of this chapter, we introduced quantum transport through a discussion of tunneling in planar barrier structures. We now extend these concepts to more complicated geometrical structures in which phase coherence is preserved and the description in terms of quantum flux is important. Such systems include quantum point contacts (introduced in Chapters 1 and 2) as well as more complicated quantum waveguide structures in 2 and 3 dimensions. To generalize our formalism for treating current flow through such systems, we go back to an early formalism due to Landauer [76], [77] which has now become standard in the parlance of nanostructure transport.

To begin, we note that in our previous derivation of the tunneling current through planar barrier structures, inconsistencies arise associated with current being injected from ideal equilibrium contacts, and with the space charge that forms as a result of current flow through the structure. In semiconductors, charge fluctuations near the barriers may substantially affect the potential near the barriers. Also, with current flow, the distribution function itself may be significantly perturbed from the equilibrium. The spatial inhomogeneity resulting

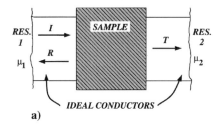

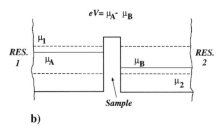

Fig. 3.15. Schematic illustration of the conductance of a one-dimensional sample: a) the ideal de-composition of the structure into the sample where scattering occurs, and ideal 1D leads connecting the sample to reservoirs on either side; b) the redistribution of charge due to the reservoirs resulting in new Fermi energies μ_A and μ_B on the left and right sides, respectively.

from current flow around obstacles was recognized early on by Landauer [76]. He formulated the problem in a somewhat different way by considering a one-dimensional system in which a constant current was forced to flow through a structure containing scatterers, and asking the question of what the resulting potential distribution will be due to the spatially inhomogeneous distribution of scatters. In this context, our planar barrier of the previous section may be considered as a general scattering center. The result in one dimension is the so-called *Landauer formula*, which is derived below for the *single-channel* case and then in Section 3.5 is extended to the general *multi-channel* case [78].

To begin, consider the general barrier problem for a 1D conductor shown in Fig. 3.15. As we discussed in Chapter 2, an ideal 1D conductor is realized by a quantum wire in the extreme quantum limit where only one subband (channel) is occupied. Ideal (i.e., without scattering) conducting leads connect the scattering region to reservoirs on the left and right characterized by quasi-Fermi energies μ_1 and μ_2, respectively, corresponding to the electron densities there. As we discussed in the previous sections, these reservoirs or contacts randomize the phase of the injected and absorbed electrons through inelastic processes such that there is no phase relation between particles. For such an ideal 1D system, the current injected from the left and right may be written as an integral over the flux (in 1D now) as we did in the previous sections:

$$I = \frac{2e}{2\pi}\left[\int_0^\infty dk\, v(k) f_1(k) T(E) - \int_0^\infty dk'\, v(k') f_2(k') T(E') \right], \qquad (3.104)$$

where the constant is the 1D density of states in k-space, $v(k)$ is the velocity, $T(E)$ is the transmission coefficient, and f_1 and f_2 are the reservoir distribution functions characterized by μ_1 and μ_2, respectively. The integrations are only over positive k and k' relative to the direction of the injected charge. If we now assume low temperatures, electrons are injected up to an energy μ_1 into the left lead and injected up to μ_2 into the right one. Converting to integrals over energy, the current becomes

$$I = \frac{2e}{2\pi}\left[\int_0^{\mu_1} dE\left(\frac{dk}{dE}\right)v(k)T(E) - \int_0^{\mu_2} dE\left(\frac{dk'}{dE}\right)v(k')T(E)\right]$$

$$= \frac{2e}{2\pi\hbar}\int_{\mu_2}^{\mu_1} dE\, T(E). \tag{3.105}$$

Note that in 1D, the energy-dependent velocity and density of states factor arising from the change of variables cancel to a simple constant, an important result in our later discussion of quantized conductance in point contacts. If we further assume that the applied voltage is small (i.e., in the linear response regime) so that the energy dependence of $T(E)$ is negligible, the current becomes simply

$$I = \left(\frac{2e}{h}\right)T(\mu_1 - \mu_2). \tag{3.106}$$

We saw in Section 3.3.3 that as a result of transmission and reflection about the barrier with current flowing, there is a reduction in carrier density on the left side of the barrier and pile-up of charge on the right side. Assuming we are still in the linear response regime, we can approximate this charge rearrangement by an average density in the ideal leads on either side of the scattering structure characterized by different Fermi energies μ_A and μ_B as shown in Fig. 3.15. The actual voltage drop V across the scattering structure is then given by

$$eV = \mu_A - \mu_B, \tag{3.107}$$

which is less than the voltage between the reservoirs, $\mu_1 - \mu_2$, the difference representing a *contact* potential drop. We need to find the potential difference (3.107) in terms of the current flowing through the structure given by (3.106). To do this, we can write the 1D density in the ideal lead on the left side as

$$n_a = \frac{2}{2\pi}\int_{-\infty}^{\infty} dk f_a(E) = \frac{2}{2\pi}\int_0^{\infty} dk[\{2 - T\}f_1(E) + Tf_2(E)], \tag{3.108}$$

where $f_a(E)$ represents the near-equilibrium distribution function in the left lead, characterized by a Fermi energy μ_A. The integral over $\pm k$ considers carriers traveling in both directions. On the right side, (3.101) (in one dimension) has been used to write the average density in terms of the injected carriers from the left and right reservoirs into the left lead, where again current conservation has been used to write $1 + R = 2 - T$. It is implicitly assumed that the Friedel oscillations of the charge density around the barrier damp out sufficiently rapidly to use the asymptotic form of the charge density there. Similarly, the density in the right lead is given by

$$n_b = \frac{2}{2\pi}\int_{-\infty}^{\infty} dk f_b(E) = \frac{2}{2\pi}\int_0^{\infty} dk[\{2 - T\}f_2(E) + Tf_1(E)]. \tag{3.109}$$

Now, taking the low temperature limit, we can subtract (3.109) from (3.108) to give

$$2\int_{\mu_B}^{\mu_A} dE\left(\frac{dk}{dE}\right) = \int_{\mu_2}^{\mu_1} dE\left(\frac{dk}{dE}\right)\{2 - T\} - \int_{\mu_2}^{\mu_1} dE\left(\frac{dk}{dE}\right)T. \tag{3.110}$$

If we now assume that the difference between the Fermi energies is sufficiently small that we may neglect the energy dependence of T and the inverse velocity, dk/dE, then (3.110) may be simply integrated to give

$$\mu_A - \mu_B = (1 - T)(\mu_1 - \mu_2). \tag{3.111}$$

It is obvious, however, that the distribution of carriers in the leads is not a Fermi-Dirac, particularly when the bias is nonzero. Carriers do not populate all the states between μ_2 and μ_1, for example, on the right side. If we use the Fermi energy μ_B only as a counting scheme for the density of carriers on the right side, it can be defined such that the number of occupied states above this level is equal to the number of unoccupied states (holes) below [78]. The number of states occupied above μ_B on the right side due to injection from the left is $T D(E)(\mu_1 - \mu_B)$, where $D(E)$ is the density of states in energy corresponding to positive k only (1/2 the total density of states). Likewise, the number of unoccupied states below μ_B is given by $2D(E)(\mu_B - \mu_2) - T D(E)(\mu_B - \mu_2)$, the states below μ_2 being completely filled. Equating the two gives

$$T(\mu_1 - \mu_B) = (2 - T)(\mu_B - \mu_2). \tag{3.112}$$

Similarly, on the left side, μ_A is defined from

$$(1 + R)(\mu_1 - \mu_A) = (2 - [1 + R])(\mu_A - \mu_2) \tag{3.113}$$

where the left side is the number of occupied states above μ_A, and the right side is the number of unoccupied states below. Combining the two equations gives (3.111) above.

Substituting (3.111) into (3.106) gives

$$I = \left(\frac{2e}{h}\right)\frac{T}{1 - T}(\mu_A - \mu_B), \tag{3.114}$$

or in terms of the conductance using (3.107),

$$G = \frac{I}{V} = \left(\frac{2e^2}{h}\right)\frac{T}{1 - T} = \left(\frac{2e^2}{h}\right)\frac{T}{R}, \tag{3.115}$$

which is finally the single-channel *Landauer formula* [76], [77]. We see that the conductance is given very simply by the product of the ratio of the transmission and reflection at the Fermi-energy and the fundamental conductance, $2e^2/h = 7.748 \times 10^{-5}$ mhos. As discussed in the introduction, these conductance quanta (corresponding to a resistance of $12,907\ \Omega$) play a fundamental role in the physics of mesoscopic systems, being the conductance associated with a single one-dimensional channel. One characteristic of mesoscopic systems is that their conductance properties may be governed by transport through a few such channels, giving rise to characteristic changes of precisely $2e^2/h$.

The factor $1 - T$ appearing in the denominator has resulted in some confusion in the literature, which has been discussed in detail in subsequent publications [79]–[84]. Basically, the problem is where you actually measure the voltage. If one applies current through a pair of contacts and then measures the voltage difference in the ideal leads noninvasively through a separate pair of contacts (4-terminal measurement), then Eq. (3.115) would provide the appropriate conductance. The catch is how one would devise leads to actually measure this voltage without influencing the conductance of the structure. In a two-terminal measurement, the voltage and current are measured through the same set of leads. In this case, the voltage measured is

$$eV = \mu_1 - \mu_2, \tag{3.116}$$

which from (3.106) clearly gives

$$G = \frac{I}{V} = \left(\frac{2e^2}{h}\right)T \tag{3.117}$$

without the factor $1 - T$. The reason is that now in addition to the potential drop across the scattering structure, we are measuring a contact potential drop across the ideal leads due to the self-consistent charge build-up characterized by a contact resistance

$$R_c = \frac{h}{2e^2}. \tag{3.118}$$

Of course, if T is very small, then the series resistance of the barrier dominates, and the two-terminal and four-terminal measurement give identical results.

3.5 The multi-channel case

We want to generalize now to the case in which multiple independent conducting channels are present [78]. This case corresponds, for example, to the low-temperature situation in which N 1D subbands (modes) are populated at the Fermi energy, all of which may contribute to the current. Alternately, one can view the situation at finite temperature as a multi-channel situation in which each channel represents a particular narrow range of energy. On the left side, the stationary solutions in the ideal lead may be written, as, for example,

$$\psi_i^l(\mathbf{r}_t, z) = \sum_{i=1}^{N} \left(A_i e^{ik_{i,z}z} + B_i e^{-ik_{i,z}z}\right)\phi_i(\mathbf{r}_t), \tag{3.119}$$

where $\mathbf{r}_t$ represents the position vector in the transverse direction, i labels the transverse solutions, and $k_{i,z}$ is the wavevector. A similar such solution may be written on the right side of the scatterer

$$\psi_n^r(\mathbf{r}_t, z) = \sum_{j=1}^{N'} \left(C_j e^{ik_{j,z}z} + D_j e^{-ik_{j,z}z}\right)\phi_j(\mathbf{r}_t). \tag{3.120}$$

A scattering matrix connects the coefficients on the left, $2N$ in this case, with $2N'$ on the right side, as shown in Fig. 3.16.

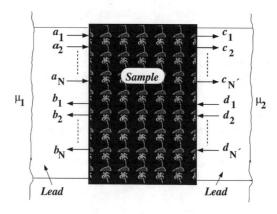

Fig. 3.16. Multi-channel system containing N input channels and N' output channels.

For simplicity, consider the case in which the input and output leads are the same, such that the number of channels on either side is the same value, N. An incoming wave in mode i on the left has a certain probability, $T_{ji} = |t_{ji}|^2$, to be transmitted into mode j on the right side, and probability $R_{ji} = |r_{ji}|^2$ of being reflected back. The transmission and reflection amplitudes, t_{ji} and r_{ji}, are the components of the $2N \times 2N$ scattering matrix

$$\mathbf{S} = \begin{bmatrix} r & t' \\ t & r' \end{bmatrix}, \tag{3.121}$$

where r, t, r', and t' all represent $N \times N$ arrays rather than individual elements as in Eq. (3.40). The scattering matrix in the case of equivalent input and output leads is again both unitary and symmetric as given by Eqs. (3.38) and (3.36).

Carriers are fed equally into all these modes in the leads from reservoirs up to the Fermi energy μ_1 on the left and up to μ_2 on the right. For a particular mode j on the left side, the current injected into channel i on the right side between μ_2 and μ_1 is again independent of the velocity and density of states as in (3.106):

$$I_{ij} = \frac{2e}{h} T_{ij}(\mu_1 - \mu_2).$$

Since each channel is assumed to be fed equally, the total current due to charge injected into mode i on the right side between μ_2 and μ_1 is now given by

$$I_i = \frac{2e}{h}\left[\sum_{j=1}^{N} T_{ij}\right](\mu_1 - \mu_2) = \frac{2e}{h} T_i(\mu_1 - \mu_2), \tag{3.122}$$

where now T_i is the shorthand notation for the sum over j on the left. Since all channels are independent, the total current is given by

$$I_{tot} = \sum_{i=1}^{N} I_i = \frac{2e}{h}(\mu_1 - \mu_2)\sum_{i=1}^{N} T_i = \frac{2e(\mu_1 - \mu_2)}{h}\text{Tr}(tt^\dagger), \tag{3.123}$$

where on the right side the sum over T_i is written in terms of the transmission submatrices

$$\sum_{i=1}^{N} T_i = \sum_{i,j=1}^{N} T_{ij} = \sum_{i,j=1}^{N} |t_{ij}|^2 = \sum_{i,j=1}^{N} t_{ij}t_{ji}^* = \text{Tr}(tt^\dagger). \tag{3.124}$$

Similarly, the current into channel i on the left side may also be expressed in terms of the reflection coefficients, R_{ij}, starting with

$$I_i = \frac{2e(\mu_1 - \mu_2)}{h}\left[1 - \sum_{i,j=1}^{N} R_{ij}\right] = \frac{2e(\mu_1 - \mu_2)}{h}[1 - R_i]. \tag{3.125}$$

Therefore, the total current becomes

$$I_{tot} = \sum_{i=1}^{N} I_i = \frac{2e(\mu_1 - \mu_2)}{h}\sum_{i=1}^{N}[1 - R_i]. \tag{3.126}$$

Comparing to Eq. (3.123), this gives the current continuity relation between all channels

$$\sum_{i=1}^{N} T_i = \sum_{i=1}^{N}(1 - R_i). \tag{3.127}$$

Again, as in the single-channel case, the ideal leads are labeled by different Fermi energies μ_A and μ_B to account for the self-consistent pile-up of charge on either side of the scatterer. To calculate these, the number counting argument in Section 3.4 is extended to look at the total number of occupied states on the right side in all channels as

$$n^r_{occ} = \sum_{i=1}^{N} T_i D_i(E)(\mu_1 - \mu_B), \tag{3.128}$$

where $D_i(E)$ is the density of states with positive velocity for channel i. The total number of unoccupied states is just the total density of states (both velocities) minus the injected density from the left,

$$n^r_{unocc} = 2 \sum_{i=1}^{N} D_i(E)(\mu_B - \mu_2) - \sum_{i=1}^{N} T_i D_i(E)(\mu_B - \mu_2). \tag{3.129}$$

Equating the two again gives μ_B in terms of the reservoir Fermi energies:

$$\sum_{i=1}^{N} T_i D_i(E)(\mu_1 - \mu_B) = \sum_{i=1}^{N} (2 - T_i)D_i(E)(\mu_B - \mu_2). \tag{3.130}$$

Rearranging to solve for μ_B,

$$\mu_B = \mu_2 + \frac{1}{2} \frac{\sum_{i=1}^{N} T_i D_i(E)(\mu_1 - \mu_2)}{\sum_{i=1}^{N} D_i(E)} = \mu_2 + \frac{1}{2} \frac{\sum_{i=1}^{N} T_i v_i^{-1}(\mu_1 - \mu_2)}{\sum_{i=1}^{N} v_i^{-1}}, \tag{3.131}$$

where the density of states in 1D has been written $D_i = 1/(\pi \hbar v_i)$. By similar generalization of Eq. (3.113), the Fermi energy on the left is determined from

$$\sum_{i=1}^{N} (1 + R_i)D_i(E)(\mu_1 - \mu_A) = \sum_{i=1}^{N} (1 - R_i)D_i(E)(\mu_A - \mu_2), \tag{3.132}$$

which upon rearranging gives μ_A as

$$\mu_A = \frac{(\mu_1 + \mu_2)}{2} + \frac{1}{2} \frac{\sum_{i=1}^{N} R_i v_i^{-1}(\mu_1 - \mu_2)}{\sum_{i=1}^{N} v_i^{-1}}. \tag{3.133}$$

The voltage drop across the scatterer is therefore given as the difference of μ_A and μ_B:

$$eV = \mu_A - \mu_B = \frac{1}{2}(\mu_1 - \mu_2) \frac{\sum_{i=1}^{N} (1 + R_i - T_i)v_i^{-1}}{\sum_{i=1}^{N} v_i^{-1}}. \tag{3.134}$$

Writing the conductance of the scatterer as I/V, using (3.123) and (3.134), we finally obtain

$$G = \frac{I_{tot}}{V} = \frac{\frac{2e^2}{h}(\mu_1 - \mu_2)\sum_{i=1}^{N} T_i}{\frac{1}{2}(\mu_1 - \mu_2)\frac{\sum_{i=1}^{N}(1 + R_i - T_i)v_i^{-1}}{\sum_{i=1}^{N} v_i^{-1}}} = \frac{2e^2}{h} \sum_{i=1}^{N} T_i \frac{2 \sum_{i=1}^{N} v_i^{-1}}{\sum_{i=1}^{N}(1 + R_i - T_i)v_i^{-1}}. \tag{3.135}$$

As before, if we performed a two-terminal measurement such that we measure the potential drop across both the scatterer and the leads, then the conductance reduces simply to

$$G_{2term} = \frac{2e^2}{h} \sum_{i=1}^{N} T_i = \frac{2e^2}{h} \text{Tr}(tt^\dagger). \tag{3.136}$$

If we had considered an unequal number of channels on either side of the scatterer (e.g., if the width of the ideal 1D conductor was different on the two sides), Eq. (3.135) may be generalized to [78]

$$G = \frac{2e^2}{h} \sum_{i=1}^{N'} T_i \frac{2}{1 + \frac{1}{g_l} \sum_{i=1}^{N} R_i v_{l,i}^{-1} - \frac{1}{g_r} \sum_{i=1}^{N'} T_i v_{r,i}^{-1}}, \qquad (3.137)$$

where $v_{l,i}$ and $v_{r,i}$ are the velocities on the left and right side, and

$$g_l = \sum_{i=1}^{N} v_{l,i}^{-1},$$
$$g_r = \sum_{i=1}^{N'} v_{r,i}^{-1}. \qquad (3.138)$$

Equation (3.137) obviously reduces to the symmetric result (3.135) when the velocities are the same on both sides. If the reflection coefficients are nearly unity and the transmission coefficients small, (3.137) reduces to the two-terminal result (3.136).

3.6 Quantized conductance in nanostructures

In the previous sections, we developed a model for transport in very small structures following the formalism developed by Landauer and Büttiker. In this picture, the conductance is determined by the number of one-dimensional channels available to carriers injected from ideal phase-randomizing contacts, and by the transmission properties of the structure for each channel. The derivation of these formulas follow from basic physical insights and certain assumptions regarding the distinction between contacts and scattering structures (if such a distinction may be made). Attempts have been made to derive them more formally from linear response theory with various degrees of success, which essentially give the same result if one is careful in making the distinction between two-terminal versus four-terminal measurements [83], [84]. A lucid discussion of the connection between the Landauer-Büttiker model and linear response theory has been given by Stone and Szafer [84]. In the present section, we now make the connection between this formulation and experimental measurements of the conductance in small semiconductor structures.

3.6.1 Experimental results in quantum point contacts

The main workhorse in regards to experimental measurements is the split-gate structure fabricated on a high-mobility electron gas structure discussed in Chapters 1 and 2 and in Fig. 3.17. For zero gate bias, the 2DEG exists essentially everywhere in the space between the two Ohmic contacts shown in Fig. 3.17a. With a negative bias applied to the two Schottky contact gates, the 2DEG is depleted underneath as well as laterally from the geometric edge of the gates as shown in Fig. 3.17b. In the narrow region between the two *split-gates*, a 1DEG is formed as discussed in Chapter 2. The density of the 1DEG decreases as the gate bias is made more negative until it eventually disappears between the split gates. The length of the 1DEG is defined by the width of the gate contacts and the shape of the depletion regions around the contacts. Such split-gate structures are often referred to *quantum point contacts* (QPCs), named after early experiments using similar structures in metallic systems [85]. Longer structures with more complicated geometries are most often referred to as

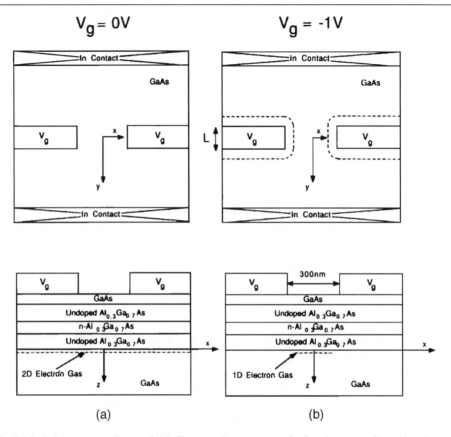

Fig. 3.17. Point contact split-gate field effect transistor structure for forming a one-dimensional quantum channel [after Timp *et al.*, in *Nanostructure Physics and Fabrication*, edited by W. P. Kirk and M. Reed (Academic Press, New York, 1989), p. 331–346, by permission].

quantum waveguides in analogy to electromagnetic waveguides, a distinction maintained throughout the rest of the book, although the terminology is often used interchangeably in the literature.

In the following subsections, we focus on phenomenon primarily associated with the phase-coherent motion of electrons through quantum point contact or quantum waveguide structures in which both the phase coherence length and the mean free path due to elastic scattering due to inhomogeneities are greater than those of the characteristic dimensions of the artificial semiconductor structure. Here we can only hope to highlight the rich experimental and theoretical body of research that has grown following the spectacular discovery of conductance quantization in QPCs in the late 1980s. We thus begin with a discussion of quantized conductance in simple QPCs and its interpretation within the Landauer-Büttiker model, and then proceed to more complicated geometries, with multiple leads, and later to the effects of adding magnetic fields.

In 1988, van Wees *et al.* [86]–[88] and Wharam *et al.* [89], [90] independently reported low-temperature conductance versus gate voltage data in split-gate structures that exhibited conductance plateaus quantized in integer multiples of $G_o = 2e/h^2$ as shown in Fig. 1.3. Such measurements are typically performed in a two-terminal configuration,

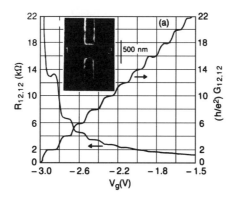

Fig. 3.18. Two terminal conductance as a function of gate bias for the Schottky gate structure shown in the inset [after Timp *et al.* in *Nanostructure Physics and Fabrication*, edited by W. P. Kirk and M. Reed, (Academic Press, New York, 1989), p. 331–346 by permission].

with the voltage and current measured through the same set of source-drain contacts while sweeping the gate voltage. The mobility of the 2DEG is generally high, often in excess of $100 \, \mathrm{m^2/Vs}$, to reduce the effects of ionized impurity scattering. The 2D sheet densities are typically in the $10^{11} - 10^{12}/\mathrm{cm^2}$ range.

Figure 3.18 shows later data by Timp *et al.* [91], [92] of the two-terminal conductance $G_{12,12}$ (using the convention introduced in Section 3.9 which signifies that the current and voltage are measured through the same pair of contacts) and resistance $R_{12,12}$ measured in the split-gate structure shown in Fig. 3.17 (the inset of Fig. 3.18 shows an SEM micrograph of the exact gate structure). As can be seen, a number of conductance plateaus are visible, all of which occur at integral multiples of the fundamental conductance, $G = NG_o$. The number N increases as the gate voltage is made less negative, corresponding to the effective 1D channel's increasing width. Eventually, the plateaus disappear as the threshold voltage for the 2DEG is reached under the gate, and the 1D channel merges into the 2D continuum. The qualitative explanation for the conductance quantization follows quite simply from our prior discussion of the multichannel model in Section 3.5, even though the experimental result was not anticipated theoretically before its discovery. The two-terminal conductance given by Eq. (3.136) predicts that the conductance is given by G_o times a sum over the occupied channels of the total transmission in each channel. In the present case, the occupied channels are nothing more than the one-dimensional subbands whose subband minima lie below the Fermi energy of the 2DEG outside of the QPC. If the coupling into and out of these channels is unimpeded, the total transmission through a given mode may approach unity, particularly if intermode scattering is weak. Thus, with unity transmission in (3.136), the conductance is nothing more than $G = 2Ne^2/h$, where N is the number of 1D subbands occupied at the Fermi energy. As the gate voltage is made more negative, the potential in the narrow region of the QPC squeezes the 1DEG, pushing successive 1D subbands through the Fermi energy (much like SdH oscillations in a magnetic field). As each 1D subband is depopulated, the conductance drops by an amount G_o until finally all the subbands are depopulated and the conductance approaches zero.

This conductance quantization in the absence of a magnetic field is reminiscent of the quantized resistance of the quantum Hall effect discussed in Section 2.7.6. However, the accuracy of the quantization in the latter is of order 10^{-7} whereas the accuracy of quantization shown for example in Fig. 3.18 is much less, on the order of a few percent accuracy. This is particularly evident in Fig. 3.18 for higher gate biases, where degradation of the plateaus is apparent. Part of the experimental uncertainty arises from the nonzero resistance of the Ohmic contacts and 2DEG between the source-drain and the QPC that includes the spreading resistance around the QPC itself in the 2DEG. Some of this series resistance may be eliminated by performing 4-terminal measurements (see Section 3.9). However, if the probes are within approximately a phase coherence length from the QPC itself, they themselves contribute to the transmission through the structure. Therefore, much of the series resistance effect cannot be eliminated other than to measure this quantity separately and to subtract its effect from the measured results, a practice which is often employed. Another contribution that degrades the ideal conductance quantization is random inhomogeneities such as impurities and boundary roughness, discussed later in Section 3.6.4. Backscattering by such inhomogeneities causes the transmission coefficient to be less than unity, thus degrading the ideal quantization of the conductance.

3.6.2 Adiabatic transport model

In the previous section, we made some hand-waving arguments for the existence of quantized conductance on the basis of the two-terminal multichannel formula (3.136). The main argument is the requirement of a smooth matching of the external 2DEG in the contact region to the narrow region of the QPC, much in analogy to impedance matching through a horn-like structure in electromagnetic waveguides. As we show later in Section 3.7, such matching is not necessary, and an abrupt constriction also allows for quantized conductance. However, to make the smooth matching assertion more quantitative, and to develop a simplified form for the quantized conductance, we first consider the local adiabatic model for transmission through a smooth point contact potential [93].

Consider the lateral geometry of the two-dimensional point contact shown in Fig. 3.19. Here it is assumed that the confining potential in the normal (growth) direction is sufficiently narrow that the width in that direction is negligibly small compared to the lateral dimensions, which is generally a good approximation in present day QPCs. It is therefore sufficient to consider the two-dimensional stationary Schrödinger equation

$$\left\{ -\frac{\hbar^2}{2m^*}\left(\frac{\partial^2}{\partial x^2} + \frac{\partial^2}{\partial y^2}\right) + V(x, y)\right\}\psi(x, y) = E\psi(x, y), \qquad (3.139)$$

where $V(x, y)$ is the lateral potential and the x- and y-directions are defined in Fig. 3.19. The adiabatic approximation amounts to assuming that the spatial variation in the x-direction is

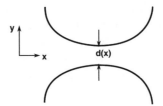

Fig. 3.19. Two-dimensional approximation for a quantum point contact that couples smoothly to the external region.

much slower than in the transverse y-direction. Thus, the second derivative with respect to x may be neglected in Eq. (3.139) to give the one-dimensional Schrödinger equation at a fixed x as

$$\left\{-\frac{\hbar^2}{2m^*}\frac{\partial^2}{\partial y^2} + V(x, y)\right\}\chi_n(x, y) = E_n(x)\chi_n(x, y), \tag{3.140}$$

where $\chi_n(x, y)$ is the transverse eigenfunction and $E_n(x)$ is the position-dependent energy eigenvalue corresponding to the transverse potential at point x. The solution of (3.139) may be constructed from the solutions of (3.140) as

$$\psi(x, y) = \sum_n \phi_n(x)\chi_n(x, y) \tag{3.141}$$

since the χ_n form a complete set. The solution (3.141) may be substituted into (3.139) multiplied by $\chi_m^*(x, y)$ and integrated over all y to get [94]

$$\left\{-\frac{\hbar^2}{2m^*}\frac{\partial^2}{\partial x^2} + E_m(x) - E\right\}\phi_m(x) = \sum_n A_{nm}\phi_n(x), \tag{3.142}$$

where A_{nm} is the operator

$$A_{nm} = \frac{\hbar^2}{m^*}\left[\int dy\,\chi_m^*(x, y)\frac{\partial}{\partial x}\chi_n(x, y)\frac{\partial}{\partial x} + \frac{1}{2}\int dy\,\chi_m^*(x, y)\frac{\partial^2}{\partial x^2}\chi_n(x, y)\right]. \tag{3.143}$$

The right side of (3.142) gives rise to terms coupling mode index m to n, which corresponds to intersubband scattering. However, these terms are proportional to spatial gradients in the longitudinal direction, which according to the adiabatic approximation are small. Therefore, to lowest order, the right side is zero ($A_{nm} = 0$), giving the simple 1D equation

$$\left\{-\frac{\hbar^2}{2m^*}\frac{\partial^2}{\partial x^2} + E_n(x)\right\}\phi_n(x) = E\phi_n(x). \tag{3.144}$$

This equation shows that for a slowly varying longitudinal potential, the envelope function for the motion in the x-direction just satisfies the 1D Schrödinger equation in an effective potential, $E_n(x)$, determined by the spatial variation of the nth energy eigenvalue of the transverse solution. The degree that this is true may be checked quantitatively by substituting back into (3.143) to determine the actual magnitude of the coefficients A_{nm}.

Now consider a simple example assuming hard wall confinement. In Fig. 3.19, the potential is assumed infinite outside of the constriction so that the transverse solution at point x is

$$\chi_n(x, y) = \sqrt{\frac{2}{d(x)}}\sin\left[\frac{\pi n\{2y + d(x)\}}{d(x)}\right], \tag{3.145}$$

where the zero of y is at the midpoint of the constriction assuming a symmetric geometry for simplicity. The corresponding effective potential is

$$E_n(x) = \frac{\hbar^2 n^2 \pi^2}{2m^* d(x)^2}. \tag{3.146}$$

If the variation of the effective potential close to the constriction has a quadratic behavior of the form

$$E_n(x) \approx V_o - \frac{1}{2}m^*\omega_x^2 x^2, \tag{3.147}$$

the transmission coefficient for mode n has a conveniently simple form [95]

$$T_n(E) = \frac{1}{1 + e^{-2[E-V_o]/\hbar\omega_x}}. \tag{3.148}$$

Thus, expanding (3.146) into quadratic form,

$$E_n(x) \approx E_n(1 - \alpha x^2), \tag{3.149}$$

where α is a parameter characterizing the variation of the potential, and $E_n = \hbar^2 n^2 \pi^2 / (2m^* d^2)$ with d the minimum separation of the channel boundaries. Large values of α correspond to a sharp point contact, similar to the structure shown earlier in Fig. 1.2 used by the Phillips-Delft group [86]. Smaller values correspond to longer structures such as that in Fig. 3.18. The resulting transmission coefficient is thus

$$T_n(E) = \frac{1}{1 + e^{-\beta_n[E-E_n]}}, \tag{3.150}$$

where

$$\beta_n = \sqrt{\frac{2m^*}{\alpha\hbar^2 E_n}}. \tag{3.151}$$

The corresponding conductance at low temperature is given by the two-terminal multichannel formula (3.136) with (3.151) above

$$G = \frac{2e^2}{h} \sum_n \frac{1}{1 + e^{-\beta_n[E_f - E_n]}}. \tag{3.152}$$

As the gate voltage changes, the Fermi energy passes through successive energy levels, E_n, giving rise to plateau-like structures as shown in Fig. 3.20. For small values of α, sharp steps are found, whereas for large values, the steps are rounded, in accordance with experiment. The rounding is due to tunneling through the constriction for a Fermi energy just below the next unoccupied level, and then approaches unity as the Fermi energy moves above the subband minimum. Note, however, that a well-defined 1DEG in a long quantum wire is not necessary for achieving conductance quantization. The important point is that the conductance is controlled by the mode spacing at the narrowest point of the channel. Thus

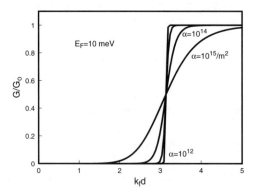

Fig. 3.20. Calculated conductance at low temperature for a fixed Fermi energy of 10 meV for various values of α as a function of the minimum constriction width, d.

a point contact or a longer wire structure will exhibit conductance quantization as observed experimentally.

The simple example above assumed hard wall boundaries for the potential, but one could equally well use some other function. Büttiker has used the two-dimensional quadratic potential [96]

$$V(x, y) = V_o - \frac{1}{2}m^*\omega_x^2 x^2 + \frac{1}{2}m^*\omega_y^2 y^2 \qquad (3.153)$$

for the potential variation around the saddle point forming the constriction, which directly gives the transmission coefficient in the form (3.150) without intermode scattering and without the necessity of expanding the potential as we did earlier. We will come back to this model potential in Section 3.10 in a discussion of the effects of magnetic field on the quantized conductance. In any case, the form of the transmission coefficient (3.150) represents a convenient empirical form for use in comparing to experiment treating β_n simply as an *adiabatic* parameter. Given that the form of the transmission coefficient is a Fermi function, it is not surprising that the effect of finite temperature is very similar to the effect of tunneling in the adiabatic model as discussed below.

3.6.3 Temperature effects

At nonzero temperature, it is qualitatively expected that the plateau structure persists until the thermal broadening, $k_B T$, is comparable to the subband spacing. Figure 3.21 shows the measured conductance plateaus for several different temperatures [97]. At $4.2K$, the conductance quantization is almost completely washed out.

In order to extend the multichannel formula to nonzero temperature, the dependence of the transmission coefficient must be explicitly taken into account. Since carriers are now injected at different energies, we can think of each incremental energy range dE around E as a separate channel into which charge is injected. Thus the incremental charge injected

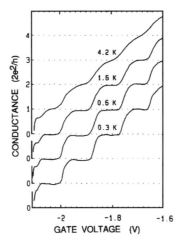

Fig. 3.21. Experimental temperature dependence of the quantized conductance in a quantum point contact structure [after van Wees *et al.*, Phys. Rev. B **43**, 12431 (1991), by permission].

into channel i on the right side (contact 2) from channel j in contact 1 becomes

$$I_{ij}dE = \frac{2e}{h}T_{ij}(E)\{f(E - \mu_1)dE - f(E - \mu_2)dE\}, \qquad (3.154)$$

where $f(E - \mu_i)$ is the quasi-equilibrium distribution function of the ith contact. The total current then summed over all channels and integrated over energy becomes

$$I = \frac{2e}{h}\int_{-\infty}^{\infty}dE \sum_i T_i(E)\{f(E - \mu_1) - f(E - \mu_2)\}. \qquad (3.155)$$

This equation represents a starting point for considering the nonlinear response in a QPC, which we will return to in Section 3.6.5. For now, we are interested in the linear response regime, so that the difference in chemical potentials may be assumed to be small. Assuming $f(E)$ to be the Fermi-Dirac distribution

$$f(E - \mu_i) = \frac{1}{1 + e^{(E-\mu_i)/k_BT}}, \qquad (3.156)$$

the current (3.155) may be rewritten [78]

$$I = \frac{2e}{h}\int_{-\infty}^{\infty}dE \sum_i T_i(E)\frac{-df}{dE}(\mu_1 - \mu_2), \qquad (3.157)$$

where

$$\frac{-df}{dE} = \frac{df}{d\mu} = \lim_{\mu_1 \to \mu_2}\frac{f(E - \mu_1) - f(E - \mu_2)}{\mu_1 - \mu_2}. \qquad (3.158)$$

Consider now the simple case that the transmission coefficient into the ith mode is unity for energies greater than the subband minimum and zero below. Then the two-terminal conductance becomes simply

$$G = \frac{I}{(\mu_1 - \mu_2)} = \frac{2e}{h}\sum_i \frac{1}{1 + e^{-(E_f - E_i)/k_BT}}, \qquad (3.159)$$

which is identical in form to (3.152) with $\beta_n = 1/k_BT$. The effect on the calculated conductance versus width is identical to that shown in Fig. 3.20 due to adiabatic tunneling. As T increases, the effective β decreases, causing increased rounding of the plateaus until they eventually disappear at sufficiently high temperature.

3.6.4 Inhomogeneous effects

In the description so far of the quantized conductance in QPCs, we have neglected the effect of unintentional inhomogeneities. Real split-gate structures such as those illustrated in Figs. 3.17 and 3.18 are formed on modulation-doped, heterojunction layers with the narrow channel formed using high-resolution lithography. The exact distribution of the impurities in the barrier region (as well as unintentional background doping) give rise to random fluctuations of the potential in addition to that defined to lowest order by the split gate. Since the ideal quantization of conductance requires unity transmission for each mode in the narrow region, a potential fluctuation in the ballistic region of the QPC contributes to backscattering which degrades the conductance below NG_o. Boundary roughness may play a dramatic role as well. The interface between lattice-matched GaAs and AlGaAs may be nearly flat on an atomic scale. However, the lateral fluctuations of the metal gates used to form the QPC may easily be on the order of several nanometers due to the lithographic process required to form narrow regions as discussed in Chapter 2. Such boundary roughness also results in backscattering and a reduction in conductance. It is important to realize that in small nanostructures such as QPCs, the total scattering matrix through the structure, including impurities and roughness, depends on the exact location of the impurities and boundary fluctuations. If such inhomogeneous effects are important, the conductance in a sample with a slightly different impurity configuration will be completely different. This behavior contrasts from that of diffusive transport over very long length scales, in which the contribution of many impurities averages to a constant contribution to the conductance that depends only on their density and not their exact location.

Given that such inhomogeneities in present-day technology exist, their role should become increasingly important as the length of the quantum point contact or waveguide is increased. Figure 3.22 shows the conductance versus gate voltage of a 600 nm long point contact structure reported by Timp *et al.* in the same material as the 200 nm long QPC of Fig. 3.18 [91], [92]. In comparing Figs. 3.18 and 3.22, it is clear that the quantized conductance of the latter is less than ideal. For all plateaus in the long waveguide, the conductance is degraded below NG_o. Some plateaus are even missing altogether. Similar behavior is observed in general by different groups, while the particular length scale where quantized conductance is degraded depends on the exact fabrication technology used.

Nixon *et al.* have calculated the explicit contribution to the conductance in a point contact due to various arrangements of random impurities in the doping layer of the AlGaAs [98]. They use a semi-classical, self-consistent potential model of the heterostructure including that due to discrete random charges projected onto the plane of the 2DEG. The transmission coefficients are calculated numerically using a coupled mode theory, which in some sense resembles the mode matching method described in Section 3.7.1 extended to an arbitrary change of potential. Figure 3.23 shows the metal gate pattern and the calculated potential distribution around the split gates for four cases corresponding to the structures of Timp *et al.* in Figs. 3.18 and 3.22. Case (a) corresponds to the ideal potential in which a uniform positive charge for the ionized donors is assumed with a density of $2.5 \times 10^{12}/cm^2$ located in a delta layer 42 nm above the channel. The ideal potential appears more like the smooth

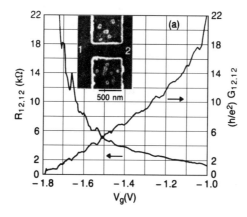

Fig. 3.22. Measured conductance in a quantum point contact structure with a 600 nm gate length (after Timp *et al.* in *Nanostructure Physics and Fabrication*, edited by W. P. Kirk and M. Reed (Academic Press, New York, 1989), p. 331–346, by permission].

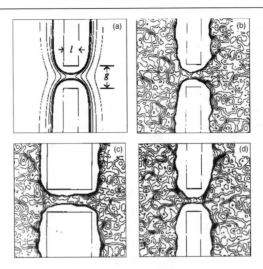

Fig. 3.23. Gate pattern and contour maps of the 2DEG electron density for 300-nm-wide split-gate constrictions with lengths 200 and 600 nm. (a) A QPC with a smooth positive background impurity distribution; (b)–(d) contours with random arrangements of impurities with areal density of $2.5 \times 10^{12}/cm^2$ [after Nixon *et al.*, Superlattices and Micro. **9**, 187 (1991), by permission].

saddle point potential used in the quasi-adiabatic model of Section 3.6.2 rather than the hard wall case used in Section 3.7.1. Cases (b) and (d) show the potential for two different realizations of the random impurity configuration in the delta plane for the 200 nm long QPC, while (c) is the case of the longer point contact of Fig. 3.22. The calculated conductance for the 200 nm case for uniform charge (top curve) and three different impurity arrangements is shown in Fig. 3.24. The ideal case shows smooth conductance plateaus without the resonances due to the abrupt transition seen in Fig. 3.20 due to the smooth tapering of the potential at the entrance and exit of the pinched region. With the discrete charge potential included, degradation of the plateaus is evident, even in the 200 nm length case. Certain impurity configurations may even give rise to resonant peaks in the conductance, as seen for example in the data for a straight wire, shown in Fig. 3.25. For the 600 nm case, the conductance is even more degraded. The calculated results shown in Fig. 3.24 predict much more variation from sample to sample than is seen experimentally, which suggests some additional role of screening in the QPC, or three-dimensional effects in reducing the fluctuations. Nevertheless, the calculations demonstrate the important role that impurities play in the conductance in QPCs.

Boundary roughness may also play a dominant role in degrading the conductance quantization in long waveguides. Takagaki and Ferry [99] have calculated the effect of fluctuations of the boundary on the quantized conductance in quantum waveguides using a mode matching theory similar to that described in Section 3.7.1. There roughness was modeled as a deviation of the local width of the wire from an average value W in terms of steps with a width uniformly distributed between $-\Delta/2$ and $\Delta/2$. The transmission is then modeled by cascading the scattering matrices for uniform sections of random length using the mode matching method, and by calculating the conductance via the multichannel formula for a given length of waveguide. Figure 3.25 shows the length dependence of the conductance for a specific realization of the boundary roughness. One can see that the reflection of the electrons from the rough surface is rapidly enhanced as L is increased. When the wire

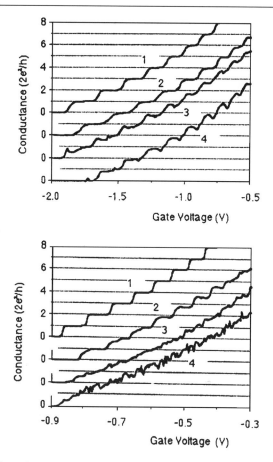

Fig. 3.24. Calculated conductance plateaus for various impurity arrangements for (a) 200 nm constriction and (b) 600 nm constriction. The upper curve corresponds in each case to the smooth impurity configuration [after Nixon *et al.*, Superlattices and Micro. **9**, 187 (1991), by permission].

length becomes longer than the elastic mean free path, univeral conductance fluctuations (UCFs) develop and dominate the behavior of the conductance. The fluctuations are not self-averaging, meaning that if the length is increased, the magnitude of the conductance fluctuations with length remains the same. This phenomenon of sample dependent UCFs will be taken up in much more detail in Chapters 5 and 6.

3.6.5 Nonlinear transport

When the source-drain bias becomes sufficiently large, the quantized conductance predicted from the equilibrium Landauer-Büttiker model is expected to break down. In particular, when the bias voltage becomes greater than the spacing of the quasi-1D subbands, different numbers of subbands become available for transport in the forward and reverse directions, giving rise to nonlinear conductance [100]. Nonlinear effects were measured experimentally by Kouwenhoven *et al.* in QPCs [101]. Later measurements by Martín-Moreno revealed oscillatory behavior in the differential conductance at nonzero source-drain bias [102]. Figure 3.26 shows typical data for the differential conductance versus source-drain bias in a QPC. For very small V_{sd}, the curves converge to integral multiples of the $2e^2/h$

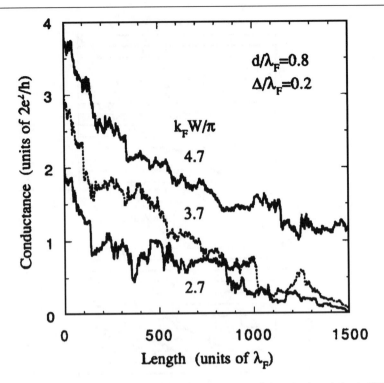

Fig. 3.25. Conductance of inhomogeneous wires as a function of the wire length for $k_f W/\pi = 2.7$, 3.7, and 4.7 [after Y. Takagaki and D. K. Ferry, J. Phys: Condens. Matter **4**, 10421 (1992)].

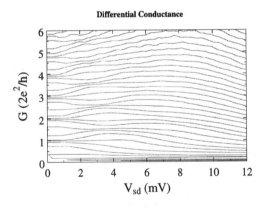

Fig. 3.26. Differential conductance versus source-drain bias for different gate biases in a quantum point contact at $1.5 K$ (after C. Berven, Ph.D. dissertation 1994).

corresponding to the linear regime. For $V_{sd} < 5$ meV, oscillatory behavior is observed, which damps out for higher bias.

We may use Eq. (3.155) to extend the Landauer-Büttiker model to the nonlinear regime. We should keep in mind that we are no longer taking into account the self-consistent pile-up of charge treated in the linear response regime in the Landauer derivation. In the nonlinear regime, we cannot simply say that Eq. (3.155) is the two-terminal simplification of a full self-consistent treatment. Rather, (3.155) in the nonlinear regime really has more relation to the Tsu-Esaki formula (3.85) in the limit that the transverse directions of motion are fully quantized, and thus subject to the limitations of that model discussed in Sections 3.3.2 and 3.3.3.

To arrive at an analytical expression for the I-V characteristics of a ballistic QPC, consider the one-dimensional model for the potential landscape of the point contact shown in Fig. 3.27. At small V_{sd}, the Fermi energies from the reservoirs on the right and left inject into the same 1D subband as shown in Fig. 3.27a. An electrostatic barrier, $\phi_o(V_g)$, exists in the point contact to electrons going from left to right due to the gate bias. The barrier is raised and lowered by the gate bias, pulling 1D levels through the Fermi energy, which results in conductance plateaus. For nonzero V_{sd}, the Fermi energy on the right is pulled down with respect to the Fermi energy on the left so that electrons are injected into higher subbands from one Fermi energy only as shown in Fig. 3.27b. The barrier height ϕ changes with respect to its zero V_{sd} value, ϕ_o. We assume that this change in barrier height is characterized by a *barrier shape* parameter, α. The barrier changes as $\phi = \phi_o - V_{sd}/\alpha$. Figure 3.27c indicates that for a symmetrical triangular barrier as shown, the barrier height changes as $V_{sd}/2$ if one pulls the right horizontal region down by V_{sd} with respect to the left horizonal region. Similarly, for the lower potential profile, the barrier height changes as $V_{sd}/3$.

Assume low temperatures such that the Fermi functions are step functions, and let the transmission coefficient be given by (3.150) for a parabolic saddlepoint potential. Then in

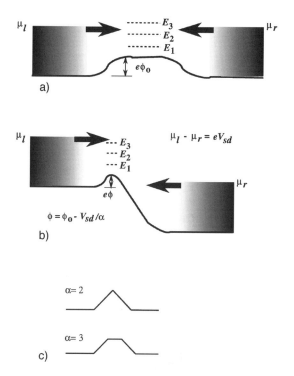

Fig. 3.27. Energy diagram of a quantum point contact with a nonzero source drain bias. (a) The point contact with zero bias; (b) the point contact with a nonzero bias; (c) the change in barrier height ϕ, with applied bias for two simple geometrical cases.

the limit of zero temperature,

$$I = \frac{2e}{h} \sum_n \left\{ \int_{-\infty}^{\mu_l} dE T_n(E) - \int_{-\infty}^{\mu_r} dE T_n(E) \right\}$$

$$T_n(E) = \frac{1}{1 + e^{-\beta_n(E - E_n)}},$$

(3.160)

where β_n is the adiabatic parameter of the nth 1D subband in the constriction. We may write E_n as the sum of the electrostatic potential in the constriction plus the confinement energy relative to this maximum, $E_n = e\phi + \varepsilon_n$. Assume that ε_n is measured relative to the maximum of the electrostatic potential, ϕ, and is fixed with respect to this reference under bias arising from either the gate or the source-drain. The Fermi-like functions are easily integrated to give

$$I = \frac{2e}{h} \sum_n \frac{1}{\beta_n} \ln \left\{ \frac{\left(1 + e^{\beta_n(\mu_l - e(\phi_o - V_{sd}/\alpha) - \varepsilon_n)} \right)}{\left(1 + e^{\beta_n(\mu_r - e(\phi_o - V_{sd}/\alpha) - \varepsilon_n)} \right)} \right\}.$$

(3.161)

If we assume that the transmission coefficient is sharp (i.e., $T_n(E) = \theta(E - E_n)$) but also assume finite temperature, the same result is obtained with $\beta_n = (k_B T)^{-1}$. If we take the limit of β_n large (i.e., low temperature and a sharp transmission probability around the subband minimum), Eq. (3.161) simplifies to

$$I = \frac{2e}{h} \sum_n \{ (\mu_l - e\phi_o - \varepsilon_n + eV_{sd}/\alpha) \theta(\mu_l - e\phi_o - \varepsilon_n + eV_{sd}/\alpha)$$

$$- (\mu_r - e\phi_o - \varepsilon_n + eV_{sd}/\alpha) \theta(\mu_r - e\phi_o - \varepsilon_n + eV_{sd}/\alpha) \},$$

(3.162)

where $\theta(x)$ is the unit step function. For subbands in which the Fermi energies of both the left and right reservoirs inject carriers, the step function is nonzero and the terms involving ϕ_o, ε_n, and V_{sd}/α cancel, leaving $\mu_l - \mu_r = eV_{sd}$ and thus a contribution of G_o to the conduction. However, under nonzero source drain bias, the upper subbands are unequally populated from the left and right (Fig. 3.27), and the current depends directly on the shape parameter, α. As an example, assume the Fermi level on the left injects into the first subband only, and V_{sd} is such that μ_r is below the subband minimum and does not contribute in (3.162). Then the current is given by

$$I = \frac{2e}{h} (\mu_l - e\phi_o - \varepsilon_1 + eV_{sd}/\alpha).$$

(3.163)

The corresponding conductance is given by

$$G = \frac{2e^2}{h\alpha}.$$

(3.164)

Since $\dot{\alpha} > 1$ by definition, the conductance drops below the fundamental conductance, G_o. This observation in part explains the oscillatory conductance shown experimentally in Fig. 3.26. As the Fermi energy on the right drops below a subband minimum with bias, the conductance decreases as this reservoir no longer contributes to the change of current in that particular subband. On the other hand, if the left chemical potential accesses a higher subband with applied bias, the conductance will increase, hence the oscillatory

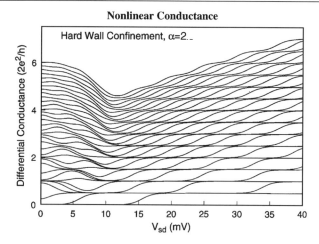

Fig. 3.28. Calculated conductance as a function of source drain bias for various widths of the constriction in a quantum point contact. The barrier shape parameter is $\alpha = 2$.

nature of the conductance with source drain bias. If the point contact is initially in the tunneling regime in which no subbands are initially occupied, application of a source drain bias lowers the potenial barrier and allows the Fermi energy on the left to inject carriers, resulting in a turn-on of the current with a conductance approaching (3.164) at low temperature.

Figure 3.28 shows the numerical calculation of the conductance given by Eq. (3.161) for energy levels given by a hard wall confinement model, where $\alpha = 2$ and the value of β is shown. Oscillatory conductance is clearly predicted, although over a much wider range of V_{sd} than observed experimentally. The absence of oscillatory conductance at higher bias may signify the breakdown of the ballistic model assumed so far. On the other hand, the simple-minded model of the barrier modification given by α may also fail as the bias voltage increases. The topic of nonlinear transport in quantum point contacts will be revisited in Section 7.1.6, where clear breakdown of the ballistic model is discussed.

3.7 Transport in quantum waveguide structures

To consider structures more complicated than a simple adiabatic point contact requires at some level a knowledge of the actual transmission characteristics of a structure in terms of its scattering matrix (see Section 3.5). In split-gate structures as well as other realizations of quantum ballistic structures, the exact treatment requires the solution of the full three-dimensional Schrödinger equation coupled with Poisson's equation for the potential (and this just within the single particle picture!). Such calculations are quite tedious and require enormous computational resources. Thus, most of the theoretical treatment of more complicated structures are based on simplified reductions of the actual geometry into two dimensions with idealized potentials and geometries. In the present chapter, we will consider two such simplified treatments. The first, the mode matching method, is the simplest to implement, requiring modest computational resources; however, it is somewhat limited in dealing with all but geometrically piecewise structures. Later, in Section 3.8, we introduce the lattice Greens function method, which allows relatively simple treatment of inhomogeneous effects as it is fully discretized, thus resembling direct numerical solution of the Schrödinger equation by finite elements or finite difference methods.

3.7.1 Mode matching analysis

The *mode matching method* is a technique used to model electromagnetic waveguides. As mentioned previously, there are a number of analogies between the solution of the wave equation(s) arising from the solution of Maxwell's equations and the wave equation in quantum mechanics. In quantum mechanics, the situation is somewhat simpler than in electromagnetics because the quantity of interest (the wavefunction) is a scalar rather than vector quantity. However, that does not prevent one from considering the effect in a ballistic quantum system of a variety of geometries used in electromagnetic waveguides. Figure 3.29 shows a schematic diagram of a variety of different geometries encountered in microwave circuits that lead to scattering of radiation characterized by scattering matrices, or *s-parameters*. In electromagnetics, the scattering matrix is represented as a function of the frequency of the radiation and may show resonant behavior at certain frequencies. In the quantum case, frequency is associated with the energy of a particle, and thus resonant behavior occurs as a function of the incident energy of a particle on the structure, as we saw for resonant tunneling diodes. Since at low temperature the conductance properties of a structure are directly determined by its transmission properties at the Fermi energy, complicated structures such as those shown in Fig. 3.29 are expected to mirror the scattering matrix of the structure as the Fermi energy or some other characteristic parameter is varied. Thus, the idea of quantum waveguides that mimic the behavior of electromagnetic waveguides has been popularized as potential nanoelectronic devices.

The technique of mode matching has been used by a number of different groups, particularly to explain the quantized conductance plateaus discussed in the previous section (see, e.g., [103]–[107]). Here we introduce one particular scheme developed by Weisshaar

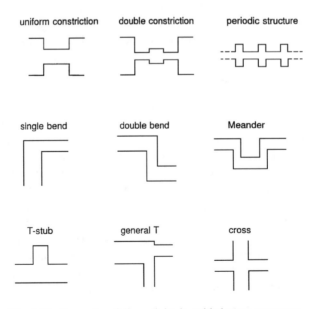

Fig. 3.29. Examples of discontinies in guided-wave structures.

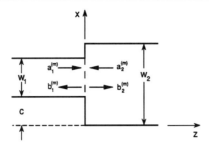

Fig. 3.30. Step discontinuity in a quantum waveguide.

et al. [108] for analyzing a variety of discontinuous waveguide structures in quantum systems. The basic idea is to break up a given geometry into sections and to expand the stationary-state solutions in terms of the transverse modes characterizing a particular section. The term "mode matching" then refers to the process of matching the usual quantum mechanical solutions of the function and its derivatives across the boundary between different sections, solving for the expansion coefficients. In essence, one matches the modes on one side of a discontinuity with those on the other side.

To start, consider a simple step discontinuity in a two-dimensional quantum wire shown in Fig. 3.30. Again we will assume that confinement in the direction normal to the plane of the figure is strong so that only the lowest confined mode in that direction is occupied. The z-direction is along the propagation direction, and the x-direction is transverse as shown. In the uniform sections of the waveguide on either side of the discontinuity, the solution of the time-independent wave equation is separable as

$$\psi_1(x, z) = \sum_{n=1}^{\infty} \left(a_n^1 e^{\gamma_n^1 z} + b_n^1 e^{-\gamma_n^1 z} \right) \phi_n^1(x)$$

$$\psi_2(x, z) = \sum_{n=1}^{\infty} \left(a_n^2 e^{-\gamma_n^2 z} + b_n^2 e^{\gamma_n^2 z} \right) \phi_n^2(x),$$

(3.165)

where $\gamma = ik_z$ is the propagation constant, n is the mode or subband index, and $\phi_n(x)$ is the transverse eigenfunction. Note the definition of a and b such that a represents incoming waves on the discontinuity and b represents outgoing waves. We now introduce a matrix notation to write (3.165) as

$$\psi_1 = \phi_1^T \left(\mathbf{P}_1 \mathbf{a}_1 + \mathbf{P}_1^{-1} \mathbf{b}_1 \right)$$

$$\psi_2 = \phi_2^T \left(\mathbf{P}_2^{-1} \mathbf{a}_2 + \mathbf{P}_2 \mathbf{b}_2 \right),$$

(3.166)

where ϕ_i^T represents a column vector containing the transverse eigenfunction, while $\mathbf{a}_i$ and $\mathbf{b}_i$ are column vectors containing the incoming and outgoing wave coefficients in region i, respectively. The square matrix $\mathbf{P}_i$ is diagonal with elements

$$P_i(n, n) = e^{\gamma_n^i z}.$$

(3.167)

The diagonal elements are phase factors for propagating modes, and attenuation factors for evanescent modes. Both have to be included in the matching. If the boundaries in Fig. 3.30 are assumed to be hard walls (infinite barriers), then the transverse eigenfunctions become particularly simple:

$$\phi_n^1(x) = \sqrt{\frac{2}{w_1}} \sin\left(\frac{n\pi}{w_1}(x - c)\right)$$

$$\phi_n^2(x) = \sqrt{\frac{2}{w_2}} \sin\left(\frac{n\pi}{w_2}x\right) \quad n = 1, 2, 3, \ldots,$$

(3.168)

where c is the offset shown in Fig. 3.30, and the wavevector in the ith region is

$$k_z^i(n) = \sqrt{\frac{2m^*}{\hbar^2}(E - V_i) - \left(\frac{n\pi}{w_i}\right)^2},$$

(3.169)

where V_i is a constant potential energy in region i.

The continuity of the wavefunction and its normal derivative at the junction $z = 0$ of the two uniform sections leads to the two matrix equations

$$\phi_2^T(\mathbf{a}_2 + \mathbf{b}_2) = \begin{cases} \phi_1^T(\mathbf{a}_1 + \mathbf{b}_1), & c \leq x \leq c + w_1 \\ 0, & otherwise \end{cases}$$

(3.170)

$$\phi_1^T \mathbf{K}_1(\mathbf{a}_1 - \mathbf{b}_1) = -\phi_2^T \mathbf{K}_2(\mathbf{a}_2 - \mathbf{b}_2), \qquad c \leq x \leq c + w_1$$

(3.171)

where the elements of the diagonal matrices $\mathbf{K}_{1,2}$ are given by Eq. (3.169) on the respective sides for hard wall boundaries. To eliminate the position dependence, multiply (3.170) by $\phi_2(m)$ and (3.171) by $\phi(m)$, and integrate over all x to get the matrix equations

$$\mathbf{a}_2 + \mathbf{b}_2 = \mathbf{H}_1(\mathbf{a}_1 + \mathbf{b}_1),$$

$$\mathbf{a}_1 - \mathbf{b}_1 = \mathbf{H}_2(\mathbf{a}_2 - \mathbf{b}_2),$$

(3.172)

where

$$\mathbf{H}_1 = \mathbf{C}$$

and

$$\mathbf{H}_2 = -\mathbf{K}_1^{-1}\mathbf{C}^T\mathbf{K}_2,$$

(3.173)

where $\mathbf{C}^T$ is the transpose of $\mathbf{C}$. The elements of the matrix $\mathbf{C}$ represent the overlap integral of the transverse modes on the left and right side of the junction:

$$C_{nm} = \int_c^{c+w_1} \phi_1(n)\phi_2(m)\,dx,$$

(3.174)

where the off-diagonal elements characterize the strength of the intermode scattering or coupling due to the abrupt discontinuity. We are now in a position to determine the scattering matrix coupling the incoming to outgoing waves, as we discussed earlier for the simpler 1D case in Section 3.2.1. From the relations (3.172), the generalized scattering parameters for the step discontinuity are given by

$$\begin{bmatrix} \mathbf{b}_1 \\ \mathbf{b}_2 \end{bmatrix} = \begin{bmatrix} \mathbf{S}_{11} & \mathbf{S}_{12} \\ \mathbf{S}_{21} & \mathbf{S}_{22} \end{bmatrix} \begin{bmatrix} \mathbf{a}_1 \\ \mathbf{a}_2 \end{bmatrix}$$

(3.175)

where from Eq. (3.172) we have

$$
\begin{aligned}
\mathbf{S}_{11} &= (\mathbf{I} - \mathbf{H}_2\mathbf{H}_1)^{-1}(\mathbf{I} + \mathbf{H}_2\mathbf{H}_1), \\
\mathbf{S}_{12} &= -2(\mathbf{I} - \mathbf{H}_2\mathbf{H}_1)^{-1}\mathbf{H}_2, \\
\mathbf{S}_{21} &= \mathbf{H}_1(\mathbf{I} + \mathbf{S}_{11}), \\
\mathbf{S}_{22} &= \mathbf{H}_1\mathbf{S}_{12} - \mathbf{I},
\end{aligned}
\tag{3.176}
$$

where $\mathbf{I}$ is the identity matrix. As was noted earlier in Section 3.2.1, the scattering matrix defined above in terms of the wave amplitudes is not unitary in general unless normalized by the characteristic wave impedances.

In general, there are an infinite number of modes that one may include in the sums (3.165). Numerically this sum is truncated, which introduces error that can be reduced by increasing the number of modes retained. The number of modes on either side of the discontinuity is in general different, which affects the convergence of the technique. For a simple step discontinuity, it is found that for optimal convergence the ratio of modes should be taken the same as the ratio of the corresponding widths on either side [109].

For a uniform-width waveguide of length L, we can define a propagation scattering matrix in analogy to the 1D case (3.43):

$$
\begin{bmatrix} \mathbf{S}_{11} & \mathbf{S}_{12} \\ \mathbf{S}_{21} & \mathbf{S}_{22} \end{bmatrix} = \begin{bmatrix} \mathbf{0} & \mathbf{P} \\ \mathbf{P} & \mathbf{0} \end{bmatrix},
\tag{3.177}
$$

where the $\mathbf{P}$ is diagonal with elements

$$
P(n, n) = e^{ik_z(n)L},
\tag{3.178}
$$

which just introduces a phase shift in the transmitted waves.

The mode matching method in and of itself would not be useful if it could not be extended to multiple discontinuities encountered in more complicated geometries. Figure 3.31 shows a block diagram of a combination of two discontinuities (A and B) separated by a uniform waveguide section C. Here there are N_A modes included to the left of A, M_A modes on the right side, etc. The scattering matrices of the two discontinuities and the uniform section may be combined using the generalized scattering matrix technique (GSM) [110]. However, this technique suffers because equal numbers of modes have to be used in each section. The main problem is that in certain regions, a large number of these modes may be evanescent, which means that they are exponentially growing or decaying depending on the direction. Such modes may cause numerical instabilities even though they contribute nothing in reality to the actual solution. Therefore, an extended GSM technique [108] is better when unequal numbers of modes are used on either side of the discontinuities, and when evanescent modes that are unimportant may be uncoupled from the solution, as illustrated schematically in Fig. 3.31. The details of the composite scattering matrix corresponding to the three blocks shown in Fig. 3.31a are somewhat involved and not repeated here, but overall the procedure is found to be numerically stable. Once the procedure for combining two discontinuities together is prescribed, then increasingly more complicated structures may be reduced as shown in Fig. 3.31, where blocks B_1 and B_2 are combined together into B.

Once the GSM of a composite structure is calculated, the transmission coefficient from a propagating mode n at the input (side 1) to all modes at the output (side 2) may be expressed

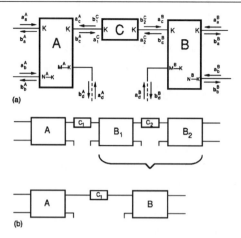

Fig. 3.31. (a) Two arbitrary discontinuities (A and B) connected by a uniform waveguide section (C). The mode amplitudes are grouped into the total number of modes K retained in the uniform waveguide section, and into the remaining modes as indicated in the figure. For the uncoupled evanescent modes, the uniform waveguide section C is treated as being infinitely long. (b) Successive GSM calculation for a quantum waveguide structure having three discontinutities [after Weisshaar et al., J. Appl. Phys. **70**, 355 (1991)].

in terms of the incoming and outgoing wave coefficients as

$$T_n(E) = \frac{\sum_{m=1}^{\infty} b_2(m) b_2^*(m) \left(k_z^{(2)}(m) + k_z^{(2)*}(m) \right)}{a_1(n) a_1^*(n) \left(k_z^{(1)}(n) + k_z^{(1)*}(n) \right)}$$

$$= \frac{\sum_{m=1}^{\infty} S_{21}(m,n) S_{21}^*(m,n) \left(k_z^{(2)}(m) + k_z^{(2)*}(m) \right)}{\left(k_z^{(1)}(n) + k_z^{(1)*}(n) \right)}, \qquad (3.179)$$

where the asterisk denotes the complex conjugate. If regions 1 and 2 are assumed to be connected to perfect phase-randomizing contacts, then the multi-terminal conductance may be defined according to Section 3.9.

Returning now to the problem of the quantized conductance in a split-gate structure, consider the geometry shown in the inset of Fig. 3.32. The split-gate QPC is modeled as a wide-narrow-wide structure with abrupt transitions between the narrow and wide regions, with hard wall boundaries. The composite scattering matrix consists of the contribution of the two discontinuities and the uniform waveguide section in the narrow region. Mimicking the effect of the gate bias in the real experiments, the width of the narrow region (and its length as well) are varied, and the total transmission coefficient at the Fermi energy is calculated using the mode matching method. The conductance is calculated from the two-terminal multichannel formula for $T = 0K$. It is immediately apparent that a plateau-like structure is obtained, even though the model QPC is nonadiabatic and has strong intermode coupling due to the discontinuities in the structure. One notes, however, that resonance-like structure is now superimposed on the conductance at the transition from one plateau to another. These resonances are cavity resonances associated with multiple reflections between the two discontinuities. As the length $L_{\min}$ is made shorter, these resonances smear out, and rounded plateaus are found. Whether such resonances are observable in real structures is still unclear. Figure 3.33 shows experimental data for a QPC in which resonance structure is evident at low temperature ($T = 40$ mK), which is suppressed with

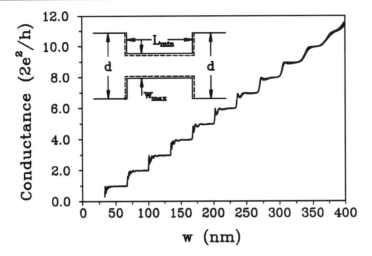

Fig. 3.32. Conductance for the split-gate geometry shown in the inset as a function of the constriction width W at low temperature. Here $L_{min} = 10$ nm while $d = 500$ nm. The Fermi energy is 5 meV in the wide region [after Weisshaar, Ph.D. dissertation (1991)].

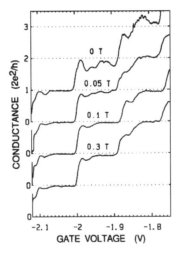

Fig. 3.33. Quantized conductance at low temperature $T = 40$ mK for a split-gate structure exhibiting oscillatory behavior. The oscillations are damped with application of a weak magnetic field [after van Wees *et al.*, Phys. Rev. B **43**, 12431 (1991), by permission].

a weak magnetic field [97]. However, it is hard to separate resonant effects due to localized impurities in the channel from geometric effects, although the magnetic field behavior is more suggestive of an impurity related effect due to reduced backscattering as discussed in Section 3.10.

3.7.2 Transport through bends

It is well known in microwave circuits that bends in a waveguide give rise to reflections and resonant behavior. Analogous behavior is expected for quantum waveguides that have

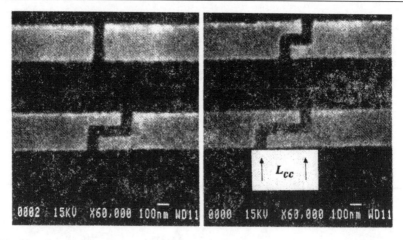

Fig. 3.34. Scanning electron micrographs of various double-bend geometries fabricated in a split-gate structure [after Wu *et al.*, J. Appl. Phys. **74**, 4590 (1993)].

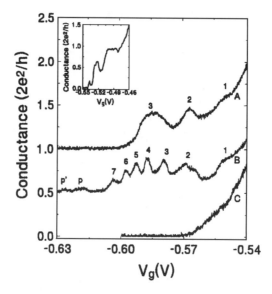

Fig. 3.35. The conductance versus gate bias of devices at $50\,mK$. Curves A, B, and C are for $L_{cc} = 177, 293$, and 377 nm. The inset shows the conductance of a device with $L_{cc} = 0$ [after Wu *et al.*, J. Appl. Phys. **74**, 4590 (1993)].

been investigated in different geometries. Wu *et al.* have investigated this phenomenon in two-terminal transport measurements for a double-bend structure shown in Fig. 3.34 [111], [112]. In this structure, electrons are forced to traverse two bend discontinuities which are separated by an effective cavity length L_{cc}. Figure 3.35 shows the measured conductance at low temperature for three different lengths as well as a straight geometry that exhibits a plateau of G_o with some resonance structure. Of particular interest is the curve labeled B which shows seven well-defined resonance peaks. The magnitude of these peaks is well below G_o, and for longer cavity lengths (curve C), the conductance plateau is

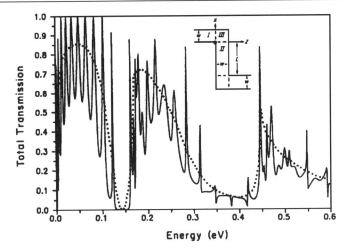

Fig. 3.36. Total transmission probability from mode 1 (T_1) as a function of energy ($w = 10$ nm, $L = 60$ nm). The dotted curve shows the transmission probability for a single right-angle bend; the solid line shows the transmission for the double bend configuration. The inset shows the modeled geometry and mode matching decomposition [after Weisshaar *et al.*, Appl. Phys. Lett. **55**, 2114 (1989)].

completely washed out. As discussed in Section 3.6.4, reduction in the conductance due to inhomogeneities becomes more severe as the waveguide length becomes longer.

From a theoretical standpoint, the double-bend structure may be modeled using the mode matching method described in the previous section. Figure 3.36 shows the idealized geometry of the double bend and the calculated transmission coefficient for both a single and double bend using the mode matching method [107]. The bend is decomposed into uniform regions I, II, III, etc., where the modal solutions are matched. The solution in the corner region III is found by superimposing solutions first for propagation in the x-direction and then in the z-direction. In comparing the single and double bend results, one sees that there is a broad resonance structure due to the presence of a single bend. The envelope of the double bend follows that of the single bend, but with additional fine resonance structure due to reflections in the cavity region between the two bends. Figure 3.37 shows the calculated conductance as a function of the width of the waveguide at low temperature and at $T = 2K$ for a structure similar to the experimental structure of Fig. 3.35. A number of resonance peaks are found in the lowest mode of the structure before pinch-off that resemble that of Fig. 3.35B. In contrast to the experiment, the calculated resonances are sharp at low temperature and approach G_o. If a temperature higher than the experiment is assumed, the calculated results more closely resemble experiment, although there is really no justification for this.

3.7.3 Lateral resonant tunneling

The previous section illustrated how the cavity due to a double bend results in resonant transmission through the waveguide, giving conductance peaks. We may consider other cavity structures, as illustrated by the double constriction in Fig. 3.38. The narrow-wide-narrow

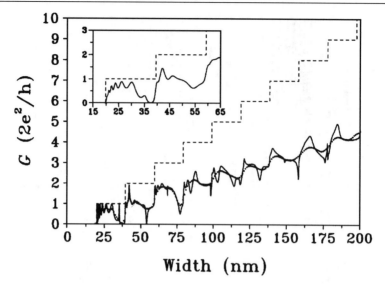

Fig. 3.37. Theoretical conductance for a double-bend constriction at zero temperature and at $4.2K$ (dotted curve). The dashed curve is the normalized conductance for an ideal split-gate structure without a bend. The inset shows an expanded view of the calculated conductance at $2K$ [after Wu *et al.*, Appl. Phys. Lett. **59**, 102 (1991)].

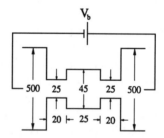

Fig. 3.38. Double constriction with a cavity region. The wide regions represent the 2DEG with a bias V_b. All dimensions are given in nanometers [after Weisshaar *et al.*, J. Appl. Phys. **70**, 355 (1991)].

structure is the generic design of a *quantum dot structure* realized using series QPCs as discussed extensively in Chapter 4. The structure is analogous to the double barrier resonant tunneling diode structure discussed in Section 3.1. In the structure of Fig. 3.38, electrons may tunnel through the narrow constrictions into quasi-bound states in the dot (wide) region, and out again. When the Fermi energy in the 2DEG outside of the dot is coincident with one of these states, a peak in the transmission occurs, giving rise to a peak in the current. Again assuming hard walls, we may cascade the scattering matrices of the discontinuities along the waveguide using the mode matching method. The calculated transmission coefficient versus energy is shown in Fig 3.39. The effect of a nonzero source drain bias, V_b, is simply modeled as a linear drop equally distributed across each narrow constriction. The transmission coefficient shows the generic onset of a single conductance plateau when multiplied by G_o, illustrating the theoretical result for two QPCs in series. Superimposed on this transition

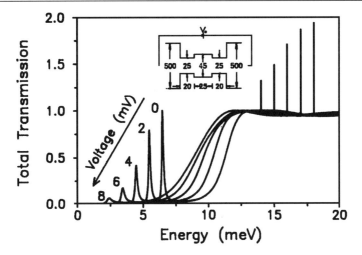

Fig. 3.39. Calculated transmission coefficient versus energy for the double constriction shown in the inset for various bias voltages as discussed in the text [after A. Weisshaar *et al.*, IEEE Elec. Dev. Lett. **12**, 2 (1991)].

from the tunneling regime to propagation in the first mode, a resonant peak is observed below the transmission plateau corresponding to tunneling from electrons outside the constriction through a quasi-bound state in the cavity (dot). The position shifts and decreases as the bias across the structure changes. The decrease is due to the increased asymmetry of the constrictions as the bias increases. If we consider nonlinear transport using (3.155), the current voltage characteristics of the double constriction exhibits a region of NDR, similar to a resonant tunneling diode. The calculated I-V is shown in Fig. 3.40 for various incremental changes in the width of the constrictions, simulating the effect of changing the gate bias in a QPC. The quasi-bound state energies, and hence the resonance position shift as the effective dimensions of the cavity change. Experimental evidence of lateral resonant tunneling through double barrier structures has been reported [113], [114]. Since lithographic limitations in these studies prevent the realization of lateral dimensions approaching the atomic dimensions of the layer structures in RTDs, NDR effects are weak, and only observed at low temperature. Further, the transport properties of double constriction structures may also be dominated by the Coulomb charging energy associated with an electron tunneling into and out of the cavity which goes beyond the simple description of lateral resonant tunneling introduced here. The interesting phenomenon of single electron charging is discussed in detail in Chapter 4.

3.7.4 Coupled waveguides

The parallels between quantum waveguides and their electromagnetic counterpart may be taken beyond single waveguide structures to consider coupled structures. Proposals have been made for realizing the optical analogues of directional couplers using phase coherent quantum waveguides [115], [116]. Figure 3.41 shows a schematic of a pair of coupled quantum waveguides and the electron probability distribution along the coupled section. The basic time-dependent behavior has already been introduced in connection with coupled quantum wells discussed earlier in Section 2.2.2. In the coupled well case for identical

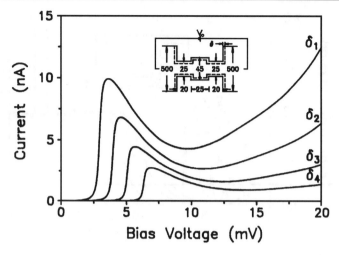

Fig. 3.40. I-V characteristics for the structure shown in the inset [after A. Weisshaar *et al.*, IEEE Elec. Dev. Lett. **12**, 2 (1991)].

wells, or for two wells biased into resonance, the electron probability density oscillates back and forth between the two wells as a function of time with a frequency proportional to the splitting of the two levels. In the case of coupled waveguides, the two waveguides have to be sufficiently close so that there is a coupling through evanescent modes. The time-dependent oscillations of the wavefunctions also correspond to spatial oscillations as shown schematically in Fig. 3.41, which may allow switching of a propagating wave from one waveguide to the other.

Coupled waveguides have been realized in practice [117] using split-gate structures consisting of three separate gates as shown by the SEM micrograph of Fig. 3.42. The middle gate acts as a tunable tunnel barrier between the two different waveguides which are independently biased. Figure 3.43 shows the measured current for a constant (small) drain bias, V_D, as the top gate bias is varied. The upper trace corresponds to the direct current from source to drain of the upper waveguide, which shows structure corresponding to quantized conductance plateaus $G = NG_o$. The lower curve is the current in the lower waveguide, which is lower in magnitude, and shows peak-like structure at bias voltages corresponding to the transition from one conductance plateau to the next. The peak-like structure of the tunneling current from one waveguide to the other is a direct result of the peaked 1D density of states for electrons in the upper waveguide. Since tunneling occurs perpendicular to the waveguide, there is no longer a cancellation of the velocity and density of states in the tunneling current, and so the current directly samples the density of occupied states. Thus the tunneling current in the second waveguide is in essence a type of tunneling spectroscopy of the occupied states of the first subband.

3.8 Lattice Green's function method

So far, in our analysis of quantum waveguides, we have considered relatively simple structures topologically that are constructed of various uniform sections of varying widths. In general, however, the modeling of real systems with three-dimensional potentials including random disorder requires one to go beyond simplified treatments such as the mode matching

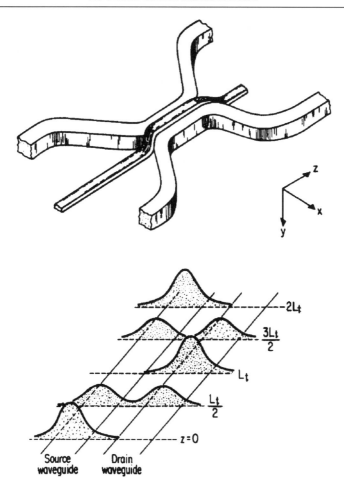

Fig. 3.41. Schematic diagram of two quantum waveguides coupled with a gate electrode (upper figure) and the corresponding propagation of a wavepacket (lower figure) along the coupled waveguides [after del Alamo and Eugster, Appl. Phys. Lett. **56**, 78 (1990), by permission].

method. Ultimately, one can take a fully numerical approach to solve the Schrödinger equation using finite differences or finite elements. Several groups have taken such an approach in using the finite elements method (FEM) to solve nanostructure problems in two dimensions [118]–[121].

In the present section, we look at a different approach to full numerical modeling of nanostructure systems based on the *lattice Green's function method*. In this approach, one solves the appropriate equations for the single-particle Green's function rather than the Schrödinger equation directly on a discretized mesh or "lattice" representing the underlying "continuous" system described in the effective mass approximation. As we show later in this section, knowledge of the Green's function allows the calculation of the transmission and reflection amplitudes which constitute the S-matrix of the system, and connects to our knowledge of transport via the Landauer-Büttiker model. The Green's function further connects us with other disciptions of transport such as linear response theory via the Kubo formula, which in the proper limits corresponds to the Landauer-Büttiker model. Thus,

(a)

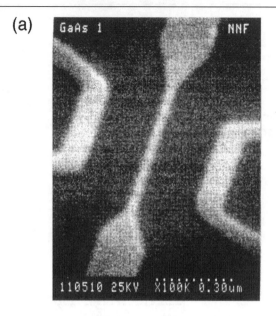

(b)

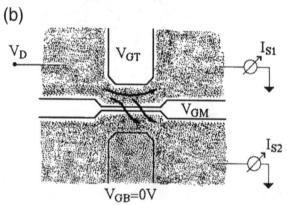

Fig. 3.42. Realization of a coupled waveguide structure using split-gate structures. The upper figure is the SEM micrograph of the sample and the lower figure is a schematic diagram showing the bias and current flow [after Eugster and del Alamo, Phys. Rev. Lett. **67**, 3586 (1991), by permission].

the popularity of Green's functions in nanostructure transport is not necessarily that they are computationally less burdensome than using FEM techniques for example, rather that they contain a rich connection with first principles transport theory and the associated theoretical heirarchy. In particular, the inclusion of various interactions such as dissipation due to phonons is handled quite readily as we discuss later in Chapter 7.

In the following, we first review some of the basic properties of single-particle Green's functions, followed by an introduction to the tight-binding Hamiltonian used to represent the discrete lattice representation of a particular nanostructure system, and the representation thereon by a lattice Green's function. Analytic forms are derived for certain special cases of interest such as infinite leads and finite width, uniform sections such as those we used in the mode matching method. The continuum limit is also investigated to determine the proper choice of the site and hopping energies. We then make the connection between the Green's

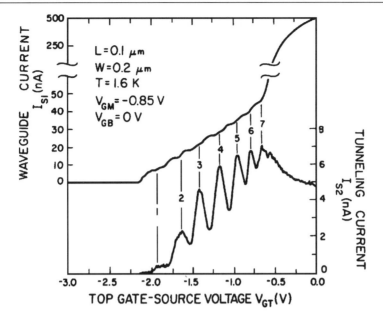

Fig. 3.43. D.C. I-V characteristics of a coupled waveguide structure [after Eugster and del Alamo, Phys. Rev. Lett. **67**, 3586 (1991), by permission].

function and the associated reflection and transmission amplitudes. The recursive Green's function method is introduced, and several example problems in 2D are investigated.

3.8.1 Single-particle Green's functions

To begin, we introduce the single-particle Green's function as discussed in many standard treatments of the subject (see, e.g., [122]). Discussion of more general Green's function formulations in connection with many-body problems (see, e.g., [123]) will be the subject of Section 5.3. In representation-independent operator notation, the time-independent Schrödinger equation may be written

$$[E - \mathsf{H}]\, \Psi = 0, \tag{3.180}$$

where H is a single-particle Hamiltonian. The Hermitian operator H obeys the following eigenvalue equation,

$$\mathsf{H}|\Phi_n\rangle = E_n|\Phi_n\rangle, \tag{3.181}$$

where the eigenfunctions $|\Phi_n\rangle$ usually form a complete set of states:

completeness: $\sum_n' |\Phi_n\rangle\langle\Phi_n| = 1,$ orthonormality: $\langle\Phi_n \mid \Phi_m\rangle = \delta_{nm},$

where the symbol $\sum_n' \equiv \sum_n + \int dn$ denotes that the eigenvalues include both a summation over discrete states as well as integration over the continuous part of the spectrum. All eigenvalues E_n are real due to the Hermiticity of H.

We may define the *Green's function (operator)*, $\mathsf{G}(E)$, corresponding to (3.180), as

$$[E - \mathsf{H}]\mathsf{G}(E) = 1, \tag{3.182}$$

where the Green's function is subject to the same boundary conditions as the wavefunction Ψ. A formal solution of this equation may be written as

$$G(E) = \frac{1}{E - H}, \tag{3.183}$$

which is defined everywhere except at the singularities, $E = E_n$. This definition differs slightly from that defined later by a factor of $\hbar$. Green's function operators are then defined by a limiting procedure from Eq. (3.183) according to

$$G^{\pm}(E) = \lim_{\eta \to 0^+} G(E \pm i\eta). \tag{3.184}$$

This limiting form is introduced when we Fourier transform the Green's function operator back to the time domain through a closed contour integration in the complex plane to obtain the time dependent Green's function or propagator. The plus and minus solutions represent different choices in avoiding the poles of Eq. (3.183) and lead to causal or anticausal solutions depending on the choice of sign as discussed below. Implicitly assuming η to represent an infinitesimal, positive real quantity, $\eta \to 0^+$, the Green's function operators are expressed as [124]

$$G^{\pm}(E) \equiv \frac{1}{E - H \pm i\eta}. \tag{3.185}$$

The $G^{\pm}$ are implicitly understood to be always replaced by $(E - H \pm i\eta)^{-1}$ and to be considered in the proper limit $\eta \to 0^+$.

We proceed by expressing the Green's functions in terms of the eigenvalues and eigenfunctions of H using the completeness relation,

$$G^{\pm}(E) = \frac{1}{E - H \pm i\eta} \sum_{n}' |\Phi_n\rangle\langle\Phi_n| = \sum_{n}' \frac{1}{E - H \pm i\eta} |\Phi_n\rangle\langle\Phi_n|. \tag{3.186}$$

Expanding $(E - H \pm i\eta)^{-1}$ into a power series, it immediately follows that

$$G^{\pm}(E) = \sum_{n}' \frac{|\Phi_n\rangle\langle\Phi_n|}{E - E_n \pm i\eta}, \tag{3.187}$$

which we will use in deriving analytic forms of certain lattice Green's functions.

Our main interest in the following will be in the G^+ operator, sometimes referred to as a *propagator*. The origin of this name becomes more apparent if we consider the Fourier transform of (3.185) above, which is easily evaluated using contour integration:

$$iG^+(t) = \frac{i}{2\pi} \int dE \, G^+(E) e^{-iEt/\hbar} = \theta(t) \, e^{-iHt/\hbar} e^{-\eta t/\hbar}, \tag{3.188}$$

where $\theta(t)$ denotes the step function. We will define the propagator more formally in Section 5.3 as the Kernel or Green's function for the Schrödinger equation. The step function is a consequence of the fact that for $t < 0$, we have to close the contour in the upper half of the complex E-plane, in which case no poles are enclosed, giving zero contribution to the integral. $G^+(t)$ is therefore the *retarded* Green's function operator

$$G^+(t) = G_r(t), \tag{3.189}$$

Since it is zero for all times less than zero. The operator $iG_r(t - t')$ above governs the "propagation" of the state vector under the influence of the full Hamiltonian H at any time

$t \geq t'$, due to the time evolution operator multiplying the step function. Likewise, we may associate $G^-(t)$ with the *advanced* Green's function, $iG_a(t - t')$, which propagates in the opposite direction in time. In the following, we will use the notation G^+ and G_r interchangeably for the retarded Green's function, which will be the central quantity of interest in connecting to the scattering matrix.

3.8.2 Tight-binding Hamiltonian

A type of Hamilton that is frequently encountered to describe a lattice (or "grid" in the numerical sense) is the *tight-binding Hamiltonian*. Let **r** denote a vector representing the spatial position of a certain lattice site. In two dimensions, this vector corresponds to a site $\mathbf{r} \equiv (m, n)$ with indices m, n in the x-, y-directions. If one calls $|\mathbf{r}\rangle$ a state that is *centered* at location **r**, the complete set of states $\{|\mathbf{r}\rangle\}$ obeys the following:

$$\text{completeness:} \quad \sum_{\mathbf{r}} |\mathbf{r}\rangle\langle\mathbf{r}| = 1, \qquad \text{orthonormality:} \quad \langle\mathbf{r}|\mathbf{r}'\rangle = \delta_{\mathbf{r},\mathbf{r}'}.$$

The tight-binding Hamiltonian is introduced as

$$H = \sum_{\mathbf{r}} |\mathbf{r}\rangle\epsilon_{\mathbf{r}}\langle\mathbf{r}| + \sum_{\mathbf{r},\mathbf{r}'} |\mathbf{r}\rangle V_{\mathbf{r},\mathbf{r}'}\langle\mathbf{r}'|, \qquad (3.190)$$

where $\epsilon_{\mathbf{r}}$ is the *on-site energy* at **r**, and $V_{\mathbf{r}\mathbf{r}'}$ is the *hopping energy* between $\mathbf{r}'$ and **r**. The most common assumption in the tight-binding model is that only *nearest neighbor* interactions are important. If $\mathbf{\Delta_r}$ represents the vectors from **r** to all its nearest neighbor sites, then this assumption of nearest neighbor interactions means that Eq. (3.190) may be written

$$H = \sum_{\mathbf{r}} |\mathbf{r}\rangle\epsilon_{\mathbf{r}}\langle\mathbf{r}| + \sum_{\mathbf{r},\mathbf{\Delta_r}} |\mathbf{r}\rangle V_{\mathbf{r},\mathbf{\Delta_r}}\langle\mathbf{r} + \mathbf{\Delta_r}|. \qquad (3.191)$$

A convenient and commonly used way of writing a tight-binding Hamiltonian is to replace the dyadic products of states $\{|\mathbf{r}\rangle\}$ by new operators using second quantization. One introduces *annihilation operators*, $a_{\mathbf{r}}$, and their adjoint *creation operators*, $a_{\mathbf{r}}^+$, with each operator acting on grid point **r**. These operators corresponding to our tight-binding lattice obey the following fundamental bosonic commutator relations:

$$\begin{aligned} \left[a_{\mathbf{r}}, a_{\mathbf{r}'}^+\right] &= \delta_{\mathbf{r},\mathbf{r}'} \\ \left[a_{\mathbf{r}}, a_{\mathbf{r}'}\right] &= 0 = [a_{\mathbf{r}}^+, a_{\mathbf{r}'}^+]. \end{aligned} \qquad (3.192)$$

The Hamiltonian (3.191) may be expressed in terms of these new operators. For simplicity, we assume a constant hopping potential V between nearest neighbor sites and obtain

$$H = \sum_{\mathbf{r}} \epsilon_{\mathbf{r}} a_{\mathbf{r}}^+ a_{\mathbf{r}} + \sum_{\mathbf{r},\mathbf{\Delta_r}} \left\{ V a_{\mathbf{r}}^+ a_{\mathbf{r}+\mathbf{\Delta_r}} + (\text{hermitian conjugate})\right\}. \qquad (3.193)$$

The Hamiltonian in this representation has diagonal contributions given by the on-site energy $\epsilon_{\mathbf{r}}$ and the hopping contributions, which create and annihilate excitations on neighboring lattice sites and annhilate or create them respectively on site **r**.

To illustrate the matrix form of a tight-binding Hamiltonian, consider a one-dimensional chain of N = 6 lattice sites that exhibit nearest-neighbor hopping (Fig. 3.44). The on-site energy of site n is denoted by ϵ_n, whereas the hopping energy between site n' and n is represented in general by $V_{nn'}$.

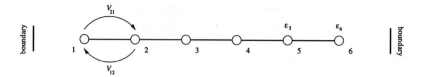

Fig. 3.44. Finite one-dimensional tight-binding chain of sites $n \in \{1, \ldots, N \equiv 6\}$.

Taking matrix elements $\langle n|H|n'\rangle$, where the indices n, n' run over all sites $\{1, \ldots, N\}$, the explicit form of the operator H corresponding to this one-dimensional problem is given by the $N \times N$-matrix

$$
H = \begin{pmatrix}
\epsilon_1 & V_{12} & 0 & 0 & 0 & 0 \\
V_{21} & \epsilon_2 & V_{23} & 0 & 0 & 0 \\
0 & V_{32} & \epsilon_3 & V_{34} & 0 & 0 \\
0 & 0 & V_{43} & \epsilon_4 & V_{45} & 0 \\
0 & 0 & 0 & V_{54} & \epsilon_5 & V_{56} \\
0 & 0 & 0 & 0 & V_{65} & \epsilon_6
\end{pmatrix}. \tag{3.194}
$$

3.8.3 Lattice Green's functions

According to Eq. (3.185), the Green's function operator $G^{\pm}(E)$ depends directly on the Hamiltonian. We refer to a Green's function represented by such a tight-binding Hamiltonian describing a discretized structure as a "*lattice Green's function*" (LGF). It is important to note that the following explicit matrix forms of the Green's function operators are given exclusively in site representation, since the tight-binding Hamiltonian was most naturally introduced in this particular representation. In this section we show in detail how the Green's functions can be transformed to any other representation.

One-dimensional lattice chains

The Green's function corresponding to a tight-binding Hamiltonian of a *finite* lattice can be obtained in general by matrix inversion of (3.185). The Green's function corresponding to the example considered in Figure 3.44 reads

$$
G^{\pm}(E) = \begin{pmatrix}
E_1 & -V_{12} & 0 & 0 & 0 & 0 \\
-V_{21} & E_2 & -V_{23} & 0 & 0 & 0 \\
0 & -V_{32} & E_3 & -V_{34} & 0 & 0 \\
0 & 0 & -V_{43} & E_4 & -V_{45} & 0 \\
0 & 0 & 0 & -V_{54} & E_5 & -V_{56} \\
0 & 0 & 0 & 0 & -V_{65} & E_6
\end{pmatrix}^{-1},
$$

where for brevity we write

$$
E_n \equiv E - \epsilon_n \pm i\eta. \tag{3.195}
$$

Generally speaking, a one-dimensional tight-binding Hamiltonian with nearest neighbor hopping in the site representation is always expressed as a *tridiagonal* matrix, that is, a matrix in which all elements vanish except for those of the principal and its two adjacent

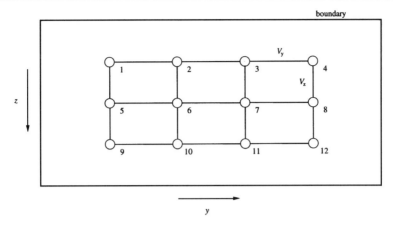

Fig. 3.45. Transverse slice of a three-dimensional grid. Uniform spacing is assumed in each transverse direction y, z.

diagonals. Not surprisingly, the same result is found in the finite difference discretization of the Schödinger equation itself. On the other hand, the Green's function in site representation is given by a *full* matrix.

Two-dimensional lattice slices

In Figure 3.45, we schematically show a two-dimensional plane (or "slice" of an underlying three-dimensional system) consisting of $N \times L = 4 \times 3$ sites in the transverse directions. If the system is uniform in both the transverse directions, hopping between the sites is described by constant energies V_y and V_z. It is inconvenient to work with two site indices for each grid point. Instead, one introduces a scheme that numbers the sites in a row-major fashion from left to right and from top to bottom. This labeling of sites is called "*natural ordering*." The tight-binding Hamiltonian corresponding to a grid of dimension $N \times L$ is given in site representation as a $(N \cdot L) \times (N \cdot L)$ matrix. As a consequence, the corresponding Green's function operator is obtained by the following matrix inversion:

$$
G^{\pm}(E) =
\begin{pmatrix}
\begin{array}{cccc|cccc|cccc}
E_1 - V_y & & & & -V_z & & & & & & & \\
-V_y & E_2 - V_y & & & & -V_z & & & & & & \\
& -V_y & E_3 - V_y & & & & -V_z & & & & & \\
& & -V_y & E_4 & & & & -V_z & & & & \\
\hline
-V_z & & & & E_5 - V_y & & & & -V_z & & & \\
& -V_z & & & -V_y & E_6 - V_y & & & & -V_z & & \\
& & -V_z & & & -V_y & E_7 - V_y & & & & -V_z & \\
& & & -V_z & & & -V_y & E_8 & & & & -V_z \\
\hline
& & & & -V_z & & & & E_9 - V_y & & & \\
& & & & & -V_z & & & -V_y & E_{10} - V_y & & \\
& & & & & & -V_z & & & -V_y & E_{11} - V_y & \\
& & & & & & & -V_z & & & -V_y & E_{12}
\end{array}
\end{pmatrix}^{-1},
$$

with the same shorthand notation for the E_n as in (3.195).

In general, this type of matrix whose inversion yields $G^{\pm}$ is called a *block-tridiagonal* matrix. It consists of $(L \times L)$ submatrices, each of dimension $(N \times N)$. Whereas the submatrices of the principal block diagonal are tridiagonal, the submatrices of the two adjacent outer block diagonals are diagonal; all other submatrices vanish.

Representations of lattice Green's functions

As a consequence of the general operator notation used, all the expressions in Section 3.8.1 are valid for *any* representation. In the site representation, the matrix elements of the Green's function operator are obtained by "sandwiching" $G^{\pm}$ between the states $|\mathbf{r}\rangle$, $|\mathbf{r}'\rangle$ that are completely determined by the site indices $\mathbf{r} \equiv (n, m, l)$, $\mathbf{r}' \equiv (n', m', l')$. Site states for different directions belong to different vector spaces. Therefore, a state of the system's entire Hilbert space can be written as a direct product of the state kets of the separate directions,

$$|\mathbf{r}\rangle = |m\rangle \otimes |n\rangle \otimes |l\rangle \equiv |m\rangle |n\rangle |l\rangle.$$

For any particular $\mathbf{r}, \mathbf{r}'$, the Green's *function* (not operator!) is written simply as

$$G^{\pm}(\mathbf{r}, \mathbf{r}'; E) = G^{\pm}(mnl, m'n'l'; E) \equiv \langle \mathbf{r}|G^{\pm}(E)|\mathbf{r}'\rangle. \qquad (3.196)$$

The *general reciprocity relation* between the Green's functions G^{+} and G^{-} follows from the operator definitions

$$G^{+}(\mathbf{r}, \mathbf{r}', E) = [G^{-}(\mathbf{r}', \mathbf{r}, E)]^{*}, \qquad (3.197)$$

where the $*$ denotes complex conjugation. Similar expressions can be found for any representation. Thus we can restrict ourselves to just one of the two species, $G^{\pm}$, since the other form in each case can always be obtained by the general relationship (3.197). As the transmission and reflection amplitudes are derived in Section 3.8.5 from G^{+}, all Green's function operators called G throughout this section are understood to be a shorthand notation for G^{+}

There are many other representations possible besides the site representation; these are based on characteristic complete sets of states. A representation of particular interest is the mixed representation as it appears in the relation between the Green's function and the transmission or reflection amplitudes discussed in Section 3.8.5. When we consider quantum waveguide structures, for example, the transverse modes form a complete set of states that may be used as a representation for the Green's function in the transverse direction (i.e., the laterally confined direction). In this case, the site index m is kept for the longitudinal propagation direction, while the transverse directions are now represented by the transverse mode numbers ν and λ. The Green's function corresponding to the state ket $|\mathbf{s}\rangle \equiv |m, \nu, \lambda\rangle = |m\rangle |\nu\rangle |\lambda\rangle$ is obtained from the site representation Green's function as

$$
\begin{aligned}
G(m\nu\lambda, m'\nu'\lambda'; E) &= \langle m, \nu, \lambda |G(E)|m', \nu', \lambda'\rangle = \langle \nu, \lambda |G_{mm'}|\nu', \lambda'\rangle \\
&= \sum_{\substack{n,l \\ n',l'}} \langle \nu, \lambda | n, l\rangle \langle n, l|G_{mm'}|n', l'\rangle \langle n', l'| |\nu', \lambda'\rangle \\
&= \sum_{\substack{n,l \\ n',l'}} \langle \lambda | l\rangle \langle \nu | n\rangle \cdot \langle n, l|G_{mm'}|n', l'\rangle \cdot \langle n' | \nu'\rangle \langle l' | \lambda'\rangle
\end{aligned}
$$

$$= \sum_{\substack{n,l \\ n',l'}} \langle l \mid \lambda \rangle^* \langle n \mid \nu \rangle^* \cdot \mathsf{G}(mnl, m'n'l'; E) \cdot \langle l' \mid \lambda' \rangle \langle n' \mid \nu' \rangle$$

$$= \underbrace{\mathsf{Z}^*_{l\lambda} \mathsf{Y}^*_{n\nu}}_{} \cdot \mathsf{G}(mnl, m'n'l'; E) \cdot \underbrace{\mathsf{Z}_{l'\lambda'} \mathsf{Y}_{n'\nu'}}_{}$$

$$= \underset{\substack{\nu,\lambda \\ n,l}}{\mathsf{U}^*} \mathsf{G}(mnl, m'n'l'; E) \underset{\substack{\nu',\lambda' \\ n',l'}}{\mathsf{U}} .$$

For the sake of brevity we have used the *Einstein summation convention* in the last line above, that is, we sum over any indices that appear at least twice in an expression.

In terms of notation and implementation, it is advantageous to express the transformation between different representations by matrices. Careful inspection of above transformation given in explicit component notation reveals that one can introduce the matrices Z and Y and their adjoints Z^+ and Y^+ by the matrix elements

$$(\mathsf{Z})_{l\lambda} = \mathsf{Z}_{l\lambda}, \qquad (\mathsf{Y})_{l\lambda} = \mathsf{Y}_{l\lambda}, \qquad (\mathsf{Z}^+)_{l\lambda} = \mathsf{Z}^*_{\lambda l}, \qquad (\mathsf{Y}^+)_{\lambda l} = \mathsf{Y}^*_{l\lambda}.$$

As an illustration the transformation matrix corresponding to a system of N transverse sites is explicitly given as

$$\mathsf{Y} = \begin{pmatrix} \langle n = 1 \mid \nu = 1 \rangle & \langle n = 1 \mid \nu = 2 \rangle & \cdots & \langle n = 1 \mid \nu = \mathsf{N} \rangle \\ \langle n = 2 \mid \nu = 1 \rangle & \langle n = 2 \mid \nu = 2 \rangle & \cdots & \langle n = 2 \mid \nu = \mathsf{N} \rangle \\ \vdots & \vdots & \ddots & \vdots \\ \langle n = \mathsf{N} \mid \nu = 1 \rangle & \langle n = \mathsf{N} \mid \nu = 2 \rangle & \cdots & \langle n = \mathsf{N} \mid \nu = \mathsf{N} \rangle \end{pmatrix}.$$

The matrices Y are made up of the two complete sets of orthonormal state vectors $\{\mid n \rangle\}$ and $\{\mid \nu \rangle\}$. Therefore, it can be easily proven that they are *unitary* matrices,

$$\text{Unitarity:} \qquad \mathsf{Y}\mathsf{Y}^+ = 1_Y = \mathsf{Y}^+\mathsf{Y},$$

by examining the matrix elements.

$$\text{diagonal elements:} \qquad (\mathsf{Y}\mathsf{Y}^+)_{ii} = \sum_{k=1}^{\mathsf{N}} \langle n = i \mid \nu = k \rangle \langle \nu = k \mid n = i \rangle \equiv 1$$

$$\text{off-diagonal elements:} \qquad (\mathsf{Y}\mathsf{Y}^+)_{ij} = \sum_{k=1}^{\mathsf{N}} \langle n = i \mid \nu = k \rangle \langle \nu = k \mid n = j \rangle \equiv 0$$

Completely analogous expressions hold for the matrices Z.

The final transformation matrix U and its adjoint U^+ are given in terms of the matrices corresponding to the two transverse directions as

$$\mathsf{U} = \mathsf{Z} \otimes \mathsf{Y}, \qquad \mathsf{U}^+ = (\mathsf{Z} \otimes \mathsf{Y})^+ = \mathsf{Z}^+ \otimes \mathsf{Y}^+.$$

Since Z and Y act on different subspaces, U is easily shown to be unitary as well:

$$\mathsf{U}\mathsf{U}^+ = (\mathsf{Z} \otimes \mathsf{Y})(\mathsf{Z} \otimes \mathsf{Y})^+ = \mathsf{Z}\mathsf{Z}^+ \otimes \mathsf{Y}\mathsf{Y}^+ = 1_Z \otimes 1_Y = 1_{Z \otimes Y}.$$

The unitary matrices U transform the Green's function from site to transverse mode representation (3.198), and vice versa (3.199). The similarity transformations in component notation read

$$G(m\nu\lambda, m'\nu'\lambda'; E) = \underset{\substack{\nu,\lambda \\ n,l}}{U^+} G(mnl, m'n'l'; E) \underset{\substack{\nu',\lambda' \\ n',l'}}{U}, \tag{3.198}$$

$$G(mnl, m'n'l'; E) = \underset{\substack{\nu,\lambda \\ n,l}}{U} G(m\nu\lambda, m'\nu'\lambda'; E) \underset{\substack{\nu',\lambda' \\ n',l'}}{U^+}. \tag{3.199}$$

If we denote the Green's function operator corresponding to the transverse mode representation by $\tilde{G}$, the transformations can be compactly written as

$$G = U\tilde{G}U^+, \qquad \tilde{G} = U^+GU.$$

The rather formal aspects of unitary transformations as shown in general terms above will reappear in a more explicit form in the next section, in which we derive analytic forms for the LGF.

3.8.4 Analytic forms of lattice Green's functions

Green's functions corresponding to certain types of lattices may be obtained in analytic form rather than by matrix inversion. In the particular case of semiinfinite lattices, which represent the ideal leads introduced earlier in the Landauer-Büttiker picture, matrix inversion *cannot* be used to obtain the Green's function due to the infinite size of the underlying lattice. Also, propagators corresponding to perfect uniform sections are taken advantage of more efficiently by using the analytic forms as well. Once we have determined the Green's function for a finite section, we can combine sections of different widths, as we did in the mode matching of Section 3.7.1 to obtain the transmission and reflection coefficients of more complicated geometries such as bends and cavity structures.

In the following, we give a detailed derivation of the lattice Green's function for a perfect semiinfinite strip, following in part [125]. While starting from expressions that are valid for a more general lattice, constraints are introduced that reflect the particular geometry. We also show how the result is readily adapted to yield the analytic lattice Green's function of a finite uniform section, and how the method may be generalized to include nonzero magnetic fields.

Semiinfinite lead

We start with the tight-binding Hamiltonian of a semiinfinite lead including nearest neighbor hopping as shown in Fig. 3.46. The lead terminates at longitudinal site, m_0, and extends infinitely to the right. The general form is given as

$$\begin{aligned}
H^{\text{semi}}(m_0) = \sum_{m=m_0}^{\infty} \sum_{n=1}^{N} \Big\{ & \epsilon_{mn} a_{mn}^+ a_{mn} + V_x^{mn} a_{mn}^+ a_{m+1,n} \\
& + V_y^{mn} a_{mn}^+ a_{m,n+1} + \left(V_x^{m-1,n} \right)^* a_{mn}^+ a_{m-1,n} \\
& + \left(V_y^{m,n-1} \right)^* a_{mn}^+ a_{m,n-1} \Big\},
\end{aligned} \tag{3.200}$$

where m and n again label the longitudinal and transverse site locations. The Hamiltonian in (3.200) is written in second-quantized form using the bosonic site annhilation operator, a_{mn},

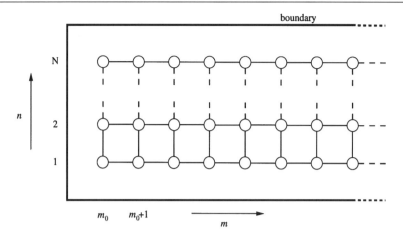

Fig. 3.46. Semiinfinite right lead with site indices $n = 1, \ldots, N$ and $m = m_0, \ldots, \infty$.

and its adjoint creation operator, a_{mn}^+. They obey the same commutator relations as (3.192):

$$[a_{kl}, a_{mn}^+] = \delta_{km} \delta_{ln}$$
$$[a_{kl}, a_{mn}] = 0 = [a_{kl}^+, a_{mn}^+]. \tag{3.201}$$

In (3.200), V_x^{mn} and V_y^{mn} denote the hopping energy onto site (m, n) from its longitudinal and transverse neighbor, respectively. In general they are complex if one includes the presence of a magnetic field in the formalism as discussed later. ϵ_{mn} represents the total site energy, which is the sum of the intrinsic lattice site energy, $\epsilon^{\text{lattice}}$, and an additional site potential, $\epsilon^{\text{potential}}$. The intrinsic site energy, $\epsilon^{\text{lattice}}$, is often neglected (see for example [126]). Such an assumption does not hold in a rigorous treatment, since the continuum limit would not yield the correct solution of the undiscretized problem.

As will be shown later in this section (p. 170), two assumptions ensure convergence to the free-particle case. First, the intrinsic site energy is set equal to the negative sum of the nearest-neighbor hopping energies,

$$\epsilon_{mn}^{\text{lattice}} = -\left(V_x^{mn} + V_y^{mn} + \left(V_x^{m-1,n}\right)^* + \left(V_y^{m,n-1}\right)^*\right), \tag{3.202}$$

which yields the same zeros of energy in the undiscretized case as in the continuous case. Second, the longitudinal hopping energy is properly chosen as

$$V_x = -\frac{\hbar^2}{2m^*} \frac{1}{a_x^2}, \tag{3.203}$$

where a_x is the longitudinal grid spacing; an equivalent expression holds for the transverse direction. The site potential $\epsilon^{\text{potential}}$ is the sum of all effects that cause a change of the otherwise constant site energy; this includes impurities, electric fields, soft-walls, and so on. In the case of a perfect lattice as considered here, $\epsilon^{\text{potential}}$ vanishes.

Next we solve the time-independent Schrödinger equation in order to obtain the eigenfunctions and eigenenergies. Knowledge of these permits us to determine the lattice Green's function according to (3.187). For the tight-binding Hamiltonian given in (3.200), the eigenvalue problem may be written

$$H^{\text{semi}} |\Psi^{\mu\nu}\rangle = E^{\mu\nu} \cdot |\Psi^{\mu\nu}\rangle, \tag{3.204}$$

where $E^{\mu\nu}$ is the total electron energy measured with respect to the conduction band edge. The eigenstates may be expanded in terms of

$$|\Psi^{\mu\nu}\rangle = \sum_{mn} c_{mn}^{\mu\nu} a_{mn}^{+} |0\rangle,$$

where $|0\rangle$ is the vacuum state, and the coefficients are given as $c_{mn}^{\mu\nu} = \langle 0|a_{mn}|\Psi^{\mu\nu}\rangle$. Substituting this expansion into (3.204) and operating using (3.200) results in a set of simultaneous equations for the expansion coefficients, $c_{mn}^{\mu\nu}$. Using the commutator relations (3.201), a comparison of coefficients yields

$$\left(V_x^{m-1,n}\right)^* c_{m-1,n}^{\mu\nu} + V_x^{mn} c_{m+1,n}^{\mu\nu} + \left(V_y^{m,n-1}\right)^* c_{m,n-1}^{\mu\nu} + V_y^{mn} c_{m,n+1}^{\mu\nu} = \left(E^{\mu\nu} - \epsilon_{mn}\right) \cdot c_{mn}^{\mu\nu}. \tag{3.205}$$

Here, as well as for the rest of the derivation, we implicitly assume $m \geq m_0$.

We now assume constant hopping energies, V_x and V_y, which correspond to an equally spaced grid in both the longitudinal and transverse directions with lattice spacing a_x and a_y, respectively. The hopping energies may be complex in general as we discuss how to incorporate the effects of a nonzero magnetic field. Introducing the separable solution

$$c_{mn}^{\mu\nu} = \phi_m^\mu \cdot \chi_n^\nu, \qquad E^{\mu\nu} = E^\mu + E^\nu,$$

splits Eq. (3.205) into longitudinal and transverse parts

$$V_x \phi_{m+1}^\mu + V_x^* \phi_{m-1}^\mu - (E^\mu - 2\mathrm{Re}V_x)\phi_m^\mu = 0 \tag{3.206}$$

$$V_y \chi_{n+1}^\nu + V_y^* \chi_{n-1}^\nu - (E^\nu - 2\mathrm{Re}V_y)\chi_n^\nu = 0. \tag{3.207}$$

These two equations are simply finite difference forms of the 1D Schrödinger equation in the longitudinal and transverse directions. We may therefore form solutions to the difference equations in terms of the well-known analytic solutions to the continuous problems that satisfy the appropriate hard wall boundary conditions. Their solutions are the following:

- Longitudinal wavefunction, ϕ_m^μ: the hard wall boundary condition at the left end of the strip, $\phi_{m_0-1}^\mu \equiv 0$ yields

$$\phi_m^\mu = \sqrt{\frac{2}{\pi}} \cdot \sin\left(\mu(m - m_0 + 1)\right) \cdot e^{im\varphi_x}, \tag{3.208}$$

  where we have written $V_x = -|V_x| \cdot e^{-i\varphi_x}$. Substituting (3.208) into (3.206), one obtains the longitudinal energy dispersion relation

$$E^\mu = -2\mathrm{Re}\, V_x - 2|V_x| \cos\mu. \tag{3.209}$$

  Here μ is the *continuous* longitudinal quantum number. In comparison with the continuum solution of the same problem, it is easily identified as the product of the longitudinal wavevector and the longitudinal grid spacing, $\mu = k_x^\mu \cdot a_x$.

- Transverse wavefunction χ_n^ν: hard wall boundary conditions at the lateral edges of the strip, $\chi_0^\nu \equiv 0 \equiv \chi_{N+1}^\nu$, yield

$$\chi_n^\nu = \sqrt{\frac{2}{N+1}} \cdot \sin\left(\frac{\pi\nu n}{N+1}\right) \cdot e^{in\varphi_y}, \tag{3.210}$$

  where we have used $V_y = -|V_y| \cdot e^{-i\varphi_y}$. The transverse phase φ_y vanishes for zero magnetic fields. At the end of this section we show how φ_y is related to a perpendicular magnetic field. Substituting (3.210) into (3.207), one obtains the transverse energy dispersion relation

$$E^\nu = -2\mathrm{Re}\,V_y - 2|V_y|\cos\left(\frac{\pi\nu}{N+1}\right), \tag{3.211}$$

where ν is the *discrete* mode number of the transverse wave function, $\nu = 1, \ldots, N$.

The total energy dispersion relation reads

$$E^{\mu\nu} = -2\mathrm{Re}(V_x + V_y) - 2|V_x|\cos\mu - 2|V_y|\cos\left(\frac{\pi\nu}{N+1}\right), \tag{3.212}$$

where the solutions of the time-independent Schrödinger equation are the eigenstates

$$|\Psi^{\mu\nu}\rangle = \sum_{mn} \phi_m^\mu \chi_n^\nu\, a_{mn}^+|0\rangle. \tag{3.213}$$

With the above definition of the eigenstates, the retarded lattice Green's function may now be defined according to (3.187):

$$G(mn, m'n'; E) \equiv G_{mm'}(n, n') \equiv \sum_{\mu\nu}{}' \frac{a_{mn}^+|\Psi^{\mu\nu}\rangle\langle\Psi^{\mu\nu}|a_{m'n'}}{E - E^{\mu\nu} + i\eta}, \tag{3.214}$$

where η is understood to be an infinitesimal, positive real number.

Inserting the known eigenstates (3.213) into (3.214), the Green's function becomes

$$G_{mm'}^{\mathrm{semi}}(n, n') = \int_0^\pi d\mu \sum_{\nu=1}^N \frac{\phi_m^\mu\left(\phi_{m'}^\mu\right)^* \chi_n^\nu\left(\chi_{n'}^\nu\right)^*}{E - E^{\mu\nu} + i\eta}. \tag{3.215}$$

In calculations involving leads attached to scattering regions, the Green's function is needed only at the left end of the semiinfinite strip. To obtain the Green's function there, we set $m \equiv m_0 \equiv m'$ and insert the longitudinal wavefunctions

$$G_{m_0}^{\mathrm{semi}}(n, n') = \sum_{\nu=1}^N \chi_n^\nu\left(\chi_{n'}^\nu\right)^* e^{i\varphi_y(n-n')} \cdot \frac{2}{\pi}\int_0^\pi d\mu \frac{\sin^2\mu}{E - (E^\mu + E^\nu) + i\eta} \equiv G^{\mathrm{semi}}. \tag{3.216}$$

The integral over the continuous variable μ,

$$\frac{2}{\pi}\int_0^\pi d\mu \frac{\sin^2\mu}{\underbrace{E + 2\mathrm{Re}(V_x + V_y) + 2|V_y|\cos\left(\frac{\pi\nu}{N+1}\right) + i\eta}_{\equiv p(\nu)\equiv p} + \underbrace{2|V_x|\cos\mu}_{\equiv q}},$$

may be evaluated by contour integration to yield [127]

$$\frac{2}{\pi}\int_0^\pi d\mu \frac{\sin^2\mu}{p + q\cdot\cos\mu} = \frac{2p}{q^2}\left(1 - \sqrt{1 - \frac{q^2}{p^2}}\right).$$

This can be further simplified by substituting $\cos\theta \equiv p/q$:

$$\frac{2}{q}\left(\frac{p}{q} + i\sqrt{1 - \left(\frac{p}{q}\right)^2}\right) = \frac{2}{q}(\cos\theta + i\sin\theta) = \frac{\exp i\theta}{|V_x|} = \tilde{G}^{\mathrm{semi}}(\nu) \equiv \tilde{G}^{\mathrm{semi}},$$

where the elements of the *diagonal* $N \times N$ matrix $\tilde{G}^{\mathrm{semi}}$ implicitly depend on the transverse mode number ν. Looking at Eq. (3.216), we see that the lattice Green's function, G^{semi}, may be written as a transformation of $\tilde{G}^{\mathrm{semi}}$, as discussed in the previous section. We introduce a transformation matrix U, whose columns are simply the transverse wave functions χ_n^ν of (3.210):

$$U \equiv \left(\chi_n^1 \middle| \chi_n^2 \middle| \cdots \middle| \chi_n^N \right).$$

The lattice Green's function may then be expressed as

$$G^{\text{semi}} = U \tilde{G}^{\text{semi}} U^+. \tag{3.217}$$

It was shown in general terms in Section 3.8.3 that the matrices U are unitary, $UU^+ = 1 = U^+U$. Therefore, one can interpret the transformation (3.217) as a similarity transformation from the *transverse mode representation* $\tilde{G}$ to the *site representation* G, or vice versa,

$$\tilde{G}^{\text{semi}} = U^+ G^{\text{semi}} U. \tag{3.218}$$

Finite uniform sections

For a uniform section of discretized size $N \times M$ surrounded by hard wall boundaries on all sides, the transverse wavefunctions are the same as in the case of a semiinfinite lead (3.210). Similarly, the longitudinal wavefunctions are given now by discrete eigenfunctions since they are bounded on both sides as well:

$$\phi_m^\mu = \sqrt{\frac{2}{M+1}} \cdot \sin\left(\frac{\pi \mu m}{M+1} \right) \cdot e^{im\varphi_x}. \tag{3.219}$$

Again, when combining finite uniform sections to form a composite structure, one needs only the Green's functions for two cases in which the two longitudinal site indices either coincide at one end of the section or represent the opposite ends. Proceeding with expressions (3.213), (3.216) and evaluating the summation over the discrete longitudinal quantum number μ yields the following matrix elements [128]

$$\tilde{G}_{11}^{\text{finite}} = \frac{\sin M\theta}{|V_x| \sin((M+1)\theta)}, \qquad \tilde{G}_{1M}^{\text{finite}} = \frac{\sin \theta}{|V_x| \sin((M+1)\theta)}, \tag{3.220}$$

both depending on the transverse mode number ν. The corresponding matrices $\tilde{G}_{11}^{\text{finite}}$, $\tilde{G}_{1M}^{\text{finite}}$ are diagonal, as in the previous section, and can be transformed from transverse mode representation to site representation as shown in (3.217)–(3.218).

Continuum Limit

It is important to note that the tight-binding lattice in the present context does *not* correspond to the crystal lattice, where each lattice site matches a single atom. Such types of *physical* lattices are commonly used to model specific properties of materials, for example, the electron band structure or phonon characteristics [129]. The tight-binding lattice should be understood as being a *mathematical* discretization that approximates the continuum behavior of the electron wavefunctions. To ensure convergence towards the desired continuum limit, one has to let the lattice spacing go to zero while keeping the geometry of the structure constant by properly increasing the number of lattice sites. It is a well-known property that uniform nearest-neighbor tight-binding lattices yield cosine-shaped electron bands [122]. However, for small wavevectors (i.e., small lattice spacing in the present model), the cosine is very well approximated by a parabolic relation

$$E(\mathbf{k}) = \frac{\hbar^2 \mathbf{k}^2}{2m^*}, \tag{3.221}$$

which is just the energy dispersion relation of the effective mass electron picture we are interested in.

To see this, consider a perfect tight-binding system in zero magnetic field, uniformly spaced in both the longitudinal and transverse directions. In this case, the hopping energies V_x and V_y are real constants. The longitudinal energy dispersion is given by (3.209) as

$$E^\mu = -2V_x + 2V_x \cos \mu. \tag{3.222}$$

Expressing the energy in terms of the longitudinal wavevector, $k_x^\mu = \mu/a_x$, and expanding the cosine into a power series this becomes

$$E^\mu = -2V_x \cdot \frac{\left(k_x^\mu a_x\right)^2}{2!} \pm \cdots = \frac{\hbar^2}{2m^*} \cdot k_x^{\mu 2} \pm \mathcal{O}\left(k_x^{\mu 4} a_x^2\right), \tag{3.223}$$

where in the last step we chose the hopping energy to be

$$V_x = -\frac{\hbar^2}{2m^*} \frac{1}{a_x^2}. \tag{3.224}$$

Similarly the transverse energy dispersion given by (3.211) for V_y real is

$$E^\nu = -2V_y + 2V_y \cos\left(\frac{\pi \nu}{N+1}\right). \tag{3.225}$$

For an infinite square-well (i.e., hard wall) potential in the transverse direction, the transverse wavevector in the continuum limit is given by $k_y^\nu = \pi \nu/(N+1)a_y$, where $(N+1)a_y \equiv W_y$ represents the width of the section. Therefore the transverse energy is readily expanded as

$$E^\nu = -V_y \cdot \left(\frac{\pi \nu}{N+1}\right)^2 \pm \cdots = \frac{\hbar^2}{2m^*} \cdot k_y^{\nu 2} \pm \mathcal{O}\left(k_y^{\nu 4} a_y^2\right), \tag{3.226}$$

and therefore V_y is given by

$$V_y = -\frac{\hbar^2}{2m^*} \frac{1}{a_y^2}. \tag{3.227}$$

As can be seen from the expressions of the longitudinal and the transverse energy dispersion relations, the tight-binding model is an accurate approximation of the free particle case as long as the lattice constants a_x and a_y are sufficiently small. The relation for the total energy $E = E^\mu + E^\nu$ introduces constraints on the ranges of wavevectors k_x^μ and k_y^ν. Therefore the tight-binding description is an excellent approximation for single electrons as long as their energy is much smaller than the characteristic hopping energies in the problem, that is, $E \ll |V_x|, |V_y|$.

We illustrate the relation of the energy dispersion (or "energy *bands*") in Fig. 3.47 for a two-dimensional square tight-binding lattice. In this particular case, the plots of the energy versus the wavevector for the longitudinal and transverse directions are identical. In general, any lattice of constant spacing in a certain direction will yield an analogous graph for the corresponding energy dispersion.

The tight-binding model under consideration yields a cosine-shaped band, whose width can be easily shown to be given as $4|V| \cdot d$, where $d = 1, 2, 3$ is the number of spatial

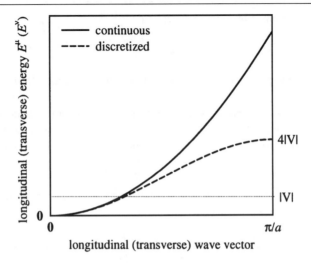

Fig. 3.47. Energy dispersion relation both for the continuous and uniformly spaced discretized case.

dimensions [122]. Furthermore, obtaining the same zeros of energy as in the free-particle case (i.e., $E(\mathbf{k} = \mathbf{0}) = 0$) requires the on-site energies $\epsilon^{\text{lattice}}$ to be set equal to the negative of half the bandwidth (which in this case is equivalent to the number of nearest neighbors times the hopping energy $|V|$).

For any typical electron energy of a problem, one obtains more accurate results by decreasing the lattice constant a. Since the lengths W of the structure have to be kept constant in turn, the number of sites N increases. Computational restrictions usually limit N, since the number of transverse sites determines the dimension of the matrices used in the recursive Green's function method. The *continuum limit* can be expressed as follows:

$$a \to 0 \quad \text{and} \quad N \to \infty \qquad \text{while} \quad W \equiv (N+1)a = \text{const.} \qquad (3.228)$$

Nonzero magnetic fields

An appealing feature of the recursive Green's function method is that some types of magnetic fields may be incorporated quite simply into the formalism, and therefore it is well suited to study various effects of magnetotransport [125], [126], [130], [131]. The basic concept is to relate the complex phase factor of the hopping energy to the vector potential $\mathbf{A}$ by a "Peierls substitution" [132], [133],

$$\mathbf{B} = \mathbf{0}: \quad |V_{\mathbf{RR'}}| \longmapsto \mathbf{B} \neq \mathbf{0}: \quad |V_{\mathbf{RR'}}| \cdot \exp\left(ie/h \int_{\mathbf{R}}^{\mathbf{R'}} \mathbf{A}(\mathbf{r})\,dr\right),$$

where the magnetic flux density $\mathbf{B}$ is determined by the vector potential as $\mathbf{B} = \nabla \times \mathbf{A}$.

Consider a configuration of particular interest which closely resembles a common experimental realization: a constant magnetic field applied perpendicular to a two-dimensional electron gas. One usually models this 2DEG in the lattice Green's function method by a square tight-binding lattice of lattice constant a with nearest-neighbor hopping only. Choosing the Landau gauge for the vector potential, $\mathbf{A} = (0, -Bx, 0)$, the matrix elements of the hopping energies become

$$V_x^{mn} = |V_x| \equiv \text{Re}\,V_x, \qquad V_y^{mn} = |V_y| \cdot \exp(i2\pi B \cdot n). \qquad (3.229)$$

Here, $B = \Phi/\Phi_0$ is given in units of h/ea^2, since $\Phi = Ba^2$ is the magnetic flux per unit grid cell and $\Phi_0 = h/e$ is the magnetic flux quantum. In accordance with (3.229), the phase arguments of Eqs. (3.208) and (3.210) can now be expressed as

$$\varphi_x \equiv 0, \qquad \varphi_y = 2\pi B.$$

3.8.5 Relation between Green's function and S-matrix

In this section we want to express the Green's function in terms of the scattering matrix S discussed in Section 3.5 of a multilead nanostructure system. Such a system is illustrated in Fig. 3.48 which is idealized to consist of N_L perfect semiinfinite leads L_a attached to the scattering region. The connection between the Green's function and the scattering matrix is critical in connecting the resent formalism to transport in the Landauer-Büttiker model, which is explicitly formulated in terms of reflection and transmission coefficients. Space is inadequate to present a proper derivation of this connection. Rather, the interested reader is referred to the work of Sols [134] where the relationship is explicitly derived. In the derivation of Sols, the scattering problem is solved by examining the evolution of wavepackets, a common approach in quantum-mechanical scattering theory (see, e.g., [135]). However, consideration of the multilead structure shown in Fig. 3.48 (as opposed to standard textbook examples) requires additional care.

By considering a nearly monochromatic wavepacket incident on lead a that is scattering into lead a' in Fig. 3.48, the transmission amplitude for a particle initially in mode ν in a to be scattered into mode ν' in lead a' is expressed in terms of the Green's function connecting these two leads as [134]

$$t_{\nu'\nu,a'a}(E) = i\hbar\sqrt{v_{E\nu'a'}v_{E\nu a}}e^{-i(k_{E\nu'a'}\hat{x}_{a'}+k_{E\nu a}\hat{x}_a)}\mathsf{G}^+(\nu'\hat{x}_{a'}, \nu\hat{x}_a; E). \qquad (3.230)$$

where $\hat{x}_{a'}$ and $\hat{x}_a$ indicate the two positions shown in Fig. 3.48, E is the energy and $k_{E\nu a}$ the wavevector. Likewise, the reflection amplitude for a particle initially in lead a in mode

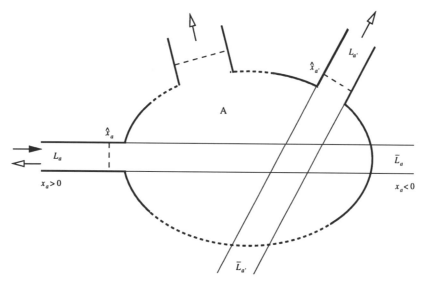

Fig. 3.48. Multilead structure consisting of a scattering region A and N_L perfect semiinfinite leads L_a attached to it. Their fictitious prolongations are called $\overline{L}_a$.

ν back into the same lead in mode ν' is given as

$$r_{\nu'\nu,aa}(E) = i\hbar \sqrt{v_{E\nu'a} v_{E\nu a}}\, e^{-i(k_{E\nu'a}+k_{E\nu a})\hat{x}_a} \cdot \left[\mathsf{G}^+(\nu'\hat{x}_a, \nu\hat{x}_a; E) - \delta_{\nu'\nu}/i\hbar v_{E\nu'a} \right]. \quad (3.231)$$

It is straightforward to adapt the results (3.230) and (3.231) obtained for a continuous problem in the previous section to a tight-binding lattice formulation, as discussed in Section 3.8.3 [136].

The basic difference that enters the transformation of the continuous expressions to the tight-binding case is the energy dispersion relation. The dispersion in the continuum case is represented by a parabolic band in the longitudinal direction

$$E = E_\nu + \hbar^2 k_{E\nu}^2 / 2m^*, \quad (3.232)$$

where $k_{E\nu}$ represents the wavevector that is available for longitudinal motion in a transverse mode ν. On the other hand, the energy dispersion for the tight-binding model (3.209) of Section 3.8.2 reads

$$E = E_\nu - 2\text{Re}V_x - 2|V_x| \cos k_{E\nu} a_x. \quad (3.233)$$

The velocities in the longitudinal direction are therefore not given as $v_{E\nu} = \hbar k_{E\nu}/m^*$, but rather are given by (3.232):

$$v_{E\nu} = \frac{1}{\hbar}\left(\frac{dE}{dk_{E\nu}}\right) = \frac{2|V_x|a_x}{\hbar} \sin k_{E\nu} a_x. \quad (3.234)$$

This expression has to be substituted for all the velocities v that appear in formulas (3.230) and (3.231).

The only other difference between the two models considered here is connected to the transformation from continuous position representation to discretized site representation. For the sake of simplicity, we assume the structure to consist of two semiinfinite leads whose axes lie parallel to a particular longitudinal direction. Let us choose the left (right) reference point to correspond to a site of longitudinal index $l < 0$ $(r > 0)$. We can omit the lead numbers a, a' from the relations (3.230) and (3.231) for this two-probe system. Further, using the abbreviation $k_{E\nu} a_x \equiv \theta_\nu$, the final relations in tight-binding language between the scattering amplitudes and the Green's function read

$$t_{\nu'\nu}(E) = i2|V_x|\sqrt{\sin \theta_{\nu'} \sin \theta_\nu}\, e^{i(\theta_\nu l - \theta_{\nu'} r)} \cdot \mathsf{G}^+(\nu'r, \nu l; E)$$

$$r_{\nu'\nu}(E) = i\sqrt{\sin \theta_{\nu'} / \sin \theta_\nu}\, e^{i(\theta_{\nu'}+\theta_\nu)l} \cdot \left[2|V_x| \sin \theta_\nu \mathsf{G}^+(\nu'l, \nu l; E) + i\delta_{\nu'\nu} \right]. \quad (3.235)$$

3.8.6 Recursive Green's function method

The idea to recursively build up the total Green's function of a tight-binding lattice was introduced in the early 1980s [137]. In such calculations, the Kubo formula was used to directly extract the transport properties from the Green's function of a structure. Since the method is used on a tight-binding grid, modeling disorder is easily incorporated by varying the on-site energies. This allows a straightforward study of localization effects in one dimension [137], [138], and two dimensions [139].

Besides the advantage of easy modeling of arbitrary potential distributions including disorder, another feature of the recursive Green's function (RGF) method is the possibility to include nonzero magnetic fields as well through the Peierl's substitution (Section 3.8.4). Several calculations on magnetotransport in two-dimensional systems have been reported

[55], [125], [126], [131]. Other extensions of the recursive Green's function method include four-terminal geometries [130], as well as superconductor junctions [140].

Two-dimensional calculations using the recursive Green's function method for semiconductor split-gate structures have been of interest primarily due to the fact that one can easily account for complicated geometries as well as disorder [128], [141], [142]. More recently, work has been reported on a three-dimensional adaption of the method [143], [144], in which the Poisson equation and the Green's function method are solved self-consistently.

Formulation of the method

The first step in applying the RGF method is to discretize the entire region of the nanostructure that is of physical interest by a tight-binding lattice. One assumes the electron wavefunctions to be zero "outside" the numerical domain, which is equivalent to imposing hard wall conditions (i.e., infinite potential energy) precisely at the boundaries where the finite lattice ends. Electron reservoirs are easily modeled by semiinfinite lattice sections.

It is not feasible to evaluate the total Green's function, G, *at once*. Instead, one divides the entire lattice into sections whose Green's functions, G^0, are *known* in a sense that they can be computed *exactly* for each separate section. Treating the nearest-neighbor hopping between two adjacent sections as a small perturbation, V, *Dyson's equation* for the case of a *single* interaction is given by (see, e.g., [145])

$$G = G^0 + G^0VG. \tag{3.236}$$

Since this represents an implicit equation for G, it leads to a system of equations that can be elegantly solved by recursion. At each step one attaches the Green's functions of a new section to the already calculated propagator of its adjacent section. Starting at one end of the discretized structure, one proceeds through the entire sample by adding the next adjacent section at each hopping location. After the last Green's function at the opposite end has been attached, the resulting total propagator between the two sides of the structure can be exploited to extract the transmission or reflection amplitudes as given by (3.235). Since new sections are attached recursively, the method is commonly called the *recursive* Green's function method, or RGF.

Attaching an extended section

In this section, we derive the expressions for coupling two adjacent lattice sections, L and R, as illustrated in Fig. 3.49. It should be noted that this figure and all associated equations are valid for any dimensionality of the problem; that is, each unfilled circle in Fig. 3.49 represents a single lattice *site*, a transverse *chain*, or a transverse *slice* of 3D lattice sites. The circles may also be associated with the corresponding Green's function at each site.

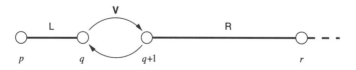

Fig. 3.49. Finite section L and semiinfinite section R are coupled by the hopping interaction V.

The two isolated sections L and R are represented by Green's functions G^L and G^R. These propagators are assumed to be already known, either by straightforward calculation or by having already used the coupling algorithm to be derived in this section. In such calculations, the right section R typically represents either a perfect semiinfinite lead (whose analytic form of the Green's function is derived in Section 3.8.4), or a semiinfinite section (whose propagator has been recursively built up to include all the interactions to the right of the current hopping site $q \rightleftharpoons q + 1$). The Green's function of the coupled system one would like to obtain is called G^{L+R}.

Since we are looking at the case where two known Green's functions (represented by G^0) are coupled by a *single* perturbation, V, Dyson's equation is of the form (3.236). As was made clear in the previous sections, we are interested only in the propagators G_{rl} from left lattice site l to right site r and in G_{ll} from l back to l, since only those contain the relevant information on transmission and reflection amplitudes t and r.

In the following we make use of the notation that was introduced in the previous sections. Matrix elements of the lattice Green's function operator G are matrices themselves. Taking matrix elements between the longitudinal site labels p and r, represented by the orthonormal state vectors $|p\rangle$, $|r\rangle$, one obtains a matrix whose indices are the transverse labels n, n'. According to the chosen representation, these indices are given by site or mode numbers, respectively

$$\langle r|G(E)|p\rangle \equiv \langle r|G(rn, pn'; E)|p\rangle \equiv G_{rp}(n, n'; E) \equiv G_{rp}. \tag{3.237}$$

"Sandwiched" between this pair of longitudinal sites, Dyson's equation (3.236) becomes

$$\langle r|G|p\rangle = \langle r|G^0|p\rangle + \sum_{a,b}\langle r|G^0|a\rangle\langle a|V|b\rangle\langle b|G|p\rangle,$$

where we have included complete sets of longitudinal site states, $|a\rangle$ and $|b\rangle$. Since the hopping acts exclusively between q and $q + 1$, and since there is no known, interaction-free propagator, G^0, from p to r or q to r, this reduces to

$$\langle r|G|p\rangle = \langle r|G^0|q + 1\rangle\langle q + 1|V|q\rangle\langle q|G|p\rangle.$$

In the notation introduced before, this matrix equation reads

$$G_{rp}^{L+R} = G_{r,q+1}^R V_{q+1,q} G_{qp}^{L+R}. \tag{3.238}$$

One proceeds by properly taking matrix elements of Dyson's equation to determine the next unknown G_{qp}^{L+R}, and so on. This procedure is terminated when one has obtained as many equations as there are unknown "variables," G^{L+R}. In the case of Fig. 3.49, the result is the set of equations (3.238)–(3.240),

$$G_{qp}^{L+R} = G_{qp}^L + G_{qq}^L V_{q,q+1} G_{q+1,p}^{L+R}, \tag{3.239}$$

$$G_{q+1,p}^{L+R} = G_{q+1,q+1}^R V_{q+1,q} G_{qp}^{L+R}. \tag{3.240}$$

In a similar way, the full propagator from p to p is obtained as

$$G_{pp}^{L+R} = G_{pp}^L + G_{pq}^L V_{q,q+1} G_{q+1,p}^{L+R}. \tag{3.241}$$

From Eqs. (3.239) and (3.240) a closed expression for the propagator of the left section L, now including the perturbation to its right-hand side, is found as

$$G_{qp}^{L+R} = G_{qp}^L + G_{qq}^L \underbrace{V_{q,q+1} G_{q+1,q+1}^R V_{q+1,q}}_{\equiv \Sigma_q^R} G_{qp}^{L+R}. \tag{3.242}$$

The unknown, full Green's function, G_{qp}^{L+R}, appears on both sides of the equation. In operator notation, this equation is again the familiar Dyson's equation

$$G = G^0 + G^0 \Sigma G. \tag{3.243}$$

In many-body physics, where the G and G^0 represent other types of Green's functions than those discussed so far in our simpler single-particle picture, Eq. (3.243) implicitly relates the full propagator G to its free propagator G^0 by including *all* the interactions with the other parts of the system in the so-called *proper self energy*, Σ. In analogy to the many-body interpretation, Σ_q^R represents the *self-energy* of an isolated section L at site q due to the hopping interaction onto R.

Recasting Eq. (3.242) as

$$G_{qp}^{L+R} = \left(1 - G_{qq}^L \Sigma_q^R\right)^{-1} G_{qp}^L,$$

finally leads to the desired expressions for the full Green's functions:

$$G_{rp}^{L+R} = G_{r,q+1}^R V_{q+1,q} \left(1 - G_{qq}^L \Sigma_q^R\right)^{-1} G_{qp}^L$$
$$G_{pp}^{L+R} = G_{pp}^L + G_{pq}^L \Sigma_q^R \left(1 - G_{qq}^L \Sigma_q^R\right)^{-1} G_{qp}^L. \tag{3.244}$$

If p still does not coincide with the longitudinal site index l on the very left, the propagator G^{L+R}, which was just calculated, becomes the new "known" Green's function G^R. The adjacent left section of longitudinal extension from site $o \geq l$ to $p - 1$ yields the new propagator G^L. This way, one recursively builds up the total propagator until the last section with left site index l has been attached. The final Green's functions from which one can extract the transmission and reflection amplitudes are then given by

$$G_{rl} \equiv G_{rl}^{L+R}, \qquad G_{ll} \equiv G_{ll}^{L+R}. \tag{3.245}$$

Attaching a section of single site extension

In cases where a section's Green's function is already known analytically, it is computationally advantageous to attach an extended section as described in the previous section. An analytic form of the propagator can only be obtained for sections, which are both geometrically uniform and without disorder, as shown in Section 3.8.4.

Whereas the method described above is sometimes refered to as the "extended recursive Green's function method" (see for example [128]), the standard way is to attach a section of unit longitudinal site extension. Following the detailed derivation of the previous section, one easily obtains the somewhat simpler recursion formulas corresponding to Fig. 3.50:

$$G_{rp}^{L+R} = G_{r,p+1}^R V_{p+1,p} \left(1 - G_{pp}^L \Sigma_p^R\right)^{-1} G_{pp}^L$$
$$G_{pp}^{L+R} = \left(1 - G_{pp}^L \Sigma_p^R\right)^{-1} G_{pp}^L. \tag{3.246}$$

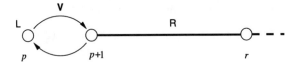

Fig. 3.50. Section L of unit longitudinal extension and semiinfinite section R are coupled by the hopping interaction V.

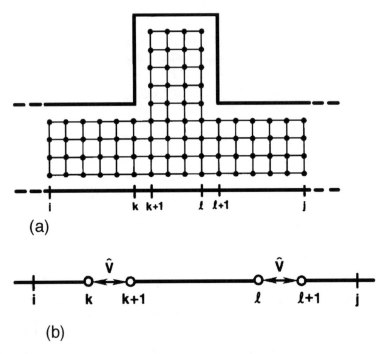

(a)

(b)

Fig. 3.51. T-stub waveguide geometry and (a) the tight-binding lattice discretization of the structure. The reduction of this problem to a representation in terms of a sequence of uniform strips is illustrated in (b) [after Sols *et al.*, J. Appl. Phys. **66**, 3895 (1989), by permission].

3.8.7 *Application to specific geometries*

T-stub waveguide

As a simple example of the application of the RGF method to the analysis of transport, consider the *T-stub geometry* illustrated in Fig. 3.51 [128]. The T-stub is a uniform waveguide with a sidearm as illustrated in this figure. The discretization in terms of a tight-binding lattice is superimposed on this geometry as shown in the figure. In a pure sight-representation calculation, each vertical slice would represent the "unperturbed" Green's function, G^0, coupled to each neighboring slice by the hopping potential. The solution for G^0 in the leads would correspond to the inversion of an $N \times N$ matrix, where N is the number of vertical sites (four in the uniform lead sections of Fig. 3.51a, nine in the sidearm section for the lattice shown). However, for an ideal T-stub with hard walls and random disorder, a simpler approach may be used based on the analytical results derived in Section 3.8.4. We have already solved the problem of the Green's function of semiinfinite leads that correspond to the left and right regions adjacent to the sidearm. We have also solved for a finite-length

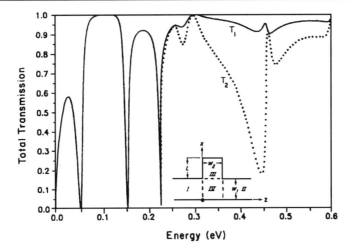

Fig. 3.52. Total transmission probabilities from mode 1 (T_1) and mode 2 (T_2) as a function of electron energy ($w_1 = w_2 = L = 10$ nm).

uniform section as well. Therefore, we can construct the full Green's function by treating the Green's function for the semiinfinite leads and a uniform section as G^0, and then recursively solving to obtain G_{kl} and G_{kk}, where k and l label the lattice sites connecting to the side-arm on the left and right as shown. The Green's functions G_{ji} and G_{ii} may then be calculated to give the transmission and reflection coefficients according to Eq. (3.235).

Figure 3.52 shows the calculated result for the transmission coefficient using this method. The solid line represents the total transmission probability from mode 1 in the left lead, and the dashed curve shows the transmission probability for the second mode (which has a cutoff in energy of 0.23 eV for the dimensions shown). The inset shows the same geometry solved using the mode matching method of Section 3.7.1, which gives the identical results for this simple geometry [107]. The transmission shows several zeros as well as regions of unity amplitude. Since the conductance is directly related to $T(E)$ at low temperature, the current should correspondingly go to zero for certain Fermi energies. This behavior has led to the proposal of fabricating a *quantum modulated transistor* (QMT) based on the T-stub structure. Essentially, if a gate bias may be employed to modulate the length of the sidearm, L, the resonances in transmission may be moved through the Fermi energy, thus modulating the conductance. Experimental realization of such a structure using split-gate QPCs has been attempted [146], which shows modulation of the conductance, although not the large deviations predicted by theory.

Quantum point contact

In comparison to the mode matching technique of Section 3.7.1, a great advantage of the RGF technique is the ability to handle arbitrary potential profiles. As discussed earlier, the electrostatic potential due to gates, impurities, and so on are assigned to the site energy, $\epsilon^{\text{potential}}$.

As an example, we reconsider the simple QPC structure discussed in Section 3.6. The actual potential due to the space charge region and the gates is complicated and requires the 3D solution of Poisson's equation. A simple analytical model of the potential in a QPC may be obtained, however, if one neglects the effects due to space charge layer formation and just treats the GaAs/AlGaAs heterostructure as a dielectric [92]. For two electrodes held at a constant voltage V_g forming a narrow constriction of length l and width w centered at the origin of the xy-plane, the potential may be expressed as

$$\phi(x, y, z) = V_g \left\{ f\left[\left(-\frac{l}{2} + \frac{x}{z}\right), \left(\frac{w}{2} + \frac{y}{z}\right)\right] - f\left[\left(\frac{l}{2} + \frac{x}{z}\right), \left(\frac{w}{2} + \frac{y}{z}\right)\right]\right.$$
$$\left. + f\left[\left(-\frac{l}{2} + \frac{x}{z}\right), \left(\frac{w}{2} - \frac{y}{z}\right)\right] - f\left[\left(\frac{l}{2} + \frac{x}{z}\right), \left(\frac{w}{2} - \frac{y}{z}\right)\right]\right\},$$

(3.247)

where f represents the auxiliary function

$$f(u, v) = \frac{1}{2\pi} \left[\frac{\pi}{2} - \arctan u - \arctan v + \arctan \frac{uv}{\sqrt{1 + u^2 + v^2}}\right].$$ (3.248)

Both a contour plot of the potential and the potential landscape in a plane $z_0 = 50$ nm below the split-gates are shown in Fig. 3.53. This plane is taken to represent the plane of the 2DEG below the AlGaAs.

To calculate the transmission using the RGF method in 2D, we assume that the thickness of the 2DEG is negligibly small, and we solve just in the plane of the electrons. We attach perfect semiinfinite leads at the extreme left and right of the finite region defining the QPC shown in Fig. 3.53. The square domain is discretized into a rectangular lattice and solved recursively in the longitudinal direction of electron wave propagation. The eigenfunctions of the Hamiltonian in the transverse direction are solved exactly and used via matrix inversion to obtain the stripe Green's function without hopping in the longitudinal direction. The RGF method is then used to build up the solution in the longitudinal direction. After that, the transmission and reflection amplitudes are calculated using (3.235) and used to calculate the conductance via the Landauer-Büttiker model.

Figure 3.54 shows the result of the conductance calculation obtained for the QPC of Fig. 3.53. The conductance versus gate voltage curve exhibits plateaus quantized in integer multiples of $2e^2/h$. This integer multiple can be related to the number of modes available

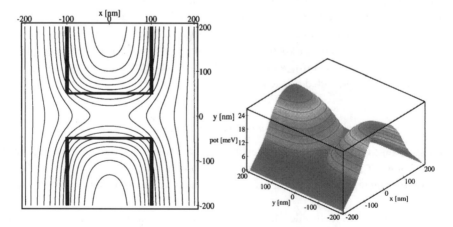

Fig. 3.53. Approximate electric potential of a quantum point contact for a distance $z_0 = 50$ nm below the gates: (a) contour plot including surface gates, (b) 3D plot.

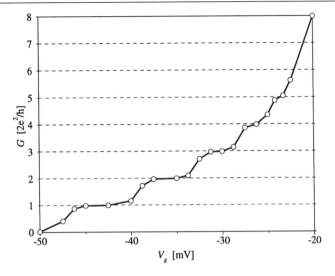

Fig. 3.54. Conductance G versus gate voltage V_g of the quantum point contact discussed in the text. The Fermi energy $E_F = 14$ meV.

for propagation in the quasi-one-dimensional conduction channel. The potential profile suggests that as the constriction couples smoothly to the wide reservoirs, intermode scattering should be negligible. No ringing or resonances are found in this fully 2D calculation as found from the mode matching method with hard walls. Thus, the assumption of adiabatic transport as discussed in Section 3.6.2 seems to provide a satisfactory explanation of the observed quantization of conductance. When the gate voltage is made less negative, the channel becomes wider, eventually the channel loses its lateral confinement, and noticeable quantization of the conductance is lost, as is observed experimentally.

3.9 Multi-probe formula

In the previous sections, we considered conductance in a system fed by two reservoirs. However, many experimental measurements such as the Hall effect discussed in Chapter 2 are performed with at least four contacts: two for injecting current, and two or more for measuring the potential drop along and across the channel. Figure 3.55 shows an example of a Hall bar structure, in which current typically flows between contacts 1 and 4, while voltages are measured with high-impedance potentiometers (through which zero current flows ideally) between contacts 2, 3, 5, and 6. In a Hall measurement, for example, the voltage measured between contacts 2 and 6 or between 3 and 5 in the presence of a magnetic field is the Hall voltage, which is used to determine the transverse conductivity, σ_{xy}. Measurement of the voltage along the channel, say between contacts 2 and 3 or between 5 and 6, determines the longitudinal conductivity, σ_{xx}. In the linear response regime (small currents), the conductivity tensor satisfies the well-known Onsager-Casimir symmetry relation [148], [149]

$$\sigma_{xy}(\mathbf{B}) = \sigma_{yx}(-\mathbf{B}), \qquad (3.249)$$

where $\mathbf{B}$ is the magnetic flux density. In a conventional Hall measurement in a homogeneous sample, for example, the voltage measured between contacts 2 and 6 changes sign upon

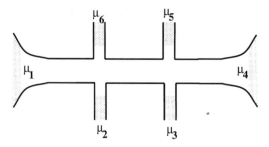

Fig. 3.55. Diagram of a Hall geometry conductor with multiple probes attached to measure voltages and currents independently. In relation to the multi-probe equation, the probes are connected via ideal leads from the sample to reservoirs with the different chemical potentials shown.

reversal of the magnetic field, whereas the longitudinal voltage drop measured between 2 and 3 remains unchanged. The corresponding Hall resistance, R_{xy}, defined in terms of the ratio of the voltage V_{26} over the current through contacts 1 and 4, changes sign; the longitudinal resistance, R_{xx}, remains the same under field reversal. As discussed below, this symmetry is preserved for nonlocal transport in mesoscopic structures as well.

We now consider the various conductances for the *multi-probe* case from the point of view of the single- and multichannel formulas of the previous two sections as derived by Büttiker [150], [151]. As a generalization of the two contact case, current probes are considered phase-randomizing agents that are connected through ideal leads to reservoirs (characterized by a chemical potential μ_i for the ith probe) which emit and absorb electrons incoherently as shown in Fig. 3.55. In general, the "probes" need not be physical objects such as shown in Fig. 3.55. They can be any phase-randomizing entity, such as inelastic scatterers distributed throughout the sample [151]. Here we will assume that voltage probes are phase randomizing, although there is some over disagreement the validity of this assumption of phase randomization.

To simplify the initial discussion, first assume that the leads contain only a single channel with two states at the Fermi energy corresponding to positive and negative velocity. The various leads are labeled $i = 1, 2, \ldots$, each with corresponding chemical potential μ_i. A scattering matrix may be defined which connects the states in lead i with those in lead j. Thus, $T_{ij} = |s_{ij}|^2$ is the transmission coefficient into lead i of a particle incident on the sample in lead j, while $R_{ii} = |s_{ii}|^2$ is the probability of a carrier incident on the sample from lead i to be reflected back into that lead. In the general case of a magnetic flux penetrating the sample, Φ, the elements of the scattering matrix satisfy the following reciprocity relations due to time reversal symmetry:

$$s_{ij}(\Phi) = s_{ji}(-\Phi). \qquad (3.250)$$

As a result, the reflection and transmission coefficients satisfy the symmetry relations

$$R_{ij}(\Phi) = R_{ji}(-\Phi), \qquad T_{ij}(\Phi) = T_{ji}(-\Phi). \qquad (3.251)$$

In order to simplify the discussion, an additional chemical potential is introduced, μ_0, which is less than or equal to the value of all the other chemical potentials such that all states below μ_0 can be considered filled (at $T_L = 0K$). With reference then to μ_0, the total current injected from lead i is given by Eq. (3.106) as $(2e/h)\Delta\mu_i$, where $\Delta\mu_i = \mu_i - \mu_0$ (again assuming the difference in chemical potentials is sufficiently small that the energy dependencies of the transmission and reflection coefficients may be neglected). A fraction R_{ii} of the current is reflected back into lead i. Similarly, according to (3.106), lead j injects carriers into lead i as $(2e/h)T_{ij}\Delta\mu_j$. The net current flowing in lead i is therefore the net difference between the current injected from lead i to that injected back into lead i due to reflection and transmission from all the other leads:

$$I_i = \left(\frac{2e}{h}\right)\left[(1 - R_{ii})\Delta\mu_i - \sum_{j \neq i} T_{ij}\Delta\mu_j\right]. \tag{3.252}$$

Conservation of flux requires that

$$1 - R_{ii} = T_i = \sum_{i \neq j} T_{ji} = \sum_{j \neq i} T_{ij}. \tag{3.253}$$

Therefore, the reference chemical potential, μ_0, drops out to give

$$I_i = \left(\frac{2e}{h}\right)\left[(1 - R_{ii})\mu_i - \sum_{j \neq i} T_{ij}\mu_j\right]. \tag{3.254}$$

The generalization of (3.274) to the multichannel case is obtained by simply considering that in each lead there are N_i channels at the Fermi energy μ_i. A generalized scattering matrix may be defined connecting the different leads and the different channels in each lead to one another. The elements of this scattering matrix are labeled $s_{ij,mn}$, where i and j label the leads and m and n label the channels. The probability for a carrier incident on lead j in channel n to be scattered into channel m of lead i is given by $T_{ij,mn} = |s_{ij,mn}|^2$. Likewise, the probability of being reflected within lead i from channel n into channel m is given by $R_{ii,mn} = |s_{ii,mn}|^2$. The reciprocity of the generalized scattering matrix with respect to magnetic flux now becomes $s_{ij,mn}(\Phi) = s_{ji,nm}(-\Phi)$. The reservoirs are assumed to feed all the channels equally up to the respective Fermi energy of a given lead. If we define the reduced transmission and reflection coefficients as

$$T_{ij} = \sum_{mn} T_{ij,mn}, \qquad R_{ij} = \sum_{mn} R_{ij,mn}, \tag{3.255}$$

then the total current in channel i becomes

$$I_i = \left(\frac{2e}{h}\right)\left[(N_i - R_{ii})\mu_i - \sum_{j \neq i} T_{ij}\mu_j\right], \tag{3.256}$$

which differs from (3.254) by the factor N_i.

Note that we may associate a voltage with each Fermi energy $eV_i = \mu_i$. Thus we can rewrite (3.256) in matrix form as

$$\mathbf{I} = \overline{\mathbf{G}}\mathbf{V}, \tag{3.257}$$

where $\mathbf{I}$ and $\mathbf{V}$ are column vectors for the probe currents and voltages respectively, and $\overline{\mathbf{G}}$ is an $M \times M$ conductance tensor

$$\overline{\mathbf{G}} = \frac{2e^2}{h}\begin{bmatrix} N_1 - R_{11} & -T_{12} & \cdots & -T_{1M} \\ -T_{21} & N_2 - R_{22} & \cdots & -T_{2M} \\ \vdots & -T_{32} & \ddots & \vdots \\ -T_{M1} & \cdots & \cdots & N_M - R_{MM} \end{bmatrix}, \tag{3.258}$$

where M is the number of probes. Due to flux conservation, both the rows and the columns add to zero. We note that the components of the conductance tensor contain the same symmetry with respect to the magnetic field as the reflection and transmission coefficients:

$$G_{ij}(\Phi) = G_{ji}(-\Phi). \tag{3.259}$$

Thus the conductance matrix is equal to its transpose under reversal of the magnetic flux. Of course, such relations are expected in linear response theory for a system governed by a local conductivity tensor satisfying the Onsager-Casimir symmetry relation (3.249). Equation (3.259), derived for the nonlocal conductance in a mesoscopic system, shows that such relationships extend to systems governed by nonlocal transport due to the underlying symmetries of the S-matrix.

The resistance matrix may be defined from the inverse of the conductance matrix

$$\overline{\mathbf{R}}\mathbf{I} = \mathbf{V}; \qquad \overline{\mathbf{R}} = \overline{\mathbf{G}}^{-1}. \tag{3.260}$$

Under magnetic flux reversal, the new resistance matrix is the inverse of the transpose of $\overline{\mathbf{G}}$. Assuming the conductance matrix is nonsingular, the inverse of the transpose of a matrix is equal to the transpose of the inverse [152]. Thus the components of the resistance matrix also satisfy

$$R_{ij}(\Phi) = R_{ji}(-\Phi). \tag{3.261}$$

The above result may be used to prove an important reciprocity relationship concerning a four-terminal transport measurement. Consider the resistance $R_{mn,kl}$, defined as the ratio of the voltage measured between leads k and l, when current $I = I_{mn}$ is driven into contact m and taken out from contact n:

$$R_{mn,kl} = \frac{V_k - V_l}{I}. \tag{3.262}$$

In terms of the resistance matrix, the voltage V_k is associated with the elements of the kth row, and V_l with the lth row of the matrix. The only nonzero current elements correspond to $I_m = I$ and $I_n = -I$, which couple to columns m and n of the resistance matrix, so that (3.262) may be written

$$R_{mn,kl}(\Phi) = \frac{(R_{km}I - R_{kn}I) - (R_{lm}I - R_{ln}I)}{I} = R_{km} + R_{ln} - R_{kn} - R_{lm} \tag{3.263}$$

in the presence of the magnetic flux Φ. If the current leads and voltage leads are exchanged (i.e., current forced between k and l while the voltage is measured between m and n), then the new resistance is defined letting $m \rightarrow k, n \rightarrow l, k \rightarrow m$, and $l \rightarrow n$ as

$$R_{kl,mn}(\Phi) = R_{mk} + R_{nl} - R_{ml} - R_{nk}. \tag{3.264}$$

If the magnetic flux is simultaneously reversed, (3.261) shows that we may write

$$R_{kl,mn}(-\Phi) = R_{km} + R_{ln} - R_{lm} - R_{kn} = R_{mn,kl}(\Phi), \tag{3.265}$$

which in the absence of magnetic flux reduces to the simpler *reciprocity theorem*

$$R_{kl,mn} = R_{mn,kl}. \tag{3.266}$$

This well-known experimental result basically states that in the absence of a magnetic field, the resistance measured by passing current through one pair of contacts and measuring the voltage between the other two is identical to that measured if the voltage and current contacts are swapped. In the presence of a magnetic field, the more general result (3.265) requires that this be done while simultaneously reversing the flux through the sample.

3.9.1 Specific examples

Two-terminal conductance

An expression for the conductance measured in a two-probe conductor was given by Eq. (3.136), where the current and voltage are simultaneously measured through the same pair of contacts. Equation (3.256) also yields the same expression if we restrict the sum over the probe indices i and j to 2. In the two-terminal case, set $I = I_1 = -I_2$ as the current flowing into terminal 1. By conservation of flux, $N_1 = R_{11} + T_{12}$ in terminal 1, and $N_2 = R_{22} + T_{21}$ in terminal 2. Using these relations in (3.256) above gives

$$I = \left(\frac{2e}{h}\right)T_{12}(\mu_1 - \mu_2) = -\left(\frac{2e}{h}\right)T_{21}(\mu_2 - \mu_1), \qquad (3.267)$$

which implies that the transmission coefficient in the two probe case, $T = T_{12} = T_{21}$, is symmetric with respect to the magnetic flux, $T(\Phi) = T(-\Phi)$. The measured voltage in the two terminal case is just the difference in Fermi energies of the two contacts, so that the conductance is identical to (3.136)

$$G = \frac{Ie}{(\mu_i - \mu_j)} = \frac{I}{V} = \frac{2e^2}{h}T, \qquad (3.268)$$

which is symmetric with respect to magnetic field reversal.

Three terminal case

Consider now the three-probe situation shown in Fig. 3.56. In this example we wish to consider the various resistances in the case that current flows between contacts 1 and 2, for which no net current flows through contact 3 (i.e., $I_3 = 0$). Thus contact 3 represents an ideal voltage probe that draws no current.

First consider the resistance $R_{12,13}$, which signifies the ratio of the voltage measured between contacts 1 and 3 for a certain current I flowing from contact 1 into contact 2. Using (3.256), the condition of zero current flowing in contact 3 gives

$$0 = \frac{2e}{h}[(N_3 - R_{33})\mu_3 - T_{31}\mu_1 - T_{32}\mu_2]. \qquad (3.269)$$

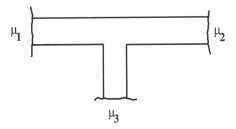

Fig. 3.56. Generic representation of a three-terminal structure.

Since $N_3 - R_{33} = T_{31} + T_{32}$ by current conservation

$$\mu_3 = \frac{T_{31}\mu_1 + T_{32}\mu_2}{T_{31} + T_{32}}. \tag{3.270}$$

The current flowing into contact 1 is again $I = I_1 = -I_2$. Starting with the equation for I_2, we may write

$$-I = \frac{2e}{h}[(N_2 - R_{22})\mu_2 - T_{21}\mu_1 - T_{23}\mu_3]. \tag{3.271}$$

Writing $N_2 - R_{22} = T_{21} + T_{23}$ and rewriting μ_2 using the result obtained in (3.270) for $I_3 = 0$, Eq. (3.271) becomes

$$-I = \frac{2e}{h}\left[(T_{21} + T_{23})\frac{(T_{31} + T_{32})}{T_{32}}\mu_3 - \frac{(T_{23} + T_{21})}{T_{32}}T_{31}\mu_1 - T_{21}\mu_1 - T_{23}\mu_3\right]. \tag{3.272}$$

The terms multiplying μ_3 may be grouped together as

$$(T_{21} + T_{23})(T_{31} + T_{32}) - T_{23}T_{32} = T_{23}T_{31} + T_{21}T_{31} + T_{21}T_{32} = D. \tag{3.273}$$

We note that upon flux reversal, $D(\Phi) = D(-\Phi)$, and so D remains invariant. This result may be proved using the symmetry relation $T_{ij}(\Phi) = T_{ji}(-\Phi)$, which implies that the last term on the left side of Eq. (3.273) is the same, while the sums in the product of the first term are equal to reflection coefficients by current conservation, which are even functions of the flux.

Similarly, the terms multiplying μ_1 are given by the same factor

$$(T_{21} + T_{23})T_{31} + T_{21}T_{32} = D. \tag{3.274}$$

Therefore, (3.272) is simply given by

$$-I = \left[\frac{2e}{h}\frac{D}{T_{32}}(\mu_3 - \mu_1)\right], \tag{3.275}$$

and the resistance $R_{12,13}$ is

$$R_{12,13} = \frac{(\mu_1 - \mu_3)}{eI} = \left(\frac{h}{2e^2}\right)\frac{T_{32}}{D}. \tag{3.276}$$

By an analogous procedure, we may calculate the resistance $R_{12,32}$ (starting from the equation for $I_1 = I$), which is the ratio of the voltage between probes 2 and 3, for a current I between 1 and 2:

$$R_{12,32} = \frac{(\mu_3 - \mu_2)}{eI} = \left(\frac{h}{2e^2}\right)\frac{T_{31}}{D}. \tag{3.277}$$

The two-terminal resistance may be calculated by combining Eqs. (3.276) and (3.277) to yield

$$R_{12,12} = \frac{(\mu_1 - \mu_2)}{eI} = \frac{(\mu_1 - \mu_3)}{eI} + \frac{(\mu_3 - \mu_2)}{eI} = R_{12,13} + R_{12,32}$$

$$= \left(\frac{h}{2e^2}\right)\frac{T_{31} + T_{32}}{D}. \tag{3.278}$$

Since $T_{31} + T_{32} = 1 - R_{33}$ is invariant with respect to flux reversal, and D was already shown to be invariant, the two-terminal resistance (or conductance) is invariant with respect to magnetic flux reversal in the presence of a third noncurrent-carrying terminal. This symmetry may be further extended to many probes as shown by Büttiker [151]. The two-terminal conductance given by the inverse of (3.278) in the presence of the third noncurrent-carrying probe is

$$G_{12,12} = \frac{2e^2}{h} \frac{T_{23}T_{31} + T_{21}T_{31} + T_{21}T_{32}}{T_{31} + T_{32}}. \tag{3.279}$$

In the limit that the transmission coefficients into and out of probe 3 from the other two probes are much smaller than the transmission coefficient from 1 to 2 (i.e., $T_{23}, T_{31} \ll T_{21}$), Eq. (3.279) reduces to the two-probe formula (3.268) with $T = T_{21}$. The fact that in general the two-terminal conductance with a third noncurrent-carrying probe is different from the simple two-terminal conductance Eq. (3.268) demonstrates that for mesoscopic systems it is difficult to separate the effect of the measurement probes from the measurement itself. Such uncertainty in separating a measured quantity from the measurement apparatus itself is of course fundamentally related to the question of measurement in quantum mechanics and phase coherence in ballistic structures.

An interesting interpretation of the effect of the passive third probe in a two-terminal measurement has been suggested by Büttiker [32], [151]. There it was proposed that the additional third terminal represent the effect of phase-randomizing inelastic scattering. Electrons that are transmitted into probe 3 lose their phase coherence by thermalizing in the reservoir. Since no net current was assumed to flow into the probe, the same flux of carriers are reemitted into the sample, but without any phase relation to the particles incident into the probe. Recalling the discussion of inelastic tunneling in connection with planar barrier structures in Section 3.3.2, the total transmission coefficient may be written as in (3.75), as the sum of the coherent (or elastic) transmission coefficient T_c and the incoherent (or inelastic) transmission coefficient T_i. Identifying the elastic contribution between probes 1 and 2 as T_{21}, then using (3.279) we can write

$$T = T_c + T_i = T_{21} + \frac{T_{23}T_{31}}{T_{31} + T_{32}}, \tag{3.280}$$

where the last term on the right side is the inelastic transmission coefficient. In the limit that the transmission coefficients into and out of probe 3 become small, T becomes $T_c = T_{21}$. In the other extreme limit of no coherent transmission (i.e. $T_{21} = 0$), the resistance becomes

$$R_{12,12} = \frac{h}{2e^2} \left(\frac{1}{T_{31}} + \frac{1}{T_{23}} \right), \tag{3.281}$$

which is nothing more than the addition of two series resistance with no coherence between them. Büttiker used this idea to develop a simple model for the effect of inelastic scattering in double barrier planar structures with similar results to those discussed in Section 3.3.2 [32].

Four-terminal conductance

As a final example, we wish to make a connection back to the discussion in Section 3.4 concerning the interpretation of the Landauer formula in terms of a two-terminal versus four-terminal measurement. As discussed earlier, if a two-terminal measurement is performed on a sample using the same pair of contacts to measure both the current and voltage, then we expect that the conductance is proportional simply to the transmission coefficient as given by (3.117) and not the ratio of transmission to reflection coefficients given by (3.115).

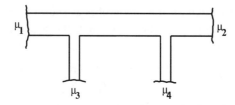

Fig. 3.57. Schematic of a four-probe experiment in which current is forced between contacts 1 and 2 while weakly coupled probes (3 and 4) measure the potential distribution along the sample.

The latter form arises if one considers the pile-up of charge on either side of the barrier, yet measures the potential drop only across the sample. Thus the conductance in the two-terminal measurement is always finite, even when transmission through the sample is unity (implying no resistance due to the contact resistance associated with connecting from a reservoir to the leads). As suggested by Engquist and Anderson [79], if noninvasive voltage probes rather than the current contacts are introduced to measure the potential distribution at the sample, one should in principle measure the Landauer conductance (3.115) through a four-terminal measurement, taking the resistance as the ratio of the voltage difference of the two voltage probes over the injected current through the other two contacts. We now consider such a measurement within the context of the multi-probe formula developed above and follow the development of Büttiker in arriving at the Landauer conductance to lowest order in the coupling of the probes to the system.

Consider the four-probe configuration shown in Fig. 3.57. A current I is passed between contacts 1 and 2 such that $I_1 = -I_2 = I$. The two probes 3 and 4 are weakly coupled to the sample. The weak coupling could be due, for example, to tunnel barriers isolating the probes from the sample. The voltage probes are assumed to draw no net current, such that $I_3 = I_4 = 0$. The condition of zero net current in probes 3 and 4 gives

$$I_3 = 0 = \frac{2e}{h}[(1 - R_{33})\mu_3 - T_{31}\mu_1 - T_{32}\mu_2 - T_{34}\mu_4]$$

$$I_4 = 0 = \frac{2e}{h}[(1 - R_{44})\mu_4 - T_{41}\mu_1 - T_{42}\mu_2 - T_{43}\mu_3].$$

(3.282)

Rewriting the above two equations we get

$$(1 - R_{33})\mu_3 - T_{34}\mu_4 = T_{31}\mu_1 + T_{32}\mu_2$$

$$(1 - R_{44})\mu_4 - T_{43}\mu_3 = T_{41}\mu_1 + T_{42}\mu_2.$$

(3.283)

Here we consider a single channel in each lead for simplicity, but the procedure is easily extendible to the multichannel, multi-probe case.

Now, following Büttiker, we assume that the coupling into the voltage probes is characterized by a small parameter, ε, which attenuates the flux into probes 3 and 4 due to tunnel barriers across the contacts for example. The transmission coefficients may thus be expanded in powers ε as

$$T_{12} = T_{12}^{(0)} + \varepsilon T_{12}^{(1)} + \cdots$$

$$T_{13} = \varepsilon T_{13}^{(1)} + \cdots$$

$$T_{34} = \varepsilon^2 T_{34}^{(2)} + \cdots,$$

(3.284)

where we interpret $T_{12}^{(0)}$ as the transmission coefficient from 1 to 2 in the absence of the other two probes, while the high order terms are small corrections arising from their presence. As such, $T_{12}^{(0)}$ is symmetric with respect to flux reversal as we saw in Section 3.9.1. T_{12} is zero order in ε as flux may go from probe 1 to 2 without crossing a tunnel barrier, T_{13} is first order in ε as flux traverses the tunnel barrier once, while $T_{34} \propto \varepsilon^2$ since the flux must traverse two such tunnel barriers.

We now proceed to calculate the two-terminal resistance, $R_{12,12}$ (where the first two indices represent the current leads, the second two the voltage leads) measured in the presence of the two voltage leads 3 and 4. To do this, we must therefore eliminate the chemical potentials μ_3 and μ_4 by expressing them in terms of the potentials μ_1 and μ_2. Eliminating μ_4 from (3.283) above gives

$$[(1 - R_{33})(1 - R_{44}) - T_{34}T_{43}]\mu_3 = [T_{31}(1 - R_{44}) + T_{34}T_{41}]\mu_1$$
$$+ [T_{32}(1 - R_{44}) + T_{34}T_{42}]\mu_2. \quad (3.285)$$

Using flux conservation, the left side of Eq. (3.285) may be written

$$(1 - R_{33})(1 - R_{44}) - T_{34}T_{43} = \overbrace{(T_{31} + T_{32})(T_{41} + T_{42})}^{\vartheta(\varepsilon^2)}$$
$$+ \overbrace{T_{34}(T_{41} + T_{42}) + T_{43}(T_{31} + T_{32})}^{\vartheta(\varepsilon^3)}, \quad (3.286)$$

which neglecting terms on the order of ε^3 gives

$$\approx (T_{31} + T_{32})(T_{41} + T_{42}). \quad (3.287)$$

Solving for μ_3 in (3.285) using (3.287) gives

$$\mu_3 \approx \frac{T_{31}(1 - R_{44}) + T_{34}T_{41}}{(T_{31} + T_{32})(T_{41} + T_{42})}\mu_1 + \frac{T_{32}(1 - R_{44}) + T_{34}T_{42}}{(T_{31} + T_{32})(T_{41} + T_{42})}\mu_2. \quad (3.288)$$

If we had started with (3.283) but eliminated μ_3 instead and solved for μ_4, an identical second order analysis yields

$$\mu_4 \approx \frac{T_{31}T_{43} + T_{41}(1 - R_{33})}{(T_{31} + T_{32})(T_{41} + T_{42})}\mu_1 + \frac{T_{32}T_{43} + T_{42}(1 - R_{33})}{(T_{31} + T_{32})(T_{41} + T_{42})}\mu_2. \quad (3.289)$$

Proceeding now to express the current in terms of the difference of the chemical potentials μ_1 and μ_2, the current $I_1 = I$ may be written from (3.254) as

$$I_1 = I = \frac{2e}{h}[(1 - R_{11})\mu_1 - T_{12}\mu_2 - T_{13}\mu_3 - T_{14}\mu_4]. \quad (3.290)$$

Substituting (3.288) and (3.289) gives

$$I = \frac{2e}{h}\left[(1 - R_{11}) - T_{13}\frac{T_{31}(1 - R_{44}) + T_{34}T_{41}}{(T_{31} + T_{32})(T_{41} + T_{42})} - T_{14}\frac{T_{31}T_{43} + T_{41}(1 - R_{33})}{(T_{31} + T_{32})(T_{41} + T_{42})}\right]\mu_1$$
$$- \left[T_{12} + T_{13}\frac{T_{32}(1 - R_{44}) + T_{34}T_{42}}{(T_{31} + T_{32})(T_{41} + T_{42})} + T_{14}\frac{T_{32}T_{43} + T_{42}(1 - R_{33})}{(T_{31} + T_{32})(T_{41} + T_{42})}\right]\mu_2. \quad (3.291)$$

We see that while the denominator of the 2nd and 3rd terms in the square brackets is of order ε^2, the numerator of both terms is at least of order ε^3. The leading terms, $T_{12} \approx (1 - R_{11}) \approx$

$T_{12}^{(0)}$, are zero order in ε and thus dominate over the other terms so that

$$I \approx \frac{2e}{h} T_{12}^{(0)}(\mu_1 - \mu_2), \tag{3.292}$$

and thus the two-terminal resistance is given by

$$R_{12,12} = \frac{(\mu_1 - \mu_2)}{eI} = \frac{h}{2e^2} \frac{1}{T_{12}^{(0)}}, \tag{3.293}$$

which is symmetric with respect to flux reversal or reversal of the measurement leads (i.e., $R_{21,21}$). Thus we have recovered to lowest order in the coupling constant the two-terminal Landauer conductance (the inverse of (3.293)) given by (3.116).

Now we want to calculate the four-terminal resistance, $R_{12,34}$, in which current is passed between terminals 1 and 2, while the voltage drop along the sample is measured independently in the weakly coupled probes 3 and 4. It is now necessary to eliminate the chemical potentials μ_1 and μ_2 from (3.282) above following an analogous procedure to that used to eliminate μ_3 and μ_4 in deriving (3.293). Eliminating μ_1 from (3.282) gives

$$\mu_2 \approx \frac{T_{41}(T_{31} + T_{32})}{T_{32}T_{41} - T_{31}T_{42}} \mu_3 - \frac{T_{31}(T_{41} + T_{42})}{T_{32}T_{41} - T_{31}T_{42}} \mu_4. \tag{3.294}$$

Likewise, eliminating μ_2 from (3.282) gives

$$\mu_1 \approx -\frac{T_{42}(T_{31} + T_{32})}{T_{32}T_{41} - T_{31}T_{42}} \mu_3 + \frac{T_{32}(T_{41} + T_{42})}{T_{32}T_{41} - T_{31}T_{42}} \mu_4. \tag{3.295}$$

Substituting into (3.290) gives

$$I \cong \frac{2e}{h} \left[-\frac{T_{12}T_{42}(T_{31} + T_{32})}{T_{32}T_{41} - T_{31}T_{42}} - \frac{T_{12}T_{41}(T_{31} + T_{32})}{T_{32}T_{41} - T_{31}T_{42}} - T_{13} \right] \mu_3$$
$$+ \left[\frac{T_{12}T_{32}(T_{41} + T_{42})}{T_{32}T_{41} - T_{31}T_{42}} + \frac{T_{12}T_{31}(T_{41} + T_{42})}{T_{32}T_{41} - T_{31}T_{42}} - T_{14} \right] \mu_4. \tag{3.296}$$

The first two terms of each of the expressions in square brackets have numerators and denominators both of order ε^2 which cancel, whereas T_{13} and T_{14} are of order ε and thus smaller compared to the first two. Neglecting terms in ε or higher then yields

$$I \approx \frac{2e}{h} T_{12}^{(0)} \frac{(T_{31} + T_{32})(T_{41} + T_{42})}{T_{31}T_{42} - T_{32}T_{41}} (\mu_3 - \mu_4), \tag{3.297}$$

which gives the four-terminal resistance

$$R_{12,34} = \frac{h}{2e^2} \frac{1}{T_{12}^{(0)}} \frac{T_{31}T_{42} - T_{32}T_{41}}{(T_{31} + T_{32})(T_{41} + T_{42})}. \tag{3.298}$$

In the four-terminal measurement, the measured resistance depends explicitly on the transmission coefficients into and out of the potential probes, which, depending on the relative values of the coefficients, may be positive, negative, or even zero!

To make a connection with the original Landauer formula (3.116), assume that the voltage probes 3 and 4 couple to perfect leads on either side of a scattering region or elastic scattering characterized by a simple transmission coefficient T and reflection coefficient R. The voltage probes still have tunnel barriers such that they are weakly coupled to the leads by an attenuation factor ε. The magnetic flux is assumed zero. Thus the transmission probability $T_{12} = T_{21} = T$ to lowest order in ε. On either side of the scattering region, electrons may be transmitted directly from contact to contact, or by reflecting from the scattering region; one may argue that

$$T_{31} = T_{13} = T_{42} = T_{24} = \varepsilon(1 + R), \qquad (3.299)$$

whereas for carriers injected to or from one of the voltage probes through the scattering region, the transmission probabilities are proportional to T:

$$T_{32} = T_{23} = T_{14} = T_{41} = \varepsilon T. \qquad (3.300)$$

Finally, carriers injected from one voltage probe to the other are attenuated twice, $T_{34} = T_{43} = \varepsilon^2 T$. Writing the four-probe resistance in this case, using (3.298), yields

$$R_{12,34} = \frac{h}{2e^2} \frac{1}{T} \frac{\varepsilon^2(1 + R)^2 - \varepsilon^2(1 - R)^2}{(\varepsilon(1 + R) + \varepsilon T)(\varepsilon T + \varepsilon(1 + R))} = \frac{h}{2e^2} \frac{R}{T}, \qquad (3.301)$$

where use has been made of $R + T = 1$. Equation (3.301) is now seen to be the inverse of the Landauer conductance (3.114). Although derived with admittedly crude approximations, one sees that the factor of T/R is recovered if the voltage probes measure the potential drop across the scattering region where charge has piled up, rather than the potential at the contacts where current is injected.

3.9.2 Experimental multi-probe measurements (B = 0)

Before turning to the topic of multi-probe measurements of magnetotransport in Section 3.10, we briefly introduce some experimental results with zero field that demonstrate the nonlocality of transport in mesoscopic structures when measured with multiple probes. A widely investigated structure is the symmetric cross shown in Fig. 3.58.

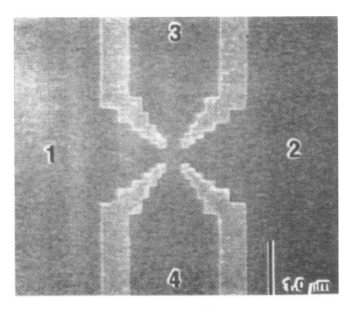

Fig. 3.58. An electron micrograph of a symmetric cross structure realized using two ballistic point contact structures in series. The regions 1–4 are separately contacted regions for making multiprobe measurements [after Timp *et al.*, in *Nanostructure Physics and Fabrication*, eds. W. P. Kirk and M. Reed (Academic Press, New York, 1989), p. 331, by permission].

Experimentally, this structure is realized using split-gate electrodes above a high-mobility 2DEG heterostructure, as shown in the SEM micrograph of this figure [91]. The ports 1 through 4 are 2DEG regions that are isolated from one another and connected to Ohmic contacts for multi-probe transport measurements. The gates are individually biased so that each of the four-point contacts may be opened or closed to realize two-, three-, and four-terminal structures.

Three-terminal measurement

With sufficiently large negative bias applied to the bottom two electrodes, region 4 is removed from the measurement, and one has an essentially three-terminal structure. Timp and coworkers [91], [92] investigated the conductance for this structure as a function of the coupling of the third electrode (3) to the system by changing the gate bias, V_{g2}, on the upper right electrode. The result is shown in Fig. 3.59 for the conductances $G_{12,12}$ and $G_{12,13}$, which theoretically are given by the inverses of (3.278) and (3.276), respectively. For $V_{g2} = 0$,

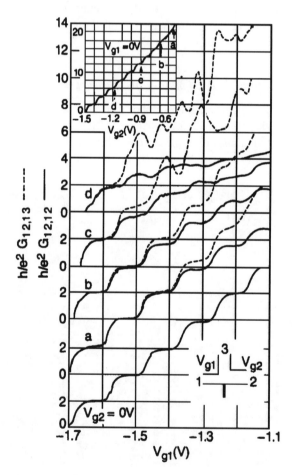

Fig. 3.59. The three-terminal conductances $G_{12,12}$ and $G_{12,13}$ for various gate biases applied to the second point contact [after Timp *et al.*, in *Nanostructure Physics and Fabrication*, eds. W. P. Kirk and M. Reed (Academic Press, New York, 1989), p. 331, by permission].

there is essentially no difference between the two-terminal and three-terminal conductance because both contacts are the same, and nearly ideal quantized conductance is measured for both conductances. As the constriction is closed, the conductance $G_{12,12}$ decreases, while $G_{12,13}$ increases, particularly for high N, where N is the number of occupied subbands in the 1D channel, illustrating that the presence of the third terminal may be invasive.

Four-terminal bend resistance

In Section 3.9.1, the theoretical four-terminal resistance in a measurement with two current probes and two ideal voltage probes is given by Eq. (3.298). Figure 3.60 shows measured data for the four-terminal resistance $R_{14,32}$, measured in a symmetric cross structure similar to that shown in Fig. 3.58. As shown, the four-terminal conductance is negative and exhibits sharp minima at certain gate biases attributable to the thresholds for occupation of 1D modes in the leads. The negative four-terminal resistance follows from the four-terminal multi-probe

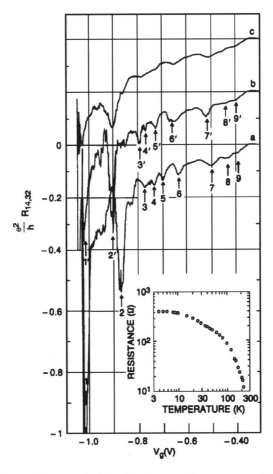

Fig. 3.60. The experimental four-terminal bend resistance, $R_{14,32}$, measured in a symmetric cross structure. The lower two traces are measured at $T = 280\,mK$ after successive cycling to room temperature and back. The upper curve is the same structure at $T = 1.2\,K$ [after Takagaki *et al.*, Solid State Comm. **71**, 809 (1989), by permission].

formula as discussed in Section 3.9.1. Such structure in the resistance is sometimes referred to as the bend resistance, as the electrons are being forced to turn 90 degrees from contact 1 to contact 4. Baranger *et al.* [153] have investigated theoretically the bend resistance through such structures, both from a classical standpoint and using a four probe recursive Green's function technique based on the theoretical approach developed in Section 3.8 including magnetic fields. The result of the calculations compares well with the bend resistance structure shown in Fig. 3.60.

3.10 Magnetic fields and quantum waveguides

The previous sections discussed ballistic transport in the absence of a magnetic field, or in the presence of a weak nonquantizing field. In the present section, we consider transport in quantum waveguide structures in the presence of strong magnetic fields. As we saw in Section 2.5, a one-dimensional quantum wire in the presence of a static magnetic field perpendicular to the longitudinal axis of the wire results in the formation of hybrid magnetoelectric subbands. Such subbands are a mixture of the quantization due to the lateral potential of the wire and the quantizing effect of the magnetic flux in terms of Landau level formation. The symmetric dispersion relations of these magnetoelectric subbands correspond to one-dimensional subbands or modes with positive and negative momentum along the wire axis. As shown in Fig. 2.21 the dispersion relationship is distinctly nonparabolic except in the particularly simple case of a parabolic confining potential. Although in general the dispersion relation is complicated, the current injected from a contact into one of the magnetoelectric modes is independent of the magnetoelectric dispersion relation, as was seen in the various formulas derived in Sections 3.4–3.9, due to cancellation of the group velocity with the density of states. Thus we may treat the magnetoelectric modes in a completely analogous way to the simple confined modes for $\mathbf{B} = 0$ within the Landauer-Büttiker formalism for transport.

A particularly important concept associated with magnetoelectric subbands is that of *edge states*, propagating states along the wire whose center coordinate is localized close to one boundary or the other depending on the direction of propagation. Semiclassically, these edge states correspond to skipping trajectories along the equipotential boundaries of the waveguide as illustrated in Fig. 2.22. An important aspect of these edge states is that states of positive and negative momentum are spatially separated from one another on opposite sides of the wire. The consequence of this is that the overlap factor for scattering from a state of forward momentum to one of backward momentum is greatly surpressed. If scattering does occur due to an impurity or boundary inhomogeneity, the particle will still remain in a state of forward scattering, which results in near unity transmission for edge states from one contact to an adjacent contact. Therefore, as the magnetic field increases, transport becomes more and more adiabatic, and backscattering due to the geometry and due to inhomogeneities is suppressed. This idea is illustrated schematically in Fig. 3.61 for a three-probe structure. The upper figure represents the edge states in an ideal geometry, and the lower figure shows the same geometry in the presence of inhomogeneities such as those discussed in Section 3.6.4. Using the representation introduced in Chapter 2, each edge state occupied at the Fermi energy (at low temperature) is indicated by a line, with the direction of propagation indicated by the arrows. The innermost lines relative to the middle of the main waveguide section are states of lower kinetic energy (higher subband index). The tendancy

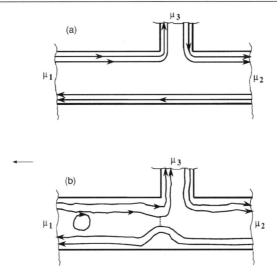

Fig. 3.61. (a) Illustration of the edge states in an ideal three-terminal waveguide structure. (b) The same system as in (a) with inhomogeneities present in the channel, showing coupling between forward and backward states, and closed orbits.

is for the edge states to follow the geometry, so edge states flowing from contact 1 follow the equipotential surfaces and flow into contact 3, while those injected from 2 flow directly to 1 in the ideal case (Fig. 3.61a). In the presence of impurities and boundary fluctuations, some of the edge states may couple to opposite flowing edge states on the opposite boundary, giving rise to a component of backscattering. Closed orbits may occur as well for localized potential fluctuations (Fig. 3.61b). As the magnetic field increases, the edge states become more localized on the boundaries, suppressing coupling across the waveguide and hence backscattering.

3.10.1 Quantized conductance in a perpendicular field

Measurement of the two-terminal conductance in the presence of a perpendicular magnetic field provides direct evidence of the current transmitted by magnetoelectric subbands. The effect of both parallel and perpendicular magnetic field on the quantized conductance was first reported by Wharam *et al.* [89]. Figure 3.62 shows data of van Wees *et al.* of the evolution of the conductance plateaus for various perpendicular magnetic field intensities [154]. The plateaus are observed to broaden and flatten as a function of increasing magnetic field. Additional plateaus are seen to arise at the highest fields at half steps of G_o. For a given gate bias as a function of increasing magnetic field, the quantized conductance remains in a given integer conductance plateau until a certain critical magnetic field intensity is reached, and then it jumps down by an integer multiple (or half integer multiple at high fields) of G_o.

A simple explanation for this behavior may be constructed by considering a 1DEG with a parabolic lateral potential in the presence of a perpendicular magnetic field, as introduced in Section 2.5.2. Considering the quantum point contact in the inset of Fig. 3.62 to be a

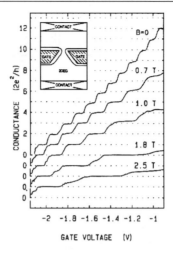

Fig. 3.62. Point contact conductance as a function of gate voltage for several magnetic field values, illustrating the transition from zero-field quantization to quantum Hall effect. The curves are offset for clarity [after van Wees *et al.*, Phys. Rev. B **38**, 3625 (1988), by permission].

waveguide, the energy eigenvalues in the constriction are given by (2.37) as

$$E_n(k_y) = \hbar\omega\left(n + \frac{1}{2}\right) + \frac{\hbar^2 k_y^2}{2M}$$

$$M = m^*\left(\frac{\omega_o^2 + \omega_c^2}{\omega_o^2}\right) \qquad \omega = \sqrt{\omega_o^2 + \omega_c^2},$$

(3.302)

where ω_o characterizes the strength of the lateral confinement, and $\omega_c = eB/m^*$ is the cyclotron frequency. In the adiabatic limit, the conductance is given by number of occupied modes below the Fermi energy in the wide 2DEG, N, times G_o. If we assume for simplicity that the Fermi energy in the 2DEG is almost constant with magnetic field (it in fact oscillates due to Landau level formation in the bulk region), the number of occupied subbands below the Fermi energy from (3.302) is given by

$$N = Int\left(\frac{E_F}{\hbar\sqrt{\omega_o^2 + \omega_c^2}} + \frac{1}{2}\right),$$

(3.303)

where $Int(x)$ means to take the integer part of x. It is clear that for low fields, $\omega_c \ll \omega_o$, and N remains constant. As ω_c approaches ω_o, increasing the magnetic field causes N to decrease by successive integer values. This decrease in N with increasing magnetic field in a quantum waveguide is referred to as *magnetic depopulation of subbands* [155]. Thus the widening of the plateaus from the zero field conductance is simply a manifestation of the increased energy separation of the subbands as the magnetic field increases, causing fewer and fewer subbands to reside below the Fermi energy, with the corresponding reduction in the conductance.

Büttiker has shown that it is not necessary to assume a waveguide geometry in explaining the transition from zero to high field magnetoconductance [96]. As discussed earlier, Eq. (3.153) gives a model form for a parabolic saddle point potential, which gives an exact form for the transmission coefficient. This potential may easily be combined with the

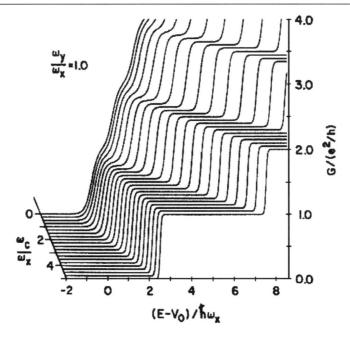

Fig. 3.63. Calculated two-terminal conductance of a point-contact constriction in a magnetic field as a function of Fermi energy. The ratio ω_c/ω_x is increased in steps of 0.25 from 0 to 5 [after M. Büttiker, Phys. Rev. B **41**, 7906 (1990), by permission].

harmonic potential due to a perpendicular magnetic field to find the transmission coefficient using the form given by Fertig and Halperin [95],

$$T_{nm} = \delta_{nm} \frac{1}{1 + e^{-\pi \varepsilon_n}}, \tag{3.304}$$

where the saddle point energies are given by

$$\varepsilon_n = \frac{E_1 - E_2 \left(n + \frac{1}{2}\right) - V_o}{E_1},$$

$$E_1 = \frac{1}{2} \frac{\hbar}{\sqrt{2}} \left(\left[\Omega^4 + 4\omega_x^2 \omega_y^2 \right]^{\frac{1}{2}} - \Omega^2 \right)^{\frac{1}{2}},$$

$$E_2 = \frac{\hbar}{\sqrt{2}} \left(\left[\Omega^4 + 4\omega_x^2 \omega_y^2 \right]^{\frac{1}{2}} + \Omega^2 \right)^{\frac{1}{2}}.$$

$\Omega^2 = \omega_c^2 + \omega_y^2 - \omega_x^2$ is an effective energy, and all other parameters are defined in Eq. (3.153). Figure 3.63 shows the calculated conductance using $T_{nm}(E_F)$ above in the two-terminal multi-channel formula. For increasing B, the curves evolve from narrow, rounded conductance plateaus to sharp, broad plateaus in qualitative agreement with the experimental results shown in Fig. 3.62.

Other features of the experimental results shown in Fig. 3.62 are of interest as well. One is the flattening of the plateaus with increasing accuracy about NG_o. It was argued earlier that backscattering is decreased as the magnetic field increases due to an increasing amount of current being carried by edge states. As discussed in Section 3.6.4, the conductance is degraded from the ideal integer multiples of the fundamental conductance due to impurities and boundary roughness which give rise to backscattering, and thus less than unity transmission. The flattening of the experimental plateaus with increasing field is thus taken as a sign of increased transmission and reduced backscattering due to these effects.

Another feature of the conductance at high fields is the appearance of half integer plateaus. This phenomenon is explained if we relax the assumption of spin degeneracy in the prescence of a field. As discussed in Section 2.5, two-fold spin-degenerate magnetosubbands are split by an amount $\pm\frac{1}{2}g^*\mu_B B$ into spin-up and spin-down states. These spin-split subbands constitute separate modes that contribute to conduction. However, they contribute only e^2/h to the conductance, since of the density of states are now reduced by a factor of two. As the magnetic field is increased, the splitting of the magnetoelectric subbands due to breaking of the spin-degeneracy increases in energy to the point that the individual contributions of spin-up and spin-down levels is resolvable in the conductance versus gate bias curve, as seen in Fig. 3.62.

3.10.2 Edge states and the quantum Hall effect

The quantum Hall effect (QHE) introduced earlier (Section 2.7) is the striking quantization of the Hall resistance at high magnetic fields measured in 2DEG structures. In the present section, we revisit this effect. However, particular emphasis will now be paid to the geometry in which the measurement is made, and to the roles that the confining boundaries play in terms of conduction through edge states. With a few exceptions, the QHE is measured in a multi-probe configuration in which current is passed longitudinally through a 2DEG sample, and the transverse voltage is measured by separate contacts on opposite sides of an imaginary line drawn from the current injecting contact to the current sink (Fig. 3.64). In this figure, a five-contact Hall geometry structure is shown in which adjacent contacts (4 and 5) are used to measure the longitudinal voltage drop, while opposite contacts (with respect to the current line) measure the transverse Hall voltage. The corresponding longitudinal and transverse resistances are found by dividing by the current. Figure 2.32 shows an example of the experimentally measured integer quantum Hall effect. The four-terminal Hall (transverse) resistance is observed to form plateaus quantized as [156]

$$R_{13,42} = \left(\frac{h}{e^2}\right)\frac{1}{n}, \qquad n = 1, 2, 3, \ldots, \tag{3.305}$$

where now we are using the nomenclature introduced in Section 3.9 to indicate that current sourced into contact 1 and contact 3 acts as the current sink, while voltage is measured between contacts 4 and 2. In the regime where the Hall resistance is quantized, the longitudinal resistance simultaneously vanishes, as shown by the SdH oscillations in Fig. 2.32.

In Section 2.7, the theoretical explanation for the QHE was explained in terms of the localized states lying between the bulk Landau level of the 2DEG in the presence of a perpendicular magnetic field. No mention was made of either the role of boundaries or the subsequent formation of edge states at such boundaries under the high magnetic field conditions in which the QHE is observed. However, as we saw in the discussion of the two-terminal conductance under high magnetic field conditions, quantization of the resistance is natural from the standpoint of the Landauer-Büttiker formalism in terms of the occupied edge states in the channel. This picture may be extended to the multiple probe case to include quantization of the the transverse Hall resistance [157]–[161]. Detailed discussion of the edge state picture of the QHE may be found in the review of Büttiker [162] as well as in references [87], [88].

In order to properly analyze the conductance, we need to invoke the multi-probe formula [151] for the four-terminal measurement of the QHE. Assume initially that the measurement structure is sufficiently small so that transport is nonlocal and that phase coherence is preserved between contacts. Further assume that the contacts are ideal in the sense that they populate edge channels equally. Then the current is given in terms of the carrier fluxes determined by the scattering matrix defining the sample, as introduced in Section 3.9. Suppose that the field is sufficiently strong to be in the edge state regime, and that back scattering is strongly suppressed. Then Fig. 3.64 gives a valid representation of the fluxes present in the system. Assume that for a given field, N single spin-degenerate subbands are occupied at the Fermi energy in all the contacts. This assumption is valid if the magnetic length is much smaller than the channel width so that the number of modes is determined by the field and not the geometry. Using Fig. 3.64, we see that N edge-state modes propagate from contact 1 to contact 2, but none propagates back in the opposite direction. Thus $T_{21} = N$ and $T_{12} = 0$. This result is true for all pairs of adjacent contacts, whereas for nonadjacent contacts, $T_{ij} = 0$, since the edge states cannot propagate without going through an intermediate contact where phase coherence is lost. The reflection coefficients are zero under these conditions due to conservation of flux. Using the multi-probe formula (3.256), we can write the current in matrix form in the five-probe structure of Fig. 3.64, assuming the above form for the various transmission and reflection coefficients as

$$
\begin{bmatrix} I \\ 0 \\ -I \\ 0 \\ 0 \end{bmatrix} = \frac{e^2}{h} \begin{bmatrix} N & 0 & 0 & 0 & -N \\ -N & N & 0 & 0 & 0 \\ 0 & -N & N & 0 & 0 \\ 0 & 0 & -N & N & 0 \\ 0 & 0 & 0 & -N & N \end{bmatrix} \begin{bmatrix} V_1 \\ V_2 \\ V_3 \\ V_4 \\ V_5 \end{bmatrix}, \tag{3.306}
$$

where the current contacts are 1 and 3 and where contacts 2, 3, and 4 are voltage contacts that draw no current. Note that if the magnetic field is reversed, the flux lines in Fig. 3.64 are reversed, and the matrix elements containing $-N$ in (3.326) are transposed. Thus the transmission coefficients satisfy the required symmetry $T_{ij}(B) = T_{ji}(-B)$. The first two

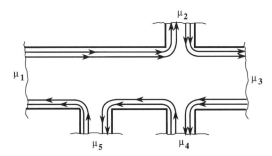

Fig. 3.64. Edge states in a five-probe structure used to measure the transverse and longitudinal resistances.

and last two equations in (3.306) give

$$I = \frac{e^2}{h} N(V_1 - V_5); \qquad V_1 = V_2,$$

$$V_3 = V_4; \qquad V_4 = V_5. \tag{3.307}$$

The four-terminal Hall resistance measured using probes 2 and 4 is therefore

$$R_{13,24} = \frac{|V_4 - V_2|}{I} = \frac{h}{e^2} \frac{1}{N}, \tag{3.308}$$

while the longitudinal resistance vanishes:

$$R_{13,45} = \frac{|V_4 - V_5|}{I} = 0. \tag{3.309}$$

Note that the Hall contacts do not have to be paired opposite one another as shown between contacts 2 and 4 in Fig. 3.64. The vanishing of the longitudinal resistance given by (3.309) signifies that $R_{13,25} = R_{13,24}$. The quantization of the Hall conductance is measured between any pair of contacts separated by the current line. Equations (3.308) and (3.309) give precisely the experimental quantized Hall resistance of (3.305) as well as the vanishing of the longitudinal resistance. However, the physical model invoked here is one of phase-coherent transport in the absence of backscattering due to spatial separation of forward and backward momentum states to opposite sides of the structures. Thus the quantization is accurate to the extent that the transmission coefficients for each mode into an adjacent contact approach unity. The result (3.308) shows that for small structures in which phase coherence exists, the quantization of the transverse conductance in the QHE has a similar origin as the quantized conductance in point contacts with no field present. Note that this transport does not have to be strictly adiabatic in the sense of no inter-mode coupling. Scattering between modes in the same momentum direction does not affect the number counting of particles transmitted from one contact to the next.

Equation (3.308) does not predict the transition from one quantized conductance plateau to next. In the regime where the transition is occuring, a new magnetoelectric subband is emerging at the Fermi energy. For such a state, the center coordinate at the Fermi energy is closer to the bulk of the sample than the other occupied levels, and therefore much more subject to backscattering. Such backscattering depends on the exact topological details of the disorder potential, which goes beyond the simple description in terms of the perfect transmission of edge states given by (3.306). Thus, transport in the transition region between plateaus is dominated by the transmission properties through the bulk 2DEG at high magnetic field when the Fermi energy is in the vicinity of one of the bulk Landau-level energies. The conductance then moves smoothly into a regime of edge-state conduction when the Fermi energy lies between Landau levels in the bulk, and Eq. (3.308) holds.

The edge state picture of the QHE derived here assumes the ideal Landauer-Büttiker picture used in Section 3.9, in which no inelastic scattering occurs in the sample, only in the contacts. However, the QHE typically is measured in macroscopic samples in which the contacts are separated by distances much longer than the expected inelastic mean free path. In fact, there is strong experimental evidence that phase coherence is preserved over very long distances [163]. However, it is clear that quantized Hall plateaus are measured in samples with dimensions exceeding the inelastic mean free path so that phase coherence is not preserved. Büttiker has demonstrated that even in the presence of inelastic scattering,

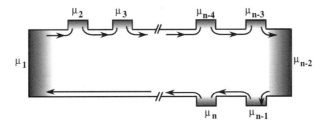

Fig. 3.65. Representation of inelastic scattering as phase-randomizing contacts in a sample under high magnetic fields.

the conductance quantization persists in the edge state picture of the QHE [161]. To show this, we can consider contacts as inelastic scattering centers as was done for RTDs [32] and in connection with the three-terminal conductance in Section 3.9 [151]. Figure 3.65 shows a sample with a random distribution of phase-randomizing contacts along the periphery that represent inelastic scattering events. Without loss of generalization, we can label these contacts in sequential order as we did for the real contacts in Fig. 3.64. Some of these contacts then represent real current or voltage probes, and the rest represent random inelastic scattering events. In the labeling scheme of Fig. 3.65, the diagonal form of the conductance matrix in (3.306) is preserved:

$$
\begin{bmatrix} I \\ 0 \\ \vdots \\ -I \\ 0 \\ 0 \end{bmatrix} = \frac{e^2}{h} \begin{bmatrix} N & 0 & 0 & \cdots & 0 & -N \\ -N & N & 0 & \cdots & 0 & 0 \\ \vdots & \vdots & \ddots & \ddots & \vdots & \vdots \\ 0 & 0 & \cdots & N & 0 & 0 \\ 0 & 0 & 0 & -N & N & 0 \\ 0 & 0 & 0 & 0 & -N & N \end{bmatrix} \begin{bmatrix} V_1 \\ V_2 \\ \vdots \\ V_{n-2} \\ V_{n-1} \\ V_n \end{bmatrix}.
\tag{3.310}
$$

A total of n contacts is assumed, where current is injected into contact 1 and taken from contact $n - 2$. One can see by example that all the voltage differences between pairs of contacts separated by the current sink contact contribute a resistance of $h/e^2 N$, while the voltage difference between any pair of contacts of the same side of the matrix with respect to the current sink contact give zero resistance. This result suggests that quantization of the transverse resistance and vanishing of the longitudinal resistance is a general property of an n-probe structure, and implies that even in the presence of inelastic scattering, the edge state model of the QHE persists up to macroscopic dimensions over which global phase coherence is lost.

3.10.3 Selective population of edge states

The edge state picture of the quantized resistances in the ballistic model shows that such quantization arises from the perfect transmission of carriers from one ideal contact to another due to the suppression of backscattering. The resistance is essentially given in terms of a fundamental constant times the number of transmitted edge states at the Fermi energy. A number of experiments have probed this picture of high-field magnetotransport through independent control of the number of transmitted edge states facilitated by a gate-induced

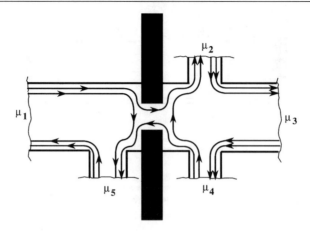

Fig. 3.66. Hall bar structure with a pinched gate-point contact structure that reflects some of the edge states and transmits others.

barrier across the channel or contact regions [164]–[168]. Such experiments usually contain a geometry that is similar to that shown in Fig. 3.66. This figure corresponds to the five-probe structure of Fig. 3.64, but with a barrier across the primary current path (Washburn *et al.* used an H-shaped sample as considered theoretically by Büttiker). Typically, this barrier may be fabricated using a split-gate-point contact structure. By similar arguments to those used in the adiabatic model for a constriction without magnetic field, a certain number of the highest occupied edge states in the wide region (in terms of their Landau level index) cannot propagate over the potential barrier of the constriction. We let that number equal an integer, K, and assume that $N - K$ states are transmitted, where N is the number of populated edge states in the wide region. Looking at the transmission coefficients now, $T_{21} = N - K$ while $T_{51} = K$. Likewise, $T_{32} = N$, $T_{43} = N$, $T_{24} = K$, $T_{54} = N - K$, and $T_{15} = N$. Modifying the conductance matrix relation (3.306) we get

$$
\begin{bmatrix} I \\ 0 \\ -I \\ 0 \\ 0 \end{bmatrix} = \frac{e^2}{h} \begin{bmatrix} N & 0 & 0 & 0 & -N \\ -(N-K) & N & 0 & -K & 0 \\ 0 & -N & N & 0 & 0 \\ 0 & 0 & -N & N & 0 \\ -K & 0 & 0 & -(N-K) & N \end{bmatrix} \begin{bmatrix} V_1 \\ V_2 \\ V_3 \\ V_4 \\ V_5 \end{bmatrix}. \tag{3.311}
$$

Using the first and last equations and eliminating V_1, the longitudinal resistance becomes

$$
R_{13,45} = \frac{h}{e^2 N} \frac{K}{N - K}, \tag{3.312}
$$

which differs from the zero resistance result (3.309) primarily by the number of K edge states reflected by the barrier. This equation thus predicts quantization of the longitudinal resistance between probes separated by the barrier. Likewise, the Hall resistance measured across opposite sides of the channel depends on whether the pair of contacts used to measure the voltage are separated by the barrier. For example, for pairs of contacts on the same side of the barrier such as contacts 2 and 4, the third and fourth equations in (3.311) give the Hall resistance h/Ne^2 as before. However, if the two contacts are on opposite sides of the

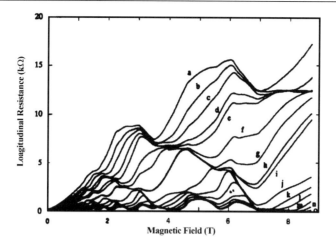

Fig. 3.67. The longitudinal resistances versus magnetic field for a variety of gate voltages [after Ryan *et al.*, Phys. Rev. B **40**, 12566 (1989)].

barrier such as contacts 2 and 5, the Hall resistance increases to

$$R_{13,42} = \frac{h}{e^2} \frac{1}{(N - K)}. \tag{3.313}$$

The evolution of the longitudinal resistance according to (3.312) was mapped out by Ryan *et al.* using gated Hall structures similar to Fig. 3.66 [169]. Figure 3.67 shows the measured longitudinal resistance versus magnetic field for various bias voltages applied to the gate traversing the channel. As predicted by (3.312), the zero resistances in the successive SdH minima become nonzero and approach plateau values due to the reflection of successive edge states by the potential barrier across the channel.

Bibliography

[1] R. H. Fowler and L. Nordheim, Proc. Roy. Soc. (London) **119**, 173 (1928).

[2] C. Zener, Proc. Roy. Soc. (London) **A145**, 523 (1943).

[3] L. Esaki, Phys. Rev. **109**, 603 (1957).

[4] I. Giaever, Phys. Rev. Lett. **5**, 147 (1960).

[5] I. Giaever, Phys. Rev. Lett. **5**, 464 (1960).

[6] J. Bardeen, Phys. Rev. Lett. **6**, 57 (1961).

[7] M. H. Cohen, L. M. Falicov, and J. C. Phillips, Phys. Rev. Lett. **8**, 316 (1962).

[8] W. A. Harrison, Phys. Rev. **123**, 85 (1961).

[9] E. O. Kane, J. Appl. Phys. **32**, 83 (1961).

[10] C. B. Duke, *Tunneling in Solids*, Solid State Physics **10** (suppl.), Academic, New York (1969).

[11] R. Tsu and L. Esaki, Appl. Phys. Lett. **22**, 562 (1973).

[12] L. L. Chang, L. Esaki, and R. Tsu, Appl. Phys. Lett. **24**, 593 (1974).

[13] T. C. L. G. Sollner, W. D. Goodhue, P. E. Tannenwald, C. D. Parker, and D. D. Peck, Appl. Phys. Lett. **43**, 588 (1983).

[14] E. R. Brown, T. C. L. G. Sollner, C. D. Parker, W. D. Goodhue, and C. L. Chen, Appl. Phys. Lett. **55**, 1777 (1989).

[15] T. J. Shewchuk, P. C. Chapin, P. D. Coleman, W. Kopp, R. Fischer, and H. Morkoç, Appl. Phys. Lett. **46**, 508 (1985).

[16] M. Tsuchiya and H. Sakaki, Appl. Phys. Lett. **49**, 88 (1986).

[17] M. Tsuchiya and H. Sakaki, Appl. Phys. Lett. **50**, 1503 (1987).

[18] M. Tsuchiya and H. Sakaki, Jpn. J. Appl. Phys. **25**, L185 (1986).

[19] P. Guéret, C. Rossel, E. Marclay, and H. Meier, J. Appl. Phys. **66**, 278 (1989).

[20] H. M. Yoo, S. M. Goodnick, and J. R. Arthur, Appl. Phys. Lett. **56**, 84 (1990).

[21] E. R. Brown, W. D. Goodhue, and T. C. L. G. Sollner, J. Appl. Phys. **64**, 1519 (1988).

[22] T. Inata, S. Muto, Y. Nakata, S. Sasa, T. Fujii, and S. Hiyamizu, Jpn. J. Appl. Phys. **26**, L1332 (1987).

[23] T. P. E. Broekaert, W. Lee, and C. G. Fonstad, Appl. Phys. Lett. **53**, 1545 (1988).

[24] L. F. Luo, R. Beresford, and W. I. Wang, Appl. Phys. Lett. **53**, 2320 (1988).

[25] J. R. Söderström, E. R. Brown, C. D. Parker, L. J. Mahoney, J. Y. Yao, T. G. Andersson, and T. C. McGill, Appl. Phys. Lett. **58**, 275 (1991).

[26] E. R. Brown, J. R. Söderström, C. D. Parker, L. J. Mahoney, K. M. Molvar, and T. C. McGill, Appl. Phys. Lett. **58**, 2291 (1991).

[27] J. R. Söderström, D. H. Chow, and T. C. McGill, Appl. Phys. Lett. **55**, 1094 (1989).

[28] K. F. Longenbach, L. F. Luo, and W. I. Wang, Appl. Phys. Lett. **57**, 1554 (1990).

[29] E. E. Mendez, Surf. Sci. **267**, 370 (1992).

[30] M. Jonson, "Tunneling Times in Quantum Mechanical Tunneling," in *Quantum Transport in Semiconductors*, eds. D. K. Ferry and C. Jacoboni (Plenum Press, New York, 1991), pp. 193–237.

[31] A. P. Jauho, "Tunneling Times in Semiconductor Heterostructures: A Critical Review," in *Hot Carriers in Semiconductor Nanostructures: Physics and Application*, ed. J. Shah (Academic Press, Boston, 1992), pp. 121–149.

[32] M. Büttiker, IBM J. Res. Develop. **32**, 63 (1988).

[33] E. R. Brown, "High-Speed Double Barrier Diodes," in *Hot Carriers in Semiconductor Nanostructures: Physics and Application,* ed. J. Shah (Academic Press, Boston, 1992), pp. 469–498.

[34] G. Breit and E. Wigner, Phys. Rev. **49**, 519 (1936).

[35] G. García-Calderón and R. E. Peierls, Nucl. Phys. A **265**, 443 (1976).

[36] L. D. Landau and E. M. Lifshitz, *Quantum Mechanics (Nonrelativistic)* (Pergamon Press, Oxford, 1977).

[37] G. García-Calderón, "Tunneling in Semiconductor Resonant Structures," in *The Physics of Low-Dimensional Semiconductor Structures,* eds. Butcher *et al.* (Plenum Press, New York, 1993), pp. 267–297.

[38] D. Bohm, *Quantum Theory* (Prentice-Hall, New York, 1951), pp. 257–261.

[39] E. H. Hauge, J. P. Falck, and T. A. Fjeldly, Phys. Rev. B **36**, 4203 (1987).

[40] J. R. Barker, Physica **134B**, 22 (1985).

[41] U. Ravaioli, M. A. Osman, W. Pötz, N. Kluksahl, and D. K. Ferry, Physica **134B**, 36 (1985).

[42] M. Büttiker and R. Landauer, Phys. Rev. Lett. **49**, 1739 (1982); IBM J. Res. Dev. **30**, 451 (1986).

[43] A. I. Baz', Sov. J. Nucl. Phys. **4**, 182 (1967); M. Büttiker, Phys. Rev. B **27**, 6178 (1983).

[44] E. R. Brown, C. D. Parker, and T. C. L. G. Sollner, Appl. Phys. Lett. **54**, 934 (1989).

[45] R. H. Good Jr. and E. W. Müller, in "Handbuch der Physik," Vol. 21, ed. S. Flugge (Springer, Berlin, 1956), p. 176.

[46] S. Collins, D. Lowe, and J. R. Barker, J. Appl. Phys. **63**, 143 (1988).

[47] H. C. Liu, J. Appl. Phys., **67**, 593 (1990).

[48] T. McGill and D. Ting, "Fluctuations in Mesoscopic Systems," in the *Proceedings of the NATO ASI on Ultrasmall Devices*, eds. D. K. Ferry, H. L. Grubin, and C. Jacoboni (Plenum Press, New York, 1995).

[49] E. E. Mendez, W. I. Wang, E. Calleja, and C. E. T. Goncalves da Silva, Appl. Phys. Lett. **50**, 1263 (1987).

[50] A. R. Bonnefoi, T. C. McGill, and R. D. Burnham, Phys. Rev. B **37**, 8754 (1988).

[51] S. Luryi, Appl. Phys. Lett. **47**, 490 (1985).

[52] F. Capasso, K. Mohammed, and A. Y. Cho, IEEE J. Quantum Elec. **QE-22**, 1853 (1986).

[53] T. Weil and B. Vinter, Appl. Phys. Lett. **50**, 1281 (1987).

[54] M. Jonson and A. Grincwaijg, Appl. Phys. Lett. **51**, 1729 (1987).

[55] A. D. Stone and P. A. Lee, Phys. Rev. Lett. **54**, 1196 (1985).

[56] M. Büttiker, Phys. Rev. B **33**, 3020 (1986).

[57] H. C. Liu and T. C. L. G. Sollner, "High-Frequency Resonant-Tunneling Devices," in *High-Speed Heterostructure Devices*, eds. R. A. Kiehl and T. C. L. G. Sollner, Semiconductors and Semimetals, Vol. **41** (Academic Press, Boston, 1994), pp. 359–419.

[58] V. J. Goldman, D. C. Tsui, and J. E. Cunningham, Phys. Rev. B **35**, 9387 (1987).

[59] E. E. Mendez, L. Esaki, and W. I. Wang, Phys. Rev. B **33**, 2893 (1986).

[60] J. F. Young, B. M. Wood, G. C. Aers, R. L. S. Devince, H. C. Liu, D. Landheer, M. Buchanan, A. J. SpringThorpe, and P. Mandeville, Phys. Rev. Lett. **60**, 2085 (1988).

[61] V. J. Goldman, D. C. Tsui, and J. E. Cunningham, Phys. Rev. Lett. **58**, 1256 (1987).

[62] T. C. L. G. Sollner, Phys. Rev. Lett. **59**, 1622 (1987); [62a] F. W. Sheard and G. A. Toombs, Appl. Phys. Lett. **52**, 1228 (1988).

[63] A. Zaslavsky, V. J. Goldman, D. C. Tsui, and J. E. Cunningham, Appl. Phys. Lett. **53**, 1408 (1988).

[64] H. Ohnishi, T. Inata, S. Muto, N. Yokoyama, and A. Shibatomi, Appl. Phys. Lett. **49**, 1248 (1986).

[65] M. Cahay, M. McLennan, S. Datta, and M. S. Lundstrom, Appl. Phys. Lett. **50**, 612 (1987).

[66] K. F. Brennan, J. Appl. Phys. **62**, 2392 (1987).

[67] H. L. Berkowitz and R. A. Lux, J. Vac. Sci. Technol. B **5**, 967 (1987).

[68] W. Pötz, J. Appl. Phys. **66**, 2458 (1989).

[69] M. O. Vassell, J. Lee, and H. F. Lockwood, J. Appl. Phys. **54**, 5206 (1983).

[70] B. Ricco and M. Ya. Azbel, Phys. Rev. B **29**, 1970 (1984).

[71] L. Eaves, G. A. Toombs, F. W. Sheard, C. A. Payling, M. L. Leadbeater, E. S. Alves, T. J. Foster, P. E. Simmonds, M. Henini, O. H. Hughes, J. C. Portal, G. Hill, and M. A. Pete, Appl. Phys. Lett. **52**, 212 (1988).

[72] P. J. Price, Phys. Rev. B **45**, 9042 (1992).

[73] E. Wigner, Phys. Rev. **40**, 749 (1932).

[74] W. R. Frensley, Phys. Rev. B **36**, 1570 (1987).

[75] N. C. Kluksdahl, A. M. Kriman, D. K. Ferry, and C. Ringhofer, Phys. Rev. B **39**, 7720 (1989).

[76] R. Landauer, IBM J. Res. Develop. **1**, 223 (1957).

[77] R. Landauer, Phil. Mag. **21**, 863 (1970).

[78] R. Büttiker, Y. Imry, R. Landauer, and S. Pinhas, Phys. Rev. B **31**, 6207 (1985).

[79] H. L. Engquist and P. W. Anderson, Phys. Rev. B **24**, 1151 (1981).

[80] Y. Imry, "Physics of Mesoscopic Systems," in *Directions in Condensed Matter Physics*, eds. G. Grinstein and G. Mazenko (World Scientific Press, Singapore, 1986), pp. 101–163.

[81] R. Landauer, IBM J. Res. Develop. **32**, 306 (1988).

[82] R. Landauer, Physica A **168**, 75 (1990).

[83] D. S. Fisher and P. A. Lee, Phys. Rev. B **23**, 6851 (1981).

[84] A. D. Stone and A. Szafer, IBM J. Res. Develop. **32**, 384 (1988).

[85] Yu. V. Sharvin, Sov. Phys. JETP **21**, 655 (1965); Yu. V. Sharvin and N. I. Bogatina, Sov. Phys. JETP **29**, 419 (1969).

[86] B. J. van Wees, H. van Houten, C. W. J. Beenakker, J. G. Williamson, L. P. Kouwenhoven, D. van der Marel, and C. T. Foxon, Phys. Rev. Lett. **60**, 848 (1988).

[87] C. W. J. Beenakker and H. van Houten, "Quantum Transport in Semiconductor Nanostructures," in Solid State Physics **44**, ed. H. Ehrenreich and D. Turnbull (Academic Press, Boston, 1991), pp. 1–228.

[88] H. van Houten, C. W. J. Beenakker, and B. J. Wees, "Quantum Point Contacts," in *Nanostructure Systems*, ed. M. Reed, Semiconductors and Semimetals **35** (Academic Press, Boston, 1992), pp. 9–112.

[89] D. A. Wharam, T. J. Thornton, R. Newbury, M. Pepper, H. Ahmed, J. E. F. Frost, D. C. Hasko, D. C. Peacock, D. A. Ritchie, and G. A. C. Jones, J. Phys. C **21**, L209 (1988).

[90] D. A. Wharam, M. Pepper, R. Newbury, D. G. Hasko, H. Ahmed, J. E. F. Frost, D. A. Ritchie, D. C. Peacock, and G. A. C. Jones, "Ballistic Transport in Quasi-One-Dimensional Structures," in *Spectroscopy of Semiconductor Microstructures*, ed. G. Fasol, A. Fasolino, and P. Lugli, NATO ASI Series B: Physics Vol. **206** (Plenum Press, New York, 1989), pp. 115–141.

[91] G. Timp, R. E. Behringer, S. Sampere, J. E. Cunningham, and R. Howard, "When Isn't the Conductance of an Electron Waveguide Quantized?" in the *Proceedings of the Int. Symp. on Nanostructure Physics and Fabrication*, ed. W. P. Kirk and M. Reed (Academic Press, New York, 1989), pp. 331–346.

[92] G. Timp, "When Does a Wire Become an Electron Waveguide?" in *Nanostructure Systems*, ed. M. Reed, Semiconductors and Semimetals **35** (Academic Press, Boston, 1992), pp. 113–190.

[93] L. I. Glazman, G. B. Lesovik, D. E. Khmel'nitskii, and R. I. Shekhter, Pis'ma Z. Eksp. Teor. Fiz. **48**, 218 (1988) [JETP Lett. **48**, 238 (1988)].

[94] F. W. J. Hekking, "Aspects of electron transport in semiconductor nanostructures," Ph.D. Dissertation, Technical University Delft, 1992.

[95] H. A. Fertig and B. I. Halperin, Phys. Rev. **36**, 7969 (1987).

[96] M. Büttiker, Phys. Rev. B **41**, 7906 (1990).

[97] B. J. van Wees, L. P. Kouwenhoven, E. M. M. Willems, C. J. P. M. Harmans, J. E. Mooij, H. van Houten, C. W. J. Beenakker, J. G. Williamson, and C. T. Foxon, Phys. Rev. B **43**, 12431 (1991).

[98] J. A. Nixon, J. H. Davies, and H. U. Baranger, Superlattices and Microstructures **9**, 187 (1991).

[99] Y. Takagaki and D. K. Ferry, J. Phys.: Condens. Matter **4**, 10421 (1992).

[100] L. I. Glazman and A. V. Khaetskii, Europhysics Lett. **9**, 263 (1989).

[101] L. P. Kouwenhoven, B. J. van Wees, C. J. P. M. Harmans, J. G. Williamson, H. van Houten, C. W. J. Beenakker, C. T. Foxon, and J. J. Harris, Phys. Rev. B **39**, 8040 (1989).

[102] L. Martín-Moreno, J. T. Nicholls, N. K. Patel, and M. Pepper, J. Phys. Condens. Matter **4**, 1323 (1992).

[103] G. Kirczenow, Solid State Comm. **68**, 715 (1988).

[104] A. Szafer and A. D. Stone, Phys. Rev. Lett. **62**, 300 (1989).

[105] E. G. Haanappel and D. van der Marel, Phys. Rev. B **39**, 5484 (1989).

[106] E. Tekman and S. Ciraci, Phys. Rev. B **39**, 8772 (1989).

[107] A. Weisshaar, J. Lary, S. M. Goodnick, and V. K. Tripathi, Appl. Phys. Lett. **55**, 2114 (1989).

[108] A. Weisshaar, J. Lary, S. M. Goodnick, and V. K. Tripathi, J. Appl. Phys. **70**, 355 (1991).

[109] R. Mittra and S. W. Lee, *Analytical Techniques in the Theory of Guided Waves* (Macmillan, New York, 1971).

[110] T. Itoh, "Generalized Scattering Matrix Technique," in *Numerical Techniques for Microwave and Millimeter-Wave Passive Structures*, ed. T. Itoh, (Wiley, New York, 1989) pp. 622–636.

[111] J. C. Wu, M. N. Wybourne, W. Yindeepol, A. Weisshaar, and S. M. Goodnick, Appl. Phys. **59**, 102 (1991).

[112] J. C. Wu, M. N. Wybourne, A. Weisshaar, and S. M. Goodnick, J. Appl. Phys. **74**, 4590 (1993).

[113] C. G. Smith, M. Pepper, H. Ahmed, J. E. F. Frost, D. G. Hasko, D. A. Ritchie, and G. A. C. Jones, Surf. Sci. **228**, 387 (1990).

[114] A. M. Bouchard, J. H. Luscombe, D. C. Seabaugh, and J. N. Randall, in *Nanostructure and Mesoscopic Systems*, eds. W. P. Kirk and M. Reed (Academic Press, New York, 1992), pp. 393–404.

[115] J. A. del Alamo and C. C. Eugster, Appl. Phys. Lett. **56**, 78 (1990).

[116] N. Tsukada, A. D. Wieck, and K. Ploog, Appl. Phys. Lett. **56**, 2527 (1990).

[117] C. C. Eugster and J. A. del Alamo, Phys. Rev. Lett. **67**, 3586 (1991).

[118] K. Kojima, K. Mitsunaga, and K. Kyuma, Appl. Phys. Lett. **55**, 882 (1989).

[119] D. J. Kirkner, C. S. Lent, and S. Sivaprakasam, Int. J. Num. Meth. in Engr., **29**, 1527 (1990).

[120] T. Kerkhoven, A. T. Galick, U. Ravaioli, J. H. Arends, and Y. Saad, J. Appl. Phys. **68**, 3461 (1990).

[121] H. K. Harbury, W. Porod, and S. M. Goodnick, J. Appl. Phys. **73**, 1509 (1993).

[122] E. N. Economou, *Green's Functions in Quantum Physics*, Vol. 7 of *Solid-State Sciences* 2nd edition (Springer-Verlag, Heidelberg, 1990).

[123] G. D. Mahan, *Many-Particle Physics* 2nd edition (Plenum Press, New York, 1990).

[124] A. Messiah, *Quantum Mechanics*, Vol. 2 (North-Holland Publishing Company, Amsterdam, 1962).

[125] R. Mezenner, *Conductivity Fluctuations in Small Structures*. Ph.D. thesis, Arizona State University, 1988.

[126] J. Skjånes, E. H. Hauge, and G. Schön, Phys. Rev. B **50**, 8636 (1994).

[127] I. S. Gradshteyn and I. M. Ryzhik, *Table of Integrals, Series, and Products* (Academic Press, Boston, 1980).

[128] F. Sols, M. Macucci, U. Ravaioli, and K. Hess, J. Appl. Phys., **66**, 3892 (1989).

[129] J. Callaway, *Quantum Theory of the Solid State*, 2nd edition (Academic Press, Boston, 1991).

[130] H. U. Baranger, D. P. DiVincenzo, R. A. Jalabert, and A. D. Stone, Phys. Rev. B **44**, 1063 (1991).

[131] T. Ando, Phys. Rev. B **44**, 8017 (1991).

[132] R. Peierls. Z. Phys. **80**, 763 (1933).

[133] A. S. Alexandrov and H. Capellmann, Phys. Rev. Lett. **66**, 365 (1991).

[134] F. Sols, Annals of Physics **214**, 386 (1992).

[135] E. Merzbacher, *Quantum Mechanics* (John Wiley & Sons, New York, 1961).

[136] A. D. Stone and A. Szafer. IBM J. Res. Develop. **32**, 384 (1988).

[137] D. J. Thouless and S. Kirkpatrick, J. Phys. C: Solid State Phys. **14**, 235 (1981).

[138] F. Guinea and J. A. Vergés, Phys. Rev. B **35** 979 (1987).

[139] P. A. Lee and D. S. Fisher, Phys. Rev. Lett. **47** 882 (1981).

[140] A. Furusaki, Physica B **203**, 214 (1994).

[141] Song He and S. Das Sarma, Phys. Rev. B **48**, 4629 (1993).

[142] T. Kawamura and J. P. Leburton, Phys. Rev. B, **48**, 8857 (1993).

[143] M. Macucci, U. Ravaioli, and T. Kerkhoven, Superlattices and Microstructures **12**, 509 (1992).

[144] M. Macucci, A. Galick, and U. Ravaioli, Phys. Rev. B **52**, 5210 (1995).

[145] A. Fetter and J. D. Walecka, *Quantum Theory of Many Particle Physics* (McGraw-Hill, New York, 1971).

[146] K. Aihara, M. Yamamoto, and T. Mizutani, Appl. Phys. Lett., **63**, 3595 (1993).

[147] L. I. Glazman, G. B. Lesovik, D. E. Khmel'nitskii, and R. I. Shekhter, JETP Lett. **48**, 238 (1988).

[148] L. Onsager, Phys. Rev. **38**, 2265 (1931).

[149] H. B. G. Casimir, Rev. Mod. Phys. **17**, 343 (1945).

[150] M. Büttiker, Phys. Rev. Lett. **57**, 1761 (1986).

[151] M. Büttiker, IBM J. Res. Develop. **32**, 317 (1988).

[152] W. Gröbner, *Matrizenrechnung* (Bibliographisches Institut AG, Mannheim, 1966), p. 47.

[153] H. U. Baranger, D. P. DiVincenzo, R. A. Jalabert, A. D. Stone, Phys. Rev. B **44**, 10637 (1991).

[154] B. J. van Wees, L. P. Kouwenhoven, H. van Houten, C. W. J. Beenakker, J. E. Mooij, C. T. Foxon, and J. J. Harris, Phys. Rev. B **38**, 3625 (1988).

[155] K. -F. Berggren, T. J. Thornton, D. J. Newson, and M. Pepper, Phys. Rev. Lett. **57**, 1769 (1986).

[156] K. von Klitzing, G. Dorda, and M. Pepper, Phys. Rev. Lett. **45**, 494 (1980).

[157] P. Streda, J. Kucera, and A. H. MacDonald, Phys. Rev. Lett. **59**, 1973 (1987).

[158] J. K. Jain and S. A. Kivelson, Phys. Rev. Lett. **60**, 1542 (1988).

[159] C. W. J. Beenakker and H. van Houten, Phys. Rev. Lett. **60**, 2406 (1988).

[160] F. M. Peeters, Phys. Rev. Lett. **61**, 589 (1988).

[161] M. Büttiker, Phys. Rev. B **38**, 9375 (1988).

[162] M. Büttiker, "The Quantum Hall Effect in Open Conductors," in *Nanostructure Systems*, ed. M. Reed, Semiconductors and Semimetals **35** (Academic Press, Boston, 1992), pp. 113–190.

[163] J. Spector, H. L. Stormer, K. W. Baldwin, L. N. Pfeiffer, and K. W. West, in *Nanostructure and Mesoscopic Systems*, eds. W. P. Kirk and M. Reed, (Academic Press, New York, 1992), pp. 107–118.

[164] R. J. Haug, A. H. MacDonald, P. Streda, and K. von Klitzing, Phys. Rev. Lett. **61**, 2797 (1988); R. J. Haug, J. Kucera, P. Streda, and K. von Klitzing, Phys. Rev. B **39**, 10892 (1989).

[165] S. Washburn, A. B. Fowler, H. Schmid, and D. Kern, Phys. Rev. Lett. **61**, 2801 (1988).

[166] H. van Houten, C. W. J. Beenakker, P. H. M. van Loosdrecht, T. J. Thornton, H. Ahmed, M. Pepper, C. T. Foxon, and J. J. Harris, Phys. Rev. B **37**, 8534 (1988).

[167] B. R. Snell, P. H. Beton, P. C. Main, A. Neves, J. R. Owers-Bradely, L. Eaves, M. Henini, O. H. Hughes, S. P. Beaumont, and C. D. W. Wilkinson, J. Phys. C **1**, 7499 (1989).

[168] S. Komiyama, H. Hirai, S. Sasa, and S. Hiyamizu, Phys. Rev. B **40**, 12566 (1989).

[169] J. M. Ryan, N. F. Deutscher, and D. K. Ferry, Phys. Rev. B **48**, 8840 (1993).

4

Quantum dots and single electron phenomena

We now turn to transport in nanostructure systems in which the electronic states are completely quantized. In Section 2.3.2 we briefly introduced quantum dots (sometimes referred to as quantum boxes) in which confinement was imposed in all three spatial directions, resulting in a discrete spectrum of energy levels much the same as an atom or molecule. We can therefore think of quantum dots and boxes as artificial atoms, which in principle can be engineered to have a particular energy level spectrum. In Section 4.1, we first consider models for the electronic states of quantum dots and boxes, and then compare these to experimental data. As in atomic systems, the electronic states in quantum dots are sensitive to the presence of multiple electrons due to the Coulomb interaction between electrons. In addition, magnetic fields serve as an experimental probe that one can use to elucidate the energy spectrum of such artificial atoms discussed below.

The primary focus of our attention in this book is on the transport properties in nanostructures; quantum dots provide some of the most interesting experiments in this regard. Transport in quantum dots and boxes implies an external coupling to these structures from which charge may be injected, as discussed in Chapter 3. Rich phenomena are observed not only because of quantum confinement and the resonant structure associated with this confinement, but also due to the granular nature of electric charge. In contrast to quantum wells and quantum wires, quantum dot structures are sufficiently small that even the introduction of a single electron is sufficient to dramatically change the transport properties due to the charging energy associated with this extra electron. One of the main effects is that of *Coulomb blockade*, where conductance oscillations are observed with the addition or subtraction of a single electron from a quantum dot, which we discuss in detail in Sections 4.2 and 4.3. The interested reader is referred to several detailed references related to the subject of single electron tunneling and Coulomb blockade [1]–[6]. Finally, in Section 4.4 we consider transport in a system of quantum dots where the dot potential is repulsive such that charge is localized outside the dots. Electrons traversing arrays of such *antidots* show interesting behavior connected to classical chaotic motion.

4.1 Electronic states in quantum dot structures

4.1.1 Noninteracting electrons in a parabolic potential

In Section 2.3.2 we introduced several methods for fabricating semiconductor quantum dot structures. Quantum dots may be realized by etching a 2DEG heterostructure, or by using an electrostatic potential due to metal gates in order to confine electrons to certain regions. Split-gate structures similar to those shown in Fig. 2.17 are used quite frequently for low-temperature transport measurements. We now look in more detail at the electron states of such a system, considering first a system of noninteracting electrons in the dot.

For split gates patterned above a 2DEG heterostructure, the confinement in the vertical growth direction is generally much stronger than that due to the lateral depletion regions under the gate. We can therefore focus on the details of the lateral confinement, assuming electrons only occupy the lowest subband associated with quantization in the vertical growth direction. Considering conduction band states in a typical type-I heterostructure system, we employ the effective mass approximation as before and assume the solution is separable as

$$\psi(\mathbf{r}, z) = \varphi_{nm}(\mathbf{r})\zeta_i(z),\tag{4.1}$$

where $\zeta_i(z)$ corresponds to the bound states in the z-direction with eigen-energies E_i characterizing the "bulk" 2DEG outside of the quantum dot. The solution in the plane of the 2DEG satisfies the effective mass equation

$$\left(-\frac{\hbar^2}{2m^*}\nabla_{\mathbf{r}}^2 + V_{eff}(\mathbf{r})\right)\varphi_{nm}(\mathbf{r}) = E_{nm}\varphi_{nm}(\mathbf{r}),\tag{4.2}$$

where m^* is the effective mass in the plane of the dot. The potential due to the space charge region under and around the split gates has a quadratic dependence. Therefore, a first approximation for the potential in a quantum dot formed with a rectangular array of gates is a parabolic potential of the form

$$V_{eff}(\mathbf{r}) = \frac{1}{2}m^*\omega_x^2 x^2 + \frac{1}{2}m^*\omega_y^2 y^2,\tag{4.3}$$

where the ωs characterize the strength of the potential. The bound states in the dot are given by the usual harmonic oscillator solutions

$$E_{n_x,n_y} = \hbar\omega_x\left(n_x + \frac{1}{2}\right) + \hbar\omega_y\left(n_y + \frac{1}{2}\right), \quad n_x, n_y = 0, 1, 2, \ldots.\tag{4.4}$$

Typical values for $\hbar\omega$ for an AlGaAs/GaAs split-gate dot are on the order of several meV, much smaller than the several 10s of meV for the vertical confinement.

By defining raising and lowering operators in the usual way for the harmonic oscillator problem, we may define the pair of adjoint operators in terms of position and momentum operators in the x-direction:

$$a_x = \sqrt{\frac{m^*\omega_x}{2\hbar}}\left(x + i\frac{p_x}{m^*\omega_x}\right),$$

$$a_x^\dagger = \sqrt{\frac{m^*\omega_x}{2\hbar}}\left(x - i\frac{p_x}{m^*\omega_x}\right).\tag{4.5}$$

An identical pair may be defined in the y-direction by replacing x with y above. The product of these two operators gives

$$a_x^\dagger a_x = \frac{1}{\hbar\omega_x}\left(\frac{p_x^2}{2m^*} + \frac{m^*\omega_x^2}{2}x^2\right) - \frac{1}{2},\tag{4.6}$$

with a similar result for $a_y^\dagger a_y$. Therefore, the Hamiltonian appearing in Eq. (4.2) may be rewritten as

$$H = \hbar\omega_x\left(a_x^\dagger a_x + \frac{1}{2}\right) + \hbar\omega_y\left(a_y^\dagger a_y + \frac{1}{2}\right). \tag{4.7}$$

Comparing (4.7) with (4.4), it is clear that $a_x^\dagger a_x$ and $a_y^\dagger a_y$ are *number operators* that, when acting on the eigenstates of the harmonic oscillator problem, give back the eigenvalues n_x and n_y respectively. The operator $a_x^\dagger$ acting on eigenstate n_x returns the eigenstate $n_x + 1$, and therefore $a_x^\dagger$ is called a *raising* or *creation operator*. The operator a_x, on the other hand, lowers the n_x to $n_x - 1$ and is therefore a *lowering* or *annihilation operator*. If we look at the product $a_x a_x^\dagger$ in terms of the position and momentum operators, we may define the commutator for these operators as

$$\left[a_x, a_x^\dagger\right] = \left[a_y, a_y^\dagger\right] = 1. \tag{4.8}$$

If the potential is equivalent in both directions such that $\omega_x = \omega_y = \omega$, we have degenerate eigenstates (in addition to spin) due to the radial symmetry of the problem:

$$E_n = (n + 1)\hbar\omega, \quad n = n_x + n_y = 0, 1, 2, \ldots. \tag{4.9}$$

The lowest state, $n_x = n_y = 0$, is nondegenerate with energy $\hbar\omega$. The next lowest level is doubly degenerate corresponding to $n_x = 1, n_y = 0$ and $n_x = 0, n_y = 1$. The third level is triply degenerate, and so forth, so that the nth level is $(n + 1)$-fold degenerate. These degeneracies correspond to different angular momentum states sharing the same energy, as one expects for a symmetric central potential.

The angular momentum is defined by the cross product of the position and linear momentum vectors,

$$\mathbf{L} = \mathbf{r} \times \mathbf{p}, \tag{4.10}$$

where quantum-mechanically $\mathbf{r}$ and $\mathbf{p}$ are the position and linear momentum operators, respectively. Since the motion in the two-dimensional problem considered here is entirely in the xy-plane, the only possible component of the angular momentum is in the z-direction:

$$L_z = xp_y - yp_x. \tag{4.11}$$

Using sums and differences of the creation and annihilation operators in (4.5) to express the position and momentum, Eq. (4.11) may be rewritten

$$\begin{aligned} L_z &= -i\frac{\hbar}{2}\left(a_x + a_x^\dagger\right)\left(a_y - a_y^\dagger\right) + i\frac{\hbar}{2}\left(a_y + a_y^\dagger\right)\left(a_x - a_x^\dagger\right) \\ &= i\hbar\left(a_y^\dagger a_x - a_x^\dagger a_y\right), \end{aligned} \tag{4.12}$$

where we have used the fact that pairs of operators with different subscripts commute, or

$$\left[a_x^\dagger, a_y\right] = \left[a_x^\dagger, a_y^\dagger\right] = \left[a_y^\dagger, a_x\right] = [a_y, a_x] = 0. \tag{4.13}$$

Using these commutation relations together with (4.8), one may show that Hamiltonian (4.7) and the z-component of the angular momentum given by (4.12) commute

$$[L_z, H] = 0, \tag{4.14}$$

and therefore L_z is a conserved quantity.

Since angular momentum is associated with the angular motion of an electron in the parabolic potential, we introduce rotating creation and annihilation operators

$$a = \frac{1}{\sqrt{2}}(a_x - ia_y),$$

$$b = \frac{1}{\sqrt{2}}(a_x + ia_y), \tag{4.15}$$

which have the associated adjoint operators

$$a^\dagger = \frac{1}{\sqrt{2}}\left(a_x^\dagger + ia_y^\dagger\right),$$

$$b^\dagger = \frac{1}{\sqrt{2}}\left(a_x^\dagger - ia_y^\dagger\right). \tag{4.16}$$

The rotational operators satisfy similar commutation relations to the linear operators,

$$[a, a^\dagger] = [b, b^\dagger] = 1,$$

$$[a, b] = [a, b^\dagger] = [a^\dagger, b] = [a^\dagger, b^\dagger] = 0, \tag{4.17}$$

which may be proved from the commutation relations (4.8) and (4.13). Effective number operators may be defined using Eqs. (4.14) and (4.16) as

$$a^\dagger a = \tfrac{1}{2}\left(a_x^\dagger a_x + a_y^\dagger a_y + ia_y^\dagger a_x - ia_x^\dagger a_y\right),$$

$$b^\dagger b = \tfrac{1}{2}\left(a_x^\dagger a_x + a_y^\dagger a_y - ia_y^\dagger a_x + ia_x^\dagger a_y\right). \tag{4.18}$$

These two expressions may be combined to realize the Hamiltonian in the symmetric case as

$$H = (a^\dagger a + b^\dagger b + 1)\hbar\omega = (n_a + n_b + 1)\hbar\omega = (n + 1)\hbar\omega, \quad n = 0, 1, 2, \ldots, \tag{4.19}$$

where n_a and n_b represent the number of excitations in the rotational states a and b. The angular momentum operator may be written similarly:

$$L_z = (a^\dagger a - b^\dagger b)\hbar = (n_a - n_b)\hbar = m\hbar, \tag{4.20}$$

where m is an integer. In this form, the Hamiltonian operator preserves its simple form, while the angular momentum operator is considerably simplified. In Eq. (4.19), n represents the principal quantum number, which determines the total energy of the state. The eigenvalues of the angular momentum are just $m\hbar$. For each $n = n_a + n_b$, there are $n + 1$ possible values of $m = n_a - n_b$:

$$
\begin{aligned}
& n_a = n & & n_b = 0 & & m = n \\
& n_a = n - 1 & & n_b = 1 & & m = n - 2 \\
& n_a = n - 2 & & n_b = 2 & & m = n - 4 \\
& \quad\vdots & & \quad\vdots & & \quad\vdots \\
& n_a = 1 & & n_b = n - 1 & & m = -(n - 2) \\
& n_a = 0 & & n_b = n & & m = -n
\end{aligned}
\tag{4.21}
$$

For a given energy level, m may therefore take on the values

$$m = n, n - 2, n - 4, \ldots, -n + 2, -n. \tag{4.22}$$

4.1.2 Dot states in a magnetic field

Using the results derived in the previous section, we are now in a position to derive the single electron states of a circular, parabolic quantum dot in the presence of a static magnetic field. This classic problem was first solved by Fock [7] and Darwin [8] and later studied in more detail by Dingle [9]. The operator approach used here follows the treatment of Rössler [10]. In the presence of a magnetic field, the Hamiltonian for the symmetric parabolic potential is generalized as

$$H = \frac{1}{2m^*}(\mathbf{p} + e\mathbf{A})^2 + \frac{1}{2}m^*\omega^2(x^2 + y^2) + \frac{g^*\mu_B}{\hbar}\mathbf{B} \cdot \mathbf{S}, \tag{4.23}$$

where $\mathbf{A}$ is the vector potential satisfying $\mathbf{B} = \nabla \times \mathbf{A}$. The last term represents the Zeeman splitting of the spin states, where $\mathbf{S}$ is the spin operator, g^* the effective Landé g factor (which is different than the free space value), and $\mu_B = e\hbar/2m_o$ is the Bohr magneton introduced in Chapter 2. For the present problem involving confinement in both lateral directions, it is more convenient to choose the symmetric gauge, rather than the Landau gauge used in Section 2.5.1 for 1D waveguides. In the symmetric gauge, $\mathbf{A} = (-By, Bx, 0)/2$, which gives $\mathbf{B} = B\mathbf{a}_z$ when the curl is taken. The kinetic energy operator may be expanded as

$$\frac{1}{2m^*}(\mathbf{p} + e\mathbf{A})^2 = -\frac{\hbar^2}{2m^*}\left(\frac{\partial^2}{\partial x^2} + \frac{\partial^2}{\partial y^2}\right) - \frac{i\hbar eB}{2m^*}\left(x\frac{\partial}{\partial y} - y\frac{\partial}{\partial x}\right) + \frac{1}{4}\frac{e^2B^2}{2m^*}(x^2 + y^2). \tag{4.24}$$

The first term is the usual kinetic energy operator. The last term may be combined with the electrostatic harmonic oscillator potential to define a renormalized oscillator frequency

$$\Omega = \sqrt{\omega^2 + \frac{\omega_c^2}{4}}, \tag{4.25}$$

where $\omega_c = eB/m^*$ is the cyclotron frequency. The second term in Eq. (4.24) is simply the product of the magnetic field and the angular momentum operator, $eBL_z/2m^*$, using (4.11). The Hamiltonian (4.23) may therefore be rewritten

$$H = \left[-\frac{\hbar^2}{2m^*}\nabla^2 + \frac{1}{2}m^*\Omega^2(x^2 + y^2)\right] + \frac{1}{2}\omega_c L_z + \frac{g^*\mu_B}{\hbar}BS_z$$

$$= H_\Omega + \frac{1}{2}\omega_c L_z + \frac{g^*\mu_B}{\hbar}BS_z. \tag{4.26}$$

The term in square brackets, H_Ω, is simply the Hamiltonian of a symmetric harmonic oscillator with frequency Ω. The spin operator S_z, operating on the spin component of the wavefunction, simply gives $\pm\hbar/2$ where the upper sign corresponds to a spin-up eigenstate and the lower sign for spin-down. This effect leads to a splitting of the two-fold spin degeneracy of the energy levels, which becomes significant at high magnetic fields. Since the angular momentum operator L_z commutes with this Hamiltonian according to (4.14) (and neglecting spin-orbit interaction), the eigenstates of H are the same as H_Ω, and we may write the Hamiltonian (4.26) in the number operator representation as

$$H = (n_a + n_b + 1)\hbar\Omega + \frac{1}{2}\omega_c L_z + \frac{g^*\mu_B}{\hbar}BS_z, \tag{4.27}$$

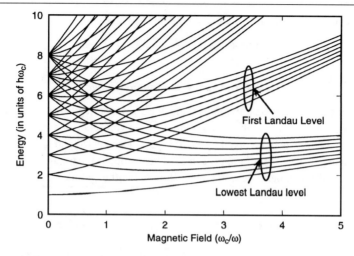

Fig. 4.1. Darwin-Fock states of a parabolic two-dimensional quantum dot in the presence of a static magnetic field. The solutions up to $n = 8$ are shown.

with the corresponding energy eigenvalues

$$E_{n,m} = (n + 1)\hbar\Omega + \frac{1}{2}\hbar\omega_c m \pm \frac{g^*\mu_B B}{2};$$

$$m = n_a - n_b = -n, -(n - 2), \ldots, (n - 2), n. \quad (4.28)$$

The result above shows that the $(n + 1)$-fold degeneracy of the angular momentum states is lifted by the magnetic field so that the total energy depends on the angular momentum itself as illustrated in Fig. 4.1, where the solutions of (4.28) are plotted as a function of magnetic field (spin splitting has been ignored for simplicity). For small fields ($\omega_c \ll \Omega$), the solutions go back to the degenerate solutions of the symmetric harmonic oscillator. As ω_c increases, the levels split towards higher and lower energy, depending on the sign of m. Each sublevel is split by an amount $\hbar\omega_c$ (since m changes by 2). In the limit of large magnetic fields ($\omega_c \gg \omega$), Eq. (4.28) may be expanded to give (neglecting the Zeeman term)

$$E_{n,m} \approx (n + m + 1)\frac{\hbar\omega_c}{2} + (n + 1)\frac{\omega^2}{\omega_c^2}. \quad (4.29)$$

In the extreme limit, (4.29) becomes $(n + m + 1)\hbar\omega_c/2$, which is the usual Landau level picture of a two-dimensional electron gas since $n + m = 2n_a$ is an even integer. Therefore, in the large magnetic field limit, the cyclotron motion dominates, and the electrons do not feel the effect of the parabolic confinement. It can be seen from Fig. 4.1 that in the large field limit, all the $m = -n$ angular momentum states combine to form the lowest level, the $m = -(n - 2)$ states the first excited level, and so on. In the intermediate regime, the splitting of the angular momentum states by the magnetic fields leads to a crossing pattern involving many levels. Such a pattern is quite evident in spectroscopic studies of quantum dots in a magnetic field, as discussed in Section 4.1.5.

In the original derivation, Fock [7] separated variables in cylindrical coordinates instead of using the operator approach developed here. The resulting energy eigenvalues appear in a slightly different form than (4.28) above (neglecting spin splitting)

$$E_{n',m} = (2n' + |m| + 1)\hbar\Omega + \frac{1}{2}\hbar\omega_c m; \quad n' = 0, 1, 2, \ldots; \quad m = 0, \pm1, \pm2, \ldots . \quad (4.30)$$

The resulting energies as a function of magnetic field are identical to those shown in Fig. 4.1. In cylindrical coordinates, the Cartesian coordinates are expressed in terms of the polar coordinates $x = r\cos\phi$ and $y = r\sin\phi$, for $r \geq 0$ and $0 \leq \phi \leq 2\pi$. The corresponding eigenstates are given by

$$\varphi_{n'm}(\mathbf{r}) = \frac{1}{\sqrt{2\pi}} e^{im\phi} \frac{1}{l_o} \sqrt{\frac{n'!}{(n' + |m|)!}} e^{-r^2/4l_o^2} (r/\sqrt{2}l_o)^{|m|} L_{n'}^{|m|}(r^2/2l_o^2), \quad (4.31)$$

where $l_o = (\hbar/m^*\Omega)^{1/2}$ is the characteristic length and $L_{n'}^{|m|}$ are generalized Laguerre polynomials. A similar set of eigenfunctions may be generated in the number operator representation using the properties of the creation and annihilation operators introduced earlier (see for example [11]).

4.1.3 Multi-electron quantum dots

The consideration of the multi-electron effects in quantum dots parallels more or less the hierarchy of approximations used in atomic and molecular physics in treating the many body problem. We need to account for not only the Coulomb interaction between particles but also the required antisymmetry of the many-body wavefunction under the exchange of identical particles. In addition, one should consider the spin-orbit interaction coupling the orbital angular momentum and spin angular momentum, although such effects are expected to be small in quantum dot systems.

One approach that has been used to calculate the electronic spectrum of many-electron quantum dots is the Hartree approximation in which the particle-particle interaction for each electron is replaced by the mean field arising from all the other electrons in the system [12]. We have already seen this approach (Sections 2.2.3 and 2.4) in connection with self-consistent calculations of the electronic states in quantum wells and quantum wires. In a quantum dot, the one-particle potential arises from the charge density in the dot, given by

$$\rho(x, y, z) = -2e \sum_n \chi_n^*(x, y, z)\chi_n(x, y, z) f((E_{Fd} - E_n)/k_B T), \quad (4.32)$$

where χ_n represent the one-electron solutions and f is the distribution function governing the occupancy of these states for a dot Fermi energy, E_{Fd}. The Hartree approximation does not directly account for the antisymmetry of the atomic-like orbitals other than through the weaker statement of the Pauli exclusion principle regarding the single occupancy of a given spin state. The exchange interaction arising from this antisymmetry property may be incorporated by including the nonlocal Fock potential in the Hartree-Fock approximation. The reduction of many-body effects in the Hartree and Hartree-Fock approximations to a one-electron problem allows the inclusion of the actual potential arising in realistic device structures rather than in the simple model we have used so far. Kumar et al. [12] used the Hartree approximation to solve self-consistently for both the eigenstates and the electrostatic potential of a quantum dot structure for an arbitrary number of particles. Needless to say, the computational requirements for the full 3D problem are large, even within the Hartree approximation. Figure 4.2 shows the calculated dot potential by Kumar et al. [12] in a plane

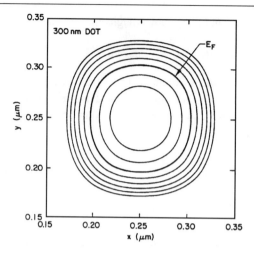

Fig. 4.2. Lateral potential contours in a quantum dot in the plane 8 nm below the GaAs/AlGaAs interface. The innermost contour is 15 meV below the Fermi level, which is indicated by the heavy line. The remaining contours are 10-meV intervals from −10 to +50 meV [after Kumar et al., Phys. Rev. B **42**, 5166 (1990), by permission].

8 nm below the interface of a square, gated GaAs/AlGaAs dot similar to that experimentally investigated by Hansen et al. [13]. Of particular interest to the present discussion is the fact that even though the geometry of the quantum structure is square, the actual equipotential surfaces close to the dot center appear more like the simple symmetric parabolic confinement we have assumed so far. This smearing out of the potential is associated with the projection of the electrostatic potential associated with a Schottky gate onto a 2DEG lying underneath, as we saw in Chapter 3. Even for ungated structures (often employed in optical studies), surface pinning due to surface states on free surfaces will result in the same effect. The resulting eigenstates calculated by Kumar et al. showed similar degeneracies for several electrons in the dot to those found for the simple parabolic potential of Section 4.1.2.

For small numbers of electrons in a dot, direct diagonalization of the many-particle Hamiltonian is computationally feasible for a fixed confinement potential [14]–[16]. A comparison of direct diagonalization versus the Hartree and Hartree-Fock approximations for a two-electron system was reported by Pfannkuche and coworkers [17]. The basis functions for such calculations may be chosen from the single-particle solutions in the absence of the Coulomb interaction between particles. Another convenient basis is to introduce center of mass (CM) and relative coordinates (RM). For the parabolic confinement potential discussed so far, this latter choice is particularly useful since there is a complete decoupling of the CM and RM coordinates. Consider as an example the case of a quantum dot "helium" atom containing two electrons, whose Hamiltonian may be written

$$\mathcal{H}(2) = H^0(1) + H^0(2) + V(\mathbf{r}_1 - \mathbf{r}_2), \tag{4.33}$$

where a pairwise potential, $V(\mathbf{r}_1 - \mathbf{r}_2)$, is assumed. For the ith particle of a multi-electron system in the presence of a magnetic field, the single particle Hamiltonian is given by Eq. (4.26):

$$H^0(i) = \frac{p_i^2}{2m^*} + \frac{1}{2}m^*\Omega^2 r_i^2 + \frac{1}{2}\omega_c L_{z,i} + \frac{g^*\mu_B}{\hbar}Bs_z^i, \tag{4.34}$$

where Ω is defined in (4.25) and s_z^i is the spin angular momentum operator of the ith particle in the z-direction. Since the interparticle potential depends only on the relative coordinates, $\mathbf{r}_1 - \mathbf{r}_2$, it is natural to introduce the CM coordinates, $\mathbf{R}$ and $\mathbf{P}$, and RM coordinates, $\mathbf{r}$ and $\mathbf{p}$:

$$\mathbf{R} = (\mathbf{r}_1 + \mathbf{r}_2)/2, \quad \mathbf{r} = \mathbf{r}_1 - \mathbf{r}_2,$$
$$\mathbf{P} = \mathbf{p}_1 + \mathbf{p}_2, \qquad \mathbf{p} = (\mathbf{p}_1 - \mathbf{p}_2)/2. \tag{4.35}$$

Equation (4.33) for a two particle system then becomes

$$\mathcal{H}(2) = \frac{P^2}{2M} + \frac{1}{2}M\Omega^2 R^2 + \frac{1}{2}\omega_c L_z^{CM} + \frac{p^2}{2m^*}$$
$$+ \frac{1}{2}m_r\Omega^2 r^2 + \frac{1}{2}\omega_c L_z^{RM} + V(r) + \frac{g^*\mu_B}{\hbar}BS_z$$
$$= H_{CM} + H_{RM} + H_{spin}, \tag{4.36}$$

where $M = m_1 + m_2 = 2m^*$, $m_r = m_1 m_2/(m_1 + m_2) = m^*/2$, and where $S_z = \sum_i s_z^i$ is the total spin angular momentum. The angular momentum operators act on their respective CM, RM, or spin coordinate systems. Thus, because of the parabolic confinement potential, the problem is separable, and one need only solve the radial equation (since angular momentum is still conserved) for the RM problem. The energy eigenvalues corresponding to the center of mass system are given by (4.28):

$$E_{CM} = (N+1)\hbar\Omega + \frac{1}{2}\hbar\omega_c M; \quad M = -N, -(N-2), \ldots, (N-2), N, \tag{4.37}$$

where N and M denote the principal and angular momentum quantum numbers of the center of mass system.

In constructing the eigenstates of the two-particle system, we have to explicitly consider the spin as well as spatial coordinates in order to form a properly antisymmetric state under the exchange of the two particles. We may write the eigenstates as

$$\psi(\mathbf{r}, \mathbf{R}) = \varphi(\mathbf{r})\psi(\mathbf{R})\chi_s, \tag{4.38}$$

where χ_s is the spin eigenfunction, which may be either antisymmetric (singlet) or symmetric (triplet) under the exchange of particles. There is only one possible singlet spin state corresponding to the combination of two electrons with essentially antiparallel spin, whereas there exist three possible combinations of essentially parallel spins in the triplet state. The total z-component of the spin angular momentum is zero in the singlet state, and $S_z = m_s\hbar$, $m_s = -1, 0$, or 1 for the triplet state. Corresponding to an antisymmetric or symmetric spin combination, the real-space part must be respectively either symmetric or antisymmetric so that the product in (4.38) remains antisymmetric. The center of mass coordinates involve the sum of the particle coordinates, $\mathbf{r}_1 + \mathbf{r}_2$, and therefore cannot change sign under the exchange of these indices. Therefore, $\psi(\mathbf{R})$ is always symmetric. On the other hand, the relative coordinate depends on $\mathbf{r}_1 - \mathbf{r}_2$ and thus changes sign under the exchange of particles. Since the relative angular momentum L_z^{RM} is still conserved, the RM wavefunction remains in the form suggested by (4.31),

$$\varphi(\mathbf{r}) = \varphi(\mathbf{r}_1 - \mathbf{r}_2) = e^{im\phi} f(r), \tag{4.39}$$

Table 4.1. *Zero-order energies, E_0 (in units of $\hbar\Omega$), in the absence of interparticle interaction, and the energies in presence of a harmonic interaction potential ($V_o = \hbar\Omega/2$; $\Omega_o = \Omega/4$). g labels the degeneracy of the state, and S is 0 for singlet and 1 for triplet states.*

E_0	E_1	**g**	**S**	(n, m)	(N, M)
2	2.75	1	0	$(0, 0)$	$(0, 0)$
3	3.5	6	1	$(1, \pm1)$	$(0, 0)$
	3.75	2	0	$(0, 0)$	$(1, \pm1)$
4	4.25	1	0	$(2, 0)$	$(0, 0)$
	4.25	2	0	$(2, \pm2)$	$(0, 0)$
	4.5	12	1	$(1, \pm1)$	$(1, \pm1)$
	4.75	1	0	$(0, 0)$	$(2, 0)$
	4.75	2	0	$(0, 0)$	$(2, \pm2)$

where $m\hbar$ is the relative angular momentum, $r = |\mathbf{r}_1 - \mathbf{r}_2|$, and $f(r)$ is a radial function. Under the exchange of the two particles, r is invariant but $\phi \to \phi + \pi$. Therefore, it is clear from (4.39) that there is a phase change of $m\pi$ following the exchange of particles. If m is odd, there is a change of sign, and the spatial function is antisymmetric. These solutions must be paired with symmetric spin functions to form a triplet state. Even values of m give symmetric solutions that must combine with the antisymmetric spin function to form a singlet state.

If we neglect the Coulomb interaction between particles, then the RM energy E_{rm} is given by (4.37) with M and N replaced with m and n. In the limit that B goes to zero, Table 4.1 gives the allowed zeroth-order energies E_0 (in units of $\hbar\omega_o$), as well as the degeneracy of the states and the classification in terms of whether the state is a singlet ($S = 0$) or triplet ($S = 1$) state. As shown above, only m odd terms give rise to triplet solutions independent of the angular momentum state of the CM system, M.

Inclusion of the Coulomb interaction raises the energy and partially lifts the degeneracies of the noninteracting case of Table 4.1. The solution requires the direct diagonalization of the Hamiltonian matrix as mentioned earlier. Johnson and Payne have reported an exactly solvable model for the n-particle system through the introduction of a harmonic potential of the form [18]

$$V(\mathbf{r}_i - \mathbf{r}_j) = 2V_o - \frac{1}{2}m_r\Omega_o^2|\mathbf{r}_i - \mathbf{r}_j|^2, \tag{4.40}$$

where V_o and Ω_o are chosen to approximate the interparticle interaction. They argue that this potential is physically reasonable if one considers effects such as the nonzero extent of the electron wavefunction and effects of image potential due to the presence of, for example, a metallic gate. In our two-body problem above, it is apparent that this potential combines with the confining potential in the equation of motion of the RM system, so that the total energy of the two-particle system may be written (ignoring the Zeeman term for simplicity)

$$E_{NM}^{nm} = 2V_o + (N + 1)\hbar\Omega + \frac{1}{2}\hbar\omega_c M + (n + 1)\hbar(\Omega - \Omega_o) + \frac{1}{2}\hbar\omega_c m. \tag{4.41}$$

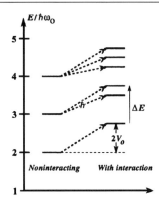

Fig. 4.3. Schematic showing the effect of a harmonic interparticle interaction potential on the two-particle states in a parabolic confinement potential. Bare energies are indicated on the left side. $\hbar\omega$ is the optical transition energy measured in FIR experiments.

Table 4.1 lists the energies, E_1, corresponding to this interaction for $V_o = \hbar\Omega/2$ and $\Omega_o = \Omega/4$ in the same $B = 0$ limit. The ground state energy is raised by the constant interaction $2V_o$; degeneracies associated with the radial motion are split by the interaction, whereas those associated with the angular momentum (for zero field) remain as illustrated in Fig. 4.3. As we will come back to in Section 4.1.5, $\hbar\omega$ on the right side is the allowed dipole transition in an optical field which essentially only measures the CM energies independent of the relative coordinates.

4.1.4 Quantum dot statistics

The quantum dot itself is not an isolated entity; rather it is coupled to the surroundings through the interaction of the electrons with elementary excitations of the lattice such as acoustic and optical phonons. The problem in the quantum dot case is that the energy of the system is strongly dependent on the actual number of particles in the dot, N, and therefore the equilibrium occupancy of the states cannot in general be simply expressed in terms of a Fermi-Dirac distribution. In equilibrium, the probability distribution governing the occupancy of the discrete states is given by the Gibbs distribution derived from the grand canonical ensemble in statistical mechanics as [3]

$$F(\{n_j\}) = Z^{-1} \exp\left[-\frac{1}{k_B T_l}\left(\sum_{j=1}^{\infty} E_j(N)n_j - NE_F \right) \right], \qquad (4.42)$$

where n_j represents the number of particles occupying a given energy state j (which for Fermions can take on only the values 0 and 1), $\{n_j\}$ represents the set of occupation values $\{n_1, n_2, \ldots, n_i\}$, and F is the probability that a chosen system from the ensemble exists in this state of occupation. N is the total number of electrons in the dot, $E_j(N)$ is the energy of state j (corresponding to some combination of n_x and n_y in Eq. (4.4)), E_F is the Fermi energy, and Z is the partition function

$$Z = \sum_{\{n_j\}} \exp\left[-\frac{1}{k_B T_l}\left(\sum_{j=1}^{\infty} E_j(N)n_j - NE_F \right) \right]. \qquad (4.43)$$

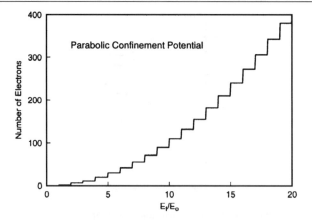

Fig. 4.4. Number of electrons versus Fermi energy for independent electrons in a parabolic confinement $E_o = \hbar\omega_o$ in both the x- and y-directions assuming a two-fold spin degeneracy.

If we assume that the occupancy of the states in the dot is governed by the Fermi energy outside of the dot in the 2DEG region, we may calculate the number of electrons in the dot occupied below a given 2D Fermi energy at zero temperature, neglecting Coulomb charging effects. For noninteracting electrons in the parabolic confining potential of Section 4.1.1, Fig. 4.4 shows the calculated number of electrons in the dot as a function of the normalized Fermi energy $E_F/\hbar\omega_o$, where for simplicity the dot has been assumed symmetric with $\omega_x = \omega_y = \omega_o$, and a spin degeneracy of two has been taken for each state.

4.1.5 Spectroscopy of quantum dots

Evidence for the existence of quantum dot states is provided by a variety of transport measurements coupling the quantum dot to external reservoirs, as well as optical spectroscopy and capacitance spectroscopy. Here we are interested primarily in experiments that probe directly the bound states existing in a quantum dot, and we defer discussion of effects associated with Coulomb charging to Section 4.2.

One of the first experiments to directly measure the excitation spectra of a quantum dot structure was reported by Reed and coworkers [19], who investigated vertical transport through resonant tunneling diodes (RTDs) (see Section 3.1 for details of RTDs). In these RTD structures (see Fig. 2.16), the smaller bandgap well region serves as the quantum dot, with electron beam lithography used to define the lateral dimensions. The lateral confinement is actually much narrower than the structural dimensions due to depletion arising from surface pinning on the free GaAs surfaces. For sufficiently small lateral dimensions, full 3D quantization of the motion occurs, which is observable in the I-V characteristics. The I-V characteristics of these structures for several different temperatures are shown in Fig. 4.5. As the temperature decreases, the main NDR peak of the high-temperature I-V resolves into a spectrum of smaller individual peaks attributed to resonant tunneling through discrete 0D states in the well region.

Far infrared (FIR) optical spectroscopy of arrays of identical quantum dots in the presence of a perpendicular magnetic field have revealed features associated with the Darwin-Fock states discussed in Section 4.1.2 for InSb [20] and GaAs [21], [22] structures. A unique feature of the parabolic confinement potential is the complete separation of the CM and

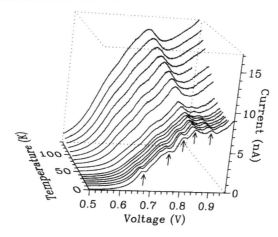

Fig. 4.5. Current-voltage characteristics of a small-area resonant tunneling diode at various temperatures [after Reed *et al.*, Phys. Rev. Lett. **60**, 535 (1988), by permission].

RM coordinates in a multi-electron dot. As a result, the *generalized Kohn theorem* holds in which the only allowed dipole transitions occur between states of the center of mass system [15], [23]. This fact may be qualitatively understood from the fact that typical wavelengths for FIR radiation far exceed the diameter of typical dot structures, and so the electric field $\mathbf{E}_o$ is approximately constant over the space of the dot. The perturbing Hamiltonian in the dipole approximation due to the optical field has the form

$$H' = \sum_{j=1}^{n_e} e\mathbf{E}_o \cdot \mathbf{r}_j \approx e\mathbf{E}_o \cdot \sum_{j=1}^{n_e} \mathbf{r}_j = Q\mathbf{E}_o \cdot \mathbf{R}, \qquad (4.44)$$

where n_e is the number of electrons in the dot, $\mathbf{R} = \sum \mathbf{r}_j / n_e$ is the CM coordinate, and $Q = n_e e$ is the total charge. The optical perturbation then couples only to the CM part of the multiparticle system. Thus, FIR spectroscopy to first-order measures only the bare one-electron properties of the dot, independent of the number of electrons and the details of the interaction between them (to the extent that the bare confinement potential is parabolic). The allowed dipole transition energies are given as [20]

$$\Delta E^{\pm} = \hbar \sqrt{\omega_o^2 + \frac{\omega_c^2}{4}} \pm \frac{\hbar \omega_c}{2}, \qquad (4.45)$$

where the upper and lower signs correspond to left and right circularly polarized light, respectively. Figure 4.6 shows a plot of the experimental resonance peak positions in FIR absorption spectroscopy for various magnetic fields illustrating this behavior for an array of rectangular, etched GaAs/AlGaAs dots. As discussed in Section 4.1.3, self-consistent calculations [12] show that in spite of the rectangular structure, the confinement potential is essentially parabolic due to lateral depletion. The resonant peaks are observed to follow the two branches predicted by (4.45). In addition, anticrossing behavior is observed close to $B = 1T$, and high-order branches are evident.

Our introduction of rotating creation and annihilation operators in Section 4.1.1 gives a natural explanation of (4.45). Circularly polarized light polarized in the plane of the quantum dot corresponds to a rotating field vector, $\mathbf{E}_o(t)$, either left-going or right-going. A natural representation for the interaction, $\mathbf{E}_o \cdot \mathbf{R}$, is to express the rotating position vector as linear combinations of the rotational operators defined in Eqs. (4.15) and (4.16). For the CM

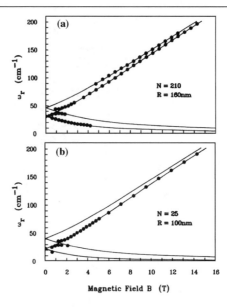

Fig. 4.6. Experimental B dispersion of resonant absorption in quantum dot structures. N is the estimated number of electrons in the dot, and R is the radius. The full lines are fits with the dispersion relation (4.45) [after Demel *et al.*, Phys. Rev. Lett. **64**, 788 (1990), by permission].

coordinates, $(A + A^\dagger)$ is proportional to the Hamiltonian for the left sense of rotation, and $(B + B^\dagger)$ for the opposite sense. (Uppercase letters are used for the operators in the CM system.) The creation operators correspond to absorption of radiation, whereas the annihilation operators correspond to emission processes. Focusing on FIR absorption, we see that absorption processes associated with left polarized light result in $N_a \to N_a + 1$, and thus according to Eqs. (4.19) and (4.20), $N \to N + 1$ and $M = N_a - N_b \to M + 1$. From the CM energy eigenvalues (4.37), $\Delta E^+ = E(N+1, M+1) - E(N, M)$, corresponding to the upper sign of (4.45). Likewise, for right polarized light, $B^\dagger$ results in $N_b \to N_b + 1$, which still increases N by one quanta but decreases the CM angular momentum $M \to M - 1$, thus accounting for ΔE^-. In this context, we understand the optical transitions as transitions between the single-electron dot levels shown in Fig. 4.1 for a given B_z subject to the above selection rules.

Transport spectroscopy has also evidenced the magnetic field evolution of states in a quantum dot [24]. Lateral structures using quantum point contacts (QPCs) were used to define a quantum dot confined by two narrow constrictions as shown in Fig. 4.7a. The transmission through a model double constriction structure was already discussed in Section 3.2, in which sharp resonance peaks were found for energies below the cutoff energy of the narrow constrictions. These resonance peaks corresponded to the quasi-bound states associated with the quantum dot which translate into conductance peaks in the Landauer-Büttiker model discussed in Chapter 3. Thus, transport spectroscopy may be performed by moving the Fermi energy of the leads through these levels, either by a gate bias (which changes the potential of the dot) or with a magnetic field. However, as we will discuss in detail in Section 4.2, this simple picture is modified by the additional charging energy associated with the capacitance in transferring an electron into or out of the dot, which superimposes an additional structure on the conductance spectra. In this case, the discrete spectra of the bare states coexist with an additional energy gap due to the effect of Coulomb blockade. Within the present context of dot spectroscopy, we are mainly concerned with

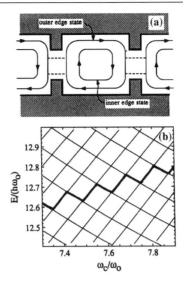

Fig. 4.7. (a) Schematic view of a split-gate double constriction quantum dot device, showing edge states traversing the periphery of the quantum dot of dimensions approximately 0.5×0.7 μm. (b) Energy levels and motion of the Fermi level (dark line) for a fixed density of carriers in the dot [after McEuen *et al.*, Phys. Rev. Lett. **66**, 1926 (1991), by permission].

the change of the discrete levels with magnetic field, and will defer discussion of Coulomb charging experiments until later.

As we saw in Section 2.5.2, magnetoelectric edge states form in a quasi-1D channel in response to a perpendicular d.c. magnetic field. As shown diagrammatically in Fig. 4.7a, this picture may be extended to the quantum dot case, where the states inside the dot form closed orbits around the periphery of the dot and couple weakly to open edge states in the leads that connect to reservoirs. Since the dot is only weakly coupled to the leads, the bare energies of these states are approximately given by the Darwin-Fock solutions (4.28) for a parabolic potential plotted in Fig. 4.1. Figure 4.7b shows an expanded cross section of the solutions shown in Fig. 4.1 (including higher n levels) in the relevant region for conductance measurements in their samples. The range of magnetic fields and energies basically corresponds to the lowest two Landau levels of the system in the large B limit. If in a simple-minded picture we filled all the states up to the maximum number of electrons in the dot at low temperature, the solid line in this figure shows the variation of the maximum occupied level ($N = 78$ in this case, neglecting spin degeneracy), which oscillates up and down while increasing with increasing B. A conductance peak occurs when this highest occupied level (including in general the Coulomb charging energy) is coincident with the Fermi energies of the leads, so that electrons can tunnel on and off the level, allowing current to flow. A gate bias applied to the structure allows the dot energies to be raised or lowered, which in turn changes the number of electrons on the dot, and gives rise to conductance peaks as the Fermi level of the dot (i.e., the highest occupied level) passes through the Fermi energies of the leads. As the magnetic field is varied, the Fermi energy of the dot shifts in accordance with the level shifts of the quantized dot states since to first order the Coulomb charging is not affected by the magnetic field. The peak positions in terms of the gate bias at which they occur are plotted as a function of magnetic field in Fig. 4.8a. The corresponding energy spectrum inferred from this figure is plotted in Fig. 4.8b assuming an arbitrary zero of energy and subtracting the constant Coulomb charging energy between

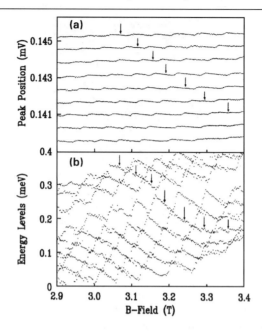

Fig. 4.8. (a) Peak position of the longitudinal conductance peak as a function of the magnetic field for a series of conductance peaks. The arrow follows a particular state in the first Landau level. (b) Energy spectrum inferred from (a), with an arbitrary zero of energy [after McEuen *et al.*, Phys. Rev. Lett. **66**, 1926 (1991), by permission].

peaks (see Section 4.2). The qualitative comparison of the energy spectrum of Fig. 4.8b and the theoretical plot of the Darwin-Fock states in Figs. 4.1 and 4.7b is striking, illustrating the single-particle energy level spectrum of the quantum dot.

An alternate technique which has been used to investigate the electronic states in semiconductor quantum dots is *capacitance spectroscopy* [13], [25]. An example of the structure used by Ashoori *et al.* is shown in Fig. 4.9a [25]. As the d.c. gate voltage to the top electrode is varied, the Fermi level in the bottom electrode may be made coincident with the Fermi energy of the dot and electron tunneling may occur through the thin AlGaAs barrier as shown in Fig. 4.9a. This results in a peak in the capacitance signal measured with a superimposed a.c. signal fed to a lock-in amplifier. High-resolution measurements were made by mounting a HEMT transistor adjacent to the dot as an input amplifier. Figure 4.9b shows the capacitance versus gate voltage for their structure. A series of peaks corresponds to single electron tunneling into the lowest unfilled state of the dot. The out-of-phase component in the lower trace shows that the electron tunneling lags the applied a.c. signal of 210 kHz for the higher states. The measurement shown in Fig. 4.9 was also performed in the presence of a static B field perpendicular to the dot, and the evolution of the dot energies versus magnetic field was measured, again illustrating the Darwin-Fock-like solutions given by (4.28). In addition, evidence for the transition from the singlet to a triplet state for the lowest energy level was observed. This transition is expected due to the Zeeman term which eventually causes the triplet state to be energetically favorable, although at much higher fields (25 T) than measured experimentally (1.5 T). Ashoori *et al.* argued that the difference is due to the electron-electron interaction which reduces the B field necessary for this transition to occur. To understand this point, consider the simple two-particle dot with a harmonic

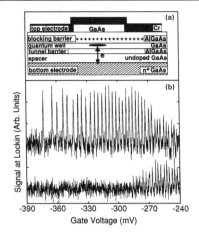

Fig. 4.9. (a) Schematic of sample used in capacitance spectroscopy measurement. (b) Capacitance data versus gate bias for the quantum dot sample in zero magnetic field. The top and bottom traces show the signal resulting from electron tunneling in phase and electron tunneling in 90° lagging phase with 210 kHz excitation voltage, respectively [from R. C. Ashoori *et al.*, Phys. Rev. Lett. **71**, 613 (1993), by permission].

interaction potential whose energy eigenvalues are given by (4.41) and listed in Table 4.1. Without including the Zeeman contribution (which was neglected in (4.41)), there is no magnetic field where the lowest two-particle singlet ground state crosses the first excited triplet level ($n = 1$, $m = -1$) for two *independent* particles (i.e., $\Omega_o = 0$). However, for $\Omega_o \neq 0$, a solution exists for $\Omega_o = \Omega - \omega_c/2$ in the absence of the Zeeman splitting. Depending on the value of the interaction potential Ω_o, the singlet-to-triplet transition may occur at an arbitrarily low magnetic field (i.e., ω_c). A more rigorous theoretical calculation for an N-electron dot was reported by Hawrylak, which supports their conjecture of such a transition with magnetic field [26].

Throughout this section, we have concentrated on the spectroscopy of semiconductor quantum single dots and quantum dot arrays fabricated through lithographic techniques. In addition, we mention in passing that there exists a large literature associated with confinement effects in microcrystalline systems (for a review, see [27], [28]). Such systems may consist of nanocrystals of a semiconductor, which grow as islands on the surface of another material or are suspended in a matrix such as a glass or a polymer or even a liquid. The dimensions of such systems may be as small as a few nanometers, and therefore confinement effects are strong. Optical spectroscopy is generally employed in studying confinement effects in exciton formation and carrier relaxation as well as the general nonlinear optical properties of such systems. The disadvantage compared to the systems discussed in this section is that control of the uniformity of the size of such nanocrystals is difficult, and so one generally measures an average over a distribution of various sizes. Recent interest has been generated by the epitaxial growth of strained InAs islands on GaAs [29]–[31]. Since InAs has a larger lattice constant than GaAs, the strain in very thin films is relieved by forming small InAs clusters or islands. Under the proper growth conditions, nanoscale islands with relatively uniform diameters may be realized. Luminescence and absorption spectroscopy have evidenced the zero-dimensional nature of the quantized states in these islands.

4.2 Single electron tunneling and Coulomb blockade

4.2.1 Introduction to Coulomb blockade and experimental studies

In Section 4.1 we discussed the electronic states of essentially isolated quantum dot and quantum box structures. The addition of an electron to the many-particle system resulted in a renormalization of the states due to the electron-electron interaction. However, we did not consider the effect on the surrounding environment of adding this electron to the dot. When we discuss transport *through* quantum dots, we are implicitly considering the coupling to this external environment, which provides the sources and sinks for electrons into and out of the dot (see Fig. 4.10). The quantum dot (also referred to as an "island") is isolated from the surroundings except for tunneling between the leads and the island. In a split-gate semiconductor quantum dot such as Figs. 2.15 and 2.17, the tunneling junction between the dot and the leads consist of point contact structures biased below the pinch-off for 1D conduction (see Section 3.6). In metal tunnel junctions, the metallic island is surrounded by an insulator such as an oxide, through which electrons tunnel to and from the metal electrodes. The principal difference in the analysis of metal islands versus semiconductor islands is in the relative number of electrons and the effects of quantum confinement on the allowed energy states. As we saw in the previous section, quantum confinement effects in semiconductor quantum dots may be quite large, leading to structures that justifiably may be considered artificial atoms consisting of just a few particles. Metallic systems, on the other hand, have much larger electron densities, and mean free paths at the Fermi energy of only a few nanometers. Therefore, metallic islands behave more or less as small bulk-like systems. However, both systems share a common feature, that the discrete nature of the electron charge becomes strongly evident when particles tunnel into and out of the structure shown in Fig. 4.10.

When an electron is transferred from the reservoir into the quantum dot system, there is a rearrangement of charge in the electrode, resulting in a change in the electrostatic potential. In a large system, this change in potential due to the injection of electrons via tunneling or over the barrier by thermionic emission is hardly noticeable and is associated with the *shot noise* in the system [32]. However, in small systems the potential change may be greater than the thermal energy, $k_B T$, particularly at low temperature. Such large changes in the electrostatic energy due to the transfer of a single charge may result in a gap in the energy spectrum at the Fermi energy, leading to the phenomenon of *Coulomb blockade,* in which the tunneling of electrons is inhibited until this charging energy is overcome through an applied bias.

In the theory of Coulomb blockade, the basic experimental results are conveniently discussed in terms of a macroscopic *capacitance* associated with the system. The change in electrostatic potential due to a change in the charge on an ideal conductor is associated

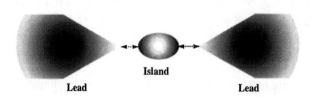

Fig. 4.10. A quantum dot coupled to two leads connected to an external circuit.

with the linear relationship

$$Q = CV \tag{4.46}$$

where C is the capacitance, Q is the charge on the conductor, and V the electrostatic potential relative to some chosen reference (e.g., ground). Since we are considering an ideal conductor, any charge added to the conductor rearranges itself such that the electric field inside vanishes, and the surface of the conductor becomes an equipotential surface. Therefore, the electrostatic potential associated with the conductor relative to its reference is uniquely defined. If we consider two conductors connected by a d.c. voltage source, a voltage $+Q$ builds up on one conductor and a charge $-Q$ on the other. The capacitance of the two conductor system is then defined as $C = Q/V_{12}$. The electrostatic energy stored in the two conductor system is the work done in building up the charge Q on the two conductors and is given by

$$E = \frac{Q^2}{2C}. \tag{4.47}$$

For a system of N conductors, the charge on conductor i may be written

$$Q_i = \sum_{j=1}^{N} C_{ij} V_j, \tag{4.48}$$

where the diagonal values C_{ii} are the capacitance of conductor i if all other conductors are grounded. The diagonal elements are commonly referred to as the *coefficients of capacitance*; the off-diagonal elements are called the *coefficients of induction*. The total electrostatic energy stored in a multiconductor system is given by the generalization of Eq. (4.47) as

$$E = \frac{1}{2} \sum_i \sum_j (C^{-1})_{ij} Q_i Q_j, \tag{4.49}$$

where C^{-1} is the inverse capacitance matrix.

It is important to note that the polarization charge on the capacitor, Q, does not have to be associated with a *discrete* number of electrons, N. This charge is essentially due to a rearrangement of the electron gas with respect to the positive background of ions, and as such it may take on a continuous range of values. It is only when we consider changes in this charge due to the tunneling of a single electron between the conductors that the discrete nature becomes apparent.

In systems of very small conductors, the capacitances approach values sufficiently small that the charging energy given by (4.47) due to a single electron, $e^2/2C$, becomes comparable to the thermal energy, $k_B T_l$. The transfer of a single electron between conductors therefore results in a voltage change that is significant compared to the thermal voltage fluctuations and creates an energy barrier to the further transfer of electrons. This barrier remains until the charging energy is overcome by sufficient bias. How small must such a structure be? A simple example is the case of a conducting sphere above a grounded conducting plane. This example approximates a metal cluster imbedded in an insulator above a conducting substrate, which is a commonly realized structure that has been extensively studied experimentally. The exact solution may be found using the method of images, which gives the capacitance of the sphere as [33]

$$C = 4\pi\epsilon a\left(1 + \alpha + \frac{\alpha^2}{1-\alpha^2} + \cdots\right), \quad \alpha = \frac{a}{2l}, \tag{4.50}$$

where a is the radius of the sphere and l is the distance above the conducting substrate. As the radius of the sphere becomes small compared to l, the capacitance becomes independent of the distance of the cluster from the substrate. An alternate example is that of a flat circular disk located parallel to and a distance d above a ground plane. This example is more closely analogous to the semiconductor quantum dots fabricated by lateral confinement of a 2DEG as discussed in the previous sections. The solution is given in a problem in Jackson's textbook [34] (which we leave as an exercise for the reader!), with the capacitance given in the limit of $d \gg R$ as

$$C = 8\epsilon R \tag{4.51}$$

where R is the radius of the disk. Equating the charging energy with the thermal energy, we see that at room temperature, $C \sim 3 \times 10^{-18}$ F. The corresponding radius for a sphere from (4.50) is on the order of $a \sim 28$ nm (assuming a relative dielectric constant of 1), and somewhat larger for the disk. The facts that $\epsilon > \epsilon_o$ in real structures and that the charging energy should be several times larger than the thermal energy imply that sub-10-nm structures need to be fabricated in order to see clear single-electron charging effects at room temperature. Although it is still somewhat challenging with today's lithographic techniques to nanoengineer such structures, it is not difficult to grow insulating films with random metallic clusters on this order in which Coulomb blockade effects are readily observed, even at room temperature. Further, if we perform measurements at cryogenic temperatures, then the size scale becomes comparably larger, allowing single-electron effects to be observed in nanofabricated quantum dot structures.

We should keep in mind that in such small systems, the concept of a lumped "capacitance" to characterize the distributed rearrangement of charges in the system may have limited validity. In semiconductor systems in particular, the capacitance in general is not linear but depends on the operating voltage. Thus the capacitance is more generally defined as the differential change in charge with voltage,

$$C(V') = \left.\frac{\partial Q}{\partial V}\right|_{V'}. \tag{4.52}$$

The capacitance associated with a depletion region (such as those used to define split-gate point contacts) is a good example of a capacitance which depends strongly on voltage due to the change of width of the depletion region with gate bias. Further, for ultra-small nanostructure systems, the nonlocality of charge itself on the quantum level introduces its own contribution to the effective capacitance of the system [35]. Despite these caveats, the simple idea of single electron charging associated with a macroscopic quantity such as the capacitance seems to work surprisingly well in describing a host of experimental results, some of which we review briefly below.

Coulomb blockade in normal metal tunnel junctions

Historically, Coulomb blockade effects were first predicted and observed in small metallic tunnel junction systems. As mentioned earlier, the conditions in metallic systems of high

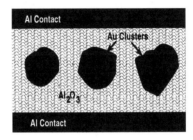

Fig. 4.11. A cross section of Coulomb island structures realized by imbedding metal clusters such as *Au* in a dielectric medium.

electron density, large effective mass, and short phase coherence length (compared to semi-conductor systems) usually allow us to neglect size quantization effects. The dominant single electron effect for small metal tunnel junctions is therefore the charging energy due to the transfer of individual electrons, $e^2/2C$. The effects of single electron charging in the conductance properties of very thin metallic films was recognized in the early 1950s by Gorter [36] and Darmois [37]. Thin metal films tend to form planar arrays of small islands due to surface tension, and conduction occurs due to tunneling between islands. Since the island size is small, the tunneling electron has to overcome an additional barrier due to the charging energy, which leads to an increase in resistance at low temperature. Such discontinuous metal films show an activated conductance, $\sigma \sim \exp(-E_c/k_B T)$, similar to an intrinsic semiconductor. Neugebauer and Webb [38] developed a theory of *activated tunneling* in which this activation energy was the electrostatic energy required to tunnel electrons in and out of the metal islands. In analogy to the semiconductor case, this activation energy resembles an energy gap and is therefore referred to as a *Coulomb gap*.

A number of studies have been conducted concerning the transport properties of metal clusters or islands imbedded in an insulator that are then contacted by conducting electrodes. A schematic of such a structure for Au particles imbedded in native or evaporated Al_2O_3 is shown in Fig. 4.11. Each metal cluster represents a Coulomb island of the sort shown schematically in Fig. 4.10 in which electrons may tunnel through the insulator from the contacts to the islands and vice versa. Giaever and Zeller [39] investigated the differential resistance of oxidized Sn islands sandwiched between two Al electrodes forming Coulomb islands down to 2.5 nm diameter depending upon the evaporation conditions. They measured one of the telltale signs of Coulomb blockade in these samples, that is, a region of high resistance for small bias voltages about the origin followed by a strong decrease in resistance past a voltage of approximately 1 mV. Shortly thereafter, Lambe and Jaklevic [40] performed capacitance-voltage measurements on structures similar to those of Giaever and Zeller. The structures were designed with thick oxides between the islands and the substrate so that tunneling of electrons occurred only through the top contact. Oscillatory behavior in the differential capacitance was interpreted in terms of the addition of charges one by one to the islands as the bias voltage increases in integral multiples of e/C_I, where here C_I is the substrate to island capacitance. Cavicchi and Silsbee [41] later investigated the frequency and temperature dependence of the capacitance and associated charge transfer in asymmetric structures similar to those of Lambe and Jaklevic.

Meanwhile, a rigorous theoretical treatment of single electron effects during tunneling was introduced by Kulik and Shekhter [42] based on the tunneling Hamiltonian method (discussed in Section 4.2.2) in order to derive a kinetic equation for charge transport. This kinetic equation approach was later improved by Averin and Likharev [43] to derive the so-called *Orthodox* model of single charge tunneling. A similar master equation approach was contemporaneously introduced by Ben-Jacob and coworkers [44] based on a semi-classical analysis. One prediction of this new theory was the occurrence of single charge effects in transport through single tunnel junctions rather than islands. Subsequent attempts to observe such effects in single tunnel junctions have not been particularly successful due to environmental effects such as parasitic capacitances. However, the renewed interest in single charge effects also led to new predictions for transport in Coulomb island structures. One effect in particular is the *Coulomb staircase* in the current-voltage characteristics, first observed by Kuzmin and Likharev [45] and Barner and Ruggiero [46] in metallic island structures of the same general structure as in Fig. 4.11. The difficulty of measuring single electron effects in metal films is that transport involves many islands, and therefore fluctuations in the size, capacitance, and tunnel resistance of individual islands tend to average out effects due to single charge tunneling. The use of a scanning tunneling microscopy (STM) tip to contact individual islands alleviates this problem and allows clean measurement of the Coulomb staircase (see Fig. 4.12) [47], [48]. The current is essentially zero about the origin, evidencing the Coulomb blockade effect, and then rises in jumps, giving a staircase-like appearance. The subsequent jumps in the I-V characteristics correspond to the stable voltage regimes in which one more electron is added to or subtracted from the island. Each plateau essentially corresponds to a stable regime with a fixed integer number

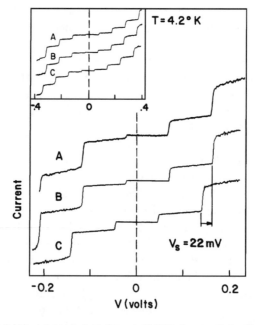

Fig. 4.12. Experimental (A) and theoretical (B and C) I-V characteristics from an STM-contacted 10 nm diameter In droplet illustrating the Coulomb staircase in a double junction system. The peak-to-peak current is 1.8 nA. The curves are offset from one another along the current axis, with the intercept corresponding to zero current. [After Wilkins *et al.*, Phys. Rev. Lett. **63**, 801 (1989), by permission.]

of electrons on the island. It is interesting to note that the I-V characteristics in Fig. 4.12 are not symmetric with respect to the origin. This offset in the I-V curve is due to the presence of unintentional background charges which contribute an additional charging energy to the Coulomb island. Such random charges are very difficult to eliminate experimentally and represent a severe problem in realizing practical device technologies based on single electron devices.

To understand the Coulomb staircase effect somewhat more quantitatively, we consider the equivalent circuit of the Coulomb island shown in Fig. 4.10 in which the "island" consists of a small metallic cluster coupled weakly through thin insulators to metal leads as shown schematically in Fig. 4.13. We have introduced a circuit element representing the tunnel junction as a parallel combination of the tunneling resistance R_t and the capacitance C. In metal tunnel junctions, the tunneling barrier is typically very high and thin, while the density of states at the Fermi energy is very high. The tunneling resistance is therefore almost independent of the voltage drop across the junction (see Section 4.2.2). In the analysis that follows, we implicitly assume a sequential tunneling model as discussed in 3.3.2. Electrons that tunnel through one junction or the other are assumed to immediately relax due to carrier-carrier scattering so that resonant tunneling through both barriers simultaneously is neglected. One has to be careful at this point in distinguishing the tunnel resistance from an ordinary Ohmic resistance. In an ordinary resistor, charge flow is quasi-continuous and changes almost instantaneously in response to a change in electric field (at least down to time scales on the order of the collision times, or picoseconds). Tunneling represents the injections of single particles, which involve several characteristic time scales as discussed in Section 3.3.2. The tunneling time (the time to tunnel from one side of the barrier to the other) is the shortest time (on the order of 10^{-14} s), whereas the actual time between tunneling events themselves is on the order of the current divided by e, which

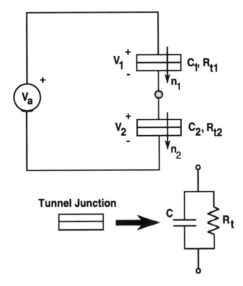

Fig. 4.13. Equivalent circuit of a metallic island weakly coupled to a voltage source through two tunnel junctions with capacitances C_1 and C_2. R_t is the tunnel resistance, n_1 is the number electrons that have tunneled *into* the island through junction 1, and n_2 is the number of electrons that have tunneled *out of* the island through junction 2.

for typical currents in the nanoamp range implies a mean time of several hundred picoseconds between events. The time for charge to rearrange itself on the electrodes due to the tunneling of a single electron will be something on the order of the dielectric relaxation time, which is also very short. Therefore, for purposes of analysis, we can consider that the junctions in the regime of interest behave as ideal capacitors through which charge is slowly leaked.

For the circuit shown in Fig. 4.13, the capacitor charges are given by

$$Q_1 = C_1 V_1,$$

$$Q_2 = C_2 V_2. \tag{4.53}$$

The net charge Q on the island is the difference of these two charges. In the absence of tunneling, the difference in charge would be zero and the island neutral. Tunneling allows an integer number of excess electrons to accumulate on the island so that

$$Q = Q_2 - Q_1 = -ne, \tag{4.54}$$

where $n = n_1 - n_2$ is the net number of excess electrons on the island (which can be positive or negative), with n_1 and n_2 defined as in Fig. 4.13. This convention is chosen such that an increase in either n_1 or n_2 corresponds to increasing either the junction charge Q_1 or Q_2, respectively, in (4.53). We follow this convention later in writing the tunneling rates for forward and backward tunneling in Section 4.2.2. The sum of the junction voltages is just the applied voltage, V_a, so that using (4.53) and (4.54) we may write the voltage drops across the two tunnel junctions as

$$V_1 = \frac{1}{C_{eq}}(C_2 V_a + ne),$$

$$V_2 = \frac{1}{C_{eq}}(C_1 V_a - ne), \tag{4.55}$$

where $C_{eq} = C_1 + C_2$ is the capacitance of the island. The electrostatic energy stored in the capacitors is given by

$$E_s = \frac{Q_1^2}{2C_1} + \frac{Q_2^2}{2C_2}, \tag{4.56}$$

which using (4.53) and (4.55) gives

$$E_s = \frac{1}{2C_{eq}}\left(C_1 C_2 V_a^2 + Q^2\right). \tag{4.57}$$

In addition, we must consider the work done by the voltage source in transferring charge in and out of the island via tunneling. The work done by the voltage source may be considered as the time integral over the power delivered to the tunnel junctions by this source,

$$W_s = \int dt V_a I(t) = V_a \Delta Q, \tag{4.58}$$

where ΔQ is the total charge transferred from the voltage source, including the integer number of electrons that tunnel into the island and the continuous polarization charge that builds up in response to the change of electrostatic potential on the island. A change in the charge on the island due to one electron tunneling through 2 (so that $n_2' = n_2 + 1$) changes the charge on the island to $Q' = Q + e$, and $n' = n - 1$. According to (4.55), the voltage across junction 1 changes as $V_1' = V_1 - e/C_{eq}$. Therefore, from (4.53) a polarization charge flows in from the voltage source $\Delta Q = -eC_1/C_{eq}$ to compensate. The total work done to pass in n_2 charges through junction 2 is therefore

$$W_s(n_2) = -n_2 e V_a \frac{C_1}{C_{eq}}. \tag{4.59}$$

By a similar analysis, the work done in transferring n_1 charges through junction 1 is given by

$$W_s(n_1) = -n_1 e V_a \frac{C_2}{C_{eq}}. \tag{4.60}$$

We may therefore write the total energy of the complete circuit including the voltage source as

$$E(n_1, n_2) = E_s - W_s = \frac{1}{2C_{eq}}(C_1 C_2 V_a^2 + Q^2) + \frac{eV_a}{C_{eq}}(C_1 n_2 + C_2 n_1). \tag{4.61}$$

We may now look at the condition for Coulomb blockade based on the change in this electrostatic energy with the tunneling of a particle through either junction. At zero temperature, the system has to evolve from a state of higher energy to one of lower energy. Therefore, tunneling transitions that take the system to a state of higher energy are not allowed, at least at zero temperature (at higher temperature, thermal fluctuations in energy on the order of $k_B T$ weaken this condition; see Section 4.2.2). The change in energy of the system with a particle tunneling through the second junction is

$$\Delta E_2^\pm = E(n_1, n_2) - E(n_1, n_2 \pm 1) = \frac{Q^2}{2C_{eq}} - \frac{(Q \pm e)^2}{2C_{eq}} \mp \frac{eV_a C_1}{C_{eq}},$$

$$= \frac{e}{C_{eq}}\left[-\frac{e}{2} \pm (en - V_a C_1)\right]. \tag{4.62}$$

Similarly, the change in energy of the system with a particle tunneling through junction 1 is given by

$$\Delta E_1^\pm = E(n_1, n_2) - E(n_1 \pm 1, n_2) = \frac{e}{C_{eq}}\left[-\frac{e}{2} \mp (en + V_a C_2)\right]. \tag{4.63}$$

According to our previous assertion, only transitions for which $\Delta E_j > 0$ are allowed at zero temperature.

Consider now a system where the island is initially neutral, so that $n = 0$. Equations (4.62) and (4.63) reduce to

$$\Delta E_{1,2}^\pm = -\frac{e^2}{2C_{eq}} \mp \frac{eV_a C_{2,1}}{C_{eq}} > 0. \tag{4.64}$$

For all possible transitions into and out of the island, the leading term involving the Coulomb energy of the island causes ΔE to be negative until the magnitude of V_a exceeds a threshold that depends on the lesser of the two capacitances. For $C_1 = C_2 = C$, the requirement becomes simply $|V_a| > e/C_{eq}$. Tunneling is prohibited and no current flows below this threshold, as evident in the I-V characteristics shown in Fig. 4.12. This region of *Coulomb*

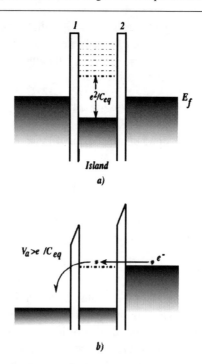

Fig. 4.14. Band diagram of an $MIMIM$ double junction structure (a) in equilibrium and (b) under an applied bias. A gap exists in the density of states of the island system due to the Coulomb charging energy which prohibits tunneling into and out of the island below the threshold voltage.

blockade is a direct result of the additional Coulomb energy, $e^2/2C_{eq}$, which must be expended by an electron in order to tunnel into or out of the island. The effect on the current voltage characteristics is a region of very low conductance around the origin, as shown in Fig. 4.12. For large-area junctions where C_{eq} is large, no regime of Coulomb blockade is observed, and current flows according to the tunnel resistance, R_t.

Figure 4.14a shows the equilibrium band diagram for an $MIMIM$ double junction structure illustrating the Coulomb blockade effect for equal capacitances. A Coulomb gap of width e^2/C_{eq} has opened at the Fermi energy of the metal island, half of which appears above and half below the original Fermi energy so that no states are available for electrons to tunnel into from the left and right electrodes. Likewise, electrons in the island have no empty states to tunnel to either until the blockade region is overcome by sufficient bias as shown in Fig. 4.14b.

Now consider what happens when the double junction structure is biased above the threshold voltage. Assume for the sake of illustration that the capacitance of the two barriers is the same, $C_1 = C_2 = C$. Suppose the threshold for tunneling, $V_a > e/2C$, has been reached so that one electron has already tunneled into the island and $n = 1$ as shown in Fig. 4.14b. The Fermi energy in the dot is raised by e^2/C_{eq} and a gap appears that prohibits a second electron from tunneling into the island from the right electrode until a new voltage, $V_a > 3e/2C$, is reached, as is apparent from Eq. (4.62). Within this range of V_a, no further charge flows until the extra electron on the island tunnels into the left electrode, taking the dot back to the $n = 0$ state. This transfer lowers the Fermi energy in the dot and allows another electron to tunnel from the right electrode, and the process repeats itself. Thus, a correlated set of tunneling processes of tunneling into and out of

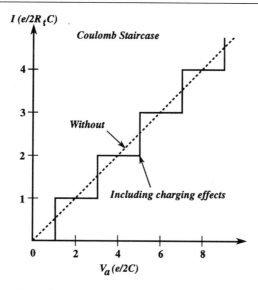

Fig. 4.15. Ideal current-voltage characteristics for an asymmetric double junction system with and without consideration of Coulomb charging effects. For this system, $C_1 = C_2 = C$ and $R_t = R_{t1} \gg R_{t2}$

the dot for the $n = 1$ configuration occur, giving rise to a nonzero current. In order to observe the Coulomb staircase of Fig. 4.12, the junctions should be asymmetric such that either the capacitances or the tunneling resistances are quite different. Assume in our present case that the capacitances are still equal but that the tunneling resistances are quite different, with $R_{t1} \gg R_{t2}$. For this situation, the limiting rate is tunneling through the first barrier, so that the island remains essentially in a charge state corresponding to the voltage range defined by (4.62) for positive voltages. As soon as an electron tunnels out of the dot through junction 1, it is immediately replenished by junction 2. Under these conditions, the current is approximately controlled by the voltage drop across junction 1, which is given by (4.55) for the equal capacitance case as $V_1 = V_a/2 + ne/C_{eq}$. The voltage across the first barrier therefore jumps by an amount e/C_{eq} whenever the threshold for increasing n is reached for junction 2. The current correspondingly jumps by an amount given by

$$\Delta I \approx \frac{\Delta V_1}{R_{t1}} = \frac{e}{C_{eq} R_{t1}} = \frac{e}{2C R_{t1}}. \tag{4.65}$$

Assuming the current does not vary much between jumps, the I-V characteristics exhibit the staircase structure shown in Fig. 4.15, which qualitatively explains the staircase structure in the experimental data of (4.12).

The existence of well-defined structure in the transport properties of a multijunction system due to Coulomb charging depends on the magnitude of the Coulomb gap, e^2/C_{eq}, compared to the thermal energy. Qualitatively it is clear that this gap must greatly exceed the thermal energy, $e^2/C_{eq} \gg k_B T$, to observe well-defined Coulomb blockade effects at a given temperature. A further constraint is that the quantum fluctuations in the particle number, n, be sufficiently small that the charge is well localized on the island. A hand waving argument is to consider the energy uncertainty relationship

$$\Delta E \Delta t > h, \tag{4.66}$$

where $\Delta E \sim e^2/C_{eq}$, and the time to transfer charge into and out of the island given by $\Delta t \approx R_t C_{eq}$, where R_t is the smaller of the two tunnel resistances. Combining these two expressions together gives

$$(e^2/C_{eq})(R_t C_{eq}) > h,$$

so that the requirement for clear Coulomb charging effects is that the tunnel resistance be sufficiently large (otherwise it would not be a tunnel junction)

$$R_t \gg \frac{h}{e^2} = 25.813\,\text{k}\Omega. \tag{4.67}$$

Up to this point, we have discussed experimental results in Coulomb islands that arise due to clustering and island formation during the growth of very thin metal films. Except for the STM experiments mentioned earlier, single electron effects were measured in an averaged way. For practical applications it is much more desirable to have the capability to fabricate nanostructure dots in an intentional fashion rather than relying on the random occurrence of such structures. Improvements in nanolithography technologies such as direct write e-beam lithography and atomic force microscopy now allow the routine fabrication of single island nanostructures exhibiting single electron phenomena.

One of the first realizations of a nanofabricated single electron structure was reported by Fulton and Dolan [49], where a shadow-mask technique using electron beam lithography was used to fabricate 30 nm × 30 nm Al − Al$_2$O$_3$ − Al tunnel junctions. The device showed clear evidence of Coulomb blockade and structure due to the Coulomb staircase (see Fig. 4.16). Besides realizing a double junction using nanolithography, the work of Fulton and Dolan was seminal in that additional gate contacts to the island were fabricated to realize a *single electron transistor* [43]. As seen in Fig. 4.16, the Coulomb staircase structure in their experiments changes as the bias applied to a substrate contact changes (the substrate contact was capacitively coupled to the island and acts as the gate in this case).

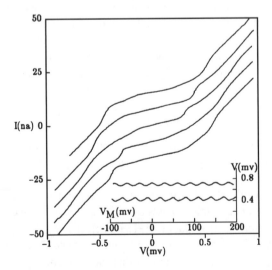

Fig. 4.16. I-V curves for a substrate-biased double junction system at $T = 1.1\,K$ for different substrate biases covering one cycle. Curves are offset by increments of 7.5 nA [after Fulton and Dolan, Phys. Rev. Lett. **59**, 109 (1987), by permission.]

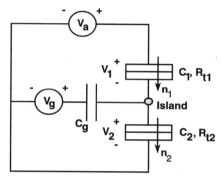

Single Electron Transistor

Fig. 4.17. Equivalent circuit for a single electron transistor.

The change in the I-V curves with gate bias was periodic with period approximately given by e/C_s, where C_s is the island-to-substrate capacitance.

To understand this behavior, we consider the topologically equivalent single electron transistor circuit shown in Fig. 4.17. There, a separate voltage source, V_g, is coupled to the island through an ideal (infinite tunnel resistance) capacitor, C_g. This additional voltage modifies the charge balance on the island so that Eq. (4.53) requires an additional polarization charge:

$$Q_g = C_g(V_g - V_2). \tag{4.68}$$

The island charge becomes

$$Q = Q_2 - Q_1 - Q_g = -ne + Q_p, \tag{4.69}$$

where Q_p has been added to represent both the unintentional background polarization charge that usually exists in real structures due to workfunction differences and random charges trapped near the junctions. The existence of such random charge was necessary in explaining the asymmetry of the experimental I-V characteristics about the origin shown in Fig. 4.12. Equations (4.53), (4.68), and (4.70) together yield the voltages across the two tunnel junctions,

$$V_1 = \frac{1}{C_{eq}}[(C_g + C_2)V_a - C_g V_g + ne - Q_p],$$

$$V_2 = \frac{1}{C_{eq}}[C_1 V_a + C_g V_g - ne + Q_p]. \tag{4.70}$$

The equivalent capacitance of the island is that obtained by grounding the independent voltage sources:

$$C_{eq} = C_1 + C_2 + C_g. \tag{4.71}$$

The electrostatic energy given by Eqs. (4.56) and (4.57) now includes the energy of the gate capacitor, $e^2/2C_g$, as

$$E_s = \frac{1}{2C_{eq}}\left[C_g C_1(V_a - V_g)^2 + C_1 C_2 V_a^2 + C_g C_2 V_g^2 + Q^2\right]. \tag{4.72}$$

The work performed by the voltage sources during the tunneling through junctions 1 and 2 now includes both the work done by the gate voltage and the additional charge flowing onto the gate capacitor electrodes. Equations (4.59) and (4.60) are now generalized to

$$W_s(n_1) = -n_1 \left[\frac{C_2}{C_{eq}} e V_a + \frac{C_g}{C_{eq}} e(V_a - V_g) \right]$$

$$W_s(n_2) = -n_2 \left[\frac{C_1}{C_{eq}} e V_a + \frac{C_g}{C_{eq}} e V_g \right]. \tag{4.73}$$

The total energy for a charge state characterized by n_1 and n_2 is given by Eq. (4.61). For tunnel events across junction 1, the change in energy of the system is now given by

$$\Delta E_1^{\pm} = \frac{Q^2}{2C_{eq}} - \frac{(Q \mp e)^2}{2C_{eq}} \mp \frac{e}{C_{eq}} [(C_g + C_2)V_a - C_g V_g],$$

$$= \frac{e}{C_{eq}} \left(-\frac{e}{2} \mp [en - Q_p + (C_g + C_2)V_a - C_g V_g] \right), \tag{4.74}$$

and for tunnel events across junction 2, the change in energy is given by

$$\Delta E_2^{\pm} = \frac{Q^2}{2C_{eq}} - \frac{(Q \pm e)^2}{2C_{eq}} \mp \frac{e}{C_{eq}} [C_1 V_a + C_g V_g],$$

$$= \frac{e}{C_{eq}} \left(-\frac{e}{2} \pm [en - Q_p - C_1 V_a - C_g V_g] \right). \tag{4.75}$$

In comparison to Eqs. (4.62) and (4.63), the gate bias allows us to change the effective charge on the island, and therefore to shift the region of Coulomb blockage with V_g. Thus, a stable region of Coulomb blockade may be realized for $n \neq 0$. As before, the condition for tunneling at low temperature is that $\Delta E_{1,2} > 0$ such that the system goes to a state of lower energy after tunneling. The random polarization charge acts as an effective offset in the gate voltage, so we may define a new voltage, $V_g' = V_g + Q_p/C_g$. The conditions for forward and backward tunneling then become

$$-\frac{e}{2} \mp \left[en + (C_g + C_2)V_a - C_g V_g' \right] > 0$$

$$-\frac{e}{2} \pm \left[en - C_1 V_a - C_g V_g' \right] > 0. \tag{4.76}$$

These four equations for each value of n may be used to generate a *stability plot* in the $V_a V_g$ plane, which shows stable regions corresponding to each n for which no tunneling may occur. Such a diagram is shown in Fig. 4.18 for the case of $C_g = C_2 = C, C_1 = 2C$. The lines represent the boundaries for the onset of tunneling given by Eq. (4.76) for different values of n. The trapezoidal shaded areas correspond to regions where no solution satisfies (4.76), and hence where Coulomb blockade exists. Each of the regions corresponds to a different integer number of electrons on the island, which is "stable" in the sense that this charge state cannot change, at least at low temperature when thermal fluctuations are negligible. The gate voltage then allows us to tune between stable regimes, essentially adding or subtracting one electron at a time to the island. For a given gate bias, the range of V_a over which Coulomb blockade occurs is given by the vertical extent of the shaded region. For the case shown in Fig. 4.18, the maximum blockade occurs when $C_g V_g' = me$, $m = 0, \pm 1, \pm 2, \ldots$. As $C_g V_g'$ approaches half integer values of the charge of a single electron, the width of

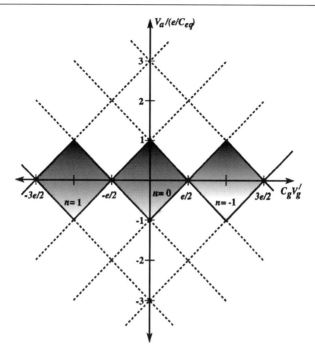

Fig. 4.18. A stability diagram for the single electron transistor for the case $C_2 = C_g = C, C_1 = 2C$, illustrating the regions of energy where tunneling is prohibited at $T = 0$ for various numbers of electrons on the island. The shaded areas correspond to regions where no tunneling through either junction may occur, and thus they represent stable regimes of fixed electron number.

the Coulomb blockade region vanishes and tunneling may occur. Therefore, for a small source-drain bias V_a across the double junction, a measurement of the current versus gate bias will exhibit peaks in the current for a narrow range of gate bias around half integer values of the gate charge, as illustrated in Fig. 4.19. The distance between peaks is given by $\Delta V_g = e/C_g$. Between peaks, the number of electrons on the dot remains a stable integer value. As long as $eV_a \ll k_B T$, the width of these peaks will essentially be limited by thermal broadening and therefore smear out at temperatures greater than the energy width of the Coulomb blockade regime. The conductance linewidth has been calculated in the limit of $e^2/C \gg k_B T$ by Beenakker as [50]

$$\frac{G}{G_{\max}} \approx \cosh^{-2}\left[\frac{e(C_g/C_{eq}) \cdot \left(V_g^{res} - V_g\right)}{2.5 k_B T}\right], \tag{4.77}$$

where V_g^{res} corresponds to a resonant value such that $C_g V_g'$ is a half integer multiple of the fundamental charge. Results using (4.77) for $k_B T = 0.05 e^2/C_{eq}$ are shown in Fig. 4.19, illustrating the expected lineshape. The existence of conductance peaks may be more clearly understood from the energy band diagram of the system shown in Fig. 4.20. The measurement of current or conductance peaks with gate bias in the linear response regime is typical of experimental results reported in semiconductor quantum dot structures discussed in the following section. In such systems, the quantum dot corresponds to the Coulomb island of the systems discussed in the present section. In the semiconductor case, however, additional complication occurs due to the discrete nature of the states in the dot, which was neglected in the qualitative discussion of metallic systems.

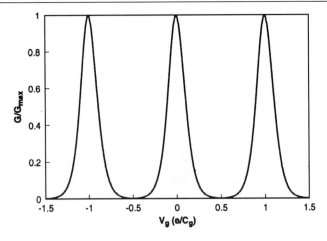

Fig. 4.19. Calculated conductance versus gate voltage in the linear response regime of a double junction single-electron transistor for $k_B T = 0.05 e2/C_{eq}$.

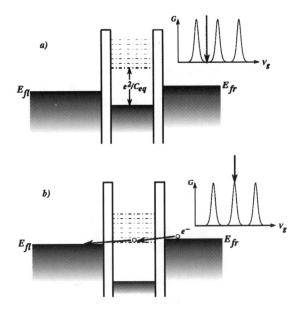

Fig. 4.20. Energy-band diagram of a double junction system, illustrating the occurance of conductance peaks due to Coulomb blockade. (a) The system biased off resonance. (b) Resonance condition when the excited state due to Coulomb charging lies between the Fermi energies on the left and right in the linear response regime.

More recent research in metallic systems has focused on increasingly more complicated metal tunnel junction circuits, both from a fundamental perspective as well as for practical applications such as metrology and single electron logic (see for example [51]). A typical technology used to realize small-area tunnel junctions is shown in Fig. 4.21. A small shadow mask is defined by using electron beam lithography and subsequently etching away the support material under the bridge. Al is evaporated from one direction and then oxidized at

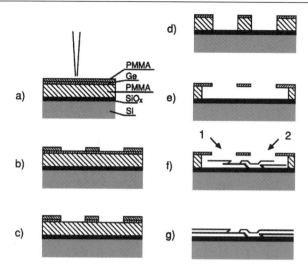

Fig. 4.21. Side view of the fabrication of a small metal tunnel junction using shadow evaporation [after Geerligs, in *Physics of Nanostructures*, eds. J. H. Davies and A. R. Long (IOP Publishing, London, 1992), p. 171, by permission].

low pressure *in situ* in the evaporation system, followed by a second Al evaporation from the other direction, forming the small overlap area junction shown in Fig. 4.21c. Typical capacitances are on the order of 10^{-15} F for a junction area of $(100\,\text{nm})^2$.

Theoretical and experimental investigations of one- and two-dimensional arrays of small tunnel junctions have demonstrated a rich field of new physical phenomena due to the coupled motion of single charges through the system, which exhibits nonlinear soliton-like behavior (for reviews see [52], [53]). From the standpoint of practical applications, a particularly interesting device structure is the *single electron turnstile* [54], which clocks electrons one by one through a series of junctions using an rf frequency a.c. gate bias. Figure 4.22 shows the operation of this structure. As shown by the circuit diagram in Fig. 4.22a, the turnstile is essentially the single electron transistor of Fig. 4.17 with the addition of two additional tunnel junctions on either side of the central island. The gate bias in this case consists of both a d.c. and an a.c. bias. The basic idea of the turnstile is to move an electron into the central island from left to right during the first half of the cycle, and then from the island to the right side during the second half, thus clocking one electron per cycle through the array. The resulting current is therefore accurately given by $I = ef$, where f is the frequency of the a.c. source. Sweeping the gate bias in the simple single electron transistor in time means moving horizontally in the stability diagram of Fig. 4.18 for a given bias voltage across the junctions. In the two-junction case, one alternately moves from stable regions of fixed charge (and no current flow) to open regions where tunneling occurs, destroying the desired turnstile effect. However, in the structure of Geerligs *et al.*, the additional junctions lead to the stability plot shown in Fig. 4.22b in which multiple stable regions exist. The bottom regions have two stable charge states, whereas the upper regions have one for the integer charges shown. An a.c. gate bias now allows the center electrode to alternate between two different stable charge states if it traverses two such regions horizontally in the diagram. Figure 4.22c shows the band diagram at various times during the tunneling process. With the gate capacitance, for example, equal to half the

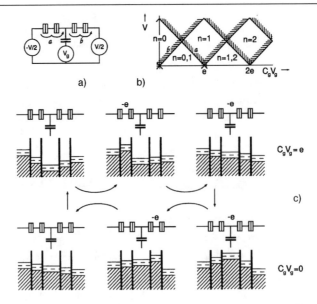

Fig. 4.22. (a) Schematic of a single electron turnstile device. (b) Charge stability diagram for the circuit in part (a). The threshold lines for events a and b are indicated. (c) Operation for a square-wave modulation; $V = 0.4e/C$, $C_g = C/2$. Top line, left to right: $C_g V_g = e$, an electron enters from the left. Bottom, right to left: $C_g V_g = 0$, an electron leaves through the right arm. [After Geerligs, in *Physics of Nanostructures*, eds. J. H. Davies and A. R. Long (IOP Publishing, London, 1992), p. 171, by permission.]

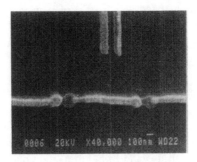

Fig. 4.23. Scanning electron microscope photograph of a four-junction turnstile system [after Geerligs, in *Physics of Nanostructures*, eds. J. H. Davies and A. R. Long (IOP Publishing, London, 1992), p. 171, by permission].

junction capacitance, the threshold for tunneling is first reached for the leftmost junction as the gate bias increases positively. This allows an electron to enter the leftmost island, from which it may subsequently tunnel to the central island, where it remains stable until the threshold from tunneling into the rightmost island is reached. The charge then leaves the central island to the right island, from which it then tunnels into the right electrode, completing the cycle.

The experimental realization of the turnstile structure using metal tunnel junctions is shown in Fig. 4.23. The tunnel junctions lie along the horizontal line in which 4 shadow-evaporated junctions are fabricated. The gate electrode is capacitively coupled to the center

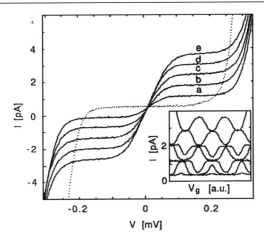

Fig. 4.24. I-V curves of the single electron turnstile without a.c. gate bias (dotted curve) and with a.c. gate bias voltage of frequency 4 to 20 MHz in steps of 4 MHz (a-e). The inset shows the $I = V_g$ curves for an a.c. gate voltage (5 MHz) of increasing amplitude (top to bottom), taken at a bias voltage of 0.15 mV. [After Geerligs and Mooij, in *Granular Nanoelectronics*, eds. D. K. Ferry, J. R. Barker, and C. Jacoboni (Plenum Press, New York, 1991), p. 393, by permission.]

metal line. Additional side gates are provided to the left and right islands, which are not shown in the circuit diagram Fig. 4.22a. These are necessary to cancel the effect of random polarization charges, Q_p, as discussed earlier. The current-voltage characteristics from an experimentally fabricated device are shown in Fig. 4.24. The tunnel resistance of the junctions was $R_t = 340$ kΩ, the junction capacitances $C = 0.5$ fF, and the gate capacitance $C_g = 0.3$ fF. Each I-V curve corresponds to different a.c. frequencies ranging from 4 to 20 MHz. The plateaus correspond to $I = \pm ef$ within the accuracy of the experiment. The inset in Fig. 4.24 shows the I-V_g curves for various values of the a.c. voltage for a frequency of 5 MHz. As the a.c. amplitude increases, it becomes possible for higher integer numbers of electrons to tunnel during one cycle. The inset shows up to 3 electrons tunneling per cycle, depending on the d.c. gate bias.

The transfer of electrons through the structure of the single electron turnstile is dissipative in that the tunneling electron gives up its excess kinetic energy when tunneling into the island. The power dissipation increases as the frequency increases. An alternate single charge transfer device referred to as the *single electron pump* was reported by Pothier *et al.* in which charge transfer is quasi-adiabatic (i.e., with little energy dissipation during the transfer process) [55]. Figure 4.25 shows a circuit schematic and stability diagram for the pump structure. It consists of three tunnel junctions, with separate gate biases capacitively coupled to each of the two Coulomb islands. For small bias voltages across the junctions ($V \approx 0$), stable configurations exist for the excess electron numbers on each island, n_1 and n_2, for various combinations of the gate bias voltages U_1 and U_2. The stability diagram showing these stable configurations has a honeycomb pattern (Fig. 4.25b). By adjusting the d.c. bias close to a "P-type" triple point, such as point P in Fig. 4.25b, one can traverse the region around the triple point in a clockwise or counterclockwise direction by using phase-shifted a.c. biases, u_1 and u_2, applied on top of the d.c. gate biases. If u_2 lags behind u_1 by a factor of $\pi/2$, the system moves from state 00 to state 10 as u_1 increases, crossing a domain boundary, meaning that an electron moves from the left electrode into island 1

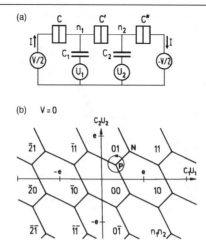

Fig. 4.25. (a) Single electron pump circuit schematic. n_1 and n_2 are the extra numbers of electrons on the two islands. (b) Stability diagram in the domain of the two gate biases for zero bias voltage across the junctions. One turn around the point P transfers one electron across the circuit, in a direction determined by the sense of rotation. [After Esteve, in *Single Charge Tunneling*, eds. H. Grabert and M. H. Devoret (Plenum Press, New York, 1992), p. 109, by permission.]

during the first positive part of the a.c. cycle. Then as u_2 increases, a second domain is crossed, taking the system from 10 to 01 (i.e., the electron moves from island one to island 2). Finally, during the negative-going part of the u_1 cycle, the system crosses back down to the original domain 00, and one electron has tunneled through the whole system from left to right. The net effect is a negative current (in terms of the sign convention in Fig. 4.25) with a magnitude given by ef. If the phase of the two a.c. voltages is reversed such that u_1 lags behind u_2, the path around the triple point is traversed in the clockwise direction, causing an electron to transit the junctions from right to left and giving a positive current. If the system is closer to the "N-type" triple point labeled N, a counterclockwise rotation produces the opposite current as an electron tunnels first into island 2 and then into island 1.

Figure 4.26 shows the experimental I-V characteristics for a pump operated close to a "P-type" triple point. Close to the origin, the current is given by a value $I = \pm ef$ depending on whether the phase difference is positive or negative, independent of the bias voltage across the junction. The quantization of the current, and its polarity dependence on the phase difference of the two a.c. sources, illustrates nicely the pump principle of Fig. 4.25. The operation of the single electron pump is a single-electron analog of charge-coupled or "bucket brigade" devices (CCDs), which are used extensively in memory and image processing applications. The pump's small size and its ability to achieve such accurate charge transfer using single electrons in both the pump and the turnstile makes such concepts attractive for very-high-density memory systems. Attendant problems in nanotechnology scaling, and particularly the elimination of random offset charges, remain to be addressed before practical applications may be considered.

Coulomb blockade in semiconductor quantum dots

Investigations of Coulomb blockade in semiconductor structures followed naturally the technological evolution of the low-dimensional systems community, which historically began with quasi-two-dimensional systems, one-dimensional quantum wires, and finally

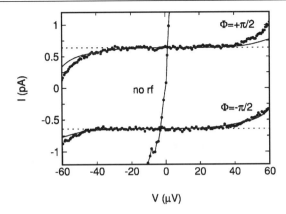

Fig. 4.26. Current-voltage characteristic of the single electron pump with and without a gate voltage modulation ($f = 4$ MHz) around a "P-type" triple point. Dashed lines mark $I = \pm ef$. [After Esteve, in *Single Charge Tunneling*, eds. H. Grabert and M. H. Devoret (Plenum Press, New York, 1992), p. 109, by permission.]

quantum dots. This contrasts to the relatively long history of investigations of single particle effects in metallic systems discussed in the preceding section which has many roots in the superconducting tunnel junction community. Both research areas converged in the late 1980s when clear demonstrations of Coulomb blockade and single electron tunneling in semiconductor quantum dot structures were reported.

Semiconductor structures offer several advantages in terms of flexibility. In particular, quantum point contact (QPC) tunnel barriers allow one to independently tune the tunnel resistance, which is not possible in thin oxide tunnel barriers. One advantage is that the tunnel barriers themselves may be used to modulate the tunneling of electrons with an a.c. bias and therefore effect turnstile-like behavior. The main difference between semiconductor quantum dot structures and metallic Coulomb islands is the coexistence of Coulomb blockade and quantization of the energy levels. This fact leads to discrete structure in the I-V characteristics due to tunneling through discrete levels in addition to the Coulomb blockade effect, which complicates the description.

We can use the equivalent circuit model of Fig. 4.17 to represent the basic structure of the relevant experiments in semiconductor quantum dot structures. The gate represents either Schottky gates on the surface of the 2DEG structure, or a bias voltage connected to the substrate that changes the electron concentration in the channel of the 2DEG. As illustrated in Fig. 4.20 for a metal system, in the linear response regime in which the applied voltage is small, the condition for lifting the Coulomb blockade is to bias the gate voltage such that the Fermi energy of the dot corresponding to the addition of one electron, $n+1$, lies between the reservoir energies on the left and right. This condition corresponds to a conductance peak in a measurement of conductance versus gate bias. In the semiconductor system, we must additionally account for the energy spacing of the discrete states comprising the allowed states in the dot, as shown in Fig. 4.27. We may write the total ground-state energy in a dot of n electrons as the sum of the filled single particle energy states plus the electrostatic energy due to the filling of the dot with electrons

$$E_t(n) = \sum_{i=1}^{n} E_i + E(n_1, n_2), \qquad (4.78)$$

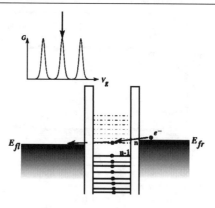

Fig. 4.27. Illustration of the condition for a conductance peak in a semiconductor double barrier island structure where discrete states co-exist with a Coulomb gap.

where E_i are the single particle energies of the quantum dot as discussed in Section 4.1, and $E(n_1, n_2) = E_s - W_s$ is the total charging energy of the system due to tunneling through the left and right barriers given by Eqs. (4.72) and (4.73). We may define the electrochemical potential in the dot as the difference $\mu_d(n) = E_t(n) - E_t(n-1)$, that is, the energy difference associated with the removal of one electron from the dot. In the linear response regime, the applied voltage V_a is small, and therefore the charging energy does not depend on which barrier the electron tunnels through. Using (4.74) and (4.75), the change in charging energy from $n \rightarrow n-1$ corresponds to $n_1 \rightarrow n_1 - 1$ and $n_2 \rightarrow n_2 + 1$, which gives for either case

$$\mu_d(n) = E_n + \frac{e^2 \left(n - \frac{1}{2}\right)}{C_{eq}} - e \frac{C_g V_g^{'}}{C_{eq}}, \tag{4.79}$$

where E_n represents the energy of the highest filled state (at zero temperature), and the prime on V_g signifies that this term may contain the random polarization charge potential Q_p/C_g if necessary. For a given gate bias $V_g^{'}$, if the chemical potential $\mu_d(n)$, corresponding to the addition of an electron to the dot with $n - 1$ electrons, lies between μ_l and μ_r of the reservoirs, tunneling may occur. The system goes from $n - 1 \rightarrow n \rightarrow n - 1$ alternately, giving rise to current. An increase in gate bias causes the state n to be stable and the dot to go off resonance. The conductance vanishes until the level $\mu_d(n + 1)$ lies between the Fermi energies of the reservoirs, and a new conductance peak appears. From (4.79) we see that the gate period corresponding to these two successive conductance peaks is

$$\Delta V_g = \frac{C_{eq}}{C_g} \left(\frac{E_{n+1} - E_n}{e} \right) + \frac{e}{C_g}. \tag{4.80}$$

In comparison to the spacing between conductance peaks shown in Fig. 4.19, which are simply $\Delta V_g = e/C_g$, there is an additional contribution due to the energy spacing between the $n + 1$ and n levels of the system, as illustrated in Fig. 4.27. In a metal system, the electron mass and the Fermi energy are large such that the spacing of levels is on the order of μeV. Therefore, the discrete spacing of the levels is negligible, and the gate voltage period is regularly spaced in e/C_g. In semiconductor quantum dots, where strong quantization may exist, the energy spacing may be comparable or exceed the Coulomb charging contribution. In the extreme limit that the Coulomb charging energy is negligible, we just return to the simple resonant tunneling picture through a dot as discussed in Section 3.1. As we saw in

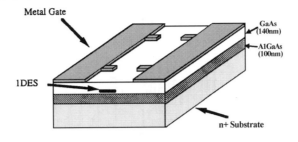

Top View:

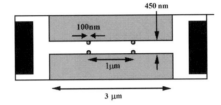

Fig. 4.28. Schematic drawing of a double-constriction split-gate structure exhibiting single electron conductance oscillations [after Kastner, Rev. Mod. Phys. **64**, 849 (1992), by permission.]

Section 4.1, the energy spacing is in general irregular, and so one signature of quantization effects is the lack of regularity of both the peak spacing and peak height. The latter arises due to the sensitive dependence of resonant tunneling on the exact nature of the transmission coefficients through a quasi-bound level as we saw in Chapter 3.

The first evidence of single electron charging in semiconductor structures was the experimental results of Scott-Thomas *et al.* where conductance oscillations versus gate voltage were observed in the low temperature I-V characteristics of side gated Si MOSFET [56]. Such structures were discussed in Chapter 2 in connection with early attempts to realize quantum wires. The oscillations were explained as Coulomb charging effects in an "accidental" quantum dot formed by impurities along the 1D inversion layer channel [57], [58].

The first definitive demonstration of Coulomb charging in deliberately defined quantum dots was reported by Meirav *et al.* using quantum point contacts to define the input and output barriers of a double junction quantum dot structure shown in Fig. 4.28 [59]. The overall heterostructure was grown on a conducting, n^+ GaAs substrate, which was biased to change the electron density in the 2DEG existing at the GaAs/AlGaAs interface. The structure was biased such that the Fermi energy in the 2DEG outside of the dot is below the lowest conducting channel in the constriction, i.e., in the tunneling regime. The resistance is thus much greater than h/e^2. Figure 4.29 shows the measured conductance versus gate voltage for two different length constrictions. The oscillations are periodic in terms of $\Delta V_g = e/C_g$, which is evidenced by the increase in spacing of the oscillations when the dot area is reduced. Although the periodic oscillations with gate bias are understood within the context of the double junction model described for metallic systems, the random modulation of the peak amplitude is not. This modulation arises from the discrete nature of the quantum dot states compared to a metal, where the tunneling probability varies strongly with energy. With increasing temperature, the peak-to-valley amplitude of the conductance

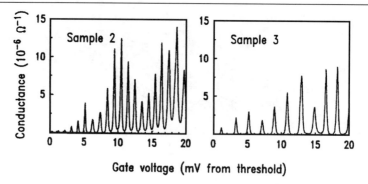

Fig. 4.29. Conductance as a function of V_g for two samples with different lengths. Sample 2 has $L_0 = 0.8\,\mu\text{m}$, and sample 3 has $L_0 = 0.6\,\mu\text{m}$. The period increases inversely with L_0. [After Kastner, Rev. Mod. Phys. **64**, 849 (1992), by permission.]

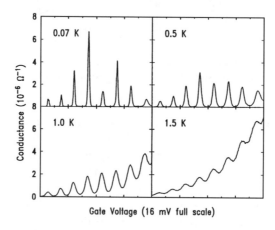

Fig. 4.30. Conductance oscillations for various temperatures [after Kastner *et al.*, in *Nanostuctures and Mesoscopic Systems*, eds. M. A. Reed and W. P. Kirk (Academic Press, Boston, 1992), p. 239, with permission.]

peaks decreases rapidly as shown in Fig. 4.30. It is interesting to note that the temperature dependence is not uniform for all the peaks. In fact, some peaks initially increase with temperature before decreasing. This is further evidence of the role of discrete states and their coexistence with Coulomb charging in these semiconductor structures.

The relative ease of fabrication of split-gate quantum dots has led to the realization of increasingly more complex structures with multiple dots and sidearm contacts [60]. Figure 4.31 shows a semiconductor realization of the single electron turnstile discussed in the preceding section. The inset shows the metal gate pattern above a 2DEG layer. In the actual experiments, gates 3 and 4 were grounded and therefore not utilized so that the system was basically a single quantum dot structure with independently biased input and output QPCs. The d.c. I-V characteristics show clear evidence of the Coulomb staircase, which is controlled by the gate contact V_c, where contact C is the plunger shown in Fig. 4.31. With phase-shifted a.c. biases applied to gates 1 and 2, electrons could be clocked through the dot one electron at a time as illustrated by the main I-V characteristics shown in the figure. These show the current at a frequency of 10 MHz. The staircase structure in this

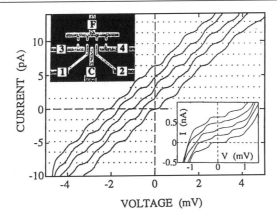

Fig. 4.31. Semiconductor turnstile device realized with a split-gate geometry on a 2DEG substrate. The main figure shows the I-V curves with 10-MHz signals applied to the QPC gates and different values of V_c. The upper inset shows the metal gate pattern, while the lower inset shows the d.c. Coulomb staircase I-V characteristics for different values of V_c. [After Johnson *et al.*, in *Nanostuctures and Mesoscopic Systems*, eds. M. A. Reed and W. P. Kirk (Academic Press, Boston, 1992), p. 267, by permission.]

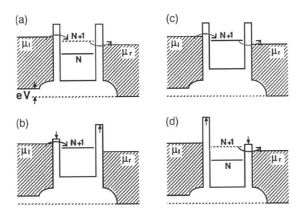

Fig. 4.32. Schematic band diagram of the quantum dot structure during various stages of an RF cycle [after Johnson *et al.*, in *Nanostuctures and Mesoscopic Systems*, eds. M. A. Reed and W. P. Kirk (Academic Press, Boston, 1992), p. 267, by permission.]

case corresponds to $I = nef$, $n = 1, 2, \ldots$. For increasing source-drain bias, the turnstile may accommodate an increasing integer number of electrons transferred per clock cycle. Figure 4.32 illustrates the operation in more detail. As the left barrier is pulled down in (b), an electron may enter the dot from the left, but not tunnel out to the right. During the opposite cycle (d), the right barrier is lowered and the electron tunnels out on the right, resulting in the transfer of a single electron per clock cycle.

4.2.2 Orthodox theory of single electron tunneling

We now want to introduce in somewhat more detail the theoretical models used for analyzing single charge transport. The most widely invoked theory is the so-called "Orthodox" theory of single electron tunneling (SET) of Averin and Likharev [43] in which a kinetic equation is derived for the distribution function describing the charge state of a junction or system of junctions. In the semiclassical limit, this theory has proved extremely valuable in analyzing the transport properties of metallic tunnel junctions where size quantization effects are negligible. The method may also be extended to the semiconductor quantum dot case by introducing quantization of the dot states in addition to Coulomb charging effects [61].

In order to establish a kinetic equation based on tunneling into and out of the Coulomb island or quantum dot, it is generally more convenient to describe tunneling in terms of transition rates using perturbation theory rather than in terms of tunneling probabilities as we did in Chapter 3. Using the transition rates for tunneling allows one to write a detailed balance for tunneling into and out of the dot, and thus an equation of motion describing the evolution of the charge with time. For this reason, the *transfer Hamiltonian method of tunneling* is employed in the Orthodox model, described in more detail in the following section.

Transfer Hamiltonian method of tunneling

In Chapter 3 we introduced tunneling using the scattering matrix formalism. Historically, an alternative method has also been successfully employed in tunneling problems based on the transfer or tunneling Hamiltonian approach [62], [63]. The transfer Hamiltonian method has been reviewed thoroughly by Duke [64]. This technique was used extensively in describing transport in superconducting tunnel junctions, and has been the basis for most models describing tunneling in small tunnel junctions including Coulomb blockade effects.

In the tunneling Hamiltonian approach, the tunneling barrier is treated as a perturbation to the (much larger) systems forming the left and right sides. The current may be investigated by calculating the rate of transfer of particles from left to right (and right to left) using time-dependent perturbation theory. As such, the applicability of the model is valid only when the perturbation is sufficiently small, which in the tunneling case usually means that the transmission coefficient is small, $T \ll 1$. However, the advantage of the method is that within perturbation theory, the powerful diagrammatic techniques of quantum field theory and many-body theory may be utilized. This allows many-body effects such as quasi-particle tunneling or phonon-assisted tunneling to be treated.

The model system for the transfer Hamiltonian approach is shown schematically in Fig. 4.33 for a simple planar rectangular barrier. The two systems, l and r, represent unperturbed systems on the left and right that are independent of one another except for the perturbation. For one particular choice of subsystems shown in Fig. 4.33, the unperturbed system on the left represents the half-space $x < d$, while the system on the right is the half-space $x > 0$; both systems contain the tunnel barrier itself. The total Hamiltonian is written

$$H = H_l + H_r + H_t, \tag{4.81}$$

where the Hamiltonians on the left and right, H_l and H_r, presumably are known with eigenvectors and eigenvalues

$$H_l \psi_l = E_l \psi_l$$
$$H_r \psi_r = E_r \psi_r. \tag{4.82}$$

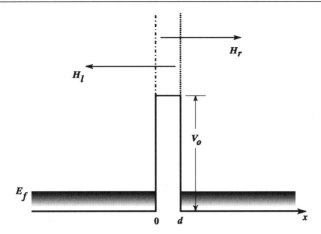

Fig. 4.33. Single barrier between two isolated systems illustrating the decomposition of the system into two independent systems in the tunnel Hamiltonian approach.

The significance of the tunneling Hamiltonian becomes particularly transparent if we write the Hamiltonian (4.81) explicitly in second quantized form:

$$H_0 = H_l + H_r = \sum_{\mathbf{k}_l} E_{\mathbf{k}_l} c^{\dagger}_{\mathbf{k}_l} c_{\mathbf{k}_l} + \sum_{\mathbf{k}_r} E_{\mathbf{k}_r} c^{\dagger}_{\mathbf{k}_r} c_{\mathbf{k}_r}, \tag{4.83}$$

$$H_t = \sum_{\mathbf{k}_l \mathbf{k}_r} T_{\mathbf{k}_l \mathbf{k}_r} c^{\dagger}_{\mathbf{k}_r} c_{\mathbf{k}_l} + \sum_{\mathbf{k}_l \mathbf{k}_r} T_{\mathbf{k}_r \mathbf{k}_l} c^{\dagger}_{\mathbf{k}_l} c_{\mathbf{k}_r}, \tag{4.84}$$

where the $c^{\dagger}_{l,r}$ and $c_{l,r}$ are the Fermion creation and annihilation operators of the independent many-body state of the left and right sides of the barrier, respectively. Such operators for bosons were introduced in Section 2.7.4 in connection with lattice scattering. In (4.83), the pairs of operators represent the occupation or number operators:

$$N_{\mathbf{k}_l} = c^{\dagger}_{\mathbf{k}_l} c_{\mathbf{k}_l}$$

$$N_{\mathbf{k}_r} = c^{\dagger}_{\mathbf{k}_r} c_{\mathbf{k}_r}. \tag{4.85}$$

For particles obeying Fermi-Dirac statistics, the result of such an operation can give only one or zero. For finite temperature, the expectation value averaged over the equilibrium ground state gives the Fermi-Dirac distribution

$$\langle N_{\mathbf{k}_{l,r}} \rangle = f(E_{\mathbf{k}_{l,r}}) = \frac{1}{1 + e^{(E_{\mathbf{k}_{l,r}} - E^{l,r}_F)/k_B T_L}}. \tag{4.86}$$

Equation (4.84) is the tunnel Hamiltonian, and its first term annihilates a particle of wavevector $\mathbf{k}_l$ on the left side, and creates it in $\mathbf{k}_r$ on the right side. This process therefore corresponds to tunneling from left to right; the second term corresponds to the reverse process. The tunneling rate is calculated using time-dependent perturbation theory for transitions from an initial state on one side of the barrier to a final state on the other side using Fermi's golden rule, in which the tunnel Hamiltonian (4.84) is the perturbation.

We now consider tunneling for a metallic tunnel junction within the single-particle picture using the transfer Hamiltonian method. To begin, a constant voltage V is applied to the left electrode relative to the right. Assume that the electrodes remain approximately in thermal equilibrium as we did in Chapter 3, so that the one-particle distribution functions are still in the form of (4.86). A positive potential applied to the left side with respect to the right lowers the Fermi energy on that side according to

$$E_F^r - E_F^l = eV. \tag{4.87}$$

Due to the capacitance of the junction, there is a certain charge buildup Q on the left electrode, and charge $-Q$ on the right electrode, associated with $CV = Q$. Tunneling from left to right increases the charge, whereas tunneling of an electron from right to left decreases the charge for the voltage convention we are using (positive voltage on the left electrode). Therefore, we will use the convention that Γ^+ is the tunneling rate from left to right which increases the electrode charge, and that Γ^- is the rate from right to left which decreases the charge.

The transition rate from an initial state $\mathbf{k}_l$ to a final state $\mathbf{k}_r$ is treated as a scattering process using Fermi's golden rule:

$$\Gamma_{k_l \to k_r}^+ = \frac{2\pi}{\hbar} |T_{k_l,k_r}|^2 [1 - f(E_r)] \delta(E_l - E_r),$$

$$T_{k_l,k_r} = \langle \mathbf{k}_r | H_t | \mathbf{k}_l \rangle, \tag{4.88}$$

where the tunnel Hamiltonian is the perturbation and the probability that the state is unoccupied has been included. Duke has discussed in detail how to define the tunnel Hamiltonian and to apply it to the simple single barrier case [64]. The results are approximately the same as those derived using the scattering matrix approach, particularly in the limit of thick barriers. The total rate from occupied states on the left to unoccupied states on the right is given by

$$\Gamma^+(V) = \frac{2\pi}{\hbar} \sum_{\mathbf{k}_l, \mathbf{k}_r} |T_{k_l,k_r}|^2 f(E_l)[1 - f(E_r)] \delta(E_l - E_r). \tag{4.89}$$

For a typical metal tunnel junction, the barrier consists of a thin native oxide with a relatively high barrier height. It is usually a reasonable approximation for such cases to neglect the variation of the tunnel matrix element with energy and momentum so that the matrix element is treated as a constant which may be taken outside of the summation. The sums over momentum are converted to sums over energy in the usual fashion to obtain

$$\Gamma^+(V) = \frac{2\pi}{\hbar} |T|^2 \int_{E_{c_l}}^\infty dE_l \int_{E_{c_r}}^\infty dE_r \, D_l(E_l) D_r(E_r) f(E_l)[1 - f(E_r)] \delta(E_l - E_r), \tag{4.90}$$

where $D_{l,r}(E)$ is the density of states in energy on the left and right sides of the barrier. Since the main contribution from the integral is for a narrow range of energies (depending on the applied voltage) around the Fermi energies on the left and right, the densities of states appearing in the integral may be taken constant as well, $D_{l,r}(E) = D_{lo,ro}$. The delta function then reduces one of the integrations such that

$$\Gamma^+(V) = \frac{2\pi}{\hbar} |T|^2 D_{lo} D_{ro} \int_{E_{cm}}^\infty dE f\left(E - E_F^l\right)\left[1 - f\left(E - E_F^r\right)\right], \tag{4.91}$$

where the lower limit is the higher of the two conduction band minima on the left and right. The difference between the Fermi energies is given by (4.87). The same calculation for the charge flow the other way gives by symmetry

$$\Gamma^-(V) = \frac{2\pi}{\hbar} |T|^2 D_{lo} D_{ro} \int_{E_{cm}}^\infty dE f\left(E - E_F^r\right)\left[1 - f\left(E - E_F^l\right)\right]. \tag{4.92}$$

The total current is written

$$I(V) = e[\Gamma^-(V) - \Gamma^+(V)]$$

$$= \frac{2\pi e}{\hbar}|T|^2 D_{lo} D_{ro} \int_{E_{cm}}^{\infty} dE [f(E - E_F^r) - f(E - E_F^l)]. \qquad (4.93)$$

Choosing the Fermi energy on the right as the zero of energy, then $E_F^l = -eV$. The integral over the Fermi functions may be performed analytically in the limit that the minimum energy is far below the Fermi energy to give

$$\lim_{E_{cm} \to -\infty} \int_{E_{cm}}^{\infty} dE [f(E) - f(E + eV)] = eV. \qquad (4.94)$$

Hence from (4.93), the I-V characteristic of the metal tunnel junction is found to be Ohmic:

$$V = IR_t; \quad R_t = \frac{\hbar}{2\pi e^2 |T|^2 D_{lo} D_{ro}}, \qquad (4.95)$$

where R_t defines the *tunneling resistance* invoked in Section 4.2.1. The resistance decreases as the density of states and the transition probability increase. Note that the density of states factors and the transition matrix element contain the cross-sectional area of the junction itself.

For a semiconductor system, the typical Fermi energies are quite small compared to metallic systems, so the limit taken in (4.94) is not strictly valid, particularly at higher temperature. Further, the approximations leading to (4.95) in terms of the neglect of the energy dependencies of the transmission matrix element and densities of states are not well satisfied. In Section 3.6.5 we showed that for a saddle-point QPC, the conductance in the tunneling regime was in fact strongly nonlinear due to the exponential dependence of the transmission coefficient with energy below the lowest subband minimum, given by

$$R_t^{QPC} = \alpha\beta_n \frac{h}{2e^2} \{1 + \exp(-\beta_n(\mu_l - e\phi_o + eV_{sd}/\alpha - \varepsilon_n))\}, \qquad (4.96)$$

where the constants are defined as before. We note from this expression that the gate bias defining the point contact directly modulates the barrier height, ϕ_o. This fact allows one to sensitively control the tunneling resistance of the tunnel junctions forming the input and output of a QPC quantum dot. This control was utilized, for example, in the turnstile device of Kouwenhoven *et al.* [60] discussed in Section 4.2.1.

We now want to consider the modification of the transition rates for nanostructure systems in which the charging energy is no longer a negligible contribution. Consider a system of N tunnel junctions that are coupled together and characterized by the number of electrons that have passed through the junction, $\{n\} \equiv \{n_1, n_2, \ldots, n_j, \ldots, n_N\}$. In Section 4.2.1, we derived the change in energy through a double junction system (including the voltage sources) due to electrons tunneling through either the first or second junction, Eqs. (4.62) and (4.63). Through a generalization of such analysis for the N-junction system, we may write the change in energy of the jth junction as

$$\Delta E_j^\pm = E\{n_1, n_2, \ldots, n_{j+1}, \ldots, n_N\} - E\{n_1, n_2, \ldots, n_j, \ldots, n_N\}, \qquad (4.97)$$

where the $\pm$ sign refers to the forward or reverse tunneling process across the junctions. Using a "golden rule" approximation [2], the rate of tunneling of electrons back and forth through the jth junction is

$$\Gamma_j^\pm = \frac{2\pi}{\hbar} \sum_{\mathbf{k}_i, \mathbf{k}_f} |T_{if}|^2 f(E_i)[1 - f(E_f)]\delta(E_i - E_f + \Delta E_j^\pm), \qquad (4.98)$$

where i and f refer to the initial and final states in the forward or reverse directions. The delta function now includes the change in the total energy of the multi-junction system due to a single electron tunneling across the jth junction. Again assuming weak energy dependencies for the transfer matrix element and the densities of states, the tunneling rate may be written

$$\Gamma_j^\pm(V) = \frac{1}{e^2 R_{tj}} \int_{E_{cm}}^\infty dE f(E)[1 - f(E + \Delta E_j^\pm)], \qquad (4.99)$$

where R_{tj} is given by (4.95) for junction j. Using the property of the Fermi function

$$f(E)[1 - f(E + \Delta E_j^\pm)] = \frac{f(E) - f(E + \Delta E_j^\pm)}{1 - e^{-\Delta E_j^\pm / k_B T}}, \qquad (4.100)$$

we may perform the integration in (4.99) using (4.94) to obtain

$$\Gamma_j^\pm(V) = \frac{1}{e R_{tj}} \frac{\Delta E_j^\pm / e}{1 - e^{-\Delta E_j^\pm / k_B T}}. \qquad (4.101)$$

The term $(\Delta E_j^\pm / e)/R_{tj}$ looks like the Ohmic current, which normally would flow for an effective bias $\Delta E_j^\pm / e$ across the jth tunnel junction. The total current flowing across the junction is still given by (4.93) as the difference of the left and right tunneling rates. The result (4.101) is central to the theoretical treatment of Coulomb charging in the Orthodox model discussed in the next section. The energetic arguments made in Section 4.2.1 now may be stated more quantitatively. We see that in the limit that $\Delta E_j^\pm$ is positive and much greater than the thermal energy,

$$\Gamma_j^\pm(V) = \frac{1}{e}\left(\frac{\Delta E_j^\pm / e}{R_{tj}}\right), \quad \Delta E_j^\pm \gg k_B T, \qquad (4.102)$$

so that tunneling is thus "allowed." On the other hand, when $\Delta E_j^\pm$ is large and negative, (4.101) shows that tunneling is "forbidden":

$$\Gamma_j^\pm(V) \simeq 0, \quad -\Delta E_j^\pm \gg k_B T. \qquad (4.103)$$

Thus the energetic arguments leading to the qualitative explanation for Coulomb blockade discussed in Section 4.2.1 are strictly valid in the limit that $|\Delta E_j^\pm| \gg k_B T$.

Considering the double junction system of Fig. 4.13 as an example, the change in energy associated with forward and backward tunneling across the second junction was given by (4.62) as

$$\Delta E_2^\pm = \frac{e}{C_{eq}}\left[-\frac{e}{2} \mp (en - V_a C_1)\right]. \qquad (4.104)$$

For zero applied bias and an initially charge-neutral island ($n = 0$), $\Delta E_2^\pm = -e^2/2C_{eq}$. From (4.103), the tunneling current is approximately zero as long as $e^2/2C_{eq} \gg k_B T$, which sets the temperature limits for observing Coulomb blockade. In the limit that the capacitances are large so that the charging energy is small compared to the applied bias, the change in energy is just proportional to the voltage drop across that junction

$$\Delta E_2^\pm \simeq eV_aC_1/C_{eq} = \pm eV_2.$$ (4.105)

The net current across the second junction is given by (4.93) and (4.101) as

$$I(V) = e\left[\Gamma_2^-(V) - \Gamma_2^+(V)\right] = \frac{V_2}{R_{t2}}\left[\frac{1}{1 - e^{eV_1/k_BT}} - \frac{1}{1 - e^{-eV_1/k_BT}}\right] = \frac{V_2}{R_{t2}}.$$ (4.106)

We see that in the limit of negligible charging effects, the simple Ohmic relation of (4.95) is recovered.

Before going on to discuss the kinetic equation governing the charge state of the junctions based on the tunneling model derived above, it is important to note that we have implicitly assumed time scales for the problem such that the charge distribution fully relaxes during the time it takes to tunnel through the junction, which is subsequently assumed to be much shorter than the time between tunnel events. We have not accounted for the coupling of the system to the electromagnetic environment surrounding the circuit, which may alter the nature of the charge relaxation and the energetics of the tunneling electron. Detailed consideration of such environmental effects have been given by Ingold and Nazarov [65] using a model system. The tunneling rates corresponding to Eq. (4.101) represent the limiting case of a low-impedance environment coupling the voltage source to the tunnel junctions.

Equation of motion for charge in single tunnel junction

To begin, we first consider the dynamics of a single tunnel junction with capacitance C biased by an ideal current source with current I shown in Fig. 4.34. The basic approach in the Orthodox model is to characterize the state of the junction in terms of its charge, $Q(t) = CV(t)$. This quantity may be treated as a quantum mechanical variable which is conjugate to the "phase" of the junction

$$\phi = \int_{-\infty}^{t} V(t')\,dt'.$$ (4.107)

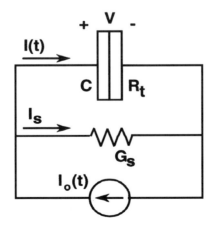

Fig. 4.34. A single tunnel junction of capacitance C and tunnel resistance R_T biased by an ideal current source. G_s is the shunt resistance of the external circuit.

Hence, the charge and the phase are conjugate variables, which ultimately may be quantized for a fully quantum-mechanical behavior [11]. Here, however, we will work with the semiclassical approach utilizing these variables.

In order to study Coulomb blockade, the equation of motion for the probability distribution governing the charge on the junction, $f(Q, t)$, must be derived. In order to do this, it is assumed that we can combine a picture of discrete random tunneling events with the continuous flow of charge associated with metallic conductors and the circuit elements characterizing the driving source and environment. For this framework to be valid, the tunneling time itself is required to be almost instantaneous, and the thermalization time of the metallic junctions themselves is assumed to be much shorter than the time between tunneling events so that we can characterize the distribution functions in the leads by quasi-equilibrium (i.e., Fermi-Dirac) functions. The major point in this approach is that tunneling of an electron from left-to-right or right-to-left changes the charge Q on the junction as $Q + e$ or $Q - e$, respectively. Further, tunneling of electrons within the same electrode is assumed to be uncorrelated.

Averin and Likharev first derived the master equation governing the evolution of the charge probability distribution in small tunnel junctions using a density matrix approach in their pioneering work [2], [43], [66]. Here, we take a simpler semiclassical kinetic equation approach to arrive at the same equation, in which Q is treated as a classical variable. The justification for this latter approach is that most of the assumptions discussed above are analogous to the same assumptions made in going from the equation of motion for the density matrix (the Liouville equation) to the Boltzmann transport equation, the semiclassical equation governing particle transport in solids (as in Chapter 2) [67]. A similar semiclassical approach based on the Kolmogorov master equation has been given by Ben-Jacob et al. [68]. For this approach, we first develop the driving force terms and then turn our attention to the "scattering" terms that represent tunneling of the individual electrons.

In the simple semiclassical picture, we can decompose the system into classical trajectories in phase space, terminated by instantaneous scattering events, in this case tunneling, which instantaneously changes the charge on the tunnel junction by an amount $\pm e$. In the simple current biased circuit of Fig. 4.34, if we neglect for the moment the effect of the shunt conductance, the change in charge between tunneling events after a short time dt is given by

$$dQ(t) = I(t) \, dt, \tag{4.108}$$

where in general $I(t)$ may have some time variation. The charge distribution function at a time $t + dt$ due to this "ballistic" motion becomes $f(Q + I(t) \, dt, t + dt)$. Expanding to first order, we may write this latter expression as

$$f(Q + I(t)dt, t + dt) = f(Q, t) + I(t) \, dt \frac{\partial f}{\partial Q} + dt \frac{\partial f}{\partial t} + \cdots, \tag{4.109}$$

so that the total rate of change of f therefore becomes

$$\frac{Df}{Dt} = \frac{f(Q + Idt, t + dt) - f(Q, t)}{dt} = \frac{\partial f}{\partial t} + I(t) \frac{\partial f}{\partial Q} = \frac{\partial f}{\partial t} \bigg|_{tunn}, \tag{4.110}$$

where it has been assumed that the total change in the distribution is balanced by the change induced by tunneling events.

The nonideality of the current source is accounted for with the shunt conductance, G_s, in parallel with the tunnel junction as shown in Fig. (4.34) as considered by Averin and Likharev. We need to be a little more careful on the left side of (4.110) in that we need to account for the existence of the shunt conductance (we want to write everything in terms of the *external* current) and also for fluctuations in the charge on the capacitor that arise from thermal fluctuations in the shunt conductance. Let $I_o(t)$ be the current provided by the current source. A certain fraction of this current is shunted through the conductance, I_s, which depends on the voltage drop across the tunnel junction. Therefore

$$I_s = G_s V = G_s Q/C, \qquad I(t) = I_o(t) - G_s Q/C, \qquad (4.111)$$

where $I(t)$ is, as before, the current through the tunnel junction. In addition, we need to account for the contribution due to Nyquist noise in the shunt conductance. As is well known, the thermal noise in a conductor contributes a fluctuating current given by $\langle i_{th}^2 \rangle = 4k_B T G_s B$, where B is the bandwidth of the noise [32]. This contribution to the junction charge distribution function gives an additional term that we have to consider. Therefore, the distribution function at time $t + dt$ is $f(Q + (I_o(t) - G_s Q/C) dt, t + dt)$, which may be expanded to higher order to give

$$f(Q + (I_o(t) - I_s) dt, t + dt) = f(Q, t) + [I_o - I_s] dt \frac{\partial f}{\partial Q} - f dt \frac{\partial I_s}{\partial Q} + dt \frac{\partial f}{\partial t}$$

$$+ \frac{1}{2} [I_o - I_s]^2 dt^2 \frac{\partial^2 f}{\partial Q^2} - dt^2 \frac{\partial I_s}{\partial Q} \frac{\partial f}{\partial Q} + \cdots.$$

$$(4.112)$$

Now, in general, all the second-order terms vanish in the limit $dt \to 0$, except the term involving the square of the shunt current. In this case, we make the connections $I_s^2 \to \langle i_{th}^2 \rangle$ and use Nyquist theorem to relate the bandwidth to the sample time interval $B dt \to 1/2$. Following the procedure leading to (4.110), this then leads to the master equation [42]

$$\frac{\partial f}{\partial t} = -I_o(t) \frac{\partial f}{\partial Q} + \frac{G_s}{C} \frac{\partial (Q f)}{\partial Q} + k_B T G_s \frac{\partial^2 f}{\partial Q^2} + \left. \frac{\partial f}{\partial t} \right|_{tunn} \qquad (4.113)$$

where the last term is given below. This equation is identical to the same equation derived by Averin and Likharev. It should be pointed out that some quantum-mechanical features are not recovered from this semiclassical derivation although they do arise in the full derivation. In particular, a fuller treatment also defines the limits of applicability of Eq. (4.113) to the limit of $R_t \gg h/2e^2$, which in the hand waving argument given in Section 4.2.1 arises from consideration of the fact that the leakage is actually due to tunneling!

To derive the rate of change due to tunneling, we look at a detailed balance between in-scattering and out-scattering events that modify the charge Q. To do this, consider a hypothetical ensemble of N tunnel junctions, $n(Q)$ of which are in charge state Q. $\Gamma^+(Q) dt$ corresponds to the number of systems that change from state Q to $Q + e$ due to tunneling of an electron from left to right in Fig. 4.34. Likewise, $\Gamma^-(Q) dt$ is the number of systems that go from Q to $Q - e$. Thus, within a small time interval dt, the change in the number of systems in state Q is given by

$$n(Q, t + dt) = n(Q, t) - \Gamma^+(Q) n(Q, t) dt - \Gamma^-(Q) n(Q, t) dt$$

$$+ \Gamma^+(Q - e) n(Q - e, t) dt + \Gamma^-(Q + e) n(Q + e, t) dt.$$

$$(4.114)$$

Letting $dt \to 0$ and $f(Q, t) = n(Q, t)/N$, the rate of change due to tunneling is written

$$\left.\frac{\partial f}{\partial t}\right|_{tunn} = \underbrace{-\Gamma^+(Q)f(Q, t) - \Gamma^-(Q)f(Q, t)}_{Outscattering}$$

$$+ \underbrace{\Gamma^+(Q - e)f(Q - e, t) + \Gamma^-(Q + e)f(Q + e, t)}_{inscattering}. \qquad (4.115)$$

In each case, it must be recalled that it is assumed that the time scale involved for the above master equation is such that both the tunneling time through the insulator and the equilibration time (such as plasmon decay) within the electrodes are small, so that these events can be assumed to occur instantaneously.

The solution of the master equation (4.113) for $f(Q, t)$ with the in- and out-scattering terms (4.115) allows one to calculate the transport characteristics of the junction. For example, the time-dependent voltage across the junction is given as

$$V(t) = \frac{1}{C} \int dQ\, Qf(Q, t) = \frac{\bar{Q}(t)}{C}, \qquad (4.116)$$

assuming $f(Q, t)$ is properly normalized. The d.c. I-V characteristics may be calculated using (4.116) from the stationary solutions of (4.113), $f(Q)$, for $\partial f/\partial t = 0$.

In the previous section, we showed that the tunneling rate itself depends on the change in energy of the system before and after charge transfer, $\Delta E^\pm(Q)$, given by (4.101). The change in electrostatic energy due to forward and backward tunneling for the current-biased single junction considered in Fig 4.34 is simply the change in electrostatic stored energy (the constant current source performs negligible work in the incremental tunneling time interval dt)

$$\Delta E^\pm = \frac{Q^2}{2C} - \frac{(Q \pm e)^2}{2C} = -\frac{e}{C}\left(\frac{e}{2} \pm Q\right). \qquad (4.117)$$

One observation from this equation is that for $|Q| < e/2$, the energy change associated with tunneling is negative which implies from the tunneling rate (4.101) at zero temperature that tunneling is prohibited. Therefore, even in the single junction case, a region of Coulomb blockade is expected in the range of voltages $-e/2C < V < e/2C$ for the d.c. current-voltage characteristics, similar to the experimental results shown in Fig. 4.12 for the double junction case.

An interesting prediction of the master equation (4.113) is that of periodic SET oscillations in the voltage across the junction. To understand this phenomena, consider the time evolution of the junction charge. In the range $-e/2 < Q < e/2$, tunneling is suppressed, and the charge varies linearly as $Q = Q_o + It$, where I is the current (ignoring the shunt resistance for the moment) and t is the time from the last tunneling event. When the charge reaches $Q = e/2$, the sign of (4.117) changes and tunneling is allowed. The first electron to tunnel then drives the junction charge back to $Q = Q_o = -e/2$, and the cycle repeats itself. The Coulomb blockade results in a correlated tunneling of single electrons with a frequency cycle of $f = e/I$, similar to the correlated tunneling in the single electron turnstile discussed earlier.

To make this slightly more quantitative, if one looks at the regime $-e/2 < Q < e/2$, the tunneling contributions may be neglected, and one just has a diffusion-like equation for the probability distribution, $f(Q, t)$. If we can neglect the shunt current contribution

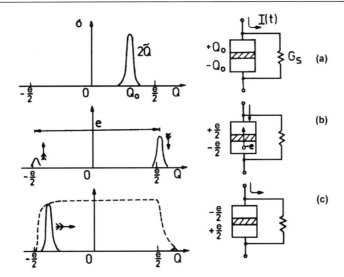

Fig. 4.35. Schematic depiction of the dynamics of the probability density (left column) and the corresponding change in the junction state (right column) in a small current-biased tunnel junction illustrating SET oscillations [after D. V. Averin and K. K. Likharev, J. Low Temp. Phys. **62**, 345 (1986), by permission].

corresponding to the second term on the right side of (4.113), this equation has precisely the form of a forced diffusion equation. For the system initially in a definite charge state $Q = -e/2$ at $t = 0$ (i.e., $f(Q, 0)$ is a delta function at $Q = -e/2$), the solution for subsequent times is

$$f(Q, t) = \frac{1}{\sqrt{4\pi k_B T G_s t}} \exp\left[-\frac{\left(Q + \frac{e}{2} - I_o t\right)^2}{4 k_B T G_s t} \right]. \qquad (4.118)$$

This solution represents a propagating Gaussian pulse centered at $Q = -e/2 + I_o t$ which spreads in time due to the Nyquist noise contribution. Figure 4.35 illustrates the solution to the master equation showing the cyclical motion of the probability density. As the average of the probability density approaches $e/2$, the tunnel rate for $\Gamma^-(Q)$ becomes large and scatters the system back to the $-e/2$ state.

Calculated results for the d.c. I-V characteristics and the time-dependent voltage across a single tunnel junction were reported by Averin and Likharev [43] under the simplifying assumption of zero temperature and $G_s = 0$ as shown in Fig. 4.36. The d.c. I-V characteristic in Fig. 4.36a exhibits a region of Coulomb blockade for small $\overline{V} < e/2C$. SET oscillations are evident in the calculated voltage versus time in Fig. 4.36b for small current values in the Coulomb blockade regime. As the current bias increases such that the average voltage exceeds the critical voltage, the oscillations disappear as illustrated in the successive curves plotted.

Clear experimental evidence of SET oscillations in single junctions has remained elusive to date. The principal problem is the realization of the ideal current source–driven junction illustrated in Fig. 4.34. In nanostructure devices such as the experimental structures discussed in Section 4.2.1, the parasitic capacitances of the leads leading to the junction far exceed that of the junction itself. An external current source charges up the lead capacitance, which tends to behaves as a voltage source with respect to the tunnel junction

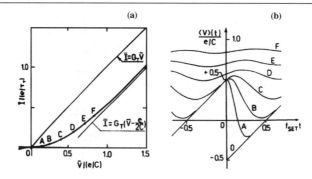

Fig. 4.36. (a) Calculated d.c. I-V characteristics for a single current-biased tunnel junction for $G_s = 0$ and $T = 0$. (b) Time-dependent voltage calculated for the different current bias points shown in part (a). [After D. V. Averin and K. K. Likharev, J. Low Temp. Phys. **62**, 345 (1986), by permission.]

rather than an ideal current source. This argument again illustrates a nontrivial problem with nanostructure systems, that of isolating the behavior of the nanostructure system from its surrounding environment (see for example [65]).

Multiple junctions

We saw earlier that double junction structures show clear evidence of single electron–like phenomena such as the Coulomb staircase and periodic conductance oscillations with a gate bias in the experimental studies of Section 4.2.1. In the double junction, single electron effects are observed with voltage rather than current bias, which circumvents the problems of observing SET oscillations in a single junction.

The master equation approach of the preceding section is easily extended to the case of multiple junctions. The state of an N junction array is characterized by the set of junction charges, $\{Q_1, Q_2, \ldots, Q_N\}$. Due to the capacitance-voltage relationships of each of these junctions, we could alternately characterize the state of such a system by the excess number of electrons in each *node* between junctions, $\{n_1, n_2, \ldots, n_{N-1}\}$. The distribution function $f(n_1, n_2, \ldots, n_{N-1})$ is then the multidimensional probability of finding the system in a state characterized by this set of node occupancies. This distribution function evolves continuously under the time-dependent influence of the external sources (voltage or current), as well as through (instantaneous) tunneling events which change the charge state of each node. For the single electron transistor structure of Fig. 4.17, there are two tunnel junctions and one node, driven by a constant voltage source. The following equation then describes the evolution of the one-dimensional distribution function:

$$\frac{\partial f(n, t)}{\partial t} = \sum_{j=1,2} \left\{ \Gamma_j^+(n-1)f(n-1, t) + \Gamma_j^-(n+1)f(n+1, t) \right.$$

$$\left. - \left[\Gamma_j^+(n) + \Gamma_j^-(n) \right] f(n, t) \right\}, \tag{4.119}$$

where j is a sum over the tunnel junctions, the tunneling rates are given by (4.101), and the change in energy during tunneling is given by (4.74) and (4.75). The right side of (4.119) is nothing more than the balance of in-scattering and out-scattering terms to the distribution function that we wrote in (4.115), now generalized to include both junctions. The effects

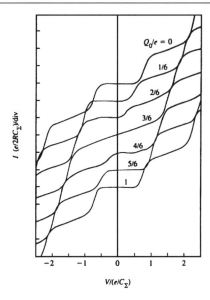

Fig. 4.37. D.c. I-V characteristics of a single electron transistor circuit for several values of gate charge $(R_1 \gg R_2, C_1 = 2C_2)$ [after K. Likharev, IBM J. Res. Develop. **32**, 144 (1988), by permission.]

of the junction and gate biases are contained in the energy changes, $\Delta E_j^{\pm}$, given by (4.74) and (4.75), which in turn determine the tunneling rates $\Gamma_j^{\pm}(n)$. Once $f(n)$ is determined, the average current through junction j is written in terms of the net rate of forward and backward tunneling:

$$\langle I_j(t) \rangle = e \sum_n f(n, t) \left[\Gamma_j^-(n) - \Gamma_j^+(n) \right], \tag{4.120}$$

where the sum is over all the possible node occupancies, n.

Figure 4.37 shows the calculated d.c. I-V characteristics due to Likharev [66] of a similar SET circuit to that of Fig. 4.17 at a temperature $k_B T = 0.1e^2/2C_{eq}$. The circuit topology used there is slightly different, such that the parameter Q_o represents the total charge injected into the central node by the gate. For the sake of discussion, we can essentially take this charge to be $C_g V_g'$ as defined in (4.76), which was plotted in the stability diagram of Fig. 4.18. The plotted curves in Fig. 4.37 represent one complete cycle of gate charge in the stability diagram (Fig. 4.18) going from $Q_o = 0$ to e. For this asymmetric example, the calculated I-V shows distinctly the Coulomb staircase discussed earlier, with a region of Coulomb blockade about the origin. As the gate charge approaches $e/2$, the width of the Coulomb blockade region vanishes as qualitatively predicted by the stability diagram, Fig. 4.18, and the conductance becomes nonzero. For small bias voltages, the range of gate voltages where the conductance is nonzero is small, giving rise to conductance peaks as illustrated in Fig. 4.19. Solutions to the master equation for a two-junction system using stochastic methods were also given by Ben-Jacob *et al.* [44]. Comparison of this calculation with the Coulomb staircase measured in STM studies by Wilkins *et al.* [48] was shown in Fig. 4.12.

4.2.3 Co-tunneling of electrons

In the prior analysis of single electron tunneling, we have considered tunneling to lowest order in perturbation theory. However, in the regime of Coulomb blockade, when the tunnel current is small, higher-order processes may become important. These higher-order processes become increasingly important when the resistances of the tunnel junction begin to approach that of the fundamental resistance, $R_K = e^2/h$, such that quantum fluctuations essentially broaden the energy levels, allowing more channels for charge transfer. In the first-order theory, an electron could not tunnel from the leads to the dot or island due to conservation of energy when biased in the Coulomb blockade regime. However, in a higher-order process, the electron can transfer from the left lead to the right lead (or vice versa) via a virtual state in the island, conserving energy for the entire process even if tunneling into a virtual state does not. The consideration of such higher-order effects is important for the proposed operation of single electron transistors and metrological structures such as the turnstile, since these effects ultimately determine the accuracy of single charge transfer.

The theoretical treatment of higher-order tunneling processes is referred to as the *macroscopic quantum tunneling of charge* (abbreviated *q-mqt*) [69], [70]. This process may be either elastic or inelastic. The elastic process corresponds essentially to the same electron tunneling into and out of the virtual state, and thus is a coherent process. Inelastic tunneling, or *cotunneling*, involves one electron tunneling in from a state below the Fermi energy of the lead into the dot, and a second electron leaving from a different state in the dot into the other lead at an energy above the Fermi energy. This latter process is illustrated schematically in Fig. 4.38. As shown, the process of tunneling of an electron from the right electrode to the left creates an electron-hole excitation in the center electrode, so that the electron appearing on the left does not necessarily have to have the same energy as the initial electron on the right. This extra energy eventually is dissipated through carrier-carrier interactions in the island. Since this process results in the creation of an electron-hole pair (with respect to the Fermi energy of the dot), it is referred to as *inelastic q-mqt*. It does not involve phase coherence as the two electrons involved are different. Since this is essentially a two-electron process, it is more popularly known as "co-tunneling." The cotunneling process usually is dominant in comparison to the elastic q-mqt, except at very small bias voltages and temperatures [70].

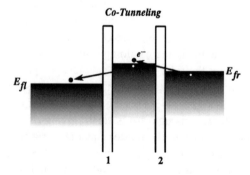

Fig. 4.38. Illustration of the inelastic macroscopic quantum tunneling of charge (q-mqt), or cotunneling.

Following the Fermi golden rule treatment of cotunneling by Averin and Nazarov [70], the matrix element for tunneling via an intermediate state may be written

$$\langle i|M|f\rangle = T^{(1)}T^{(2)}\left(\frac{1}{\Delta E_1} + \frac{1}{\Delta E_2}\right), \tag{4.121}$$

where $T^{(i)}$ represents the tunneling amplitude through barrier i, and ΔE_i is the energy barrier difference including the charging energy between the initial state and the virtual state in the island for tunneling through the ith barrier. These states are different due to the different energies in the dot of the two tunneling electrons. The two terms represents the fact that the process could either occur due first to the electron tunneling out of the island into the left electrode and then filled from the right, or due to the first electron tunneling into the island from the right electrode and then the second one tunneling out. The transition rate is given by the usual expression

$$\Gamma = \frac{2\pi}{\hbar}|\langle i|M|f\rangle|^2\delta(E_i - E_f), \tag{4.122}$$

where E_i and E_f are the initial and final energies of the system after tunneling, which says that the energy lost during the tunneling process by the electron motion from right to left electrode in Fig. 4.38 is equal to that given up in creating the electron-hole excitation in the island. The squares of the tunneling amplitudes are inversely proportional to the respective tunnel resistances, as shown in (4.95). The total rate of cotunneling is found by summing (4.122) over all possible initial and final states for the two processes indicated in Fig. 4.38. The interested reader is referred to [70] for more detail of the general relationship. Under the limiting case that the applied voltage is much less than the charging energy of the island, the I-V relationship resulting from this tunneling rate is given by

$$I_c(V) = \frac{\hbar}{12\pi e^2 R_{t1} R_{t2}}\left(\frac{1}{\left|\Delta E_1^\pm\right|} + \frac{1}{\left|\Delta E_2^\pm\right|}\right)^2 [e^2 V^2 + (2\pi k_B T)^2]V, \tag{4.123}$$

where V is the voltage across the double junction, and R_{ti} is the tunnel resistance of junction i. The energies $\Delta E_i^\pm$ are the change in energies due to forward and backward tunneling across the ith tunnel junction given by (4.74) and (4.75) in the Coulomb blockade regime (i.e., $\Delta E_i^\pm < 0$). This equation shows that the co-tunneling current goes as the inverse of the square of the tunneling resistance due to the second-order nature of the process. It becomes significant when the tunnel resistance approaches the fundamental resistance h/e^2. The current for low temperature has a characteristic power-law dependence, V^3, and a quadratic temperature dependence. In contrast, we may argue from the form of the first-order tunnel rate (4.101) that the current decreases exponentially as the voltage decreases in the Coulomb blockade regime, and has an activated temperature dependence. Thus, while the first-order tunnel rate is usually dominant, in the Coulomb blockade regime the first-order current goes rapidly to zero, and the higher-order process given by (4.123) may dominate.

Experimental evidence for the V^3 dependence of the current measured in the Coulomb blockade regime of double junction structures was reported by Geerligs et al. using metal tunnel junctions such as those shown in Fig. 4.23 [71]. The temperature dependence evident in (4.123) was measured in metal double junctions [72]. Investigation of the dependence of the cotunneling current on the tunnel resistance was measured in a double barrier semiconductor quantum dot structure by the Saclay/CNRS group [73]. In these quantum-point contact structures similar to those discussed earlier, the tunnel resistance is independently controlled by the gate potential, which allowed the observation of Coulomb blockade oscillations from the strong tunneling regime $R_t \gg R_K$ to the regime where the resistance is less than R_K.

Because the q-mqt or cotunneling represents a current that flows in a regime in which to first order it should be suppressed, it is viewed as a parasitic or undesirable effect in proposals for utilizing single electron charging for practical applications such as logic or memory elements. Understanding how to control this effect is important. If one extends the calculation outlined above for the double junction to include multiple junctions, the tunneling current at $T = 0$ is given by [70]

$$I_c(V) = \frac{2\pi e}{\hbar} \left(\prod_{i=1}^{N} \frac{R_K}{4\pi^2 R_{ti}} \right) \frac{S^2}{(2N-1)!} (eV)^{2N-1}, \qquad (4.124)$$

where N is the number of junctions and S is the generalized sum of the energy denominators. For identical capacitances, with no stray or self-capacitance on the islands, this term may be written

$$S = N! \left(\prod_{i=1}^{N-1} |\Delta E_i| \right)^{-1}. \qquad (4.125)$$

For $N = 2$, (4.124) reduces to (4.123) in the limit of $T = 0$. As the number of junctions increases, the power-law dependence becomes stronger, and the magnitude of the current decreases as roughly α^N, where $\alpha = R_K/R_t$ (keeping the tunnel resistances the same for simplicity). For $R_t < R_K$, the current due to inelastic q-mqt is increasingly suppressed as the number of junctions increases. This fact implies that the stability of single electron circuits can be increased by increasing the number of junctions per element.

4.2.4 Theoretical treatment of semiconductor quantum dots

A complete theory of Coulomb blockade and single electron charging in semiconductor structures should take into account the discrete nature of the electronic states due to quantum confinement in the dot, as discussed earlier. Extension of the Orthodox model presented in the preceding sections to include discrete levels in the dot was given by Averin, Korotkov, and Likharev [74]. Beenakker developed a similar kinetic equation model to solve for the nonequilibrium probability distribution governing the occupancy of states in the quantum dot coupled to leads, including the charging energy [50], [75]. Space does not permit a presentation of such approaches, unfortunately. In the linear response regime, conductance formulas are derived that basically give the proper lineshape and peak amplitude with phenomenological choices of the tunneling rates, charging energies, etc. A similar linear response result for the conductance was derived independently by Meir *et al.* [76]. A full nonequilibrium treatment of transport through quantum dots based on the nonequilibrium Green's function method discussed in Chapter 8 was given by Groshev and coworkers [77].

4.3 Coupled dots and quantum molecules

When we begin to talk about coupled dots (or antidots, as discussed in the next section), the system under discussion can vary from two coupled dots to very large arrays of dots. The

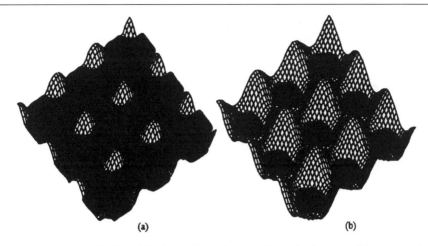

Fig. 4.39. A schematic display of the imposition of a lateral superlattice potential on a quasi-two-dimensional electron gas.

latter are often referred to as *lateral surface superlattices* [78]. The conditions necessary to observe superlattice effects can be easily defined. We recognize that there are several length scales at work here, just as in any mesoscopic system. The important factor is that the electrons must have sufficient phase memory to recognize that the periodic (super-) potential is present. Thus, the phase coherence length must be larger than the superlattice period a. If we are to be able to observe the superlattice structure, then we must also have minigaps that are large compared to the thermal energy $k_B T$. However, at this point there are several possibilities. First, the periodic potential could merely be a perturbation that opens gaps in an otherwise nearly-free electron spectrum characterized by the bulk effective mass. Then, for a weak periodic potential, the nth minigap is proportional to the nth Fourier component of the potential [79]. This minigap is centered around the effective mass energy at the minizone boundary. On the other hand, the periodic potential can localize the electrons within each quantum dot potential minimum. Consider Fig. 4.39. If the carrier density is high (a relatively high value of the Fermi energy) so that the Fermi energy lies well above the saddle-point energies of the two-dimensional periodic potential (only the potential peaks rise out of the Fermi sea in Fig. 4.39a), then the electron gas is mainly that of the two-dimensional system and the potential is a small perturbation. However, if the density is low (a relatively low value of the Fermi energy), the Fermi energy will lie below the saddle-point energies, and the carriers will be trapped in the potential minima, as shown in Fig. 4.39b. In this latter case, the transport through the array of dots is mostly that of single electron transport, as discussed in the previous sections of this chapter. As the Fermi energy increases toward the saddle-point energy, the coupling between the wells becomes much stronger, and the superlattice effects can begin to arise. This results in the formation of minibands and the spreading of the isolated dot energy levels as the coupling increases.

In this section, we will discuss the latter result, in which the dots are basically coupled weakly to each other in the array. In the next section, we will discuss the opposite configuration, in which the perturbation of the periodic potential is often called an *antidot system*.

As discussed in Section 4.2.1, transport through arrays of metal tunnel junctions has been investigated. In such systems, the tunnel resistance is relatively difficult to modify, particularly after fabrication. On the other hand, gated semiconductor structures such as QPCs allow one to tune continually between the isolated dot limit and strongly coupled systems such as the superlattice discussed above. More complicated chains of semiconductor quantum dots may be realized using split-gate structures [80]. Most of the experimental and theoretical work in recent years in coupled quantum dot systems has focused on the near-equilibrium conductance peaks versus gate voltage. As discussed in Section 4.2.1, conductance peaks occur in the Coulomb blockade regime whenever the Fermi energy of the dot is aligned between that of the left and right reservoirs. The period of the peaks with gate bias was simply given by $\Delta V_g = e/C_g$, where C_g is the gate capacitances coupling to the dot. Quantization of the states in the dot adds an additional contribution. In a coupled dot system, this condition must occur simultaneously in both dots. If the gate capacitances coupling the two dots are unequal, then one has two incommensurate ladders of different periods that coincide only infrequently (compared to the period of either dot), giving aperiodic peaks in conductance referred to as "stochastic" or quasi-periodic Coulomb blockade as predicted theoretically [81]. Experimental verification of this effect was reported in coupled QPC dots by Kemerink and Molenkamp [82]. In systems where the gate capacitances are almost identical, the gate periodicity of the conductance oscillations is periodic in e/C_g. However, if the coupling between the dots is increased by opening the QPCs separating the dots, splitting of the conductance peaks occurs [83] as illustrated in Fig. 4.40. In the experiments of Waugh *et al.*, the independently biased split-gate structure shown in Fig. 4.40a was used to form two and three coupled dots. The QPCs separating the dots were used to adjust the degree of coupling between the dots. The parabolas plotted in Figs. 4.40c and 4.40d represent the charging energies versus gate voltage for different occupancies of electrons in the dots for identical dots. With no coupling, the parabolas for unequal occupancy are degenerate. When the parabolas

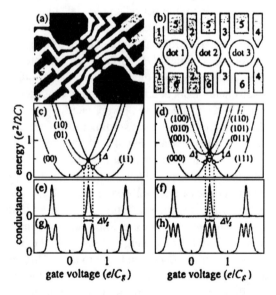

Fig. 4.40. (a) SEM micrograph of three coupled quantum dots in a GaAs/AlGaAs heterostructure. (b) Schematic of the dot structure. (c) Double-dot and (d) triple-dot charging energy versus gate voltage for the indicated numbers $(N_1, N_2, \ldots)$ of electrons on each dot for identical dots. (e) and (f) are the schematic conductance versus gate voltage without coupling. (g) and (h) are the schematic conductance with coupling in which peak splitting occurs due to the splitting of the degenerate dot energies [after F. R. Waugh *et al.*, Phys. Rev. Lett. **75**, 705 (1995), by permission.]

cross, the energy for having either 1 or 0 electrons in each of the dots is the same, and the system can alternate between these states, giving rise to tunneling and a conductance peak as discussed in Section 4.2.1. Coupling between the dots gives rise to a splitting in the degenerate energies, Δ, of the (10) and (01) states in the double dot case, lowering one with respect to the crossing point. Therefore, tunneling may occur between the (00) and (01) states or the (11) and (01) states, giving two conductance peaks separated in voltage by ΔV_s. A similar argument holds for three dots as shown in Figs. 4.40d, f, and h. Experimentally, such behavior was observed with single peaks evolving into two or three peaks for double and triple dots. In the extreme limit that the overlap of dots was strong, it was further observed that the splitting evolved into single peaks characteristic of one large dot molecule.

4.4 Transport in antidot systems

The antidot system is considered to be the case in which the carrier density is relatively high, so that the Fermi energy lies above the saddle-point energy of the dot potential. In this case, the transport is nearly that of the quasi-two-dimensional gas, as shown in Fig. 4.39a. A major complication arises, however, in that the periodic potential antidots, which protrude through the Fermi sea, cannot be treated simply as isolated scattering centers. Rather, it is important to note that the basic region between four antidot potentials (Fig. 4.41) forms what is called a Sinai billiard [84]. The classical trajectories that strike the various potential barriers are chaotic in nature, and this chaos governs the transport by providing new resonances in the magnetotransport (see the review by Fleischmann *et al.* [85]). In this section, we will review the experimental and theoretical understanding of the transport through these antidot systems, primarily for the two-dimensional periodic potential.

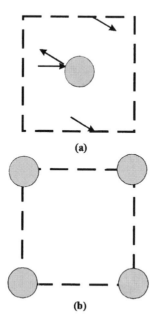

Fig. 4.41. The Sinai billiard is an antidot superlattice. (a) The unit cell, centered on one antidot, with typical trajectories that are periodic in the cell. (b) The superlattice cell centered on the positions between the antidots.

4.4.1 The low-magnetic field regime

Some years ago, a new type of periodic conductance oscillation was observed in the magnetotransport of electrons in a quasi-two-dimensional electron gas that was subjected to an additional linear superlattice (one-dimensional potential) along the surface [86]. The periodic modulation generally was thought to perturb the Landau levels [87], [88], although it might not be expected that these are well formed at the low magnetic fields used (<1 T). However, the discussion of Section 2.5 shows that the electrostatic energy levels in a quasi-one-dimensional quantum wire move smoothly into the Landau levels as the magnetic field is increased, so the argument is only one of the size of the thermal broadening of the levels.

In the case of the two-dimensional (grating) potentials, the effects are modified somewhat [89]. The oscillations in one dimension arise from commensurability of the Landau orbits with the width of the "wire" induced by the potential. In two dimensions, however, this commensurability condition becomes much more stringent. As we discussed earlier in this chapter, the magnetic levels of the quantum dots that now exist are derived from the electrostatic levels, but are split into distinct magnetic levels separated by $\hbar\omega_c$. The natural extension of the Weiss experiments [86] to two dimensions, in which the periodicity of the potential is $a = 365$ nm, is shown in Fig. 4.42 (actually both one-dimensional and two-dimensional cases are shown). The oscillations that arise in the case of the grid potential are considerably weaker, and maxima are found in the longitudinal conductivity where minima are found in the quantum wire (one-dimensional) case. However, the arrows in the figure correspond to resonances predicted by the commensurability condition. It is presumed that the periodic potential actually destroys the band conductivity oscillations seen in the grating case. If the collisional broadening for the two-dimensional case is small compared to the gaps in the spectrum (or the superlattice-induced broadening is small compared to the gaps in the quantum dot spectrum), the transport contributions will be greatly suppressed compared to the one-dimensional case.

The argument for commensurability is based upon a simple perturbation treatment of the transport. It is clear that these new oscillations, the Weiss oscillations, are different from the Shubnikov–de Haas oscillations and have a completely different magnetic field behavior. The period can be found from relatively simple arguments [88] by considering the energy in the nth Landau level as

$$E' = \left(n + \frac{1}{2}\right)\hbar\omega_c + \langle n|V(x)|n\rangle, \tag{4.126}$$

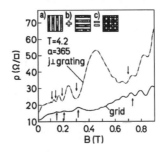

Fig. 4.42. Magnetoresistance in a grating (current normal to the grating) and grid potential [after D. Weiss in *Localization and Confinement of Electrons in Semiconductors* (Springer-Verlag, Heidelberg 1990), p. 247, by permission.]

where the last term is the perturbation of the periodic potential. This term can be calculated by averaging the periodic potential over the extent of the electron's motion about the guiding center x_0. The cyclotron orbit has a radius, which in the quantum limit is given by $\sqrt{(2n + 1)\hbar/eB}$. This is used to calculate the average of the second term in (4.126) with a classical average [88], [90]:

$$\langle n|V(x)|n\rangle \sim \int_{x-x_c}^{x+x_c} \frac{V}{\pi y} \cos\left(\frac{2\pi x}{a}\right) dx = V J_0\left(\frac{2\pi x_c}{a}\right) \cos\left(\frac{2\pi x_c}{a}\right). \qquad (4.127)$$

In this latter expression, J_0 is a Bessel function. This Bessel function can be expanded in its limiting form, as the results are for a rather high Landau index at these low magnetic fields, so that

$$J_0\left(\frac{2\pi x_c}{a}\right) \sim \frac{1}{\pi} \cos\left(\frac{2\pi x_c}{a} - \frac{\pi}{4}\right) \qquad (4.128)$$

so that the resulting current, which is proportional to the square of the product of the two cosine terms, is given by

$$\cos^2\left(\frac{2\pi(x_0 \pm x_c)}{a} - \frac{\pi}{4}\right). \qquad (4.129)$$

The rapid variation with the magnetic field is produced primarily by the x_0 term rather than the x_c term. Using the positive sign on the last term, this leads to extrema when

$$\frac{1}{B} = \frac{ea}{2\hbar k_F}\left(\nu - \frac{1}{4}\right), \qquad (4.130)$$

where ν is an integer. In essence, this leads to the condition of $2R_c = (\nu - 0.25)a$ for the extrema. The minima of the longitudinal magnetoresistance are given by (4.130), while the maxima are usually found with the numerical factor of -0.25 replaced by $+0.17$ in the perpendicular transport [91]. In the two-dimensional case, it may be noted that the roles of maxima and minima are reversed, a result that is thought to be due to the tensor behavior of the conductivity in the magnetic field.

Slightly different results have been obtained by Alves *et al.* [92]. These authors created a two-dimensional array of dots in a similar manner to that of Weiss; that is, an antidot gate structure is fabricated by evaporating a nichrome gate metal over PMMA in which an array of circular holes has been created. This leaves an array of circular metal gates, isolated from one another but creating the superlattice potential by varying the surface potential itself. The period of these dots was 145 nm. In Fig. 4.43, the low-field magnetoresistance is shown for both a grating (one-dimensional potential) and a grid (two-dimensional periodic potential). At fields below 0.4 T, the expected series of new oscillations, which are periodic in $1/B$, appear. Contrary to the above results of Weiss, however, the amplitudes here are comparable for both samples. For fields above 0.4 T, the normal Shubnikov–de Haas oscillations begin to appear. From analysis of the oscillations, these authors are able to determine the period in $1/B$, which is related to the density through the magnetic radius at the Fermi energy. By illuminating the sample and using the persistent photoconductivity, the sheet density can be varied. In Fig. 4.44, the relationship between the density and the inverse of the magnetic periodic of the oscillation (B_f) deduced from the low magnetic field series of oscillations is plotted. The solid curves are given by

$$B_f = 2\sqrt{2\pi n_s} \frac{\hbar e}{a}, \qquad (4.131)$$

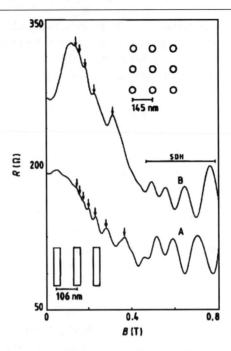

Fig. 4.43. Resistance versus magnetic field for a two-dimensional electron gas subjected to a weak periodic potential. Curve A is for a grating, and curve B is for a grid potential [after E. S. Alves *et al.*, J. Phys. Cond. Matter **1**, 8257 (1989), by permission].

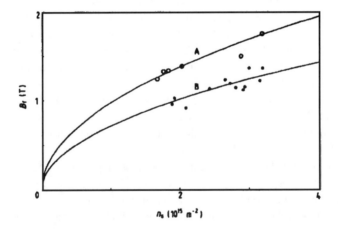

Fig. 4.44. Inverse period of the low-field 1/B oscillations, plotted against electron density. Curves A and B correspond to the grating and grid potentials, respectively [after E. S. Alves *et al.*, J. Phys. Cond. Matter **1**, 8257 (1989), by permission].

which is obtained from (4.130), with a set equal to the grating period of 106 nm in Fig. 4.44a and the grid period of 145 nm in Fig. 4.44b. Clearly, within the experimental error in the determination of the period of the oscillations, the relationship between this period and the sheet density is the same for both one-dimensional and two-dimensional periodic potentials. In both cases, the so-called Weiss oscillations are confirmed to arise from a commensurability condition between the semiclassical cyclotron orbit diameter and the period of the superlattice.

In other results, these latter authors indicate that by varying the gate voltage, the strength of the perturbation caused by the potential may be increased to suppress the formation of the Landau levels and push the onset of the Shubnikov–de Haas oscillations to higher magnetic fields. They do not comment on this further, but it may be expected from their work on the one-dimensional grating potential that a stronger potential will actually weaken, and eliminate, the Weiss oscillations by deepening the potential wells and converting the structure to a dot superlattice rather than an antidot superlattice. In one dimension, the cyclotron orbit is broken up by a strong superlattice potential, and the motion is confined mostly to a single quantum wire. The same effect in a two-dimensional potential will confine the electrons mostly to the dots themselves. The problem in this is the slow variation of the electrostatic potentials. In order to avoid these localization effects in the stronger potential, Weiss et al. [93] set out to create the antidot potentials in a *hard-wall* manner. In this case, the periodic potential is created by etching a series of holes completely through the two-dimensional electron gas region, as shown in the inset to Fig. 4.45. By comparing the magnetotransport in the unpatterned regions with the patterned regions, the effect of the superlattice potential can be isolated and studied. The dots are fabricated with a period of 300 nm, and the transport mean free path is in the range of 4.4–9.6 microns, so that $l/a \gg 1$. The imposed array of antidots dramatically affects the magnetotransport. First, the Shubnikov–de Haas oscillations (Fig. 4.45a) is shifted slightly to *higher* density. Secondly, a series of low-field anomalies appears in the transport, and it is this latter effect that is the result of the antidot potential. In Fig. 4.46, the low-field anomalies are plotted for three different structures. The arrows mark the magnetic field for which the cyclotron orbit is

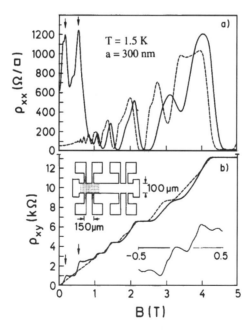

Fig. 4.45. (a) Magnetoresistance and (b) Hall resistance in patterned (solid line) and unpatterned (dashed line) sample segments. The arrows mark magnetic field positions where $R_c/a \approx 0.5$ and 1.5 [after D. Weiss et al., Surf. Sci. **305**, 408 (1994), by permission].

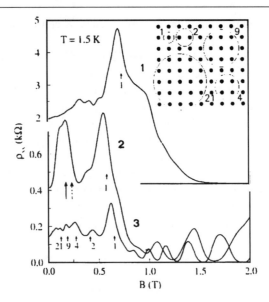

Fig. 4.46. The low-magnetic-field anomalies for three different samples as discussed in the text [after D. Weiss *et al.*, Surf. Sci. **305**, 408 (1994), by permission].

one-half the period (i.e., $r_c = a/2$). The magnetoresistance peaks correspond to trapping orbits enclosing 1, 2, 4, 9, or 21 antidots as indicated in the inset. That is, for these commensurate magnetic fields, a significant number of the carriers are trapped in these orbits and do not contribute to the conductivity. These orbits are sketched in the inset of the latter figures.

The fact that some of the peak positions in the magnetoresistance can be associated with these trapped orbits has stimulated an explanation based upon a modified Drude picture, in which the particles move as if billiards, as discussed above; this is supported by the excellent fit to the experimental data. However, not all the samples fit nicely into this pattern, and the variation of the Hall effect is not well explained [94]. More to the point, the variation of the scattering potential is quite like the Sinai billiard studied by Jalabert *et al.* [84]. Fleischmann *et al.* [95] used a model of the potential and carefully studied the classical billiards in such a structure. The phase space is generally divided into regions in which the motion is either regular (cyclotron-like) or chaotic, and the detailed understanding of this behavior must be found in such studies (we will see this role of chaos again in the next chapter). In Fig. 4.47 (the figure is due to Fleischmann *et al.* [94]), the results of such studies are illustrated. In panel (a), the model potential is shown, and panel (b) illustrates one Poincare section for the reduced phase space (y, v_y). Panels (c) and (d) illustrate typical trajectories for incommensurate (c) and commensurate (d) values of the magnetic field. It is clear that the latter case will result in higher resistance. Such studies have led this group to a new model for Bloch electrons in a magnetic field, which is exact although quite reminiscent of Harper's equation solutions for periodic potentials [96], [97] (see also [98]).

4.4.2 The high-magnetic-field regime

We generally refer to the high-magnetic-field regime as the one in which the cyclotron radius is small compared to the periodicity of the superlattice. For this to occur, we want

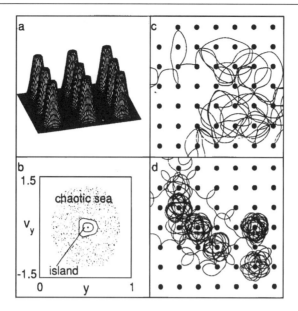

Fig. 4.47. Results of the study of chaotic trajectories in the antidot potential. The various panels are discussed in the text [after D. Weiss *et al.*, Surf. Sci. **305**, 408 (1994), by permission].

the cyclotron orbit at the Fermi surface $k_F l_m^2$ to be smaller than a, which leads to

$$\frac{\Phi}{\Phi_0} = \sqrt{\frac{n_s a^2}{2\pi}} > 1, \tag{4.132}$$

which, for a density of 10^{11} cm^{-2} and $a = 0.16$ mm, leads to $B > 0.35$ T. For higher densities, the required magnetic field increases, and generally a changeover at about 1 T is expected. In this region, the effects of the superlattice compete with those expected from the quantum Hall effect.

Most of the measurements reported in this section were obtained on high-electron-mobility samples, which are fabricated from modulation-doped AlGaAs/GaAs heterostructures. Typically, the carrier density is 4.7×10^{11} cm^{-2}, and the mobility is greater than 100 m^2/Vs. A grid gate of 170 nm periodicity produces a periodic potential, but the combination of the Fermi energy and the magnetic field leads to transport dominantly in the antidot regime. Weiss *et al.* [99] observed an unusual extra minimum in the Shubnikov–de Haas effect, near $\nu = 8/3$, which they attributed to the first onset of van Hove singularities arising from a periodic potential. In Fig. 4.48, the Hall resistance and longitudinal resistance are shown for a sample with a grid-gate superlattice potential applied [101]. Here, the destruction of the Shubnikov–de Haas peak around $\nu = 2$ is accompanied by a series of weak oscillations, not all of which are clearly discernible because they ride on top of another structure (the anomaly in the Hall resistance in the transition from the $\nu = 3$ to the $\nu = 2$ plateau is not real). Extra structure is also beginning to be observed in the next two lower Shubnikov–de Haas peaks as well. The periodicity of the oscillation seems to be about 0.29 T, corresponding to an energy of 0.5 meV.

It is tempting to associate these oscillations with depopulation of the minibands (with the consequent modulation expected from the van Hove singularities). However, this would have to be in the regime where the superlattice potential is a perturbation upon the quantum

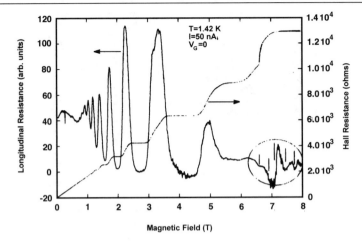

Fig. 4.48. The Hall resistance and magnetoresistance for a sample with a grid-gate superlattice potential applied [after E. Paris *et al.*, J. Phys. Cond. Matter **3**, 6605 (1991)].

dot states. The periodicity in the magnetic field can be assumed to arise from the fact that each quantum dot level, in the presence of the magnetic field and the superlattice potential, is broadened into a miniband. If the number of electrons per miniband is taken as $2/a^2 = 6.9 \times 10^9$ cm^{-2}, there are approximately 50 levels occupied and each miniband has a nominal width of 0.25 meV. This value of the miniband width leads to a $\Delta B = 0.27$ T, which is close to the observed periodicity. This suggests that the oscillations arise from the movement of the Fermi energy through the minibands that have broken up the Landau level quantization. In this regime, the oscillations should be periodic in magnetic field (as opposed to $1/B$), and this is observed in the experiments. There is extra structure also apparent between the $\nu = 4$ and $\nu = 6$ peaks that also has this spacing in magnetic field, but this is quite weak, and the number of oscillations is insufficient to determine whether this is a true periodicity.

Further information can be obtained by varying the gate voltage on the grid gate. In Fig. 4.49, the behavior of the longitudinal magnetoresistance and the Hall resistance is shown for a magnetic field of 3 T. In Fig. 4.49a, it is clear that a new series of oscillations, periodic in the gate voltage, are now seen near the pinchoff point (at –0.32 V), where the Fermi energy presumably drops below the saddle-point energy of the periodic potential, at which point the carriers become strongly localized in the quantum dots. As the negative bias is increased, successive Hall plateaus are seen in the Hall resistance corresponding to a transition from the $\nu = 3$ to the $\nu = 2$ plateau (the former is not resolved in the zero gate voltage case), consistent with a reduction of sheet density with gate voltage. The region between these two plateaus is expanded in Fig. 4.49b to show that the oscillations are also present in this regime, with essentially the same periodicity. These oscillations occur whenever the Fermi energy lies within a given Landau level and are not well formed when the Fermi energy traverses the gap between the Landau levels (corresponding to the Hall plateaus). This reinforces the view that the oscillations are connected with the quantum dot–derived minibands, as these levels coalesce into the Landau levels. While the oscillations can be seen within the edges of the plateaus, they are heavily damped when the Fermi energy is within the localized states between the Landau levels. Of course, the quantum dots spread the Landau levels into a much broader energy range which presumably

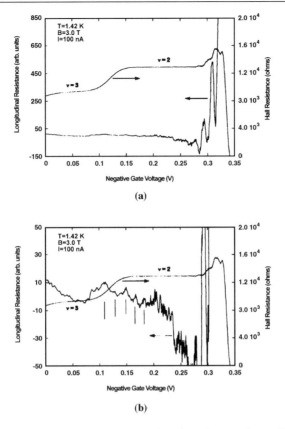

Fig. 4.49. (a) Hall and longitudinal resistances as a function of gate voltage. (b) Expansion of the signals showing the oscillations. [After D. K. Ferry, Prog. Quantum Electron. **16**, 251 (1992).]

allows the observation of the minibands. Moreover, as the Fermi energy moves to lower values, the periodic potential is less well screened, the coupling becomes stronger and the oscillations increase in amplitude, but not in period. The regularity of the oscillations, and their reproducibility, exclude as a possible explanation a parasitic charging of the grid gate. The fact that the periodicity is not a function of the strength of the periodic potential further suggests that they are derived from the quantum dot levels themselves, with the minibands forming from these levels. The oscillations occur for all values of the magnetic field (in the high field limit) near the pinchoff of the channel, and the period of the oscillations does not seem to be a strong function of the magnetic field [100]. Over the range of gate voltages and magnetic fields for which these oscillations are observed, the periodicity is almost constant with a value of 15.7 ± 1.0 meV. Using the density variation with the gate voltage (determined by the Hall effect near $B = 0$), this corresponds to a density variation of 2.12 ± 0.15 electrons per unit cell of the superlattice per oscillation, which clearly indicates that the oscillations arise from the passage of the Fermi energy through the minibands.

Toriumi *et al.* [102] have observed structure like the above in a similar structure fabricated by X-ray lithography (200 nm period), and observe the oscillations down to zero magnetic field. Their oscillations, however, are observed only quite near threshold, with a period of about 0.1 V and an amplitude that is attenuated by the magnetic field. They attribute these oscillations to resonant tunneling through the superlattice potential, with the magnetic field attenuation assumed to be due to reduction of the off-resonant tunneling contributions to the current. This would involve tunneling to bound states within the quantum dots, an observation that also seems to be similar to that observed by Hansen *et al.* [103], in which charging into fractional quantum Hall states in quantum dots containing very few electrons (1–40). In these latter structures, the capacitance spectra have clear minima at gate voltages and magnetic fields for which the filling factors are 1/3 and 2/3.

Bibliography

[1] *Granular Nanoelectronics*, eds. by D. K. Ferry, J. R. Barker, and C. Jacoboni, NATO ASI Series B **251** (Plenum Press, New York, 1991).

[2] D. V. Averin and K. K. Likharev, "Single Electronics: A Correlated Transfer of Single Electrons and Cooper Pairs in Systems of Small Tunnel Junctions," in *Mesoscopic Phenomena in Solids*, eds. B. L. Altshuler, P. A. Lee and R. A. Webb (Elsevier Science Publishers, Oxford), pp. 173–271 (1991).

[3] *Single Charge Tunneling, Coulomb Blockade Phenomena in Nanostructures*, eds. H. Grabert and Michel H. Devoret, NATO ASI Series B **294** (Plenum Press, New York, 1992).

[4] M. A. Kastner, Rev. Mod. Phys. **64**, 849 (1992).

[5] *Quantum Transport in Ultrasmall Devices*, eds. D. K. Ferry, H. L. Grubin, C. Jacoboni, and A. Jauho, NATO ASI Series B **342** (Plenum Press, New York, 1995).

[6] L. P. Kouwenhoven and P. L. McEuen, "Single Electron Transport Through a Quantum Dot," in *Nano-Science and Technology*, ed. G. Timp (AIP Press, New York, 1996).

[7] V. Fock, Z. Phys. **47**, 446 (1928).

[8] C. G. Darwin, Proc. Cambridge Philos. Soc. **27**, 86 (1930).

[9] R. B. Dingle, Proc. Roy. Soc. London A **211**, 500 (1952).

[10] U. Rössler, in *Quantum Coherence in Mesoscopic Systems*, eds. B. Kramer (Plenum Press, New York, 1991), pp. 45–62.

[11] D. K. Ferry, *Quantum Mechanics: An Introduction for Device Physicists and Electrical Engineers* (IOP Publishing Ltd., Bristol, UK, 1995).

[12] A. Kumar, S. E. Laux, and F. Stern, Phys. Rev. B **42**, 5166 (1990).

[13] W. Hansen, T. P. Smith, K. Y. Lee, J. A. Brum, C. M. Knoedler, J. M. Hong, and D. P. Kern, Phys. Rev. Lett. **62**, 2168 (1989).

[14] G. W. Bryant, Phys. Rev. Lett. **59**, 1140 (1987).

[15] P. A. Maksym and T. Chakraborty, Phys. Rev. Lett. **65**, 108 (1990).

[16] U. Merkt, J. Huser and M. Wagner, Phys. Rev. B **43**, 7320 (1991).

[17] D. Pfannkuche, V. Gudmundsson, and P. A. Maksym, Phys. Rev. B **47**, 2244 (1993).

[18] N. F. Johnson and M. C. Payne, Phys. Rev. Lett. **67**, 1157 (1991).

[19] M. A. Reed, J. N. Randall, R. J. Aggarwal, R. J. Matyi, T. M. Moore, and A. E. Wetsel, Phys. Rev. Lett. **60**, 535 (1988).

[20] Ch. Sikorski and U. Merkt, Phys. Rev. Lett. **62**, 2164 (1989); U. Merkt, Ch. Sikorski, and J. Alsmeier, in *Spectroscopy of Semiconductor Microstructures*, eds. G. Fasol, A. Fasolino and P. Lugli, (Plenum Press, New York, 1989), pp. 89–114.

[21] T. Demel, D. Heitmann, P. Grambow, and K. Ploog, Phys. Rev. Lett. **64**, 788 (1990); D. Heitmann, in *Physics of Nanostructures,* eds. J. H. Davies and A. R. Long (Institute of Physics Publishing, Ltd., London, 1992), pp. 229–256.

[22] A. Lorke, J. P. Kotthaus, and K. Ploog, Phys. Rev. Lett. **64**, 2559 (1990).

[23] F. M. Peeters, Phys. Rev. B **42**, 1486 (1990).

[24] P. L. McEuen, E. B. Foxman, U. Meirav, M. A. Kastner, Y. Meir, and N. S. Wingreen, Phys. Rev. Lett. **66**, 1926 (1991).

[25] R. C. Ashoori, H. L. Störmer, J. S. Weiner, L. N. Pfeiffer, K. W. Baldwin, and K. W. West, Phys. Rev. Lett. **71**, 613 (1993).

[26] P. Hawrylak, Phys. Rev. Lett. **71**, 3347 (1993).

[27] L. E. Brus, Appl. Phys. **A53**, 465 (1991).

[28] A. D. Yoffe, Adv. Phys. **42**, 173 (1993).

[29] D. Leonard, M. Krishnamurthy, C. M. Reaves, S. P. Denbaars, and P. M. Petroff, Appl. Phys. Lett. **63**, 3203 (1993).

[30] J. M. Moison, F. Houzay, F. Barthe, and L. Leprince, Appl. Phys. Lett. **64**, 196 (1994).

[31] A. Madhukar, Q. Xie, P. Chen, and A. Konkar, Appl. Phys. Lett. **64**, 2727 (1994).

[32] A. Van der Ziel, *Noise in Solid State Devices and Circuits* (Wiley, New York, 1986).

[33] D. K. Cheng, *Field and Wave Electromagnetics*, 2nd edition (Addison-Wesley Publishing, Reading, Massachusetts, 1989), pp. 172–174.

[34] J. D. Jackson, *Classical Electrodynamics*, 2nd edition (John Wiley and Sons, New York, 1975), p. 133.

[35] M. Macucci, K. Hess, and G. J. Iafrate, Phys. Rev. B **48**, 17354 (1993).

[36] C. J. Gorter, Physica **17**, 777 (1951).

[37] E. Darmois, J. Phys. Radium **17**, 210 (1956).

[38] C. A. Neugebauer and M. B. Webb, J. Appl. Phys. **33**, 74 (1962).

[39] I. Giaever and H. R. Zeller, Phys. Rev. Lett. **20**, 1504 (1968).

[40] J. Lambe and R. C. Jaklevic, Phys. Rev. Lett. **22**, 1371 (1969).

[41] R. E. Cavicchi and R. H. Silsbee, Phys. Rev. Lett. **52**, 1453 (1984); Phys. Rev. B **37**, 706 (1987).

[42] I. O. Kulik and R. I. Shekhter, Sov. Phys.-JETP **41**, 308 (1975).

[43] D. V. Averin and K. K. Likharev, in the *Proceedings of the Third International Conference on Superconducting Quantum Devices (SQUID)*, Berlin, 1985, eds. H.-D. Hahlbohm and H. Lubbig (W. de Gruyter, Berlin, 1985), p. 197; D. V. Averin and K. K. Likharev, J. Low Temp. Phys. **62**, 345 (1986).

[44] E. Ben-Jacob, Y. Gefen, K. Mullen, and Z. Schuss, in the *Proceedings of the Third International Conference on Superconducting Quantum Devices (SQUID)*, Berlin, 1985, eds. H.-D. Hahlbohm and H. Lubbig (W. de Gruyter, Berlin, 1985), p. 203; E. Ben-Jacob, D. J. Bergman, B. J. Matkowsky, and Z. Schuss, Phys. Rev. B **34**, 1572 (1986).

[45] L. S. Kuz'min and K. K. Likharev, JETP Lett. **45**, 495 (1987).

[46] J. B. Barner and S. T. Ruggiero, Phys. Rev. Lett. **59**, 807 (1987).

[47] P. J. M. van Bentum, R. T. M. Smokers, and H. van Kempen, Phys. Rev. Lett. **60**, 2543 (1988).

[48] R. Wilkins, E. Ben-Jacob, and R. C. Jaklevic, Phys. Rev. Lett. **63**, 801 (1989).

[49] T. A. Fulton and G. J. Dolan, Phys. Rev. Lett. **59**, 109 (1987).

[50] C. W. J. Beenakker, Phys. Rev. B **44**, 1646 (1991).

[51] D. V. Averin and K. K. Likharev, in *Single Charge Tunneling, Coulomb Blockade Phenomena in Nanostructures*, eds. H. Grabert and Michel H. Devoret, NATO ASI Series B **294** (Plenum Press, New York, 1992), pp. 311–332.

[52] P. Delsing, in *Single Charge Tunneling, Coulomb Blockade Phenomena in Nanostructures*, eds. H. Grabert and Michel H. Devoret, NATO ASI Series B **294** (Plenum Press, New York, 1992), pp. 249–273.

[53] J. E. Mooij and G. Schön, in *Single Charge Tunneling, Coulomb Blockade Phenomena in Nanostructures*, eds. H. Grabert and Michel H. Devoret, NATO ASI Series B **294** (Plenum Press, New York, 1992), pp. 275–310.

[54] L. J. Geerligs, V. F. Anderegg, P. A. M. Holweg, J. E. Mooij, H. Pothier, D. Esteve, C. Urbina, and M. H. Devoret, Phys. Rev. Lett. **64**, 2691 (1990); L. J. Geerligs and J. E. Mooij, in *Granular Nanoelectronics*, eds. D. K. Ferry, J. R. Barker, and C. Jacoboni, NATO ASI Series B **251** (Plenum Press, New York, 1991), pp. 393–412; L. J. Geerligs, in *Physics of Nanostructures*, eds. J. H. Davies and A. R. Long (Institute of Physics Publishings, Ltd., London, 1992), pp. 171–204.

[55] H. Pothier, P. Lafarge, C. Urbina, D. Esteve, and M. H. Devoret, Europhysics Lett. **17**, 249 (1992); E. Esteve, in *Single Charge Tunneling, Coulomb Blockade Phenomena in Nanostructures*, eds. H. Grabert and Michel H. Devoret, NATO ASI Series B **294** (Plenum Press, New York, 1992), pp. 109–137.

[56] J. H. F. Scott-Thomas, S. B. Field, M. A. Kastner, H. I. Smith, and D. A. Antoniadis, Phys. Rev. Lett. **62**, 583 (1989).

[57] L. I. Glazman and R. I. Shekhter, J. Phys.: Condens. Matter **1**, 5811 (1989).

[58] H. van Houten and C. W. J. Beenakker, Phys. Rev. Lett. **63**, 1893 (1989).

[59] U. Meirav, M. A. Kastner, and S. J. Wind, Phys. Rev. Lett. **65**, 771 (1990); M. A. Kastner, Rev. Mod. Phys. **64**, 849 (1992).

[60] L. P. Kouwenhoven, C. T. Johnson, N. C. van der Vaart, A. van der Enden, C. J. P. M. Harmans, and C. T. Foxon, Z. Phys. B **85**, 381 (1991).

[61] D. V. Averin, A. N. Korotkov and K. K. Likharev, Phys. Rev. B **44**, 6199 (1991).

[62] J. Bardeen, Phys. Rev. Lett. **6**, 57 (1961).

[63] M. N. Cohen, L. M. Falicov and J. C. Phillips, Phys. Rev. Lett. **8**, 316 (1962).

[64] C. B. Duke, *Tunneling in Solids*, Chapter VII, Solid State Physics **10**, eds. F. Seitz, D. Turnbull, and H. Ehrenreich (Academic Press, New York, 1969).

[65] G.-L. Ingold and Yu.V. Nazarov, "Charge Tunneling Rates in Ultrasmall Junctions," in *Single Charge Tunneling, Coulomb Blockade Phenomena in Nanostructures*, eds. H. Grabert and Michel H. Devoret, NATO ASI Series B **294** (Plenum Press, New York, 1992), pp. 21–108.

[66] K. K. Likharev, IBM J. Res. Develop. **32**, 144 (1988).

[67] W. Kohn and J. M. Luttinger, Phys. Rev. **108**, 590 (1957).

[68] E. Ben-Jacob, Y. Gefen, K. Mullen, and Z. Schuss, Phys. Rev. B **37**, 7400 (1988).

[69] D. V. Averin and A. A. Odintsov, Phys. Lett. A **140**, 251 (1989).

[70] D. V. Averin and Yu. V. Nazarov, Phys. Rev. Lett. **65**, 2446 (1990); D. V. Averin and Yu. V. Nazarov, in *Single Charge Tunneling, Coulomb Blockade Phenomena in Nanostructures*, eds. H. Grabert and Michel H. Devoret, NATO ASI Series B **294** (Plenum Press, New York, 1992), pp. 217–247.

[71] L. J. Geerligs, D. V. Averin, and J. E. Mooij, Phys. Rev. Lett. **65**, 3037 (1990).

[72] T. M. Eiles, G. Zimmerli, H. D. Jensen, and J. M. Martinis, Phys. Rev. Lett. **69**, 148 (1992).

[73] D. C. Glattli, C. Pasquier, U. Meirav, F. I. B. Williams, Y. Jin, and B. Etienne, Z. Phys. B **85**, 375 (1991); C. Pasquier, U. Meirav, F. I. B. Williams, D. C. Glattli, Y. Jin, and B. Etienne, Phys. Rev. Lett. **70**, 69 (1993).

[74] D. V. Averin, A. N. Korotkov, and K. K. Likharev, Phys. Rev. B **44**, 6199 (1991).

[75] H. van Houten, C. W. J. Beenakker, and A. A. M. Staring, in *Single Charge Tunneling, Coulomb Blockade Phenomena in Nanostructures*, eds. H. Grabert and Michel H. Devoret, NATO ASI Series B **294** (Plenum Press, New York, 1992), pp. 167–216.

[76] Y. Meir, N. S. Wingreen, and P. A. Lee, Phys. Rev. Lett. **66**, 3048 (1991).

[77] A. Groshev, T. Ivanov, and V. Valtchinov, Phys. Rev. Lett. **66**, 1082 (1991).

[78] D. K. Ferry, Prog. Quantum Electron. **16**, 251 (1992).

[79] N. W. Ashcroft and N. D. Mermin, *Solid State Physics* (Holt, Rinehart, and Winston, New York, 1976).

[80] R. J. Haug, J. M. Hong, and K. Y. Lee, Surf. Sci. **263**, 415 (1992).

[81] I. M. Ruzin, V. Chandrasekhar, E. I. Levin, and L. I. Glazman, Phys. Rev. B **45**, 13469 (1992).

[82] M. Kemerink and L. W. Molenkamp, Appl. Phys. Lett. **65**, 1012 (1994).

[83] F. R. Waugh, M. J. Berry, D. J. Mar, R. M. Westervelt, K. L. Campman, and A. C. Gossard, Phys. Rev. Lett. **75**, 705 (1995).

[84] R. A. Jalabert, H. U. Baranger, and A. D. Stone, Phys. Rev. Lett. **65**, 2442 (1990).

[85] R. Fleischmann, T. Geisel, R. Ketzmerick, and G. Petschel, Physica D **86**, 171 (1995).

[86] D. Weiss, K. von Klitzing, K. Ploog, and G. Weimann, in *High Magnetic Fields in Semiconductors II*, eds. G. Landwehr (Springer-Verlag, Heidelberg, 1989), p. 357.

[87] R. W. Winkler, J. P. Kotthaus, and K. Ploog, Phys. Rev. Lett. **62**, 1177 (1989).

[88] R. Gerhardts, D. Weiss, and K. von Klitzing, Phys. Rev. Lett. **62**, 1173 (1989).

[89] D. Weiss, in *Localization and Confinement of Electrons in Semiconductors*, eds. F. Kuchar, H. Heinrich, and G. Bauer (Springer-Verlag, Heidelberg, 1990), p. 247; Physica Scripta **T35**, 226 (1991).

[90] C. W. J. Beenakker, Phys. Rev. Lett. **62**, 2020 (1989).

[91] D. Weiss, K. von Klitzing, K. Ploog, and G. Weimann, Europhys. Lett. **8**, 179 (1989).

[92] E. S. Alves, P. H. Beton, M. Henini, L. Eaves, P. C. Main, O. H. Hughes, G. A. Toombs, S. P. Beaumont, and C. D. Wilkinson, J. Phys. Cond. Matter **1**, 8257 (1989).

[93] D. Weiss, M. L. Roukes, A. Menschig, P. Grambow, K. von Klitzing, and G. Weimann, Phys. Rev. Lett. **66**, 2790 (1991).

[94] D. Weiss, K. Richter, E. Vasiliadou, and G. Lütjering, Surf. Sci. **305**, 408 (1994).

[95] R. Fleischmann, T. Geisel, and R. Ketzmerick, Phys. Rev. Lett. **68**, 1367 (1992).

[96] R. Fleischmann, T. Geisel, R. Ketzmerick, and G. Petschel, Semicond. Sci. Technol. **9**, 1902 (1994).

[97] R. Fleischmann, T. Geisel, R. Ketzmerick, and G. Petschel, Physica D **86**, 171 (1995).

[98] T. Yamada and D. K. Ferry, Phys. Rev. B **47**, 1444, 6416 (1993), **48**, 8076 (1993).

[99] D. Weiss, K. von Klitzing, K. Ploog, and G. Weiman, Surf. Sci. **229**, 88 (1990).

[100] E. Paris, J. Ma, A. M. Kriman, D. K. Ferry, and E. Barbier, in *Nanostructure and Mesoscopic Systems*, eds. W. P. Kirk and M. A. Reed (Academic Press, New York, 1992), pp. 311–322.

[101] E. Paris, J. Ma, A. M. Kriman, D. K. Ferry, and E. Barbier, J. Phys. Cond. Matter **3**, 6605 (1991).

[102] A. Toriumi, K. Ismail, M. Burkhardt, D. A. Antoniadis, and H. I. Smith, in the *Proc. 20th Intern. Conf. Physics of Semicond.*, Vol. 2, eds. E. M. Anastassakis and J. D. Joanopoulos (World Scientific Press, Singapore, 1990), p. 1313.

[103] W. Hansen, T. P. Smith III, K. Y. Lee, J. M. Hong, and C. M. Knoedler, Appl. Phys. Lett. **56**, 168 (1990).

5

Interference in diffusive transport

The phase interference between two distinct electron (or hole) waves was treated in the Introduction to this book. Whereas the past few chapters dealt largely with the quasi-ballistic transport of these waves through mesoscopic systems, in this chapter we want to begin to treat systems in which the transport is more diffusive than quasi-ballistic. How do we distinguish between these two regimes? Certainly the existence of scattering is possible in both regimes, but we distinguish the *diffusive* regime from the *quasi-ballistic* regime by the level of the scattering processes. In the diffusive regime, we assume that scattering dominates the transport to a level such that there are no "ballistic" trajectories that extend for any significant length within the sample. That is, we assert that $l = v_F \tau \ll L$, where L is any characteristic dimension of the sample. Typically, this means that the material under investigation is characterized by a relatively low mobility, certainly not the mobility of several million that can be obtained in good modulation-doped heterostructures. In a sense, the transport is now considered to be composed of short paths between a relatively large number of impurity scattering centers. Thus, we deal with the smooth diffusion of particles through the mesoscopic system. To be sure, the Landauer formula does not distinguish ballistic from diffusive transport, but its treatment in multimode waveguides is more appropriately considered a ballistic transport. To illustrate the difference, consider the Aharonov-Bohm effect and the presence of weak localization. In the former, the wavefunction (particles) splits into two parts that propagate around opposite sides of a ring "interferometer," as illustrated in Fig. 1.4. The phase difference between the two paths leads to interference effects that can be modulated by a magnetic field enclosed within the ring. On the other hand, in weak localization particles *diffuse* by a sequence of scattering events around a path that closes upon itself. If we examine such a diffusive path, such as that shown in Fig. 1.8, it is clear that the particle could go around the ring in either direction. In the discussion of Chapter 1, it was asserted that the two possible paths were *time-reversed* equivalents (by which it is implied that we cannot distinguish which direction the electron goes around the ring), and that interference between these two paths creates the effect known as "weak localization." It doesn't require much creativity to recognize that the paths in Fig. 1.8 form a ring, as in the case of the Aharonov-Bohm effect. The difference here is that each path goes completely around the ring, rather than just halfway around. Thus, the flux enclosed is twice that of the Aharonov-Bohm effect, and one expects oscillations in flux of $h/2e$ [1] rather than h/e, or a magnetic field periodicity of $h/2eA$ rather than h/eA. In the Aharonov-Bohm effect, we expect the particle to travel ballistically around the ring, whereas in weak localization the particles diffuse around the ring and it is the interference in the two time-reversed paths that is important. (One thinks of this interference as occurring uniformly around the paths

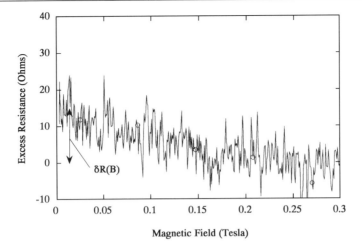

Fig. 5.1. Weak localization in a quantum wire formed between two depletion gates in a GaAs/AlGaAs heterostructure.

rather than at the exit slit, as in the former case.) Another way to think of this is that in the normal Aharonov-Bohm effect, a *single* incident wave splits coherently between the two paths and then interferes upon recombining at the output. On the other hand, the diffusive transport could be a single particle that scatters around the closed path and interferes with itself. Since the particle can go in either direction, these two possibilities lead to the observed dependence upon twice the area. In the first, the oscillations are primarily in the transmission coefficients that appear in the Landauer equation. In the second process, the self-interference is a modulation of the input resistance, or conductance (we will not indulge in the fact that these two methods of interference can be the same in a philosophical sense). In fact, the latter effect is small and leads to a small additional resistance (reduction in conductance), which is the observed weak localization effect.

The presence of weak localization was clearly exhibited in the typical data that were shown in Fig. 1.9 (repeated as Fig. 5.1). One may also note that this curve also exhibits considerable fluctuations in the actual conductance, by which it is meant that the conductance exhibits a smooth part and a fluctuating (in magnetic field or Fermi energy) part. The fluctuating part has come to be known as *universal conductance fluctuations*. The oscillations in conductance that can be seen in this figure are not time-dependent, but rather are steady conductance values that change dramatically with a small change in the chemical potential of the sample. It should be pointed out that, in general, these universal conductance fluctuations are not observed in high-mobility material, where the transport is quasi-ballistic. They are observed only in relatively low-mobility material, in which the transport is primarily diffusive. These oscillations tend to depend critically upon the *exact* configuration of the impurities (which are the dominant scattering process at low temperatures). These two processes, weak localization and universal conductance fluctuations, are not isolated, independent processes. Both arise from the correlation between different particle trajectories for the carriers in a mesoscopic device. Although one calculates the two effects in slightly different manners from the detailed quantum-mechanical formulation of the conductance, they can most effectively be treated in parallel, which is the path we shall adopt here.

In this chapter, we shall develop the basic quantum-mechanical (and some semiclassical) treatments for these two effects, using the simplest zero-temperature formalisms. (We leave the more exact temperature variations to the next chapter.) First, however, the experimental observables and scaling behaviors are discussed along with the simplest semiclassical approaches needed to convey the understanding of the effects. We then turn to the quantum treatments, complete with the complicated behavior of such quantities as diffusions and cooperons. Finally, we treat the presence of fluctuations within quantum dots (as introduced in the last chapter), where these may arise from universal conductance fluctuations in diffusive dots or chaotic behavior within quasi-ballistic dots.

5.1 Weak localization

The presence of weak localization results in a reduction in the actual conductivity of the sample, that is, weak localization is a negative contribution to the conductivity. The basic idea was illustrated briefly in Chapter 1 (and discussed above) as a result of interference around a diffusive loop between the two trajectories. Let us repeat part of that discussion for convenience. Consider, for example, Fig. 5.1. What is shown in the figure is just the weak localization contribution to the overall resistance of a 2.0-μm-long quantum wire, defined in a GaAs/AlGaAs heterostructure by a pair of "split gates" which define the wire by depletion of the region under the gates. This leaves a narrow channel open in the region between the gates. The device of interest here has a width of about 0.5 μm and a background resistance of 590 ohms, so the peak conductance correction ΔG is about 43 μS ($\Delta G = -\Delta R/R^2$, which leads to the observed value; the same value can be obtained by taking the reciprocal of the background resistance and subtracting the reciprocal of the total resistance). A second device is shown in Fig. 5.2. Here, a low-mobility GaAs MESFET structure (a heavily doped epitaxial layer on top of a semiinsulating substrate) has been biased nearly to depletion, and the resulting strong impurity scattering leads to the weak localization in the sample. (Also shown is the Hall voltage, but this is not of interest to us here.) The normal magnetoresistance has not been subtracted out of this plot. The weak localization contribution here is about 30 μS (the structure observed will be discussed in a later section).

In both cases, the weak localization correction is damped by the application of a magnetic field. We will see below that a critical magnetic field can be defined as that at which the weak localization correction has been reduced to half of its peak value, and this defines a critical magnetic field. It will be argued that this value of critical magnetic field is just

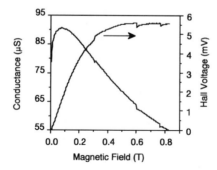

Fig. 5.2. Weak localization in a MESFET layer subjected to a superlattice potential.

enough to enclose one flux quanta ($\hbar/e$) within the area of the phase-coherent loop. In this case, these field values are about 0.1 T for the heterostructure (Fig. 5.1) and about 0.01 T for the MESFET, which is quite small. (Here, as in most cases, one can only make a rough guess of the actual value due to the conductance fluctuations present in most samples.) These values will be used later in the semiclassical theory discussed below.

Before beginning to talk about the quantum treatment of the conductance fluctuations and weak localization, we need to refresh our memory of the semiclassical transport theory in terms of correlation functions. The reason for this is that the primary measurement on the fluctuations is to determine their correlation function as the field or density is varied. Thus, we need to understand that the correlation functions are a general property of transport and not a new phenomenon for describing these effects.

5.1.1 Semiclassical treatment of the conductance

The conductivity can be generally related to the current-current correlation function, a result that arises from the Kubo formula, but it also is dependent upon an approach involving retarded Langevin equations [2]. In general, the conductivity can be expressed simply as

$$\sigma = \frac{ne^2}{m\langle v^2(0)\rangle}\int_0^\infty \langle v(t), v(0)\rangle\, dt = \frac{ne^2}{m}\int_0^\infty e^{-t/\tau}\, dt = \frac{ne^2\tau}{m}, \tag{5.1}$$

where it has been assumed that the velocity decays with a simple exponential, and where the conductivity is measured at the Fermi surface (although this is not strictly required for the definitions used in this equation). The exponential in Eq. (5.1) is related to the probability that a particle diffuses, without scattering, for a time τ, which is taken as the mean time between collisions. Now, in one dimension, we can use the facts that $D = v_F^2\tau$ (or more generally, $D = v_F^2\tau/d$, where d is the dimensionality of the system), $l_e = v_F\tau$, and $n_1 = k_F/\pi$ to write (5.1) as

$$\sigma_1 = \frac{e^2}{\pi\hbar}2D\int_0^\infty \frac{1}{2l_e}e^{-t/\tau}\, dt. \tag{5.2}$$

In two dimensions, we use the facts that $D = v_F^2\tau/2$ and $n_2 = k_F^2/2\pi$ to give

$$\sigma_2 = \frac{e^2}{\pi\hbar}2D\int_0^\infty \frac{k_F}{2l_e}e^{-t/\tau}\, dt. \tag{5.3}$$

We note that the integrand now has the units L^{-d} in both cases. It is naturally expected that this scaling will continue to higher dimensions. We note, however, that the quantity inside the integral is no longer just the simple probability that a particle has escaped scattering. Instead, it now has a prefactor that arises from critical lengths in the problem. To continue, one could convert each of these to a *conductance* by multiplying by L^{d-2}. This dimensionality couples with the diffusion constant and time integration to produce the proper units of conductance. To be consistent with the remaining discussion, however, this will not be done.

In weak localization, we will seek the correlation function that is related to the *probability of return* to the initial position. We define the integral analogously to the above as a *correlator*, specifically known as the *particle-particle correlator*. Hence, we define

$$C(\tau) = \int_0^\infty W(t)\, dt \sim \int_0^\infty \frac{1}{2L^d}e^{-t/\tau}dt, \tag{5.4}$$

where $W(t)$ is the time-dependent correlation function describing this return, and the overall expression has exact similarities to the above equations (in fact, the correlator in the first two equations is just the integral). What is of importance in the case of weak localization is that we are not interested in the drift time of the free carriers. Rather, we are interested in the diffusive transport of the carriers and in their probability of return to the original position. Thus, we will calculate the weak localization by replacing C or W by the appropriate quantity defined by a diffusion equation for the strongly scattering regime. Hence, we must find the expression for the correlation function in a different manner than simply the exponential decay due to scattering.

Following this train of thought, we can define the weak localization correction factor through the probability that a particle diffuses some distance and *returns to the original position*. If we define this latter probability as $W(t)$, where T is the time required to diffuse around the loop, we can then define the conductivity correction in analogy with (5.3) to be

$$\Delta\sigma = -\frac{2e^2}{h}2D\int_0^\infty W(t)\,dt, \ x(t) \to x(0), \tag{5.5}$$

and the prefactor of the integral is precisely that occurring in the first two equations of this section. (The negative sign is chosen because the phase interference *reduces* the conductance.) In fact, most particles will not return to the original position. Only a small fraction will do so, and the correction to the conductance is in general small. It is just that small fraction of particles that actually does undergo backscattering (and reversal of momentum) after several scattering events that is of interest. In the above equations, the correlation function describes the decay of "knowledge" of the initial state. Here, however, we use the "probability" that particles can diffuse for a time T and return to the initial position while retaining some "knowledge" of that initial condition (and, more precisely, the phase of the particle at that initial position). Only in this case can there be interference between the initial wave and the returning wave, where the "knowledge" is by necessity defined as the retention of phase coherence in the quantum sense. Now, while we have defined $W(t)$ as a probability, it is not a true probability since it has the units L^{-d}, characteristic of the conductivity in some dimension (DWt is dimensionless). In going over to the proper conductance, this dimensionality is correctly treated, and in choosing a properly normalized probability function, no further problems will arise. This discussion has begun with the semiclassical case, but now we are seeking a quantum-mechanical memory term, exemplifying the problems in connecting the classical world to the quantum mechanical world. There are certainly other approaches to get to this latter equation, but the approach used here is chosen to enhance its connection with classical transport; what we have to do is describe the phase memory in such an approach, and this is traditionally ascribed to the WKB method in quantum mechanics.

It has been assumed so far that the transport of the carriers is diffusive, that is, that the motion moves between a great many scattering centers so that the net drift is one characterized well by Brownian motion. By this we assume that quantum effects cause the interference that leads to (5.5), and that the motion may be dominantly described by the classical motion (the quantum treatment is taken up below). This means that $kl_e < 1$, where k is the carrier's wavevector (usually the Fermi wavevector) and l_e is the mean free path between collisions, which is normally the elastic mean free path (which is usually shorter than the inelastic mean free path). This means that the probability function will be Gaussian (characteristic of diffusion), and this is relatively easily established by the fact that $W(t)$

should satisfy the diffusion equation for motion away from a point source (at time $t = 0$), since the transport is diffusive. This means that [3]

$$\left(\frac{\partial}{\partial t} - D\nabla^2 \right) W(\mathbf{r}, t) = \delta(\mathbf{r})\delta(t), \tag{5.6}$$

which has the general solution

$$W(\mathbf{r}, t) = \frac{1}{(4\pi Dt)^{d/2}} \exp\left(-\frac{r^2}{4Dt} \right). \tag{5.7}$$

In fact, this solution is for unconstrained motion (motion that arises in an infinite d-dimensional system). If the system is bounded, as in a two-dimensional quantum well or in a quantum wire, then the modal solution must be found. At this point we will not worry about this, but we will return to this in the formal theory below. Our interest is in the probability of return, so we set $r = 0$. There is one more factor that has been omitted so far, and that is the likelihood that the particle can diffuse through these multiple collisions without losing phase memory. Thus, we must add this simple probability, which is an exponential, as in (5.1). This leads us to the probability of return after a time T, without loss of phase, being

$$W(t) = \frac{1}{(4\pi Dt)^{d/2}} e^{-t/\tau_\varphi}. \tag{5.8}$$

Here, the phase-breaking time τ_φ has been introduced to characterize the phase-breaking process. We note at this point that the dimensionality of $W(t)$ is L^{-d} (Dt has the dimensions of L^2), which is the dimensionality of the integrand for the conductivity, not the conductance. Thus, this fits in with the discussion above.

One further modification of this simple semiclassical treatment has been suggested by Beenakker and van Houten [4]. This has to do with the fact that we do not expect to find these diffusive effects in ballistic transport regimes. Thus, it can be expected that on the short-time basis, these effects go away. Here, "short time" is appropriate in that collisions must occur before diffusive transport can take place. If there are no collisions, there is little chance for the particle to be backscattered and to return to the original position. Thus, these authors suggest modifying (5.8) to account for this process. This gives the new form for the probability of return to be

$$W(t) = \frac{1}{(4\pi Dt)^{d/2}} e^{-t/\tau_\varphi}(1 - e^{-t/\tau}). \tag{5.9}$$

At this point, we have slipped in the only quantum mechanics in the current approach. This quantum mechanics is connected with the phase of the electrons and is described by the phenomenological phase relaxation time τ_φ, which has been introduced. We have not actually carried out a quantum-mechanical calculation, yet we have introduced all the necessary phase interference through this phenomenological term. The actual quantum calculations are buried at this point, but it is important to recognize where they have entered in the discussion.

The dimensionality correction to the probability of return will go away if we work with the total conductance, rather than the conductivity. However, as above, we will not make this change. We can use (5.9) in (5.5). This gives the conductivity corrections for weak localization to be

$$\Delta\sigma = -\frac{e^2}{\pi\hbar} \begin{cases} \frac{1}{2\pi l_\varphi}\left(\sqrt{1 + \frac{\tau_\varphi}{\tau}} - 1\right), & d = 3, \\ \frac{1}{2\pi}\ln\left(\frac{\tau_\varphi}{\tau} + 1\right), & d = 2, \\ l_\varphi\left(1 - \sqrt{\frac{\tau}{\tau + \tau_\varphi}}\right), & d = 1. \end{cases} \qquad (5.10)$$

It is clear that the important length in this diffusive regime is the phase coherence length $l_\varphi = \sqrt{D\tau_\varphi}$. To be sure, this result is the most simple one that can be obtained within reasonable constraints. Nevertheless, the results are quite useful to point out that the weak localization reduction of conductance is relatively universal in its amplitude, but also that it has an adjustment depending upon the ratio of the important time scales in the transport problem. Nevertheless, the quantum mechanics is buried in the *ad hoc* introduction of the phase coherence time τ_φ. Without this introduction, none of the above formulas would be meaningful. The detailed quantum treatment described in a later section, while being more formal, will basically return these same values for the weak localization corrections to the conductance.

Let us now compare with the experiments presented above. In the GaAs/AlGaAs heterostructure wire, the conductance reduction was about $43\,\mu$S. In this sample, the (quasi-two-dimensional) density was about $6.63 \times 10^{11}\,\text{cm}^{-2}$, and the mobility was about $9.2 \times 10^4\,\text{cm}^2/\text{Vs}$. These lead to values of $\tau = 3.5 \times 10^{-12}$ s and $D = 2180\,\text{cm}^2/\text{s}$. The observed value of the weak localization correction suggests that (iterating the above quasi-one-dimensional result) l_φ is about $2.6\,\mu$m, which is longer than the wire length itself, which suggests that the two-dimensional regions adjoining the wire will be important as well, a point to which we return below. The quasi-two-dimensional GaAs MESFET device, on the other hand, had a density of $4 \times 10^{11}\,\text{cm}^{-2}$, and the mobility was only $2.0 \times 10^4\,\text{cm}^2/\text{Vs}$. Using these values, it is estimated that the value for the phase coherence length is about $0.16\,\mu$m.

5.1.2 Effect of a magnetic field

In the presence of a magnetic field, the diffusive paths that return to their initial coordinates will enclose magnetic flux. As discussed above, the situation is slightly different than that of the Aharonov-Bohm effect, in that each of the two interfering paths cycles completely around the loop, doubling the enclosed flux. Thus the expected periodicity is in $h/2e$ (rather than h/e). To examine this behavior is only slightly more complicated than that of the above treatment, and we will take the magnetic field in the z-direction and in the Landau gauge $\mathbf{A} = (0, Bx, 0)$. We will also consider only a thin two-dimensional slab with no z variation at present, so that (5.6) can be written as

$$\left(\frac{\partial}{\partial t} - D\left(\nabla - i\frac{2e\mathbf{A}}{\hbar}\right)^2 + \frac{1}{\tau_\varphi}\right)W(\mathbf{r}, t) = \delta(\mathbf{r})\delta(t), \qquad (5.11)$$

or

$$\left[\frac{\partial}{\partial t} - D\frac{\partial^2}{\partial x^2} - D\left(\frac{\partial}{\partial y} - \left(i\frac{2eBx}{\hbar}\right)^2\right)^2 + \frac{1}{\tau_\varphi}\right]W(\mathbf{r}, t) = \delta(\mathbf{r})\delta(t). \qquad (5.12)$$

Before proceeding, the two equations above should be justified. The earlier approach was contained in (5.6). In the equations here, an extra term appears related to the phase coherence time, and this is just the expected term that will give rise to the exponential decay introduced earlier. However, the gradient operator has also been replaced by a term combining the gradient and the vector potential (divided by $\hbar$). It is this replacement, which seems somewhat questionable in the classical approach, that has been followed so far. Here, we are introducing another aspect of the quantum behavior, first studied for band electrons in terms of the susceptibility. The connection is that the ∇^2 term is related to the square of the momentum in the total energy operator (as related to the Schrödinger equation in Chapter 1). Classically, when one introduces a magnetic field described by the vector potential $\mathbf{A}$, the proper conjugate "momentum" becomes $\mathbf{p} + e\mathbf{A}$ [5]. Quantum mechanically, we expect both the vector potential, which is the source of the magnetic field, to modify the momentum as $\hbar\mathbf{k} \to \mathbf{p} + e\mathbf{A} \to -i\hbar\nabla + e\mathbf{A}$, where it is the relation to quantum mechanics that allows us to replace the total conjugate momentum $\mathbf{p}$ by $-i\hbar\nabla$, and the replacement of the square of the simple momentum in the diffusion equation that has been used above. In addition, the vector potential is modified by the factor of 2 for the special circumstances of the problem of interest. This is the second adjustment that arises from quantum mechanics. We have used the connection between the diffusion equation and the Schrödinger equation discussed in Chapter 1 to guide this replacement, so now our quantum treatment lies in the phase-breaking process and in the modification of the diffusion equation itself. In the following, we go essentially completely to the quantum treatment but leave the proper derivation of the above equation to a later section.

This approach is of course quite similar to that for Landau levels. The major difference is that we are developing the response with the diffusion equation, and the decay due to the inelastic (phase-breaking) time represented by (5.8) has been included explicitly. However, we will proceed in exactly the same manner as in the case of Landau levels by assuming that the y variation is characterized by a free momentum plane wave in this direction with wavevector k, so that the last equation can be rewritten as

$$\left[\frac{\partial}{\partial t} - D\frac{\partial^2}{\partial x^2} - D\left(ik - i\left(\frac{2eBx}{\hbar}\right)^2\right)^2 + \frac{1}{\tau_\varphi}\right]W(\mathbf{r}, t) = \delta(\mathbf{r})\delta(t). \tag{5.13}$$

At this point, we shall diverge from a Landau level treatment and recognize that in the final process the limit $x = 0$ will be introduced. Moving directly to this, it may be recognized that the inelastic decay time of (5.8) may be replaced as

$$\frac{1}{\tau_\varphi} \to \frac{1}{\tau_\varphi} + Dk^2 = \frac{1}{\tau_\varphi} + (2n + 1)\frac{2eBD}{\hbar}, \tag{5.14}$$

where it has been assumed in the last expression that the relevant momentum is the Fermi momentum with $E_F = (n + \frac{1}{2})\hbar\omega_c$. The simplest assumption is to replace the phase-breaking time in Eq. (5.10) by the combination of the phase-breaking and magnetic *times* (with $n = 0$) defined in (5.14). Of course, this treatment is oversimplified, and a more extensive exact treatment is required. However, the important point is that the role of the magnetic field is to dramatically increase the phase-breaking *rate*, thus reducing the *effective* phase-breaking time. As a consequence, the magnetic field breaks up the correlation of the time-reversed paths that led to the weak localization correction, thus reducing this term. (On a formal basis, the magnetic field breaks the *time-reversal symmetry* of the two counter-propagating waves in the ring. Thus they are no longer equivalent to one another, and the weak localization interference is destroyed.) By estimating the magnetic field at which the weak localization correction is reduced by a factor of two, one has an estimate of the value of the phase-breaking time, since it will be equal to the magnetic time defined in the second term of (5.14)

$$B_c \sim \frac{h}{el_\varphi^2},\tag{5.15}$$

where it is assumed that D is known. This leads us to conclude that the critical magnetic field is that necessary to couple one flux quantum (h/e) through a phase-coherent area defined by the diffusive coherence length l_φ. Hence, one can thus determine the phase coherence length by two measurements: the amplitude of the weak localization correction and the magnetic field at which this correction has decayed by a factor of 2. In actual fact, the equations (5.10) give a slightly different value for the effective phase-breaking time to reduce the correction by a factor of two, but this is a detail correction, not a fundamental variation.

To compare, we note that the quantum wire discussed above had the weak localization correction reduced by a factor of 2 for a magnetic field of 0.1 T. To reduce the phase coherence by a factor of two (which is the major factor in this case), we need to reduce the inelastic time by a factor of 4. In this case, the magnetic dependence suggests a phase coherence length of only 0.33 μm, which is a strange occurence suggesting that not all the flux is effective in cancelling the weak localization. We return to this point below. In the quasi-two-dimensional GaAs MESFET structure, on the other hand, the reduction of the weak localization by a factor of two at 0.01 T (which is a crude estimate given the scales of the data in the figure), suggests a value of 0.11 μm, which compares reasonably well with the above estimate of the phase coherence length. As will be seen below, there are a number of correction terms that need to be included before these estimates can be taken to be accurate.

The one point left out of the latter development, but alluded to earlier, is the fact that the particle may return to its initial point many times due to the Landau quantization of the orbits. This is an equivalent effect to the Aharonov-Bohm oscillations. The integration in (5.5) does not include the possibility of a periodicity in the probability for return. This fact must be taken out by hand at this point by noting that the time of interest should be reduced by the period of the orbit around the closed path. This orbit may be due to the Landau quantization, or just due to the fact that the circumference of the path is smaller than the coherence length. It is the latter that produces the competition to the Aharonov-Bohm effect. Moving completely around the loop produces a phase correction of twice that of the Aharonov-Bohm effect, so that there is a correction to the probability of return in the form of

$$\sum_{n'=-\infty}^{\infty} \exp\left(-2\pi i \frac{2eBa^2}{h} n'\right),\tag{5.16}$$

where a^2 is the enclosed area of the loop. Thus, this factor (using only the case of $n' = \pm 1$) produces a periodicity in $h/2e$. This is shown in Fig. 5.3 for metal rings (similar effects are also observed in semiconductor samples) [6]. For small magnetic fields, the $h/2e$ oscillations dominate the conductance correction, whereas for large magnetic fields, where the weak localization effects are damped, the Aharonov-Bohm oscillations dominate the conductance. On the other hand, the latter are heavily damped (actually exponentially attenuated) as the number of rings increases, while the weak-localization $h/2e$ oscillations are attenuated only as $1/N$, where N is the number of rings in the structure. Equivalently, N is the number of equivalent regions of size dimension comparable to the inelastic mean free path.

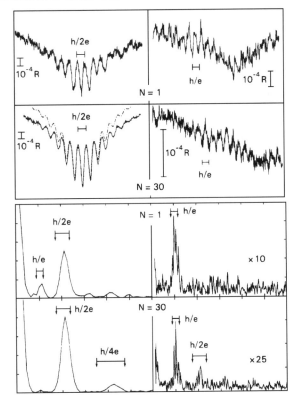

Fig. 5.3. The oscillations in the upper frame show the resistance oscillations, and the lower frame shows the Fourier transforms, for arrays of 1 and 30 rings, which are discussed in the text. [After Umbach *et al.*, Phys. Rev. Lett. **56**, 386 (1986), by permission.]

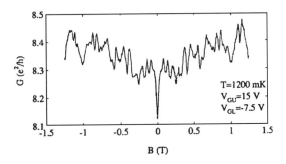

Fig. 5.4. Typical magnetoresistance curve for a short quantum wire in Si. The negative going peak near $B = 0$ is the weak localization. [After de Graaf *et al.*, Phys. Rev. B **46**, 12814 (1992), by permission.]

Because these are well-formed rings, the oscillations remain coherent. In a bulk region, where the "rings" are phase-coherent regions of random size, the oscillations are expected not to be coherent and to produce interfering effects leading to universal conductance fluctuations.

In Fig. 5.4, the weak localization measurements for a short "wire" formed in a Si MOS-FET are shown [7]. The devices used are dual-gate Si MOSFETs, in which a short one-dimensional channel is formed by a split pair of lower gates (biased negatively), while

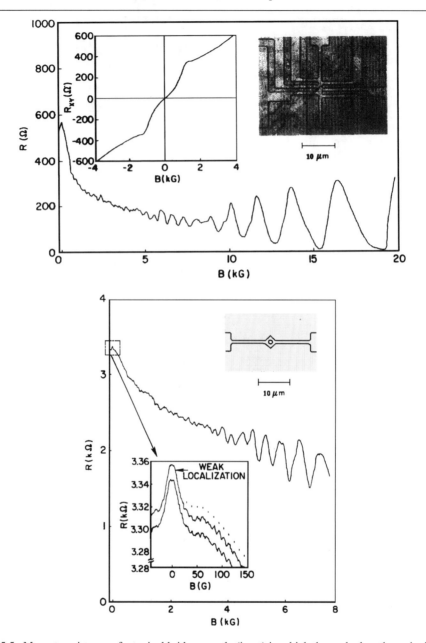

Fig. 5.5. Magnetoresistance of a typical bridge sample (inset) in which the probe lengths and width are less than the elastic mean free path. The Hall voltage is also shown in the inset. [After J. Simmons *et al.*, Surf. Sci. **196**, 81 (1988), by permission.]

the overall inversion layer density is controlled by a large upper gate (biased positively to attract carriers into the inversion channel). The width of the gap in the split lower gates is $0.35\,\mu$m, but the width of the conducting channel is considerably smaller due to the fringing fields of the gate. The length of the split-gate defined wire is about $0.73\,\mu$m. The real one-dimensional channel will be somewhat longer than this, again due to the fringing fields around the actual metal gates. The low-field magnetoresistance of this device is given in the figure. The strong dip in the conductance around zero magnetic field is due to weak localization of the carriers. In addition, universal conductance fluctuations are seen in the conductance at higher fields. First, in these one-dimensional wires, surface scattering from the side walls of the wire strongly affect the conductance and actually increase the amplitude of the weak localization. Second, the shortness of this wire means that the two-dimensional areas to which the wire is attached (the areas that are not covered by the lower split gates) play an important role in the size and field dependence of the weak localization. We will turn to these effects below, but the analysis presented there suggests that the phase coherence lengths for this device (at the measuring temperature of 1.2 K) are about 0.7 μm in the two-dimensional regions and about 0.2 μm in the wire itself.

The results of the weak localization in a wire formed in a GaAs/AlGaAs heterostructure are shown in Fig. 5.5 [8]. The structure of the wire and its measuring wires is shown in the inset to this figure. The magnetoresistance for this structure is for a wire segment of length $L = 2 \mu$m at 0.3 K. The wires, in this case, are formed by etching them to a width of 1 μm. Universal conductance fluctuations and the onset of the quantum Hall effect are shown at higher fields for this sample. The large resistance peak near $B = 0$ is the weak localization. The phase coherence length in this sample has been estimated from the universal conductance fluctuations to be about 12 μm. It should be noted, however, that the basic heterostructure has very high mobility so that both the length and width of the wire segments are less than the *elastic* scattering mean free path. The mobility ($\sim 10^6$ cm^2/Vs) was such that the latter was on the order of 6 μm in this wire structure.

5.1.3 Size effects in quantum wires

In general, the weak localization seems to be stronger in the quantum wires. This is because the longitudinal resistivity (along the magnetic field) in a quasi-two-dimensional system remains independent of the magnetic field, at least within the semiclassical treatment. However, in a high-mobility quantum wire, the transverse motion causes the carriers to interact strongly with the "walls" of the quantum wire. If the transport is primarily diffusive (lower mobility), this is not an important effect. On the other hand, in the high-mobility wires, where the transport can be quasi-ballistic, this effect is a natural extension of the properties of boundary scattering in the wire itself.

In Chapter 3, it was shown how boundary scattering changes the basic resistivity of the region contained within the quantum wire. Let us review that discussion. We consider that the channel (or wire) is defined by the presence of hard walls at $x = \pm W/2$, at which the carriers are diffusely scattered. By diffusive scattering we mean that the carriers lose all information of the incoming angle of propagation and are reflected uniformly in an angle of π radians. If they have some memory of their initial momentum direction, they are then said to have been scattered specularly [9]–[11]. The stationary distribution function for carriers in the wire satisfies the simple Boltzmann equation, within the relaxation time approximation [12],

$$\mathbf{v} \cdot \frac{\partial f(\mathbf{r}, \alpha)}{\partial \mathbf{r}} = -\frac{f(\mathbf{r}, \alpha)}{\tau} + \frac{1}{\tau} \int_0^{2\pi} \frac{f(\mathbf{r}, \alpha)}{2\pi} d\alpha, \qquad (5.17)$$

where $\mathbf{r}$ is the two-dimensional position vector in the quantum wire and α is the angle that the velocity makes with the transverse x-axis. The second term on the right is a balancing term so that integrating over all angles α produces no net flow term on the left side (as required by current continuity along the wire). With diffuse scattering, the distribution function should be independent of the direction of the velocity for those velocities directed *away* from the boundary. Current continuity then leads to the boundary conditions

$$f(\mathbf{r}, \alpha) = \begin{cases} \frac{1}{2} \int_{-\pi/2}^{\pi/2} f(\mathbf{r}, \alpha') \cos(\alpha') \, d\alpha' & \text{for } x = \frac{W}{2}, \frac{\pi}{2} < \alpha < \frac{3\pi}{2}, \\ \frac{1}{2} \int_{\pi/2}^{3\pi/2} f(\mathbf{r}, \alpha') \cos(\alpha') \, d\alpha' & \text{for } x = -\frac{W}{2}, -\frac{\pi}{2} < \alpha < \frac{\pi}{2}. \end{cases}$$

(5.18)

Since there is no magnetic field, it may be assumed that the density is uniform in α, so that the integral over the angle in (5.17) vanishes. Beenakker and van Houten [4] show that the solution is given by (except for a normalization constant)

$$f(\mathbf{r}, \alpha) = -y + l_e \sin(\alpha) \left[1 - \exp\left(-\frac{W}{2l_e |\cos \alpha|} - \frac{x}{l_e \sin \alpha} \right) \right],$$

(5.19)

where $l_e = v_F \tau$ is the elastic mean free path at the Fermi surface. For $W \ll l_e$, this leads to

$$\rho = \rho_0 \frac{\pi l_e}{2W \ln(l_e/W)} = \frac{m v_F}{n e^2 W} \frac{\pi}{2 \ln(l_e/W)}.$$

(5.20)

In a magnetic field, it is possible for skipping orbits to form discussed in Section 2.5.2. Even when the magnetic field is directed along the axis of the wire, it is possible for particles, whose motion would normally be along the wire, to be deflected into the surface and to undergo backscattering. Our major interest, however, is for the case in which the magnetic field is normal to the quasi-two-dimensional layer. Then the motion is much clearer. In the presence of this z-directed magnetic field, Eq. (5.17) becomes

$$\mathbf{v} \cdot \frac{\partial f(\mathbf{r}, \alpha)}{\partial \mathbf{r}} + \omega_c \frac{\partial f(\mathbf{r}, \alpha)}{\partial \alpha} = -\frac{f(\mathbf{r}, \alpha)}{\tau} + \frac{1}{\tau} \int_0^{2\pi} \frac{f(\mathbf{r}, \alpha)}{2\pi} \, d\alpha,$$

(5.21)

where $\omega_c = eB/m$ is the cyclotron frequency. If we have *specular* scattering, then we require $f(\mathbf{r}, \alpha) = f(\mathbf{r}, \pi - \alpha)$ at $x = \pm W/2$. This leads to the result [12]

$$f(\mathbf{r}, \alpha) = -y - \omega_c \tau x + l_e \sin(\alpha).$$

(5.22)

The interesting aspect is that the diffusive current along the wire $I_y = \pi W v_F l_e$ is the same whether or not there is a magnetic field. This implies that, in the presence of specular scattering at the surface, the motion along the wire is not affected by the magnetic field. Longitudinal motion is converted to edge-state skipping orbits, as discussed in previous chapters, but the overall effect of the magnetic field on the resistance is a zero effect. In the presence of diffusive scattering at the boundaries, however, this is no longer the case. As we have seen above, diffusive scattering increases the resistivity. This is further increased in a small magnetic field, in which trajectories that would normally move directly along the wire are diverted into the surface. In a high magnetic field, the cyclotron motion inhibits the backscattering that can occur from the diffuse scattering by bending the orbits back into the forward direction. Thus, when $W > 2\ell_{cy}$, where $\ell_{cy} = r_c$ is the cyclotron orbit at the Fermi surface, the resistivity increase due to diffusive backscattering is dramatically reduced. This is apparent in Fig. 5.6, where experiments from Thornton *et al.* [13] are shown. Data for several different widths are shown for comparison. First, it is clear that the resistance increase at zero magnetic field is much larger than that which arises from the simple reduction of the wire width (decreasing the width by a factor of two increases the resistance by a factor of 2.5–3.5). There is a rise in the resistance at small magnetic

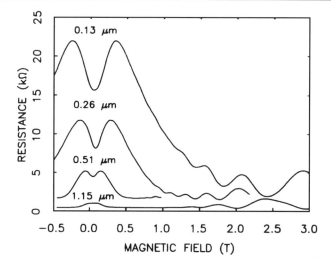

Fig. 5.6. The role of boundary scattering in thin wire structures is indicated here for a set of wires of varying width [after Thornton *et al.*, Phys. Rev. Lett. **63**, 2128 (1989), with permission].

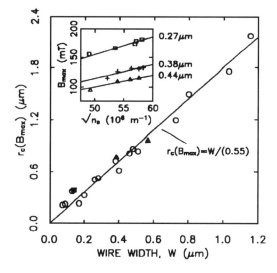

Fig. 5.7. The cyclotron radius at the peak position of the resistance, plotted versus wire width. The inset shows the dependence upon the carrier density. [After Thornton *et al.*, in "Granular Nanoelectronics," eds. D. K. Ferry *et al.* (Plenum Press, New York, 1991), pp. 165–179.]

field, which is the effect of deflecting the longitudinal trajectories into the walls. Finally, at higher magnetic fields, the diffusive scattering is greatly reduced due to the formation of the edge states and their skipping orbits. Finally, at the highest magnetic fields, the normal Shubnikov–de Haas effects are seen. The position of the anomalous resistance peak, B_{max}, is determined by the ratio of the width of the wire to the cyclotron radius. Calculations by Forsvoll and Holwech [14] and by Pippard [10] suggest that $W/\ell_{cy} = 0.55$ at the maximum. In Fig. 5.7, the peak position (in magnetic field) of the resistance maximum is plotted for a variety of wire widths [15]. The straight line is a fit to this theory. The dependence on the cyclotron radius also suggests a connection to the carrier density and this is illustrated in the inset. In general, the fit to the theory is quite good.

That this is a high magnetic field effect in relatively high-mobility material is evident from the fact that the peak is found to occur at $B = 0.2$–0.4 T, which suggests that $\omega_c \tau = \mu B > 2$, or that $\mu > 5 - 10 \times 10^4$ cm^2/Vs at low temperature. No conductance fluctuations are seen in the data, which is further indication that the impurity concentration is sufficiently low that diffusive scattering within the bulk of the wire is not significant.

It is important to note that this diffusive edge scattering effect is different from the weak localization discussed above. Weak localization is broken up by the magnetic field so that the enhanced resistance decays already from zero magnetic field. This leads to a reduction in resistance with magnetic field. Diffusive edge scattering, however, is *enhanced* in a small magnetic field, which causes an increase in the resistance with magnetic field; this increase is much larger than normally expected from the magnetoresistance. Thus, weak localization and diffusive edge scattering show opposite behavior with magnetic field in the small field regime. However, in many cases it is not clear that the two effects can be easily separated, particularly in quantum wires. Note that for the boundary scattering, the overall change in the resistance at low magnetic fields is opposite to that of the weak localization. However, the role of boundary scattering is decreased with much stronger magnetic fields than those at which the coherence of the time-reversed orbits is broken. Consider Fig. 5.5, discussed earlier, as an example. There is a general decay of the resistance with magnetic fields in the range around 1–3 T. This decay could be due to the decay of surface scattering as well as the general onset of field quantization in the quantum wire. The weak localization, however, is observed clearly only at the lowest magnetic fields and has been totally broken up by a field value of about 1 T. Although it can not be seen clearly in this figure, one would expect the weak localization to be gone at fields of the order of 0.1 T or so. This was clear in Fig. 1.8 where, although for a sample with lower mobility, the weak localization was gone for fields above 0.2 T.

5.1.4 The magnetic decay "time"

As discussed above, the presence of surface (or edge) scattering will modify the behavior of the magnetoresistance, and particularly the weak localization, in the presence of a magnetic field. In general, we may consider that the effectiveness of the magnetic field in suppressing the weak localization contribution to the resistance lies in the amount of magnetic flux enclosed in the loop formed by the two time-reversed trajectories. If the bulk trajectories are modified by scattering from the boundary of the quantum wire, then the amount of flux will be modified. Thus there are different regimes, characterized by the relative sizes of the important lengths: the mean free path $l_e = v_F \tau$, the coherence length for diffusive transport $l_\varphi = (D\tau_\varphi)^{1/2}$, the magnetic length $l_m = (\hbar/eB)^{1/2}$, and the width of the wire W (as well as the length of the wire L). The case for $l_e \ll W$, in which the elastic mean free path is much less than the width of the wire, corresponds to a narrow quasi-two-dimensional structure rather than to a proper quantum wire. This limit has been called the *dirty metal regime*, but it is sometimes found in Si quantum wires. In this limit, transport is mostly characterized by that of a disordered material. The opposite limit, for which $l_e \gg W$, usually called the *pure metal regime*, is one in which the differences between specular and diffusive scattering from the boundaries can be delineated. The latter regime is the one that we have called the quasi-ballistic regime.

Table 5.1. *Magnetic relaxation time and critical magnetic field for weak localization in a channel.*

Length Scales	τ_B	B_c
$l_e, l_\varphi \ll W$	$\dfrac{l_m^2}{2D}$	$\dfrac{\hbar}{2el_\varphi^2}$
$l_e \ll W \ll l_\varphi$	$\dfrac{3l_m^4}{W^2 D}$	$\dfrac{\hbar 3^{1/2}}{eWl_\varphi}$
$W \ll l_e, Wl_e \ll l_m^2$	$\dfrac{C_1 l_m^4}{W^3 v_F}$	$\dfrac{\hbar}{eW}\left(\dfrac{C_1}{W v_F \tau_\varphi}\right)^{1/2}$
$W \ll l_e, W^2 \ll l_m^2 \ll Wl_e$	$\dfrac{C_2 l_m^2 l_e}{W^2 v_F}$	$\dfrac{\hbar}{eW}\left(\dfrac{C_2 l_e}{W v_F \tau_\varphi}\right)$

Another relationship between the various lengths determines whether we are concerned with quasi-two-dimensional corrections to the conductivity or truly quantum wire corrections. This relationship concerns the coherence length and the width of the wire. If $l_\varphi \ll W$, it is generally felt that the weak localization corrections are those corresponding to a quasi-two-dimensional medium. On the other hand, if $l_\varphi \gg W$, it is generally felt that the system is properly a quasi-one-dimensional wire. The key factors in the magnetic field dependence of the weak localization are the magnetic "lifetime" τ_B, given for example by the last term in (5.14), and the magnetic field B_c, at which τ_B and τ_φ produce comparable effects in the weak localization. Several theories have been presented for these quantities, and the predictions are summarized in Table 5.1 [4], [16]–[18]. The two constants C_1 and C_2 depend upon the nature of the boundary (surface) scattering. For specular scattering, it is generally found that $C_1 = 9.5$ and $C_2 = 4.8$, while for diffusive scattering, $C_1 = 4\pi$ and $C_2 = 3$ [4].

To compare with the results for the pseudo-wire case, one can consider a quasi-two-dimensional device for which $l_\varphi \ll W$. Then, the magnetic length is given by the result of (5.14), in which $\tau_B \sim \hbar/2eBD = l_m^2/2D$, and the critical magnetic field is found from $\tau_B \sim \tau_\varphi$, so that $B_c \sim \hbar/2eD\tau_\varphi = \hbar/2el_\varphi^2$. These values are for the lowest Landau level, and smaller values are expected for higher levels. We note that it is just this value that is used to estimate the quantum wire data above, and apparently this estimate was made in the wrong limits of lengths, since this wire clearly had a width smaller than the phase coherence length.

The full expression for the magnetoconductance correction due to weak localization is found by expanding the solution to (5.13) with a summation over the index n, leading to [3]

$$\delta G_{WL}^{2D}(B) - \delta G_{WL}^{2D}(0) = \frac{W}{L}\frac{ge2}{4\pi^2\hbar}\left[\Psi\left(\frac{1}{2} + \frac{\tau_B}{2\tau_\varphi}\right) - \Psi\left(\frac{1}{2} + \frac{\tau_B}{2\tau}\right) + \ln\left(\frac{\tau_\varphi}{\tau} + 1\right)\right],$$

(5.23)

where g is a degeneracy factor that accounts for spin and valley degeneracies, and $\Psi(x)$ is the digamma function. Normally, τ_B will be given by the value in the first row of Table 5.1. In the absence of the magnetic field, Eq. (5.10) is recovered. The weak localization will be completely destroyed by the magnetic field when the magnetic field is such that $\tau_B < \tau$, which occurs for $\hbar/2el_e^2 < 1$. These fields are still much weaker than the usual quantum limit.

In the one-dimensional case $W \ll l_\varphi$, the time-reversed trajectories are squeezed by the finite width of the wire. This compression of the orbits leads to the second line of Table 5.1

[17]. The full expression for the magnetic field dependence of the conductivity correction in this limit is given by

$$\delta G_{WL}^{1D}(B) = \frac{ge^2}{hL}\left(\frac{1}{D\tau_\varphi} + \frac{1}{D\tau_B}\right)^{-1/2}. \tag{5.24}$$

It should be noted that one expects to see a crossover from quasi-one-dimensional to quasi-two-dimensional results at a magnetic field for which $l_m \sim W$. The reason is that the lateral confinement is replaced by skipping orbit formation and the dominance of edge-state conduction in the wire. We will see more of this result in the discussion of the universal conductance fluctuations below.

In semiconductor nanostructures formed in high-mobility material, the elastic mean free path is usually quite long. For example, in GaAs with a mobility of 10^5 cm^2/Vs, it was shown in Chapter 1 that the average elastic mean free path is about 1 μm. Now it is possible in heterostructures to have the mobility 10–100 times larger than this, which means that the mean free path can reach 10–100 μm. For wire widths of less than 1.0 μm, it is easy to obtain $l_e \gg W$. This is then the "pure metal" regime (or pure quasi-ballistic regime), in which scattering is relatively weak and surface scattering dominates the diffusive process. This regime seems to have been first treated by Dugaev and Khmelnitskii [19]. There is a major new factor that becomes important in this regime, and that is the possibility of *flux cancellation* [4] by multiple reflections from the wire boundary. By this we mean that the diffusive scattering from the boundaries causes trajectories to fold back upon themselves. Consider Fig. 5.8 as an example. Here, a trajectory is indicated that actually closes upon itself. If the magnetic field is directed out of the paper, then that part of the trajectory with the dark shading is closed in a positive manner, while that part with the light shading is closed in a negative manner. Thus, these two contributions to the enclosed flux may cancel one another. Then, the consideration of weak localization is more complex than that discussed above. Now, for example, a single impurity scattering can break up the flux cancellation and lead to the onset of weak localization with a much greater effect than expected from a single scattering center. On the other hand, in trajectories that normally do not have flux cancellation, a single impurity scattering can cause the onset of weak localization by the influence of the boundaries. In this regime, one can distinguish the difference between a weak field (third row of Table 5.1) and a strong field (last row of the table) regime. In the former, many impurity scattering events are required to achieve weak localization. In the latter regime, on the other hand, a single impurity may be sufficient to cause this effect. Both of these cases are for the situation in which the magnetic length is large compared to the width, however, so that edge states and skipping orbits are not important. These effects

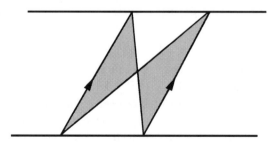

Fig. 5.8. Illustration of the manner in which diffusive scattering can cause cancellation of enclosed flux in an orbit.

lead to the numerical coefficients for the magnetic quantities in Table 5.1. In numerical simulations of quantum wires, it is generally found that neither limit is well formed, and an average of the two magnetic times is often preferred as [4]

$$\tau_{B,eff} = \tau_B^{weak} + \tau_B^{strong}. \tag{5.25}$$

The conduction correction for weak localization is now given for these 1D channels as

$$\delta G_{WL}^{1D}(B) = \frac{ge^2}{hL}\left[\left(\frac{1}{D\tau_\varphi} + \frac{1}{D\tau_B}\right)^{-1/2} - \left(\frac{1}{D\tau_\varphi} + \frac{1}{D\tau_B} + \frac{1}{D\tau}\right)^{-1/2}\right]. \tag{5.26}$$

This is, of course, the same result as (5.24) but with the addition of the term arising from elastic scattering, and the change in the definition of the effective magnetic relaxation time.

The use of these latter expressions improves the estimates of the coherence length from the magnetic decay for the short quantum wires discussed above. These now give numbers near $0.5\,\mu$m, but it is clear that these are nowhere near the value estimated from the amplitude of the weak localization correction at zero magnetic field. The shortness of the wire itself is the last culprit to be examined.

5.1.5 Extension to short wires

In the above discussions, it has been inherently assumed that the length L of the wire was large compared to any other characteristic length. In short wires, different effects may be observed. One needs to consider how the coherence may be broken within the regions at the ends of the wire. In other words, one needs to question whether the wire is terminated in two quasi-two-dimensional contact regions or by a network of other quantum wires, such as in the inset of Fig. 5.5. The question that must be addressed in the experiments is just what is being measured. When samples have characteristic lengths that are smaller than the coherence length, the amplitude of the conductance fluctuations often are found to be larger than the universal value of e^2/h [20], [21]. Second, the magnetoconductance is found to be asymmetrical in magnetic field [22]. More important, the values of the coherence length l_φ are smaller than those obtained from measuring long wires, even in the same material. Thus, it seems that in measurements in these short wires, the nature of the "contacting" regions becomes very important in determining just what is being measured. This problem was examined by Chandrasekhar et al. [23]. In steady state, the weak localization contribution to the conductivity is given by (5.5) to be

$$\Delta\sigma(\mathbf{r}) = -\frac{2e^2 D}{\pi\hbar}C(\mathbf{r}, \mathbf{r}')\delta(\mathbf{r} - \mathbf{r}'). \tag{5.27}$$

Here, $C(\mathbf{r}, \mathbf{r})$ is the time-integrated probability of return (e.g., when $\mathbf{r}' \to \mathbf{r}$) and therefore is given by the steady-state form of (5.11), in a magnetic field, as the solution to the equation

$$D\left(\left(-i\nabla - \frac{2e\mathbf{A}}{\hbar}\right)^2 + \frac{1}{l_\varphi^2}\right)C(\mathbf{r}, \mathbf{r}') = \delta(\mathbf{r} - \mathbf{r}'). \tag{5.28}$$

The quantity $C(\mathbf{r}, \mathbf{r}')$ is also called the particle-particle propagator because it describes the spatial correlation of the particle, and in this form will be our connection to a more basic quantum approach. On an insulating boundary, $C(\mathbf{r}, \mathbf{r}')$ is required to satisfy the condition of zero flow across the boundary, or

$$\left(-i\nabla_n - \frac{2e\mathbf{A}_n}{\hbar}\right)C(\mathbf{r}, \mathbf{r}') = 0, \tag{5.29}$$

where the subscript n indicates the normal component of the vector operator. For a simple wire of length L, this leads to the weak-localization contribution to the resistance to be ($\Delta G \ll G$)

$$\left.\frac{\Delta R}{R}\right|_{loc} = -\left.\frac{\Delta G}{G}\right|_{loc} = \frac{2e^2 D}{\pi\hbar\sigma_0 L} \int_0^L C(\mathbf{r}, \mathbf{r}) \, d\mathbf{r}. \tag{5.30}$$

It is quite usual to set the particle-particle propagator to zero, as in this equation. Yet in short wires, one cannot rule out the possibility that the coherent propagation extends well into the quasi-two-dimensional contacting regions. Here, $G = \sigma_0 L$ is the conductance for a one-dimensional wire. The above equation provides a simple approach to determine the weak-localization correction for any one-dimensional sample (one-dimensional in the fact that there is only an average of the particle-particle propagator over the length of the wire; higher dimensions require a more complete averaging process).

Chandrasekhar *et al.* [23] consider the general case in which the wire is terminated by a network of other one-dimensional wires, as well as the case in which it is terminated by two-dimensional regions. Here, we demonstrate only the latter situation. We work first in the absence of a magnetic field and count upon the substitutions developed above for the insertion of the field. In the one-dimensional wire, $C(\mathbf{r}, \mathbf{r}')$ depends only on the coordinates along the wire, which is taken to be the x-axis. Following these authors, the solution to (5.29) can be formulated in terms of Green's functions (here, the solutions will be hyperbolic sines and cosines) and connected to the two-dimensional regions. This leads to

$$\left.\frac{\Delta R}{R}\right|_{loc} = \frac{e^2 R_\square}{\pi\hbar} \frac{l_\varphi}{W} \left[\frac{(\eta^2 + \alpha^2)\coth\left(\frac{L}{l_\varphi}\right) - \frac{l_\varphi}{L}(\eta^2 - \alpha^2) + 2\alpha\eta}{(\eta^2 + \alpha^2) + 2\alpha\eta\coth\left(\frac{L}{l_\varphi}\right)} \right], \tag{5.31}$$

where $\alpha = W/l_\varphi$, and where

$$\eta = \pi \ln\left(\frac{2l_\varphi^{2D}}{l_e}\right) \tag{5.32}$$

describes the effects of the two-dimensional boundary layers. Here, $R_\square = 1/\sigma_\square$ is the sheet resistance of the two-dimensional layer. It is clear that in the case for which $L \gg l_\varphi$, the result reduces to the normal quantum wire result

$$\left.\frac{\Delta R}{R}\right|_{loc} = \frac{e^2 R_\square}{\pi\hbar} \frac{l_\varphi}{W}. \tag{5.33}$$

In the opposite limit, $l_\varphi \gg L$, the fluctuations are dominated by the two-dimensional regions, and

$$\left.\frac{\Delta R}{R}\right|_{loc} = \frac{e^2 R_\square}{\pi\hbar\eta}. \tag{5.34}$$

It should be mentioned that if the wire is measured in a four-probe configuration with narrow probes attached at the junction of the two-dimensional region and the wire of interest, then $\eta = \eta_1 + \eta_2$, where η_1 corresponds to (5.32) with the angle π replaced by the appropriate acceptance angle of the two-dimensional region, and $\eta_2 = W_p/l_\varphi^p$, which describes the nature of the voltage probe wires. In the presence of the magnetic field, the magnetic relaxation time is introduced through the replacement of the appropriate τ_φ according to the same procedure as above,

$$\frac{1}{\tau_\varphi} \rightarrow \frac{1}{\tau_\varphi} + \frac{1}{\tau_B}, \tag{5.35}$$

with the appropriate τ_B introduced into each of the coherence lengths used in (5.31).

The major effect of the results obtained in this section is that the propagator in the wire does not end (is not terminated to a zero value) at the contacts but rather transitions smoothly into the appropriate one- or two-dimensional propagators for the "contacting regions." These differ by having either one- or two-dimensional values for the coherence lengths. The latter differ in the wire and in the two-dimensional regions due to the additional effect of the side-wall scattering as discussed above. Careful analysis of the data in Fig. 5.1, with the formulas of this section, now leads to an estimate for the coherence length of $2.5\,\mu$m in the wire section and about $5.0\,\mu$m in the two-dimensional sections to which the wire is connected. The actual fit is not particularly sensitive to the latter quantity, however, so the value for this quantity is not well determined by such an analysis. The value for the wire is now quite close to that estimated from the amplitude of the weak-localization correction at zero magnetic field.

5.2 Universal conductance fluctuations

It is clear from the above discussion for weak localization that the existence of time-reversed paths can lead to quantum interference that causes a correction to the conductivity. These paths are modified by the change in the electrochemical potential (changes in the wavephase velocity) and in the magnetic field (similar changes in the momentum occur through the vector potential). If a sample is composed of a great many such loops (and not necessarily time-reversed loops), with each contributing a phase-dependent correction to the conductivity, then the summation over these loops may or may not ensemble average to zero, depending upon the number of such loops contained within the sample (and, hence, upon the size of the sample). On the other hand, in mesoscopic systems, where the number of such loops is relatively small, this is not the case, and these loops lead to the presence of universal conductance fluctuations. Since the impurity distribution is reasonably fixed for a given sample configuration, the particular oscillatory pattern is often thought of as a *fingerprint* of an individual sample. That is, the impurity distribution is different from one sample to another, and therefore the detailed nature of the interference pattern seen in any one sample is a characterization of its unique impurity distribution. Perhaps the most remarkable feature of these oscillations is that, for $L \gg l_e$, the amplitude of these oscillations seems to satisfy a universal scaling with a nominal value of $\delta G = e^2/h$, regardless of the size of the sample (but notice the discussion of this below) and the degree of disorder [24], [25]. One view of this universality in amplitude arises from the Landauer formula, in which the conductance is quantized for each possible channel through the sample. Here, the conductivity is given by (see Chapter 3) [26]

$$G = \frac{e^2}{h} \sum_{i,j=1}^{N} |t_{ij}|^2, \tag{5.36}$$

where t_{ij} is the transmission from incoming channel i to outgoing channel j. In this regard, one conclusion that may be drawn from the above is that the fluctuation in conductance arises from the random turning on and/or off of a single channel of the many channels that are occupied. Generally, one would now construct the variance of G, but the problem lies in the fact that there will be correlation between the individual t_{ij}, especially for similar pairs of ingoing and outgoing waves. However, since we are considering channels that are primarily transmitting, the correlation between the reflection coefficients can be much smaller (and therefore more easily averaged) [27]. Thus, we rewrite (5.36) as

$$G = \frac{e^2}{h}\left[N - \sum_{i,j=1}^{N} |r_{ij}|^2 \right], \tag{5.37}$$

where N is the number of channels, and the variance of the conductance is given simply by

$$var(G) = \left(\frac{e^2}{h}\right)^2 var\left(\sum_{i,j=1}^{N} |r_{ij}|^2 \right) = \left(\frac{e^2}{h}\right)^2 N^2 var\left(|r_{ij}|^2\right), \tag{5.38}$$

with uncorrelated reflection coefficients. The exact evaluation of the variance is quite difficult and will be deferred until later. However, one simple approximation is to assume that Wick's theorem carries over to this problem, by which we mean that there are no high-order correlation functions. Thus we can use the simple approximation

$$\langle |r_{ij}|^4 \rangle = 2\langle |r_{ij}|^2 \rangle^2. \tag{5.39}$$

To lowest order, the last term in the angular brackets is simply $1/N$, so that

$$var(G) \sim \left(\frac{e^2}{h}\right)^2 N^2 \frac{2}{N^2} \sim 2\left(\frac{e^2}{h}\right)^2. \tag{5.40}$$

The numerical prefactors are more problematic and have to be treated carefully [28], [29]. However, the general result can be written in the form [12], [25]

$$\delta G = [var(G)]^{1/2} = \frac{g}{2}\frac{C}{\sqrt{\beta}}\frac{e^2}{h}, \tag{5.41}$$

where g is a factor for spin and valley degeneracies, C is a constant of order unity (C is about 0.73 for a narrow channel with $L \gg W$, but on the order of $\sqrt{W/L}$ in the opposite limit of a wide channel), and β is unity in zero magnetic field but takes a value of 2 when the magnetic field lifts the time-reversal symmetry of the system. (It should also be noted that the lifting of the spin degeneracy by the magnetic field leads to this contribution to the degeneracy factor to be reduced from 2 to $\sqrt{2}$.)

The simplest experiments were discussed in Fig. 1.7, which was measured for a wide wire formed in a Si MOSFET. These measurements have been studied for a large range of effective wire lengths L. Similar universal conductance fluctuations are also apparent in the data of Fig. 5.1. If we calculate the variance of this data, where the amplitude is relatively constant down to zero field, we arrive at a value of C of about 0.4.

In general, most measurements are made under constant current conditions (as was the latter data), so that fluctuations in the voltage couple to *resistance* fluctuations (a similar effect appeared in the discussion of the previous section). If the resistance fluctuations are independent of the sample length, then it is expected that $\delta G \approx \delta R/R^2 \sim L^{-2}$, since the resistance is expected to scale linearly with the length of the wire. The results of an

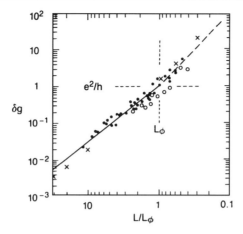

Fig. 5.9. Variation of the conductance fluctuation with the wire length, demonstrating the scaling behavior. [After Skocpol, Phys. Scripta **T19**, 95 (1987), by permission.]

extensive study of Si wires are shown in Fig. 5.9, where the reduced conductance fluctuation (in units of e^2/h) is plotted as a function of L/l_φ [30]. For $L < l_\varphi$, this behavior is clearly observed; that is, for lengths smaller than the coherence length, the resistance fluctuation is independent of the length of the sample. This result indicates that the voltages on the individual probes are fluctuating with some coherence to one another (the measuring length is still l_φ, which can reach around from one voltage probe to another). The net voltage fluctuation is then independent of L, since the critical length for coherence is l_φ. On the other hand, when $L > l_\varphi$, the measurement is over many phase-coherent regions (approximately L/l_φ such regions), so that the fluctuation in the resistance is proportional to the number of these resistances. In this case, the two voltage probes are considered to be oscillating independently from one another, with the amplitude of the variation proportional to the square of the number of such phase-independent cells. Thus, $\delta R \sim \sqrt{N} \sim L^{1/2}$, which leads to a variation in the conductance fluctuation as $L^{-3/2}$, and this behavior is also clearly seen in the figure. Finally, we note that the estimate of the data from Fig. 5.1 lies within the spread of experimental points in Fig. 5.9 when we consider that the value of the phase coherence length determined previously is slightly larger than the length of the wire itself.

The behavior of the scaling of the voltage (or the resistance) with the length is also illustrated in Fig. 5.10 [20]. Here, it is important to note that under the application of a magnetic field, there is a part of the fluctuation that is symmetric in magnetic field and a part that is asymmetric. Only the symmetric part satisfies this behavior of relative independence for $L > l_\varphi$. In Fig. 5.10a, the voltage fluctuation itself is plotted as a function of the relative length of the sample. Here, the symmetric and asymmetric parts are separated and plotted. In Fig. 5.10b, the corresponding conductance fluctuation is plotted. Although one might think that there should not be a part of the voltage fluctuation that is asymmetric in magnetic field (most of us are taught that magnetic field effects should be symmetric in the field), it should be remembered that these mesoscopic devices behave like small *circuits* rather than as discrete devices. As circuits, they exhibit nonlocal properties, whether simple circuit theory or the equivalent microwave modal theory (see Chapter 3) is introduced. Consider, for example, the device of Fig. 5.11. If we feed the current through the contacts labeled 1 and 4, the voltage measured between contacts 2 and 3 will depend upon the manner in which

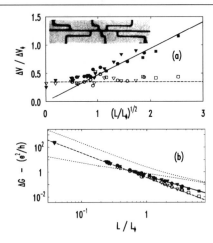

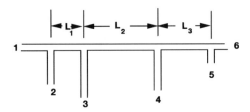

Fig. 5.10. (a) The measured voltage fluctuations, and (b) the conductance fluctuations as a function of wire length. [After Benoit *et al.*, Phys. Rev. Lett. **58**, 2343 (1987), with permission.]

Fig. 5.11. A typical multi-lead sample geometry with the leads numbered for reference.

leads 5 and 6 are terminated. If these latter two leads are grounded, they act as current sinks. If they are left open-circuited, then they can have floating voltages. The principle that this will strongly affect the measurements is well known from introductory circuit theory through Thevenin's equivalent circuit principles. In fact, if an Aharonov-Bohm ring is attached between lead 5 and ground, the voltage measurements between leads 2 and 3 will show oscillations that arise from the varying conductance in lead 5 to ground. (This is true even if the end of the ring is left "open," since there really is no such thing in the semiconductor circuits due to parasitic resistances and capacitances.) These effects are seen in mesoscopic devices [31]. The reversal of the magnetic field, and the consequent asymmetry, brings another basic circuit principle into play, the principle of reciprocity, in which measurements are required to be equal only when the current and voltage leads are interchanged in a multi-loop/node circuit (this was discussed in Chapter 3). In reversing the magnetic field, the symmetry of the problem requires the invoking of reciprocity, so that [32]

$$V_{ij,kl}(B) = V_{kl,ij}(-B), \tag{5.42}$$

where the first set of indices refers to the current probes and the second set of indices refers to the voltage probes. The full importance of this relation was not appreciated when mesoscopic systems were first studied, and only after several experiments and the explanation of Büttiker [32] was it realized that (5.42) was the proper form in which the Onsager relation would

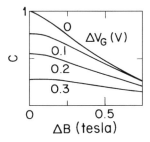

Fig. 5.12. The correlation function for the fluctuations measured in Fig. 1.6 are plotted as a function of the change in magnetic field, with the change in gate voltage as a parameter. [After Skocpol *et al.*, Phys. Rev. Lett. **56**, 2865 (1986), by permission.]

appear. Since the current is a constant value in the experiments, the voltages indicated are precisely just a constant value different from the resistances, so that $R_{ij,kl} = R_{kl,ij}$.

It was indicated in the earlier treatment that the magnetic field would cause the weak-localization correction to decay. But, we have pointed out that the universal conductance fluctuation arises from similar considerations, and that any variation in the position of the electrochemical potential in the density of states, whether caused by a gate bias or by a magnetic field, would also cause the correlation in the loops to decay (or at least to change). This is treated in the universal conductance fluctuation by the presence of a correlation function

$$C(\Delta E, \Delta B) = \langle G(E + \Delta E, B + \Delta B)G(E, B)\rangle - \langle G(E, B)\rangle^2. \qquad (5.43)$$

It is clear that the magnetic field variation causes a change in the correlation function. The energy variation arises from the facts that the impurities cause an inhomogeneous energy surface and that different current paths will see different local variations in the chemical potential. This means that small changes in the overall chemical potential, due to an applied gate voltage, for example, will cause larger variations in the local chemical potential. This will lead to an equivalent correlation "length" in the energy, or in the gate voltage. In Fig. 5.12, the correlation function is plotted for several values of ΔV_G (which corresponds to ΔE) as a function of the magnetic field for the measurements shown in Fig. 1.7. The correlation function has a normal decay and can be characterized by correlation energies ΔE_c and magnetic fields ΔB_c. (The tendency is to use a different term than the correlation field B_c, which has been used previously to define the value of the field at which the magnetic decay time is comparable to the phase-breaking time in weak localization, although the two terms arise from the same physical processes and are both used often with the same symbolism.) At least for low magnetic fields, these correlation fields and energies are universal quantities related to the size of phase-coherent regions.

The correlation "lengths" that can be defined from the correlation function (5.43), and measurements such as those of Fig. 5.12, are in terms of magnetic fields $\Delta B = \Delta B_c$ and energies $\Delta E = \Delta E_c$, which may formally be defined theoretically, as will be done later in this chapter. The magnetic field correlation length ΔB_c, the correlation field, is the critical field that characterizes the magnetic field–induced dephasing of the interference phenomena. The correlation field is defined as that field at which C is reduced to its half-height, so that the correlation field is the half-width at half-height of the correlation function. As in weak localization, the correlation function is made up of three factors [3]: (1) the probability of return of the trajectory from $\mathbf{r}$ to $\mathbf{r}'$, in a time T, independent of the magnetic field, (2) a phase relaxation factor, and (3) the average phase factor itself. The averages in the correlation function are taken over all classical trajectories that can contribute to the diffusive transport being measured. In fact, therefore, the critical correlation field corresponds to coupling flux quanta through the area of a so-called phase-coherent region corresponding to some product of length and width within the sample. One can expect that these length scales will then be the smaller of the characteristic length of the sample (width or length) and the phase coherence length l_φ. Also, as in the case of weak localization, the dephasing is caused by the magnetic field in a manner in which the effective inelastic scattering process is given by (we use a common notation here, which will be explained further below)

$$\frac{1}{\tau_\varphi} \to \frac{1}{\tau_\varphi} + \frac{1}{\tau_{\Delta B/2}}, \tag{5.44}$$

so that the magnetic field actually does produce dephasing of the correlation function in the same manner as it reduces the weak-localization correction to the conductance. It is for this reason that these phenomena are termed *universal*. The magnetic field at which the correlation function is reduced to half-height is the same quantity given in Table 5.1, but with the substitution $B \to \Delta B/2$. The factor of 2 that enters this latter substitution has to do with the nature of the averaging process that appeared in (5.39), in that we are no longer dealing with *two* time-reversed interfering paths. Since we are not using two time-reversed paths, but only a *single net loop* around the enclosed area, the magnetic field coupled to the loop is reduced by a factor of 2 from that in the weak-localization correction. Thus, for UCF, we replace $2B$ in the magnetic decay time by ΔB (or B by $\Delta B/2$), where ΔB is the shift in the magnetic field used in the correlation function (5.43). This means that for a wire in the dirty metal regime

$$\tau_{\Delta B/2} = \frac{12 l_m^4}{W^2 D} = 12 \left(\frac{\hbar}{e \Delta B} \right)^2 \frac{1}{W^2 D}, \quad l_e \ll W \ll l_\varphi, \tag{5.45}$$

and this leads to the correlation field

$$\Delta B_c = 0.55 \frac{h}{e} \frac{1}{W l_\varphi}, \quad l_e \ll W \ll l_\varphi. \tag{5.46}$$

In fact, the numerical pre-factor can vary somewhat with the measurements. It should be noted that this value differs from the equivalent value obtained from the weak localization. For a wire in the pure metal regime,

$$\tau_{\Delta B/2} = 4 C_1 \left(\frac{\hbar}{e \Delta B} \right)^2 \frac{1}{W^3 v_F} + 2 C_2 \left(\frac{\hbar}{e \Delta B} \right) \frac{l_e}{W^2 v_F}, \quad W \ll l_e, \tag{5.47}$$

where the weak and strong magnetic field results have been combined. The quantities C_1 and C_2 were discussed below in Table 5.1. This now leads to

$$\Delta B_c = \frac{C_2}{4\pi} \frac{h}{e} \frac{l_e^2}{W^2 l_\varphi^2} \left[1 \mp \sqrt{1 \mp \frac{8 C_1}{C_2} \frac{W l_\varphi^2}{l_e^3}} \right], \quad W \ll l_e. \tag{5.48}$$

It should be remarked, however, that this version of the correlation magnetic field differs from that obtained in the earlier detailed theories, where it was found that [28]

$$\Delta B_c = \beta \frac{h}{e} \frac{1}{W l_\varphi}, \quad W \ll l_\varphi, l_e. \tag{5.49}$$

Here, $\beta \sim 0.25$ is found for a variety of long wires in high-mobility material [33]. A similar behavior can be found in the low-magnetic-field limit of (5.48) only if $W \sim l_e$ is invoked and much smaller values of the constants are used. In this situation, the elastic mean free path is that determined from the surface scattering, rather than from the basic mobility of the material. Thus, one should use some care in actually using this great variety of formulas for computing any one quantity.

While it can not easily be seen from Fig. 5.1, the correlation field has been determined to be about 0.8 mT at a width of 0.5 μm. If we use the previously determined values for l_φ, and this correlation field, it is found that $\beta \sim 0.4$ for this wire. On the other hand, if we use (5.46), then a good fit to the earlier value of coherence length is found. This suggests that the constants C_1 and C_2 are somewhat high for this experimental wire. It should be apparent that obtaining a unique value for the coherence length is difficult, and only when a number of determinations from different sets of measurements is achieved with some consistency is the end result believable.

The values of the constants depend on the nature of the surface scattering at the edge of the wire. In actual samples being measured, it may not be clear just which regime is appropriate for using a formula. This has complicated the evaluation quite often, and sometimes it is simpler to use numerical coefficients in these expressions in a manner that allows them to be adjusted for a particular sample. However, the above discussion leads to the conclusion that, once the regime of operation is known and the values for the width, length, elastic scattering length, and coherence length are known, the amplitude of the fluctuations and the correlation functions are universal for any value of the magnetic field and bias. This is what is meant by universal behavior. This is found to be the case so long as the magnetic field remains relatively small, that is, so long as the motion remains diffusive in the bulk of the sample. We discuss the high-magnetic-field behavior in a later section.

5.3 The Green's function in transport

At this point, we want to begin to try to understand the standard methods by which weak localization and universal conductance fluctuations are treated by the theoretical world in a quantum-mechanical manner. This involves some rather complicated mathematical approaches, because the long-range correlations are important for phase interference, and these are usually difficult to evaluate. The approach we follow here is based on the Green's functions introduced in Chapter 3. There, however, the concentration was on generating a lattice representation that could be used for numerical simulations. Here, on the other hand, the concentration is on the analytical approach. Some repetition of the previous material will be given in order to facilitate the understanding. In this section, we will basically review the use of the Green's functions in analytically calculating the conductivity. The following sections will examine weak localization and universal conductance fluctuations.

The beginning point is the recognition that any initial wavefunction $\Psi(\mathbf{r}', t')$, which can be considered as an arbitrary initial condition, can be determined at any other point at a (not necessarily) later time from the relationship

$$\Psi(\mathbf{r}, t) = \int d^3 r' \int_{t'}^{t} dt' \, K(\mathbf{r}, t; \mathbf{r}', t') \Psi(\mathbf{r}', t'). \tag{5.50}$$

Here, $K(\mathbf{r}, t; \mathbf{r}', t')$ is the propagator, or kernel, or Green's function $G(\mathbf{r}, t; \mathbf{r}', t') = -iK(\mathbf{r}, t; \mathbf{r}', t')$ for the Schrödinger equation. These quantities may be related to the actual Green's functions that one uses, the retarded and advanced Green's functions (we use the zero-temperature Green's functions in this chapter).

The kernel in (5.50) describes the general propagation of any initial wavefunction at time t' to any arbitrary time t (which is normally $>t'$, but not necessarily so). There are a number of methods of evaluating this kernel, either by differential equations (which will be pursued here) or by integral equations known as path integrals [34]. In general, the form shown here is developed for a system characterized by a well-defined set of basis functions, which are characteristic of the entire problem. We will see in the next chapter that one can just as easily take a thermodynamic equilibrium basis by passing to imaginary time with the substitution $(t - t') \rightarrow -i\hbar\beta$, where $\beta = 1/k_B T$ is the inverse temperature. We will see still later that a further approach is to use real-time Green's functions in a manner that will require us to actually solve for the distribution function for the states in a nonequilibrium system. We work here with the first basis, in which it is assumed that the temperature $T = 0$. Thus, all states up to the Fermi energy are completely full, and all states above this energy are normally empty, in the absence of any perturbation to the equilibrium situation.

In general, one separates the kernel in the wavefunction (5.50) into forward and reverse time-ordering in order to have different functions for the retarded (forward in time) and advanced (backward in time, in the simplest interpretation) behavior. We do this by introducing the retarded Green's function as (for fermions)

$$G_r(\mathbf{r}, \mathbf{r}'; t, t') = -i\Theta(t - t')\langle K(\mathbf{r}, t; \mathbf{r}', t')\rangle = -i\Theta(t - t')\langle\{\Psi(\mathbf{r}, t), \Psi^\dagger(\mathbf{r}', t')\}\rangle, \tag{5.51}$$

where the angle brackets have been added to symbolize an ensemble average, which is also a summation over the proper basis states, and where the curly brackets indicate an anticommutator for the anticommuting electrons with which we are working. On the other hand, the advanced Green's function is given by

$$G_a(\mathbf{r}, \mathbf{r}'; t, t') = i\Theta(t' - t)\langle K(\mathbf{r}, t; \mathbf{r}', t')\rangle = i\Theta(t' - t)\langle\{\Psi^\dagger(\mathbf{r}', t'), \Psi(\mathbf{r}, t)\}\rangle, \tag{5.52}$$

and one can write the kernel itself as

$$\langle K(\mathbf{r}, t; \mathbf{r}', t')\rangle = i[G_r(\mathbf{r}, \mathbf{r}'; t, t') - G_a(\mathbf{r}, \mathbf{r}'; t, t')]. \tag{5.53}$$

This particular form allows a symmetrized kernel to be used for (5.50) for which both the field operator and its adjoint satisfy this integral equation. Here, these field operators satisfy the anticommutation relations [35]

$$\{\Psi(\mathbf{r}, t), \Psi^\dagger(\mathbf{r}', t')\} = \delta(\mathbf{r} - \mathbf{r}')\delta(t - t'). \tag{5.54}$$

The Green's functions include two time variables and two spatial variables, thus describing the propagation between two different points in the four-variable space $x = (\mathbf{r}, t)$ (this four-variable notation will occasionally be used to simplify the number and complexity of integrations). Thus, temporal correlation processes can be incorporated in a fundamental manner. With this added complication – far more equations will be required to solve for the Green's functions than for the Schrödinger wavefunction – comes the benefit of a much

more direct incorporation of dissipative processes. The equations of motion for the Green's functions can be developed for the simple propagation in the absence of any interactions from the basic Schrödinger equation. This leads to the pair of equations

$$\left(i\hbar\frac{\partial}{\partial t} - H_0(\mathbf{r}) - V(\mathbf{r})\right)G_0(\mathbf{r}, \mathbf{r}'; t, t') = \hbar\delta(\mathbf{r} - \mathbf{r}')\delta(t - t'), \qquad (5.55)$$

$$\left(-i\hbar\frac{\partial}{\partial t'} - H_0(\mathbf{r}') - V(\mathbf{r}')\right)G_0(\mathbf{r}, \mathbf{r}'; t, t') = \hbar\delta(\mathbf{r} - \mathbf{r}')\delta(t - t'). \qquad (5.56)$$

The Green's function here is either the retarded or the advanced function, with the difference determined by the ordering of the two time variables. In general, we can invoke the causality of the functions and make them arguments of $t - t'$, rather than of t, t' separately. Similarly, we can assert that these functions have properties that depend only on the difference in spatial variables (there is no particular value to any one point in space, and *thus the system is basically homogeneous*, for such an approximation). If we then Fourier transform in space and time, we find that

$$G_0(\mathbf{k}, \omega) = \frac{\hbar}{\hbar\omega - E(\mathbf{k})}, \qquad (5.57)$$

where $E(\mathbf{k})$ is the energy eigenvalue of $H_0 + V$. The latter potential is a confining potential that leads to a set of basis states so that the actual Green's function must be a sum over these states. In general, to invert the Fourier transform for the time variable, one must worry about the closure of the contour integration. Either the upper or lower half-plane must be taken in the complex ω-space. One half-plane leads to the retarded function, and the other to the advanced function. Simply speaking, these are separated by introducing an infinitesimal energy η for which

$$G_0^{r,a}(\mathbf{k}, \omega) = \frac{\hbar}{\hbar\omega - E(\mathbf{k}) \pm i\eta}, \qquad (5.58)$$

where the upper sign is used for the retarded function and the lower sign is used for the advanced function. This small imaginary quantity moves the poles from the real axis into the lower half-plane for the retarded function and into the upper half-plane for the advanced function. This then gives the closure conditions to assure inclusion of the poles in the integration contour. This complexity becomes unnecessary once dissipation is incorporated in the problem.

We note that, in the case of a simple parabolic energy band for nearly free electrons $(V \to 0)$, $E(\mathbf{k}) = \hbar^2 k^2/2m$. This leads to the form

$$G_0^{r,a}(\mathbf{k}, \omega) = \frac{\hbar}{\hbar\omega - \frac{\hbar^2 k^2}{2m} \pm i\eta}. \qquad (5.59)$$

This is also the Fourier-space propagator for the diffusion equation if we replace $\hbar/2m$ by the diffusion constant D, so that the retarded and advanced Green's functions are often referred to as the results from *diffusion poles*. It should be noted, in particular, that the units of $\hbar/2m$ are cm^2/s, which are the same as those of the diffusion constant. The actual value has no connection to real transport numbers, but it should be recalled that this will provide a renormalization of the crucial frequencies (energies) when real numbers are used. We will use this fact later to replace the free-particle Green's function with a dressed *diffusion* particle.

5.3.1 Interaction and self-energies

In the classical case, and certainly before many-body theories were fully developed, the theory of electrical conduction was based upon the Boltzmann transport equation for a one-particle distribution function. Certainly this approach is heavily used still, particularly in the response of semiconductor systems in which the full quantum response is not necessary [2]; this was the case in the last chapter for Coulomb blockade and in Chapter 2. The essential assumption of this theory is the Markovian behavior of the scattering processes, that is, each scattering process is fully completed and independent of any other process. Coherence of the wavefunction is fully destroyed in each collision. On the other hand, we are talking about coherence between a great many scattering events. Certainly, replacements for the Boltzmann equation exist for the non-Markovian world, but the approach of interest here is that of the Kubo formula discussed earlier. This linear response behavior describes transport with only the assumption that the current is linear in the applied electric field. The result, the Kubo formula, is that the current itself (or an ensemble averaged current) is given by

$$\langle \mathbf{J}(\mathbf{r}, t) \rangle = \langle \mathbf{j}(\mathbf{r}) \rangle + \int dt' \int d^3 \mathbf{r}' \hat{\mathbf{G}}(\mathbf{r} - \mathbf{r}', t - t') \cdot \mathbf{A}(\mathbf{r}', t'), \tag{5.60}$$

where $\hat{\mathbf{G}}$ is the conductance tensor and $\mathbf{A}$ is the vector potential. In the diffusive limit, we can ignore the ballistic response (which leads to plasma oscillation effects in the dielectric function), and the conductance is given by the current-current correlation function,

$$\hat{\mathbf{G}}(\mathbf{r} - \mathbf{r}', t - t') \simeq -i\Theta(t - t')\langle [\mathbf{j}(\mathbf{r}, t), \mathbf{j}(\mathbf{r}', t')] \rangle. \tag{5.61}$$

If we recognize that the current operator is given by

$$\mathbf{j}(\mathbf{r}) = -i\frac{e\hbar}{2m}\{\Psi^\dagger(\mathbf{r})\nabla\Psi(\mathbf{r}) - [\nabla\Psi^\dagger(\mathbf{r})]\Psi(\mathbf{r})\} \rightarrow \frac{e\hbar k}{m}\Psi^\dagger(\mathbf{r})\Psi(\mathbf{r}), \tag{5.62}$$

we can Fourier transform everything and arrive at the solution

$$\sigma_{\alpha\beta} = \frac{e^2\hbar^2}{m^2}\sum_{\mathbf{k},\mathbf{k}'} k_\alpha k_\beta' G^{(2)}(\mathbf{k}, \mathbf{k}; \mathbf{k}', \mathbf{k}'; t' = t)\delta(E_k - E_F), \tag{5.63}$$

where $G^{(2)}(\mathbf{k}, \mathbf{k}; \mathbf{k}', \mathbf{k}')$ is the two-particle Green's function, and the last delta function insures that the evaluation is done at the Fermi level, since only those particles at the Fermi surface can contribute to transport at $T = 0$. The latter form is arrived at rather simply by Fourier-transforming (5.60) in both space and time, noting that the commutator in (5.61) will lead to four terms, and then by using (5.62) to replace each current term. The four field operators combine to define the *two-particle Green's function*.

While the above has been written in momentum space and in time, we can examine the behavior somewhat differently in real space. The two-particle Green's function is defined in terms of four wavefield operators. It can be written (using the time-ordering operator T) as

$$G^{(2)}\left(\mathbf{r}, t, \mathbf{r}_1, t_1; \mathbf{r}', t', \mathbf{r}_1', t_1\right) = -\langle T[\Psi^\dagger(\mathbf{r}, t)\Psi(\mathbf{r}_1, t_1)\Psi^\dagger(\mathbf{r}_1', t_1')\Psi(\mathbf{r}', t')] \rangle. \tag{5.64}$$

There are various approximations to evaluate this two-particle Green's function. The most usual are the Hartree and Hartree-Fock approximations. For the Hartree approximation, we neglect possible exchange interactions, and write (5.64) as

$$G^{(2)}\left(\mathbf{r}, t, \mathbf{r}_1, t_1; \mathbf{r}', t', \mathbf{r}_1', t_1\right) = G(\mathbf{r}, t; \mathbf{r}', t')G(\mathbf{r}_1, t_1; \mathbf{r}_1', t_1'). \tag{5.65}$$

Under the conditions for the above derivation (homogeneous conductance and the zero-frequency static conductivity), we may rewrite this in Fourier transform form as the product of two time functions, giving a convolution integral in Fourier space; since we are interested in the static result, only the frequency integral survives. The momentum integration is already contained in (5.63) and so does not appear. This result is then

$$G^{(2)}(\mathbf{k}, \mathbf{k}; \mathbf{k}', \mathbf{k}') = \int \frac{d\omega}{2\pi} G^r(\mathbf{k}, \mathbf{k}', \omega) G^a(\mathbf{k}, \mathbf{k}', \omega). \tag{5.66}$$

It is important to reiterate that this last result is a d.c. result and the factor ω in the integral is not the applied frequency, but the energy $E/\hbar$. The result is actually the d.c. (not the a.c.) conductivity, as has been stated several times. The interpretation of this term is that an electron is excited across the Fermi energy (at $T = 0$, all states are filled up to the Fermi energy), which creates an electron-hole pair in momentum state $\mathbf{k}'$. The creation is assumed to be accomplished by a nonmomentum-carrying process, such as a photon (whose momentum is considerably smaller than that of the electron and/or hole). This pair then propagates, scattering from charged impurities, phonons, and other centers, to state $\mathbf{k}$, where it recombines, again giving up the excess energy to a momentum-less particle of some type. This is then a particle-hole propagator, since the electron excited above the Fermi energy is called a quasi-particle with its characteristic energy measured from the Fermi energy itself. Similarly, the hole is a quasi-particle existing below the Fermi energy, and its energy is measured downward from the Fermi energy.

This notation very simply arises from the assertion that the absolute energy of the system is unimportant, and only the relative energies have sense. Since the total energy of these two particles remains $2E_F$, there is no confusion if we take the zero of energy at the Fermi energy. Changing the relation between the energy and momentum, which is often described by the mass in semiclassical systems, is more problematic but handled by talking about a quasi-particle mass. This treatment works well when the energy of the quasi-particles is only slightly different from the Fermi energy (i.e., in the linear response regime). Here, we retain the Fermi energy as a nonzero quantity for the moment.

Finally, let us connect the Green's functions in (5.66) with the noninteracting Green's functions of (5.59). This can be done as

$$G_0^{r,a}(\mathbf{k}, \mathbf{k}', \omega) = G_0^{r,a}(\mathbf{k}, \omega)\delta_{\mathbf{k}\mathbf{k}'}. \tag{5.67}$$

While there are some interactions which raise this conservation of the momentum in the Green's function of the interacting system, these will not be dealt with here. The Hartree-Fock approximation adds another term in which the two $\mathbf{k}$-states (of the electron and hole) are interchanged due to electron exchange. This is of more interest in the electron-electron interaction and will be treated later.

5.3.2 Impurity scattering

As the first introduction of the methodology to be used here, we consider the contribution of impurity scattering to the resistivity (or conductivity, as the case may be). The impurity has an associated Coulomb potential that is long-range in nature. This long-range interaction is usually cut off either by assuming that it is screened by the electrons (through the electron-electron interaction) or by some arbitrary distance. We will choose the latter approach as a crude approximation in that we assume the integrals (or summations) converge, but this should not be construed as any limit on the process; it is only done to avoid dealing with the divergences in some integrals that accompany the long-range interaction if it is not cut off. In any real physical system, the long-range interaction is certainly cut off by at least the screening process. The impurity potential is given by

$$V_{imp} = \int d^3\mathbf{r} \sum_j U(\mathbf{r} - \mathbf{R}_j)\bar{n}(\mathbf{r})$$

$$= \frac{1}{\Omega} \sum_{\mathbf{q},j} U(-\mathbf{q})e^{i\mathbf{q}\cdot\mathbf{R}_j} \sum_{\mathbf{k},\mathbf{k}'} \Psi_{\mathbf{k}'}^\dagger \Psi_{\mathbf{k}} \delta_{\mathbf{k}',\mathbf{k}-\mathbf{q}} \delta_{ss'}, \qquad (5.68)$$

where the last Kronecker delta function conserves the spin of the particle, and Ω is the volume. Usually, the wavefunctions (taken here to be field operators) reduce to the creation and annihilation operators for plane-wave states. Although the primary role may well be the scattering of the electron (or hole) from one plane-wave state to another, there is also a recoil of the impurity itself, which can ultimately couple into local modes of the lattice. This latter complication will not be considered here, although it can be a source of short-wavelength phonons that can cause impurity-dominated intervalley scattering. Generating the perturbation series usually relies upon the S-matrix expansion of the unitary operator

$$\exp\left(-\frac{i}{\hbar}\int_{t'}^{t} dt'' V_{imp}(t'')\right). \qquad (5.69)$$

The treatment to be handled first is one in which we treat the two Green's functions in (5.66) independently. We will look at their interaction through the impurities later.

The expansion of the S-matrix in the scattering operator leads to an infinite series of terms, which change the equilibrium state. The higher-order terms are usually broken up by the use of Wick's theorem [36], and this leads to a diagram expansion. The formation of the Green's functions in the S-matrix expansion is accompanied by an averaging process over the equilibrium state (this is actually coupled to an average over the impurity configuration as well, which is discussed below). The equilibrium state must be renormalized in this process, and this causes a cancellation of all disconnected diagrams [37]. The result is an expansion in only the connected diagrams. Typical diagrams for the impurity scattering are shown in Fig. 5.13a. One still must carry out an averaging process over the position of the impurities, since the end result (at least in macroscopic samples) should not depend upon the unique distribution of these impurities.

The impurity averaging may be understood by noting the summation over the impurity positions that appear in the exponential factors in (5.68). Terms like those in Fig. 5.13a involve the average

$$\left\langle \sum_{\mathbf{R}_i} e^{i\mathbf{q}\cdot\mathbf{R}_i} \right\rangle \to \delta(\mathbf{q}), \qquad (5.70)$$

where the angular brackets denote the average over impurity positions. If the number of impurities is large, then the averaging of these positions places the important contribution of the potential from the impurities as that of a regular array, which can be thought of as creating a superlattice. The vectors $\mathbf{R}_i$ are then the basis vectors for this lattice, and the summation is over all such vectors. In short, the summation then represents the closure of a complete set, which yields the delta function shown on the right side of the arrow

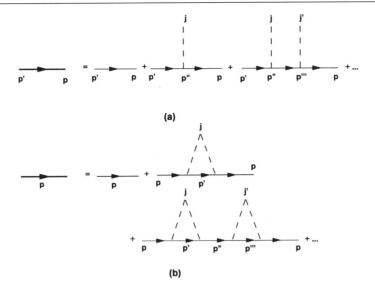

Fig. 5.13. (a) Typical impurity scattering single interactions involved in impurity averaging. (b) The second-order terms resulting in the dominant interactions after impurity averaging.

[37]. Thus, only the series of events in which the impurity imparts zero momentum to the electrons is allowed in the scattering process after the impurity averaging. Now, consider the double scattering processes shown in Fig. 5.13b, in which a single impurity interacts to second order with the propagating electron. Now, there are two momenta imparted by the impurity, and the averaging in (5.70) becomes

$$\left\langle \sum_{\mathbf{R}_i} e^{i(\mathbf{q}_1 + \mathbf{q}_2)\cdot\mathbf{R}_i} \right\rangle \to \delta(\mathbf{q}_1 + \mathbf{q}_2), \tag{5.71}$$

so that one arrives at $\mathbf{q}_1 = -\mathbf{q}_2$. Thus, the interaction matrix element contained in the resultant expansion term from (5.69) for the second interaction is the complex conjugate of that for the first interaction, and the overall scattering process is proportional to the magnitude squared of the matrix element. This is the obvious result expected if one had started with the Fermi golden rule rather than Green's functions. Impurities can also interact with three lines and four lines, and more, as obvious extensions of the two situations shown in Fig. 5.13. The terms from (5.70) produce only an unimportant shift in the energy that arises from the presence of the impurities in the real crystal lattice, and the second-order interaction is the dominant scattering process. In general, the impurity interaction is sufficiently weak so that all terms with more than two coupled impurity lines usually can be safely ignored.

It may be noticed first that after the impurity averaging, we always have $\mathbf{k}' = \mathbf{k}$. Thus, as described above, each of the Green's functions, for which the spatial variation is in the *difference* of the two coordinates, is described by a single momentum state. This is as expected for normal Fourier transformation. Thus,

$$G^{r,a}(\mathbf{k}, \mathbf{k}', \omega) = G^{r,a}(\mathbf{k}, \omega)\delta_{\mathbf{kk}'}, \tag{5.72}$$

just as for the case of the noninteracting Green's functions. The diagrams in Fig. 5.13a may be visualized as propagation, interaction, propagation, interaction, and so on. The labels indicate which impurity is involved in the process. The impurity averaging groups

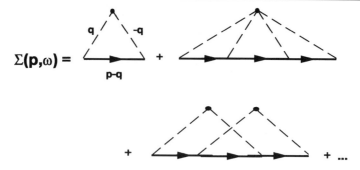

$$\Sigma(\mathbf{p},\omega) =$$

Fig. 5.14. The leading terms in the self-energy.

the terms according to the number of times that the same impurity occurs in the average. The dotted lines are connected to show this in Fig. 5.13b, as was described just above. The light lines in each of these figures represent the unperturbed propagator G_0, and the heavy line corresponds to the actual propagator G. There is also the possibility that the intermediate state has zero momentum and is connected by impurity lines of zero momentum. This subset of diagrams is not shown, for it contributes only an arbitrary shift of the energy scale, and this can be incorporated in the overall Hartree energy. Now the third term in Fig. 5.13b is a replication of the second term. There could also be terms in which the two double lines overlap one another, as well as terms in which they are nested. The resummation is defined by the bringing together of all those diagrams that are topologically distinct and not simply replications of each other. These contributions are termed the self-energy. The replications are handled by replacing the output bare Green's function by the full Green's function. This leads us to be able to write the expansion as

$$G^{r,a}(\mathbf{k}, \omega) = G_0^{r,a}(\mathbf{k}, \omega) + G_0^{r,a}(\mathbf{k}, \omega)\frac{1}{\hbar}\Sigma^{r,a}(\mathbf{k}, \omega)G^{r,a}(\mathbf{k}, \omega), \qquad (5.73)$$

which is *Dyson's equation* for the Green's functions. The self-energy expansion is shown in Fig. 5.14. Note that the internal Green's functions are the full Green's functions. In many cases, where the number of impurities is small, the scattering is weak, and the self-energy can be approximated by keeping only the first term and using the bare Green's function. Another, slightly better, approach is to keep only the lowest-order diagram but to go ahead and use the full Green's function, solving the overall problem by iteration. This is often called the *self-consistent Born approximation*.

The equation for the full Green's function can be rewritten as

$$G^{r,a}(\mathbf{k}, \omega) = \frac{1}{\left[G_0^{r,a}(\mathbf{k}, \omega)\right]^{-1} - \frac{1}{\hbar}\Sigma^{r,a}(\mathbf{k}, \omega)}. \qquad (5.74)$$

This form allows us to see just what the self-energy represents. In general, $\Sigma^{r,a}(\mathbf{k}, \omega)$ will have both a real and an imaginary part. If we compare the forms of (5.74) and (5.57), we note that the real part of the self-energy can be considered a correction to the single particle energy $E(\mathbf{k})$. This represents the *dressing* of (or change in) the energy due to the interaction with the impurities. This dressing can cause a general overall momentum-dependent shift in the energy, which also causes a change in the effective mass of the particle. On the other hand, the imaginary part of the self-energy represents the dissipative interaction that was included in an *ad hoc* manner by the insertion of the parameter η. Clearly the sign of the imaginary part of the self-energy is important to determine whether (5.74) represents the retarded or the advanced Green's function.

Let us now proceed to compute the scattering and conductance for the simplest case. This means that the task is really to evaluate the self-energy. However, if the full Green's function is retained in the latter term, then the process must be iterated. Here, we retain only the lowest-order term in the self-energy and treat the included Green's function with the bare equilibrium Green's function. According to (5.68), we need to sum over the momentum variable contained in the impurity interaction. For this, we will keep only the first term in Fig. 5.14, so that we assume the impurity scattering is weak. Then,

$$\Sigma^{r,a}(\mathbf{k}, \omega) = \frac{N_i}{\Omega} \sum_{\mathbf{q}} |V(\mathbf{q})|^2 \frac{1}{\hbar\omega - E(\mathbf{k} - \mathbf{q}) \pm i\eta}, \tag{5.75}$$

where N_i is the total number of impurities. The impurity average has replaced the scattering by the various assortment of impurities with a single interaction, the averaged interaction, and this is multiplied by the number of impurities to arrive at the total scattering strength. At this point, it is pertinent to note that we expect the impurity potential to be reasonably screened by the free carriers, which means that the potential is very short range. It is easiest to assume a δ-function potential (in real space), which means that the Fourier-transformed potential is independent of momentum, or $V(\mathbf{q}) = V_0$. This is equivalent to assuming that the screening wavevector is much larger than any scattering wavevector of interest, or that $V_0 \simeq e^2/\varepsilon q_s^2$. However, we will retain the form shown in Eq. (5.75) since we need to produce a higher-order correction that modifies this term.

We are primarily interested in the imaginary parts of the self-energy, and this can be obtained by recognizing that η is small, so that we retain only the imaginary parts of the free Green's function. The real parts of the self-energy are no more than a shift of the energy scale, and we expect this to be quite small in the weak scattering limit. Then, we take the limit of small η and ignore the principal part of the resulting expansion (since it leads to the real part of the self-energy, which we have decided to ignore), so that

$$\lim_{\eta \to 0} \frac{1}{\hbar\omega - E(\mathbf{k} - \mathbf{q}) \pm i\eta} \to \mp i\pi \delta(\hbar\omega - E(\mathbf{k} - \mathbf{q})). \tag{5.76}$$

We now convert the summation into an integration in $\mathbf{k}$-space, so that

$$\sum_{\mathbf{q}} \to \iint \frac{\sin\theta \, d\theta \, d\phi}{4\pi} \int \rho(E_q) \, dE_q, \quad E_q = \frac{\hbar^2 q^2}{2m}. \tag{5.77}$$

Here, the first two integrals represent an integration over the solid-angle portions of the overall three-dimensional integral (or an equivalent in two dimensions). If the integrand is independent of these angular variables, then these integrals yield unity. It may be noted that the impurity scattering conserves energy, so that the final state energy is the same as the initial energy, or $E(\mathbf{k} - \mathbf{q}) = E(\mathbf{k})$. Thus, the energy integral can be evaluated as well, and the angular integral is at most an angular averaging of the scattering potential. This allows us to evaluate the self-energies, with $n_i = N_i/\Omega$ (Ω is the volume of the crystal) the impurity density, as

$$\Sigma^{r,a}(\omega) = \pm i\pi n_i \rho(\omega) \int \frac{\sin\theta \, d\theta \, d\phi}{4\pi} |V(\mathbf{q})|^2 \equiv \pm i \frac{\hbar}{2\tau(\omega)}. \tag{5.78}$$

It should be noticed that with the approximations used, the self-energies are independent of the momentum and are only functions of the energy $\hbar\omega$. We can now evaluate the conductivity as

$$\sigma_{\alpha\beta} = \frac{e^2\hbar^4}{m^2} \sum_{\mathbf{k}} k_\alpha k_\beta \int \frac{d\omega}{2\pi} \frac{1}{\hbar\omega - E(k) - i(\hbar/2\tau)} \frac{1}{\hbar\omega - E(k) + i(\hbar/2\tau)} \delta(E - E_F).$$

(5.79)

With these approximations, and in the absence of a magnetic field, the conductivity is diagonal and isotropic (assuming that the energy band is isotropic). Either the frequency integral or the energy integral can be evaluated quickly by residues, with the other being replaced by the use of the delta function at the Fermi surface. Then, using $k_x^2 = k^2/d$ and $\rho(E_F)E_F = dn/2$, where $\rho(E)$ is the density of states (per unit volume) and d is the dimensionality, the conductivity is finally found to be

$$\sigma = \frac{e^2\hbar^4}{m^2} \sum_{\mathbf{k}} k_\alpha k_\alpha \frac{\tau}{\hbar^2} \delta(E - E_F) = \frac{e^2\hbar^2\tau}{dm^2} \rho(E)k_F^2 = \frac{ne^2\tau}{m}.$$

(5.80)

This is the normal low-frequency result of the Drude formula, and it is the usual conductivity one arrives at in transport theory. It must be noticed, however, that most semiclassical treatments of the impurity mobility include a factor $(1-\cos\theta)$, which is missing in this formulation, so that τ is a scattering time and not a relaxation time. It also must be noted that this result, which leads to a part of the normal Drude conductivity, is a result in which the two Green's functions are evaluated in isolation from one another, and there is no interaction between the two. We proceed to the more complicated case next.

5.3.3 Beyond the Drude result

In the previous paragraph, we treated only the interaction of the impurities with a single Green's function "line." We now consider higher-order corrections. In general, the impurity averaging still requires that the number of impurity "lines" joined by a single scattering site is an even number, with the two-line case the normal dominant term. The higher-order corrections that are most important are those in which the impurities connect the two Green's functions that appear in Eq. (5.66). The typical types of new diagrams that appear are shown in Fig. 5.15. The diagrams shown all have four interaction lines from either a single or a pair of impurities. To handle this complex situation, let us rewrite part of (5.63) as

$$L(\mathbf{k}', \mathbf{k}''', \omega; \mathbf{k}, \mathbf{k}'', \omega) = G^{(2)}(\mathbf{k}, \mathbf{k}', \omega; \mathbf{k}'', \mathbf{k}''', \omega).$$

(5.81)

(As an aside, we note that the frequencies in the above equation are most often seen as $\omega_\pm = \omega \pm \omega_a/2$, where ω_a is the applied a.c. frequency at which the conductivity is evaluated. Here, however, we are interested in the d.c. conductivity so that $\omega_a = 0$ and we can ignore this complication.) In general, the two-particle Green's function is frequency dependent as shown, although the one we need for the static conductivity does not have any frequency dependence. We define the kernel of the conductivity (for the static conductivity) in the isotropic limit as [38], [39]

$$\sigma_{\alpha\beta} = \frac{e^2\hbar^2}{dm^2} \int \frac{d\omega}{2\pi} F_{\alpha\alpha}(\omega),$$

(5.82)

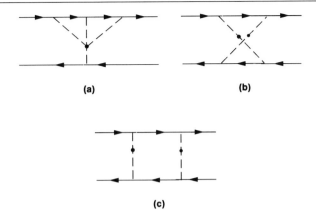

Fig. 5.15. Some typical interactions at fourth order in the impurity term.

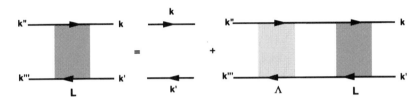

Fig. 5.16. The diagram for the two-particle interacting Green's function and the Bethe-Salpeter equation.

with

$$F_{\alpha\alpha}(\omega) = \sum_{\mathbf{k}\mathbf{k}'} \mathbf{k} \cdot \mathbf{k}' L(\mathbf{k}, \mathbf{k}', \omega; \mathbf{k}', \mathbf{k}, \omega)$$

$$= \sum_{\mathbf{k}} G^r(\mathbf{k}, \omega) G^a(\mathbf{k}, \omega) \bigg\{ \mathbf{k} \cdot \mathbf{k}$$

$$+ \sum_{\mathbf{k}'} \mathbf{k} \cdot \mathbf{k}' \Lambda(\mathbf{k}', \mathbf{k}, \omega; \mathbf{k}, \mathbf{k}', \omega) G^r(\mathbf{k}', \omega) G^a(\mathbf{k}', \omega) + \cdots \bigg\}. \quad (5.83)$$

In going from the tensor form of the conductivity to the form here, where the two wavevectors form a dot product, we note the previous choice of k_x^2 is replaced by k^2, which leads to the factor of d in the denominator. The first term in the curly brackets leads to the results obtained in the previous section. The new term is the second one in the curly brackets. Again, the terms can be resummed into the form shown in Fig. 5.16, in which the last pair of Green's functions in the second term in curly brackets is replaced by the two-particle Green's function. This particular form, illustrated in the figure, is known as the Bethe-Salpeter equation. In this, the quantity Λ is referred to as the irreducible scattering vertex.

The key contributions to the scattering vertex arise from the second-order interaction, which spans the two Green's functions. This is shown doubled in Fig. 5.15c, so that the latter is a reducible term, in that it appears as a higher-order term in the series of Eq. (5.83). The replication of the lines that lead to Fig. 5.15c are called *ladder diagrams*. The terms of parts (a) and (b) of Fig. 5.15 are irreducible, as is the first part of (c). The impurity lines do not transport energy (the scattering from impurities is elastic), so the upper and lower Green's functions are at the same energy throughout; for the static case, these two frequencies are the same. We keep only the lowest-order correction in these ladder diagrams, which is the iterated diagrams such as those of Fig. 5.15c. Now, we note that the kernel for the single scattering pair is

$$\Gamma_0 = \frac{n_i}{\hbar^2} |V(\mathbf{k} - \mathbf{k}')|^2, \tag{5.84}$$

where the two momenta are those on either side of the scattering line. We want to form a series for the set of iterated iteractions and have the two integrations over $\mathbf{k}$ and $\mathbf{k}'$. The scattering kernel (5.84) can be rewritten in terms of the scattering wavevector $\mathbf{q} = \mathbf{k} - \mathbf{k}'$, but this will make the second set of Green's functions depend upon $\mathbf{k}' + \mathbf{q}$, and the integration is then taken over $\mathbf{q}$. However, since energy is conserved in the interaction, the argument of the potential is a function only of $\mathbf{k}$ and a scattering angle, which is not part of the integration over the last Green's functions. This means that the kernel can be treated separately from the last pair of Green's functions. We can write the series as

$$\Lambda = \Gamma_0 + \Gamma_0 \Pi \Gamma_0 + \Gamma_0 \Pi \Gamma_0 \Pi \Gamma_0 + \cdots = \frac{\Gamma_0}{1 - \Pi \Gamma_0}. \tag{5.85}$$

The set of Green's function integrals for Π can be integrated (with the exception of the angular integration, and with the $\mathbf{k}'$ term replacing $\mathbf{k} \cos \theta$) to give

$$\Pi(\omega) = \int \rho(E) \, dE \frac{\hbar}{\hbar\omega - E(\mathbf{k}' + \mathbf{q}) - i(\hbar/2\tau)} \frac{\hbar}{\hbar\omega - E(\mathbf{k}' + \mathbf{q}) + i(\hbar/2\tau)}$$

$$= -2\pi\hbar\rho(\hbar\omega)\tau. \tag{5.86}$$

The last pair of Green's functions in the second term of (5.83) just adds another factor of Π to the numerator of this last expression, if we take the kerm in $\mathbf{k}'$ out of the integral. This is done by the argument that the Green's functions will change this to a quantity $(2m\omega/\hbar)^{1/2}$, which will be then changed to $\mathbf{k}$ in the integration over the frequency. What remains from the dot product $\mathbf{k} \cdot \mathbf{k}'$ is $k^2 \cos \theta$, where θ is the scattering angle. The integration of the last two Green's functions then produces just an additional factor of Π. Using these last three equations, the kernel can be rewritten as

$$F_{\alpha\alpha}(\omega) = \sum_{\mathbf{kq}} G^r(\mathbf{k}, \omega) G^a(\mathbf{k}, \omega) k^2 \left\{ 1 + \cos \theta \frac{\Gamma_0 \Pi}{1 - \Gamma_0 \Pi} \right\}$$

$$= \sum_{\mathbf{kq}} G^r(\mathbf{k}, \omega) G^a(\mathbf{k}, \omega) k^2 \left\{ 1 - \cos \theta \frac{|V(\theta)|^2}{\langle |V|^2 \rangle + |V(\theta)|^2} \right\}, \tag{5.87}$$

where $\langle |V|^2 \rangle$ is the angle-averaged potential appearing in (5.78). The difference in the last fraction is that the first term in the denominator has already been angle averaged, whereas the last term and the numerator term retain their angular variation. If the scattering is isotropic, such as for a delta-function scattering potential, then there is no angular variation and the last term vanishes, leaving only the scattering time in the conductivity. That the angular variations are inherent within the second term in the curly brackets becomes important for creating a difference between a scattering time and a relaxation time. Carrying out the frequency integration in (5.83) leads to the result (with an inferred angular variation from the momentum summation)

$$\sigma = \frac{e^2\hbar^2}{dm^2} \sum_{\mathbf{k}} \tau k^2 [1 - f(\theta)\cos\theta]\delta(E - E_F) = \frac{ne^2\tau_m}{m}, \qquad (5.88)$$

where $f(\theta)$ is the angular variation resulting from the ratio of terms in (5.87), and τ_m is a momentum relaxation time obtained by the angular averaging process.

The result obtained here, with the approximations used, provides a connection with the semiclassical result for impurity scattering obtained from the Boltzmann equation discussed in Chapter 2. The complications are greater here, but this result also allows us to create a formalism that can be moved forward to treating other diagrammatic terms for new effects that are not contained in the Boltzmann equation. The angular dependence of the fraction that appears as the second term in the curly brackets of (5.87) provides the needed conversion from a simple scattering time to a momentum relaxation time. On the other hand, if we have a matrix element that really does not depend upon $\mathbf{q}$, such as a delta-function scattering potential or a heavily screened Coulomb potential where $V(\mathbf{q}) \rightarrow e^2/\varepsilon q_{sc}^2$ (where q_{sc} is the screening wavevector, either for Fermi-Thomas screening or for Debye screening), then the ladder correction terms do not contribute as the angular variation integrates to zero. Thus, scattering processes in which the matrix element is independent of the scattering wavevector do not contribute the $(1-\cos\theta)$ correction. This simple fact is often overlooked in semiclassical treatments, as this latter angular variation is put in by hand, and its origin is not fully appreciated.

5.4 Weak-localization correction to the conductance

The important phenomenon that we have used in the introductory sections above to discuss weak localization is that of phase coherence; that is, the phase memory is maintained through several scattering events in the time-reversed paths. This requires the scattering to be elastic, in which only the direction of the momentum is changed. Moreover, we require that this coherence be maintained through a series of scattering events by which the particle returns (is scattered back) to its original position, or to a momentum directly opposite to its original momentum state (in momentum space). Weak localization itself is the interference between the two time-reversed paths around the scattering ring. The resistivity corrections that were calculated above assume that there is no coherence through the scattering process, so that the impurity scattering introduces *disorder* into the system. Strong disorder, of course, will localize part or all of the states in the system [40]. As discussed above with the experimental studies, the most remarkable sign of weak localization is a backscattering peak at zero magnetic field. This backscattering is a representation of multiple elastic scattering. Consider the construction of Fig. 5.17. A momentum vector is gradually rotated in momentum space by elastic scattering until it points in the opposite direction. In the figure, the various momentum vectors have been translated to a common origin, so that the momentum imparted by the impurity ($\mathbf{q}_i$) is clearly seen. There are, of course, two directions in which the momentum vector can be rotated, and these are the two time-reversed paths. We note that the momentum, which starts out as $\mathbf{k}_0$, is rotated to $\mathbf{k}_1$ by $\mathbf{q}_1$ and then to $\mathbf{k}_2$ by $\mathbf{q}_2$, and so on. The primed vectors denote the opposite sense of rotation. The important point is that the set of momentum vectors can be made precisely the same for each sense of rotation, and it is this coherence that leads to the interference of the time-reversed paths;

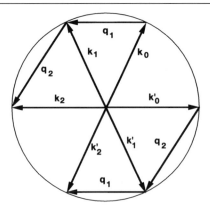

Fig. 5.17. The multiple scattering can be linked with this ring for time-reversed paths.

the wavevectors are matched by their complements, which makes the motion difficult. The coherent backscattering is the result of this constructive interference in the set of multiple scatterings. If we consider that each process of scattering involves an amplitude a_i, then we note that each a_i is matched by the equivalent a_i'. Thus, the coherence that this entails leads to [38]

$$\frac{\left|a_i + a_i'\right|^2}{\left|a_i\right|^2 + \left|a_i'\right|^2} \sim 2, \tag{5.89}$$

if there is perfect coherence between the two scattering amplitudes. If there were no coherence, the ratio in (5.89) would be unity. This leads to scattering almost a factor of two stronger, and hence to a significant reduction in the conductivity. From this consideration, the quantity $\langle a_i a_i'^* \rangle_{imp}$, where the brackets denote the impurity averaging, gives us the degree of coherence in this process and represents the coupled scattering processes. This is just the effect that we discussed in the previous section, since this coupled impurity (two interaction lines) spans across propagators that are rotating in different directions. We note that the energy-conserving delta function is not in the basic impurity interaction Hamiltonian (5.68) but occurs in the transition probability when we compute the matrix elements and the Green's functions. (The delta function arises from the Green's function itself in the absence of interactions; it is replaced here by the broadened spectral density.) Thus, one can evaluate $\langle a_i a_i'^* \rangle_{imp}$ without concern about this, showing that there is a singularity in this quantity when the backscattering condition is fulfilled. Since the propagators are rotating around the loop in opposite directions, the interactions in which we are interested are the set of maximally crossed diagrams first studied by Langer and Neal [41].

At the end of the previous section, we considered those interactions that spanned the two Green's functions but did not cross. These formed the ladder diagrams. As mentioned here, our interest is now in that set of diagrams that are maximally crossed. This raises an important general point about the perturbation series. In two cases above, only a particular set of diagrams – certainly not all possible diagrams that can be drawn – were chosen for consideration. Here, still a third set of possible diagrams is being considered. In truth, one decides *a priori* just what the nature of the physics will require and tries to find a subset of all possible diagrams that first fits the physics and second can be conveniently resummed

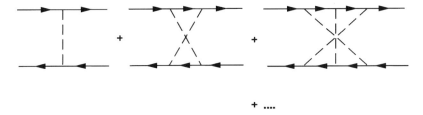

+ ....

Fig. 5.18. The set of maximally crossed diagrams that lead to weak localization.

into a simple analytic expression. Only experience can guide the practioner as to the manner in which the terms are chosen, and hopefully other series contributions are small enough that the chosen diagrams are the dominant ones.

As mentioned, we now want to evaluate the maximally-crossed diagrams. These are shown in Fig. 5.18. The first diagram was used earlier as part of the ladder diagram series; the next two are the next-higher orders of the interaction, which are the terms of interest. The approach that we follow is mainly due to Bergmann [42], although it builds on earlier work of Altshuler *et al.* [16] and others [18], [43].

5.4.1 The cooperon correction

The value of this group of electron-hole propagators is an additional term to Λ of Eq. (5.81). Consider the third diagram of Fig. 5.18. The arrangement of the Green's functions and matrix elements in the diagram is usually changed by reversing the direction of the Green's functions on one of the horizontal lines, usually the hole line, so that it becomes an electron-electron propagator and the interactions create a new ladder diagram. This changes $\mathbf{k}' \to -\mathbf{k}'$ and changes the integrals that result from the new ladder diagram. The maximally crossed diagrams now become the normal set of ladder diagrams in this new description. Such a diagram reversal was first used in discussing superconductivity, so that the set of ladder diagrams with the electron-electron propagators is often called a *cooperon*. Rather than the entire two-particle Green's function, we are interested only in the correction term. Previously we have let the applied frequency $\omega_a \to 0$, for the static conductance, but here we will find a divergence as q, $\omega_a \to 0$. Normally, one enters the two frequencies through $\omega \pm \omega_a/2$, as discussed below (5.81). Here, we will continue to let $\omega_a \to 0$ but will introduce an arbitrary decay function for the phase as the phase-breaking self-energy $i\hbar/2\tau_\varphi$ ($\omega_\pm = \omega \pm i\hbar/2\tau_\varphi$) in the critical polarization terms of the ladder diagram series. Following Bergmann, we will adopt an approach suitable for a d-dimensional system. First, it should be noted that Λ depends only on $\mathbf{k} + \mathbf{k}' = \mathbf{q}$, which is different from the result in the preceding section (where we were interested in $\mathbf{k} - \mathbf{k}' = \mathbf{q}$), since the direction of $\mathbf{k}'$ has been reversed but has little internal structure apart from the divergence already mentioned. Therefore, we can write this term as $\Lambda(\mathbf{q}; \omega)$. The two integrations (summations) that are indicated in Eq. (5.83) become integrations over $\mathbf{q}$ and $\mathbf{k}$. We further note that since we are interested in the low q limit, we can make the approximation (which is on the Fermi surface, so that $k \to k_F = mv_F/\hbar$)

$$E(k') = \frac{\hbar^2}{2m}(\pm\mathbf{k} + \mathbf{q})^2 \simeq E(k) \pm \hbar\mathbf{v}_F(k) \cdot \mathbf{q}. \tag{5.90}$$

Now, the integration over **k** in the Green's functions can be performed as

$$I = -\int \frac{d\omega}{2\pi} \int \frac{d^d k}{(2\pi)^d} k_x^2 G^r(\mathbf{k}, \omega_+) G^a(\mathbf{k}, \omega_-) G^r(-\mathbf{k}+\mathbf{q}, \omega_+) G^a(-\mathbf{k}+\mathbf{q}, \omega_-) \delta(E - E_F)$$

$$= \int \frac{d\omega}{2\pi} \int \rho(E) \frac{k^2}{d} dE \frac{1}{\hbar\omega - E + (i\hbar/2\tau)} \frac{1}{\hbar\omega - E - (i\hbar/2\tau)} \delta(E - E_F)$$

$$\times \frac{1}{\hbar\omega - E + \hbar\mathbf{v}_F(k)\cdot\mathbf{q} + (i\hbar/2\tau)} \frac{1}{\hbar\omega - E + \hbar\mathbf{v}_F(k)\cdot\mathbf{q} - (i\hbar/2\tau)}. \tag{5.91}$$

The contour of integration (the energy runs from $-\infty$ to ∞ to account for both electrons and holes in the diagrams) is closed in the upper half-plane, giving two poles at

$$\hbar\omega = E + (i\hbar/2\tau),$$

$$\hbar\omega = E - \hbar\mathbf{v}_F(k)\cdot\mathbf{q} + (i\hbar/2\tau), \tag{5.92}$$

which leads to

$$I = -i\frac{k_F^2}{d}\rho(E_F)\frac{1}{i/\tau} \frac{1}{\hbar\mathbf{v}_F\cdot\mathbf{q}} \frac{2\mathbf{v}_F\cdot\mathbf{q}}{(i/\tau)^2 - (\mathbf{v}_F\cdot\mathbf{q})^2}. \tag{5.93}$$

For small q, this becomes

$$I \simeq 2\frac{k_F^2}{d}\rho(E_F)\tau^3. \tag{5.94}$$

This now leads to the conductivity correction

$$\delta\sigma = -\frac{2e^2\hbar^2 k_F^2 \rho(E_F)\tau^3}{dm^2} \int \frac{d^d q}{(2\pi)^d} \Lambda(\mathbf{q}, \omega), \tag{5.95}$$

where we have used the coefficient from earlier results.

Now, we must calculate the kernel $\Lambda(\mathbf{q}, \omega)$. Neglecting the terminal Green's functions which have been included already in the above calculation, we can write the third-order ladder interaction as

$$\frac{1}{\hbar^6}\sum_{\mathbf{r},\mathbf{s}} V_\mathbf{r} G^r(\omega_+, \mathbf{k}+\mathbf{r}) V_\mathbf{s} G^r(\omega_+, \mathbf{k}+\mathbf{r}+\mathbf{s}) V_{\mathbf{s}'}^* V_\mathbf{r}^* G^a(\omega_-, \mathbf{k}'-\mathbf{r}) V_\mathbf{s}^* G^a(\omega_-, \mathbf{k}'-\mathbf{r}-\mathbf{s}) V_{\mathbf{s}'}^*. \tag{5.96}$$

With the reversal of one of the particle lines, the crossed diagrams are converted to a simple ladder, so that a series is more easily generated. For simplicity, we assume that the scattering is isotropic (independent of the scattering wavevector, as discussed earlier), since the angular dependence of the potential is not crucial to the presence of these terms in the series. This leads to

$$\frac{n_i}{\hbar^2}|V_\mathbf{r}|^2 = \cdots = \frac{n_i}{\hbar^2}|V_0|^2 = \frac{1}{2\pi\hbar\rho(E_F)\tau} \equiv \Gamma_0, \tag{5.97}$$

where we have used the definitions of the scattering time from the previous section. Further, we make the definition

$$\Pi(\mathbf{k}+\mathbf{k}'; \omega, \omega') = \sum_\mathbf{g} G^r(\omega_+, \mathbf{k}+\mathbf{g}) G^a(\omega_-, \mathbf{k}'+\mathbf{g}), \tag{5.98}$$

which depends on only the sum of the two wavevectors, as discussed previously. Then, the term (5.96) can be written as

$$\Gamma_0 \Pi \Gamma_0 \Pi \Gamma_0. \tag{5.99}$$

Each of the terms in the series can be written in this way, and the series can be resummed as before, which gives

$$\Lambda = \Gamma_0 + \Gamma_0 \Pi \Gamma_0 + \Gamma_0 \Pi \Gamma_0 \Pi \Gamma_0 + \cdots = \frac{\Gamma_0}{1 - \Pi \Gamma_0}. \tag{5.100}$$

Because there are no internal integrations (the terms Π are functions only of the scattering wavevector), the result (5.100) is equivalent to a Dyson's equation for the two-particle propagator. Thus, we need only to evaluate the vertex functions, and for this we will set $\mathbf{k}'' = \mathbf{k} + \mathbf{g}$ and $\mathbf{k} + \mathbf{k}' = \mathbf{q}$ in (5.100). Then, the reversal of the hole line takes $\mathbf{k}'' \to -\mathbf{k}''$, and the polarization function between the elements of the ladder, when we use (5.90) for small q, becomes

$$\Pi(\mathbf{q}; \omega) = \int \frac{d^d \mathbf{k}''}{(2\pi)^d} G^r(\omega_+, \mathbf{k}'') G^a(\omega_-, -\mathbf{k}'' + \mathbf{q})$$

$$= \rho(E_F) \int \frac{dS_{k''}}{S_{k''}} \int dE \left\{ \frac{\hbar}{\hbar\omega - E + (i\hbar/2\tau) + (i\hbar/2\tau_\varphi)} \right.$$

$$\left. \times \frac{\hbar}{\hbar\omega - E + \hbar \mathbf{v}_F(\mathbf{k}'') \cdot \mathbf{q} - (i\hbar/2\tau) - (i\hbar/2\tau_\varphi)} \right\}, \tag{5.101}$$

where we have introduced the additional phase-breaking term to the self-energy. The integration over the energy can be performed as

$$\Pi(\mathbf{q}; \omega) = -\rho(E_F) \int \frac{dS_{k''}}{S_{k''}} 2\pi i \frac{\hbar^2}{\hbar\mathbf{v}'' \cdot \mathbf{q} - i\hbar/\tau - i\hbar/2\tau_\varphi}$$

$$\simeq 2\pi\hbar\rho(E_F)\tau \left[1 - \frac{\tau}{\tau_\varphi} - Dq^2\tau - \cdots \right], \tag{5.102}$$

where $D = v_F^2 \tau/d$. We note that the density of states is inside the above integral over frequency and energy, so that it is actually evaluated at the Fermi surface. This can then be used with (5.97) to give the kernel

$$\Lambda(\mathbf{q}; \omega') = \frac{1}{2\pi\hbar\rho(E_F)\tau} \frac{1}{Dq^2\tau + \tau/\tau_\varphi}, \tag{5.103}$$

and the conductivity correction is simply

$$\delta\sigma = -\frac{e^2}{\pi\hbar}(D\tau) \int \frac{d^d\mathbf{q}}{(2\pi)^d} \frac{1}{Dq^2\tau + \tau/\tau_\varphi}. \tag{5.104}$$

We see that it does indeed diverge as $q \to 0$ and for $\tau_\varphi \to \infty$. This must be examined with some care. Normally, one ignores the last factor in the denominator, but we shall carry it along here. We note here that the first fraction in the integral looks like a Green's function but involves the diffusion constant D. This replacement for the Green's function, which has naturally arisen, was discussed much earlier in this chapter and is termed the *diffusion pole*.

The main contribution to the integral (5.104) arises from small values of the momentum, hence the finite size of the system will appear here. For example, in a film of thickness t

and a mean free path l_e, which should be less than t, the first integration over the energy arose from a three-dimensional integration. However, in the integration left in (5.104), the finite thickness of the film can limit the integration to just two dimensions if there is quantization in the direction normal to the film. Here, we will explicitly deal with only the two-dimensional case, which is the one normally encountered when studying mesoscopic effects in semiconductors. Then, we will further limit the integration to values of q below those defined by the length given by the diffusion process (the diffusion length), so that $q < 1/\sqrt{D\tau}$. In essence, this limit says that we are interested only in the small q divergence of the integral, and

$$\delta\sigma = -\frac{e^2}{4\pi\hbar}(D\tau)\int_0^{1/\sqrt{D\tau}} dq^2 \frac{1}{Dq^2\tau + \tau/\tau_\varphi} = -\frac{e^2}{4\pi\hbar}\ln\left[\frac{1+\tau/\tau_\varphi}{\tau/\tau_\varphi}\right]. \qquad (5.105)$$

Thus, we finally arrive at the two-dimensional correction to the conductivity due to weak localization as

$$\delta\sigma = -\frac{e^2}{4\pi\hbar}\ln\left(1+\frac{\tau_\varphi}{\tau}\right), \qquad (5.106)$$

where $\tau < \tau_\varphi$ is needed for the entire approach of this section to be fully valid. If $\tau = \tau_\varphi/10$, then the logarithm term is about 2.4.

The integrand of Eq. (5.104) is often called the diffusion propagator. It has a form quite similar to that of the Green's functions themselves, even though it is real. It often appears with the fraction in the denominator expressed as $i\omega\tau$, which introduces the Fourier representation. In this form, it appears more evenly in the form of the Green's function. We note that this diffusion propagator did not appear in the treatment of the particle-hole ladder of the previous section, since the small q terms are irrelevant to that development, whereas here they are critical to the development. Following the same line of argument, it can be shown that the q-dependent diffusion treated in the previous section gives the same polarization (5.103) as obtained here, provided that there is no magnetic field. This is a result of the time-reversal symmetry of the diagrams used for this purpose. Consequently, the diffusion propagator is often used for both the particle-hole and the particle-particle ladders, and these are termed the *diffuson* and the *cooperon*, respectively. It may also be noted that the leading Green's functions in each of the forms of the conductivity used in these sections form a basic polarization term but are weighted by a term in k^2. The fact that the two Green's functions give essentially the same weight as a polarizataion term is important, and the overall diagram of the left side of Fig. 5.16 can be redrawn as in Fig. 5.19. Here, the two Green's function lines on the left side have been pulled together, since they represent the same point $\mathbf{r}$. Similarly, the two Green's function lines on the right side have

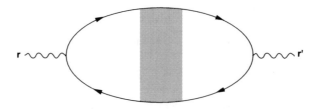

Fig. 5.19. Bubble form of the current arising from the Green's functions on the left side of the general conductivity.

been drawn together since they represent the same point $\mathbf{r}'$. The two squiggly lines represent the current carried by the factors k (one for each side of the diagram which gives the k^2), as $\mathbf{k}$ is the Fourier transform variable for $\mathbf{r} - \mathbf{r}'$. The overall diagram of Fig. 5.19 is termed the "current bubble." In the absence of either the diffuson or the cooperon contributions, the interactions spanning the two Green's function branches (top of the diagram and bottom of the diagram) are absent, and one has just the two simple Green's functions. These are of course the full Green's functions with the interaction important to the single lines, but with no interactions spanning the individual electron and hole (or electron line in the cooperon) lines.

5.4.2 Role of a magnetic field

If we now apply a magnetic field normal to the two-dimensional electron (and hole) gas, this has a nonnegligible effect on the conductance correction, as was discussed in the opening sections of this chapter. The vector potential of the magnetic field modifies the phase of the wavefunctions. In the treatment here, we must assume that the elastic mean free path is much smaller than the cyclotron radius, since we cannot have closed orbits for our diffusive motion. The main effect of the vector potential, and the magnetic field, is then to change the relative phase between any two points on the path more than would normally arise from the propagation. Hence, the phase of one wavefunction relative to the other one in the Green's function is shifted by the vector potential according to

$$G(\mathbf{r}, \mathbf{r}', B) = G(\mathbf{r}, \mathbf{r}', 0) \exp\left[\frac{ie}{\hbar} \int_{\mathbf{r}}^{\mathbf{r}'} \mathbf{A}(\mathbf{r}'') \cdot d\mathbf{r}'' \right]. \tag{5.107}$$

The form of the additional term is easiest to conceptualize when we think of plane waves. Here, one replaces the momentum by the proper conjugate momentum discussed earlier in this chapter, so that $\mathbf{k} \to \mathbf{k} + e\mathbf{A}/\hbar$, which is then integrated over the path. This vector potential breaks the translational invariance of the Green's function but does not change the basic approach that we have used. The major change will be in the summation of the maximally crossed diagrams. For the moment, this calculation is done in real space, but it will be transferred to the Fourier space after some preliminary considerations. The polarization (5.98) is now given by

$$\Pi(\mathbf{r}, \mathbf{r}'; \omega; B) = G^r(\mathbf{r}, \mathbf{r}'; \omega; B) G^a(\mathbf{r}, \mathbf{r}'; \omega; B)$$

$$= G^r(\mathbf{r}, \mathbf{r}'; \omega; 0) G^a(\mathbf{r}, \mathbf{r}'; \omega; 0) \exp\left[\frac{i2e}{\hbar} \int_{\mathbf{r}}^{\mathbf{r}'} \mathbf{A}(\mathbf{r}'') \cdot d\mathbf{r}'' \right]$$

$$= \Pi(\mathbf{r}, \mathbf{r}'; \omega; 0) \exp\left[\frac{i2e}{\hbar} \int_{\mathbf{r}}^{\mathbf{r}'} \mathbf{A}(\mathbf{r}'') \cdot d\mathbf{r}'' \right]. \tag{5.108}$$

The factor of 2 in the phase arises from the fact that each of the two lines, the electron line and the hole line, contribute a factor equivalent to that in Eq. (5.107). This is quite similar to the Aharonov-Bohm effect. The path indicated in the integral is just one-half of the overall loop, and each side of the loop contributes one-half of the total phase integral leading to the Aharonov-Bohm phase. (We will see later that the universal conductance fluctuations do not constitute a complete loop, but that the two sides of the loop tend to cancel one another, so that it is only the difference in the coupled field from the two sides that will remain in the phase integral.) Another way of seeing this is that while the first Green's function is a function of $\mathbf{r} - \mathbf{r}'$, the second is a function of the reverse of this quantity, and is a complex conjugate, so that it leads to a doubling of the phase shift. Now, we need to show that this is equivalent to the Peierl's substitution in the scattering wave vector $\mathbf{q} \to \mathbf{q} + 2e\mathbf{A}/\hbar$, with the factor of 2 coming from the phase-doubling inherent in the last equation. Moreover, it is important to show that it arrives as an eigenvalue of the diffusive operator for the polariztion.

To show the important property of the eigenvalues, we note that the polarization is an operator that can operate on an arbitrary wavefunction and produce an eigenvalue equation according to

$$\int d^3\mathbf{r}' \, P(\mathbf{r}, \mathbf{r}'; \omega; B)\psi_i(\mathbf{r}') = \lambda_i \psi_i(\mathbf{r}), \tag{5.109}$$

where P is an operator expression that will be determined below. To proceed, we introduce (5.108) into the integral and expand the exponential function and the wavefunction up to second order in a Taylor series about $\mathbf{r}$. These terms are then integrated. The expansion gives

$$\int d^3\mathbf{r}' \left\{ P(\mathbf{r} - \mathbf{r}'; \omega; 0) - \frac{(\mathbf{r} - \mathbf{r}')^2}{2} P(\mathbf{r} - \mathbf{r}'; \omega; 0)\left(-i\nabla + \frac{2eA}{\hbar}\right)^2 \right\}\psi_i(\mathbf{r}) = \lambda_i \psi_i(\mathbf{r}). \tag{5.110}$$

The wavefunction can be brought outside the integration (but carefully, as it is still subject to the operators). Then, the first term in the integral is recognized as the Fourier transform if we let $\mathbf{q} \to 0$. Similarly, if we use this Fourier transform connection, the second term becomes the second derivative with respect to the Fourier variable, and we may rewrite the integral in transformed variables as

$$\left[P(\mathbf{q} = 0; \omega; 0) + \frac{1}{2}\frac{\partial^2 P(\mathbf{q}; \omega; 0)}{\partial q^2}\bigg|_{\mathbf{q}=0}\left(-i\nabla + \frac{2eA}{\hbar}\right)^2 \right]\psi_i(\mathbf{r}) = \lambda_i \psi_i(\mathbf{r}). \tag{5.111}$$

The terms in the square brackets should be recognized as the first two terms in a Taylor series for the polarization, but expanded around zero momentum. Then, the second term tells us that we connect the momentum with $(-i\nabla + 2eA/\hbar)$, and it is clear that the eigenfunctions of the operator in momentum space are identical with wavefunctions for particles of charge $2e$ in a magnetic field (which is another connection with the cooperon). Comparison with the development much earlier in this chapter relates the operator P to the diffusion operator, which is just the polarization above that gives rise to the diffusion pole type of "Green's function." Hence, in the magnetic field, we can replace the momentum by a Peierl's substitution in vector potential for a doubly charged particle, but this leads to the replacement

$$q^2 \to q_n^2 = \frac{4eB}{\hbar}\left(n + \frac{1}{2}\right), \tag{5.112}$$

where we have used the cyclotron energy relation as $\hbar^2 q_n^2/2m = \hbar\omega_c(n + 1/2)$ and doubled the charge in the cyclotron frequency. Now, using Eq. (5.102), the eigenvalue can be written as

$$\lambda_i = 2\pi\hbar\rho(E_F)\tau\left[1 - \frac{\tau}{\tau_\varphi} - D\frac{4eB}{\hbar}\left(n + \frac{1}{2}\right)\tau - \cdots\right]. \tag{5.113}$$

One could now proceed to compute the actual value of the polarization in real space and Fourier-transform it, but the important point is that the only change in the summation for the kernel is in the quantization of the momentum. Thus, Eq. (5.103) becomes

$$\Lambda(\mathbf{q}_n;\omega) = \frac{1}{2\pi\hbar\rho(E_F)\tau} \frac{1}{Dq_n^2\tau + \tau/\tau_\varphi}, \qquad (5.114)$$

and the conductivity may be found from the first line of (5.105) as

$$\delta\sigma = -\frac{e^2}{\pi\hbar}(D\tau)\frac{eB}{\pi\hbar} \sum_{n=0}^{\hbar/4eDB\tau} \frac{1}{D\tau(4eB/\hbar)(n+\tfrac{1}{2}) + \tau/\tau_\varphi}$$

$$= -\frac{e^2}{4\pi^2\hbar}\left[\Psi\left(\frac{1}{2} + \frac{\hbar}{4eDB\tau}\right) - \Psi\left(\frac{1}{2} + \frac{\hbar}{4eDB\tau_\varphi}\right)\right], \qquad (5.115)$$

where a different normalization has been used to take care of the degeneracy of the quantized state that appears in the polarization, and $\Psi(x)$ is the digamma function. Quite often the first digamma function is replaced by its large argument limit, which is the natural logarithm of the second term in the argument.

The first line of (5.115) carries an important message. The denominator in the summation is basically a sum over $(\lambda_i' - 1)$, where λ_i' is λ_i reduced by the prefactor on the right side of (5.113). That is, the conductivity is given basically by a summation over the reciprocal of the eigenvalues of the diffusion equation. This same behavior will arise in a later section for the universal conductance fluctuations. Clearly, the reduction of the conductivity correction that causes weak localization is because of an increase in the eigenvalues with magnetic field. This increase is explicit in Eq. (5.113) through the third term in the large brackets. Understanding this behavior, in which the weak-localization amplitude depends upon the eigenvalues of the diffusion equation describing the electron-electron correlation, is crucial and will open the door to some interesting effects in the next section.

5.4.3 Periodic eigenvalues for the magnetic effects

It is important to note that each term in the series of (5.115) decays as the magnetic field is increased. This is the rationale for why the weak-localization correction decays with the magnetic field. Recall, however, that the terms that arise come from the eigenvalues of the basic diffusion equation for the polarization operator. If there is some part of the system that makes these eigenvalues periodic, then some unusual effects can occur. One such unusual behavior is found in periodic superlattices. Among the earliest to study the magnetic field and periodic potential in two dimensions was Harper [44]. Azbel [45] subsequently studied the problem and found that the magnetotransport could show oscillations periodic in the magnetic field, and also periodic in the reciprocal of the magnetic field [46], depending on the relative strengths of the periodic potential and the Landau quantization energy $\hbar\omega_c$. The most extensive studies, at least up to a few years ago, were these of Rauh et al. [47] and Hofstadter [48]. One major problem is the need to have the various lengths in the problem (the magnetic length, the periodicity of the lattice, etc.) rationally related. Through the use of Harper's equation, Hofstadter developed the discrete energy levels that arise when the ratio of the lengths are rationally related;

$$\phi(m+1) + \phi(m-1) + 2\cos[2\pi m\alpha - \nu]\phi(m) = \varepsilon\phi(m), \qquad (5.116)$$

where

$$\alpha = \frac{ea^2 B}{\hbar} \tag{5.117}$$

is the normalized flux coupled through each unit cell of the periodic potential of basis vector a, and $\psi(x, y) = \phi(ma)e^{iky}$, with $v = k_y a$. Here, ϕ is the x portion of the wavefunction. It is clear from Eq. (5.116) that the energy eigenvalues are periodic in the magnetic field, a result that is induced by the periodic potential. While Hofstadter's result is reasonable for small magnetic fields, it is only valid for a single Landau level in high magnetic fields. Geisel has given a more complete derivation for the high-magnetic-field case that accounts for coupling between the Landau levels [49], but here we are primarily interested in the low-magnetic-field regime, where the motion has not been quantized into Landau levels. In this regime, the periodicity in magnetic fields of the energy "bands" formed from the above equation is important. In the normal case of semiconductor materials, the small lattice constants force this periodicity to occur for megagauss magnetic fields, a clearly untenable and unreachable level of field. On the other hand, in artificial superlattices, for which the lattice constant can be relatively large, the magnetic fields are readily attainable.

Of interest in this discussion is the possibility of really connecting the weak-localization behavior with the eigenvalues of the diffusive nature of the polarization function. For this, the first requirement is sufficient impurity scattering so that the transport is diffusive. A structured gate electrode written by electron beam lithography is shown in Fig. 5.20. Here, the gate forms a two-dimensional periodic potential that is induced upon the electron gas under it. The potential is created by the depletion under the actual gate lines, leaving higher carrier density in the open regions. Just before pinch-off of the entire conducting region, this potential is induced in an effective two-dimensional electron gas, even in a normal MESFET structure. The structures in which the best results are seen have epitaxial layers only 50 nm thick grown on semiinsulating GaAs. The grown layers are also GaAs, but doped to $1.5 \times 10^{18}\,\mathrm{cm}^{-3}$. Near pinch-off, most of the dopants are ionized and lead to significant impurity scattering within the electron gas. The resultant conductance curves (actually current through the device) are shown in Fig. 5.21 for a variety of source-drain

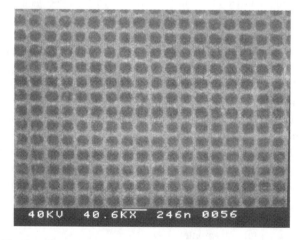

Fig. 5.20. A metallized gate that produces a periodic potential on a two-dimensional electron gas by depletion under the gate fingers. Here the lines are 40 nm wide with a periodicity of 165 nm.

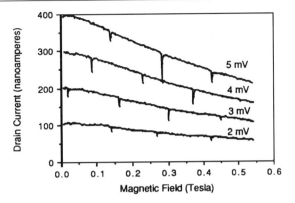

Fig. 5.21. Measured current through a device in which a superlattice potential has been applied. The parameter is the drain bias. [After Ma *et al.*, Surf. Sci. **229**, 341 (1991).]

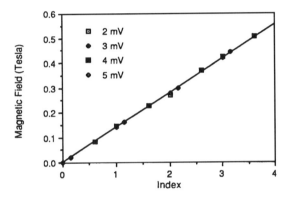

Fig. 5.22. The magnetic field at which dips in conductance occur can be plotted against an arbitrary index, which relates to the flux coupled in each unit cell.

biases [50]. Each of the curves shows a series of dips in conductance, with the dips essentially equally spaced in magnetic field. (A 2-mV source-drain potential corresponds to essentially $11.4\,\mu$V across each cell of the superlattice, a relatively small potential drop that is orders of magnitude smaller than the estimate of the minibands resulting from the superlattice.) The onset of the magnetic field–induced dips, interpreted as replicas of the weak localization, generally does not occur at zero magnetic field, except for the lowest values of the drain bias (i.e., at and below 2 mV). In Fig. 5.22, the locations of these dips in conductance are plotted versus the magnetic field. (Each dip is assigned an arbitrary index, and the magnetic field at which the dip is seen is plotted as a function of this index number.) The various data lie on a single curve. The required shift of the data for each bias is defined as an index shift. It may be seen from this latter figure that all conductance dips essentially line up with a single periodicity in magnetic field. The slope of the curve, if we are to believe the periodicity in Eq. (5.116), should correspond to the coupling of integer numbers of flux quanta per unit cell. The slope of this curve indicates a periodicity of the superlattice of about 170 nm, which is to be compared with the estimate from the electron micrographs of 165 nm.

Using the theory for weak localization around zero magnetic field, the inelastic mean free path can be estimated for each of the dips that occur in Fig. 5.21. The value obtained in this manner is relatively independent of the magnetic field and bias applied and has a value near $0.55 \pm 0.1 \, \mu m$. The conductance of the sample, at zero magnetic field, rises from $40 \, \mu S$ at 2 mV bias to $80 \, \mu S$ at 5 mV bias. The conductance expected from a fully conducting channel with the Landauer formula is about $77 \, \mu S$, so it is clear that the conductance is not a free electron channel but rather a diffusive (hopping) transport through the array of quantum dots. The data given in the figures were measured at 5.7 K, and the effects persist up to about 15 K. It is also interesting that the increase of the drain potential from 2 mV to 5 mV has shifted the spectrum almost exactly one unit cell, bringing the spectra back into commensurability (with a weak localization dip at zero magnetic field).

The shifts in the dips induced by the source-drain potential can be understood within the context of Harper's equation. With a source-drain potential applied to the device, the energy levels are not only shifted along the channel, in reference to their values at the source end, but also distorted slightly within each unit cell. It is this latter effect that can give rise to the measured shifts, but this is within the context of the derivations provided by Hofstadter. Here, both the electric field and the magnetic field must be treated in the vector potential, so that

$$A_y = Bx + e \int_0^\infty e^{-t/\tau} E \, dt, \tag{5.118}$$

where the elastic scattering time has been introduced to limit the time range of the integral. The addition of the second term in the vector potential modifies (5.116) to the form

$$\phi(m + 1) + \phi(m - 1) + 2 \cos[2\pi m\alpha - \omega_B \tau - \nu]\phi(m) = \varepsilon \phi(m), \tag{5.119}$$

where

$$\omega_B = \frac{eEa}{\hbar} \tag{5.120}$$

is the Bloch frequency. The presence of the electric field, and the resulting change in the drift velocity shifts the y-momentum that appears in ν. This shifts the zero of the cosine function and hence the value of the magnetic field that corresponds to "zero." However, we would not expect any shift until the drift momentum becomes comparable with the Fermi momentum, and for these samples this means biases above 2 mV. In fact, the total shift of the spectrum by one index unit in moving from 2 mV bias to 5 mV bias leads to an estimate of the mobility within 25% of the measured value of $2 \times 10^4 \, cm^2/Vs$. These measurements, and the above discussion, strongly suggest an interpretation of the dips as replicas of the zero-magnetic-field weak localization, an interpretation in keeping with the dependence of weak localization upon the eigenvalues, as found in the basic theory above.

5.5 Quantum treatment of the fluctuations

In this section, we want to extend the Green's function approach to the calculation of the correlation function for the universal conductance fluctuations. This will give us both the amplitude of these fluctuations and the manner in which they decay with magnetic field. As above, we will build this up in parts, by discussing the basic approach, which is similar to that above, and then look at the energy variation of the correlation function. We then look at the inclusion of the magnetic field. Most of this work is built around the

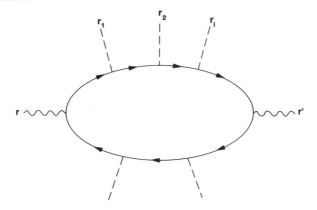

Fig. 5.23. The real-space electron-hole propagator with unconnected impurities.

electron-hole propagator, which has become known as the diffuson, and the cooperon or electron-electron propagator. Also, as above, the approach that we follow is to work with the impurity-averaged diagrams, which is a method of constructing ensemble averages for the Green's functions, and the approach to be adopted here closely follows the seminal paper of Lee, Stone, and Fukuyama [28]. The crucial assumption is that the ergodic hypothesis adopted here implies that $rms(g) = [var(g)]^{1/2}$ is a good measure of the typical amplitude of the fluctuations that are found in the conductance.

In this approach, we want to work with real-space Green's functions rather than momentum-space Green's functions. That is, we want to work with $G(\mathbf{r}, \mathbf{r}', \omega)$, as originally described above. The diagram for the electron-hole propagator is shown in Fig. 5.23. This should be compared with Fig. 5.13a for the momentum-space representation and with Fig. 5.19. The diagram in Fig. 5.23 represents the creation of an electron-hole pair (excitation of an electron above the Fermi energy leaving a hole below the Fermi energy) at $\mathbf{r}$, and the subsequent propagation of this pair to $\mathbf{r}'$, where it recombines. The dashed lines are the impurity interactions, and the "wiggly" lines are the position connections for the diagram (which couple in the k^2 term in the momentum representation). Impurity averaging connects the impurity lines, just as discussed in the treatment of conductance above. For the full Green's functions, the impurity lines connect to other impurities on the same propagator, while for higher-order terms (beyond the Drude approximation for the diffuson and the cooperon) the connections span across the two Green's functions. Here, however, we are interested in the quantity $\langle \delta g(E, B) \delta g(E + \Delta E, B + \Delta B) \rangle$, which is a conductance-conductance correlation function, so that we want to consider diagrams in which there are two loops, each of which is like that of Fig. 5.23. That is, we shall make a "double bubble" diagram, and our interest lies in those interaction terms in which the impurity lines span *between these two bubbles*, thus building in correlations between the two δg terms in the correlation function. This basic premise lies in the Kubo formula (5.63) (which is in momentum space), in which the two-particle Green's function is represented by the electron-hole propagator discussed here (the lowest-order term in the electron-hole propagator). It is assumed that each of the Green's functions in this figure is already the total Green's function including the self-energy corrections due to impurities that average together for these terms. Thus, only those new interactions that span the two pairs of propagators are of interest in this treatment and give rise to the fluctuations that are of interest.

One reason for this approach is that, if we consider only the terms that lie on one of the two bubbles and do not couple the two bubbles, then these terms are already included in the proper calculation of $g(E, B)$ and cancel when we compute the correlation of the *fluctuation* in the conductivity. Thus, the correlation function describes small changes in the conductivity at one energy due to interactions with a second conductivity bubble. This means that such an interaction must arise from terms that span the two bubbles in our double bubble diagram.

5.5.1 The correlation function in energy

Let us now turn to the evaluation of the correlation function for the fluctuations in the conductance. For simplicity in this section, the magnetic field effects will be ignored, and we will worry only about the variation with the energy. This variation is introduced by varying, for example, the Fermi energy of the two-dimensional electron gas. The diagrams are two nested polarization loops, as discussed above; each pair of loops represents one of the two conductivity bubbles given in the product of the fluctuations in $\langle \delta g(E) \delta g(E + \Delta E) \rangle$. One loop is at the energy E, and the second one is at the energy $E + \Delta E$. Since the quantity of interest is the correlation of the *fluctuation* in the conductance, the first term in the coupling of the two loops, which corresponds to no coupling between the loops (the isolated loops), cancels with the term in $\langle g(E) \rangle \langle g(E + \Delta E) \rangle$; that is, $\langle \delta g(E) \delta g(E + \Delta E) \rangle = \langle g(E)g(E + \Delta E) \rangle - \langle g(E) \rangle \langle g(E + \Delta E) \rangle$. When there are no impurity lines coupling the two loops, there is no correlation between them, and the contribution cancels the last term. Thus, after impurity averaging, the only diagrams of interest in computing the correlation in the fluctuations are those in which the impurity lines span the two loops, as discussed above. The most important diagrams are those in which the impurity lines do not cross, such as those diagrams that would normally contribute to ladder diagrams. There are a variety of ways in which these two nested loops can be arranged. For example, one loop consists of the position vectors $\mathbf{r}$ and $\mathbf{r}'$. The second loop consists of position vectors $\mathbf{r}_1$ and $\mathbf{r}'_1$. In the most logical case, $\mathbf{r}$ and $\mathbf{r}_1$ are coincident while $\mathbf{r}'$ and $\mathbf{r}'_1$ are coincident (the electron-hole pairs propagate close to one another). In another case, $\mathbf{r}$ and $\mathbf{r}'$ are coincident, while $\mathbf{r}_1$ and $\mathbf{r}'_1$ are at different locations (the electron-hole pairs diverge from one another in their propagation). A third possibility is that all four positions are unrelated. A fourth possibility is that each loop comes back upon itself so that $\mathbf{r}$ and $\mathbf{r}'$ are coincident. Still there is another possibility in which the latter case is complicated by impurity lines spanning the bubble. The latter two, however, are unimportant because they tend to be canceled by higher-order terms. Only the first three terms in which two, three, and four diffusons are involved need to be considered, since they tend to produce contributions of the same order. The first three terms are shown in Fig. 5.24a–c, respectively.

The above diagrams are in fact generated by the adoption of the two current loops (the two polarization bubbles), with the two irreducible vertices (the factors Λ, which represent the scattering plus polarization) inserted in all possible manners. Another set of diagrams arises from the maximally crossed impurity lines, in which the rotation of the "hole" propagator is reversed, giving a set of electron-electron (cooperon) bubbles. In the presence of normal impurity scattering (no spin-dependent scattering), these latter diagrams give exactly the same contribution to the correlation function. Thus, a factor of two will be added to the final result. The feature that leads to the recognition that these diagrams are the important contribution is tied to the basic singularity at small $\mathbf{q}$ and the frequency of the diffusion

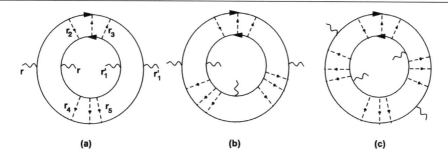

Fig. 5.24. The three diagrams that correspond to the first three contributions to the conductance correlation function.

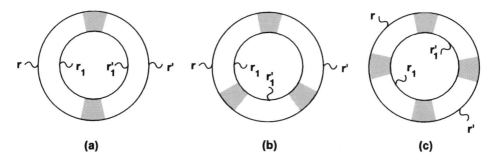

Fig. 5.25. A rearrangement of Fig. 5.24a,b showing the positions of the polarizations.

pole itself, as discussed in a previous section. When the arrows on the Green's functions are oppositely directed (this assumes a particular orientation of the two bubbles), as is normal in the particle-hole two-particle Green's function, the diagrams represent density fluctuations and have the characteristic diffusion pole. The set of diagrams shown in Fig. 5.24 are the only ones in which the diffusion poles all occur with the same value of momentum transfer.

Each of the bubble-pair combinations has two end groups corresponding to the connections to the currents, and a central group corresponding to the impurity-polarization combination, just as in the previous sections. (For example, the central impurity-polarization part led to Λ, and the end groups added an additional pair of Green's functions to the overall two-particle Green's functions.) Consider Fig. 5.24a. Each bubble has both an electron and a hole propagator (and the equivalent cooperon propagator), leading to four possible combinations of Green's functions in which the upper part of the pair of rings is represented by G^rG^r, G^aG^r, G^rG^a, and G^aG^a, respectively. Coupled to this are two current connections, each of which has a pair of Green's functions. Now consider one of these, as shown in Fig. 5.25. In (a), the pair of rings is shown with a set of coordinates, while (b) indicates the current connections. In essence, the current connections span the coordinates of the conductivity two-particle Greeen's function and couple to the diffusion coordinates.

The overall conductance fluctuation contribution can be written in the form of the conductance found earlier, which we now write

$$F_a = \frac{e^4\hbar^4}{m^4} \int d^d\mathbf{r}_2 \int d^d\mathbf{r}_3 \int d^d\mathbf{r}_4 \int d^d\mathbf{r}_5 \Lambda(\mathbf{r}_2, \mathbf{r}_3)\Lambda(\mathbf{r}_4, \mathbf{r}_5) j(\mathbf{r}_2, \mathbf{r}_4) j(\mathbf{r}_3, \mathbf{r}_5). \quad (5.121)$$

The terms Λ are the diffuson (or cooperon) contributions and are given by (5.91), and the terms j are the current connections. We have previously found the cooperon, or polarization, terms including the inelastic decay rate, so that

$$\Lambda(\mathbf{q}, \omega) = \frac{1}{2\pi\hbar\rho(E_F)\tau} \frac{1}{Dq^2\tau + \tau/\tau_\varphi}, \tag{5.122}$$

which is in Fourier transform mode. The current kernel is given equivalently by (5.91) modified for the present purpose. That is, we need to pull Green's function lines together to make a current vertex as done in the calculation of the conductivity above. This may be accomplished by letting the current operator be

$$j(\mathbf{r}_3, \mathbf{r}_5) = j_0\delta(\mathbf{r}_3 - \mathbf{r}_5), \tag{5.123}$$

and

$$j_0 = \int \frac{d^d\mathbf{k}}{(2\pi)^d} k_d^2 [G^r(\mathbf{k})G^a(\mathbf{k})]^2 = 2\rho(E_F)\tau^3 \frac{k_F^2}{d}. \tag{5.124}$$

This is well and good for infinite samples. However, when we deal with finite samples, the polarization functions must be modified to account for the boundary conditions. We work with the reduced polarization functions, represented by the terms in the last fraction in (5.122). When there is no current flow through a lateral boundary, then one has the normal derivative of Π equal to zero. On the other hand, when there are good conducting metallic contacts at the boundary, the flow is ballistic (nondiffusive) and the boundary condition must be to set Π equal to zero. In general, the polarization can be found from the differential equation in (5.111), but in real space as

$$\Pi(\mathbf{r}, \mathbf{r}', \Delta E) = \sum_m \frac{Q_m^*(\mathbf{r})Q_m(\mathbf{r}')}{\lambda_m}, \tag{5.125}$$

where $Q_m(\mathbf{r})$ and λ_m are the eigenfunctions and eigenvalues of

$$\tau\left[-D\nabla^2 - i\frac{\Delta E}{\hbar} + \frac{1}{\tau_\varphi}\right]Q_m(\mathbf{r}) = \lambda_m Q_m(\mathbf{r}), \tag{5.126}$$

subject to the boundary conditions discussed above, and where an effective diffusion constant has been introduced in place of the bare $\hbar/2m$. For a sample of cross section $L_x \times L_y$ and length (to the metallic contacts) L_z in three dimensions,

$$Q_m(\mathbf{r}) = \sqrt{\frac{8}{L_x L_y L_z}} \sin\left(\frac{m_z \pi z}{L_z}\right) \cos\left(\frac{m_y \pi y}{L_y}\right) \cos\left(\frac{m_x \pi x}{L_x}\right), \tag{5.127}$$

and

$$\lambda_m = \tau D\pi^2\left[\left(\frac{m_z}{L_z}\right)^2 + \left(\frac{m_y}{L_y}\right)^2 + \left(\frac{m_x}{L_x}\right)^2\right] - i\frac{\tau\Delta E}{\hbar} + i\frac{\tau}{\tau_\varphi}. \tag{5.128}$$

Thus, we can now write, after integration in the transverse directions,

$$F_a = \frac{e^4\hbar^4}{m^4}\left[2\rho(E_F)\tau^3\frac{k_F^2}{d}\right]^2\left[\frac{1}{2\pi\hbar\rho(E_F)\tau}\frac{4}{L_z}\right]^2\sum_m\frac{1}{\lambda_m^2}$$

$$= \left(\frac{e^2}{\pi\hbar}\right)^2\left(\frac{4}{\pi}\right)^2\sum_m\frac{1}{\tilde{\lambda}_m^2}, \tag{5.129}$$

where $D = v_F^2 \tau / d$, and where

$$\tilde{\lambda}_m = \lambda_m \frac{1}{D\tau} \left(\frac{L_z}{\pi} \right)^2 \tag{5.130}$$

produces a dimensionless eigenvalue. The other three possible arrangements of the advanced and retarded Green's functions for this simple diagram lead to changes in the sign of the wavevector (for the cooperon terms), and the three taken together lead to a factor of two multiplier on Eq. (5.129).

The terms that arise from Fig. 5.24b,c can be computed in the same manner, but they add extra summations over the eigenvalues since they have more diffusons (or cooperons) in the overall diagram. We will not go through the full derivation, as it follows clearly the outline above. The end result is that

$$F(\Delta E) = \left(\frac{e^2}{\pi \hbar} \right)^2 \left(\frac{4}{\pi^2} \right)^2 \sum_{m_x, m_y = 0} \sum_{m_z = 1,3,5,\ldots} \left\{ 2Re \left[\frac{1}{\tilde{\lambda}_m} \right]^2 \right.$$

$$- 8Re \sum_{n_z = 2,4,6,\ldots} \frac{f_{mn}^2}{\tilde{\lambda}_m \tilde{\lambda}_n} \left[\frac{1}{\tilde{\lambda}_m} + \frac{1}{\tilde{\lambda}_n} \right]$$

$$+ 24Re \sum_{p_z = 1,3,5,\ldots} \sum_{n_z q_z = 2,4,\ldots} \left. \frac{f_{mn} f_{np} f_{pq} f_{qm}}{\tilde{\lambda}_m \tilde{\lambda}_n \tilde{\lambda}_p \tilde{\lambda}_q} \right\}, \tag{5.131}$$

where

$$f_{mn} = \frac{4m_z n_z}{\pi \left(m_z^2 - n_z^2 \right)}. \tag{5.132}$$

Clearly, the amplitude of the universal conductance fluctuation is given by the square root of F and has a value of the order of $e^2/\pi \hbar$, with a numerical factor on the order of unity depending on the number of modes excited in the system and the extent to which the lateral dimensions are meaningful. These numerical factors have been computed to be [28] 0.729 in 1D, 0.862 in 2D, and 1.088 in 3D, when the current flows in the z-direction.

Just as in the previous section for weak localization, the key factor in the dependence of the correlation function on the energy difference is the reduced eigenvalue of the diffusion propagator, which is given by (5.128) and (5.130) as

$$\tilde{\lambda}_m = \left[m_z^2 + m_y^2 \left(\frac{L_z}{L_y} \right)^2 + m_x^2 \left(\frac{L_z}{L_x} \right)^2 \right] - i \frac{\Delta E L_z^2}{\pi^2 \hbar D} + \frac{L_z^2}{\pi^2 \tau_\varphi D}. \tag{5.133}$$

For a two-dimensional system, in which the current path is short compared to the lateral dimensions but large compared to the coherence length, the second and third terms in the square brackets are negligible, and the eigenvalue is dominated by the last two terms. Then, the critical value of the energy change required to reduce the correlation function to one-half its peak value is just

$$\Delta E_{c,2D} = \hbar/\tau_\varphi, \quad L_x, L_y \gg L_z \gg L_\varphi. \tag{5.134}$$

On the other hand, for a truly short current path, in which the length is much smaller than the phase coherence length $(D\tau_\varphi)^{1/2}$, the last term is also negligible, and the critical correlation energy is when the imaginary term is equal to m_z^2, which we take to be unity for the lowest eigenstate, and

$$\Delta E_{c,2D} = \frac{\pi^2 \hbar D}{L_z^2}, \quad L_z \ll L_\varphi, L_x, L_y. \tag{5.135}$$

If we are dealing with a quantum wire, in which $L_z \gg L_x, L_y$, then the transverse modes dominate the eigenvalue. The correlation energy is then defined by the condition

$$\frac{\Delta E_{c,1D} L_z^2}{\pi^2 \hbar D} = \left[m_y^2 \left(\frac{L_z}{L_y} \right)^2 + m_x^2 \left(\frac{L_z}{L_x} \right)^2 \right] + \frac{L_z^2}{\pi^2 \tau_\varphi D}, \tag{5.136}$$

or

$$\Delta E_{c,1D} = \frac{\pi^2 \hbar D}{L_x L_y} \left[\frac{m_y^2 L_x^2 + m_x^2 L_y^2}{L_x L_y} + \frac{L_x L_y}{\pi^2 L_\varphi^2} \right] \sim \frac{\pi^2 \hbar D}{L_x L_y}, \quad L_x L_y \ll L_\varphi^2. \tag{5.137}$$

5.5.2 Correlation function in a magnetic field

When we now turn to the magnetic field variation, the same diagrams will come into play. Now, however, the Green's functions that contribute to these diagrams are modified by the magnetic field (actually, by the vector potential) according to Eq. (5.107). This leads to the general changes as discussed in the previous section on weak localization. There are some differences, however. There, we were discussing the cooperon, or particle-particle channel, for which the actual value of the magnetic field appears, since the contribution of interest is of order $G(\mathbf{r}, \mathbf{r}')^2$. Thus, for the particle-particle terms, the important equation of motion (5.126) is changed by

$$-i\nabla \rightarrow -i\nabla - e(2A + \Delta A). \tag{5.138}$$

The eigenvalues for the particle-particle channel will decay away very quickly with magnetic field, as does the weak localization (unless there is a periodicity in the eigenvalues, as discussed in the previous section). For all practical purposes, the particle-particle channel can be ignored except at the lowest magnetic fields. Thus, the behavior of the correlation function near $B = 0$ will be somewhat different than at higher magnetic fields, where the particle-particle channel has been damped out. Nevertheless, where the magnetic field clearly is not large, the particle-particle channel must still be evaluated but its contribution is no longer equal to that of the particle-hole channel.

In distinction to the particle-particle channel, the particle-hole channel relies upon the Green's function product $G(\mathbf{r}, \mathbf{r}')G(\mathbf{r}', \mathbf{r})$, so that it is dependent only on the *difference* in the vector potential over the two paths. As discussed earlier in this chapter, the particle-particle channel is similar to the Aharonov-Bohm effect in that each propagator corresponds to one branch and the two add their effects together to measure the coupled flux. With the particle-hole channel, however, one propagator *cancels* the phase of the other, so that only the *difference* in phase between the two paths is important. Thus, in this case the equation of motion (5.126) is modified to depend only on the difference in the magnetic field between the two Green's functions, or

$$-i\nabla \rightarrow -i\nabla - e\Delta A. \tag{5.139}$$

Here, only the difference in the magnetic field appears in the eigenvalues for the diffuson. The equation of motion must still be solved, and this is considerably more difficult now since the magnetic field, taken to be in the x-direction, couples the y- and z-motions. This leads to a complicated coupled-mode solution for the wavefunction.

In the high-magnetic-field limit, and for a mainly two-dimensional system, where the Landau levels are fully formed, we have $BL_zL_y \gg \Phi_0 = h/e$. In this case, the eigenvalues for the particle-particle channel are just those of the Landau levels formed in the two-dimensional system, because of the role played by the absolute value of the magnetic field. This leads to [28]

$$\tilde{\lambda}_m^{pp}(B, \Delta B) = \frac{4}{\pi}\left(n + \frac{1}{2}\right)\frac{(2B + \Delta B)L_z^2}{\Phi_0}. \tag{5.140}$$

Hence, the lowest value of this eigenvalue is still much greater than unity, regardless of the value of ΔB. Certainly, this will not be the case for the diffuson contributions, which depend on only this increment in the field, and the particle-particle channel contribution will be negligible. Thus, the amplitudes of the fluctuations are reduced by a factor of two over the values quoted above.

The eigenvalues for the particle-hole channel can be computed in two limits. The first is where the difference magnetic field can be treated as a small perturbation on the normal eigenvalues (5.133). (We generally take $\Delta E = 0$ in this discussion.) If we take the vector potential as $\mathbf{A} = \Delta By\mathbf{a}_z$, where $\mathbf{a}_z$ is a unit vector pointing in the z-direction, the perturbation has two terms:

$$V' = -2ie\Delta By\frac{\partial}{\partial z} + (e\Delta By)^2. \tag{5.141}$$

For consistency, one needs to calculate the effect of the first term to second order and the effect of the second term to first order (thus giving a correction to order B^2 in both cases). This is much easier for the case of a quantum wire, in which $L_z \gg L_x, L_y$. In the case in which only the lowest lateral mode is chosen, the first term has no diagonal corrections due to the symmetry of the system, and the second term leads to the correction

$$\tilde{\lambda}_m(\Delta B) \simeq m_z^2 + \frac{1}{3}\left(\frac{\Delta BL_yL_z}{\Phi_0}\right)^2 + \frac{L_z^2}{\pi^2\tau_\varphi D}. \tag{5.142}$$

If the last term is negligible, this leads to a correlation magnetic field (for $m_z = 1$)

$$\Delta B_c \sim \sqrt{3}\frac{\Phi_0}{L_yL_z}, \quad L_z \ll l_\varphi. \tag{5.143}$$

It must be recalled that the magnetic field is in the x-direction, so that the area in this equation is the sample area normal to the magnetic field. In the other limit, in which the length is much larger than the coherence length, the correlation magnetic field is given by

$$\Delta B_c = \frac{\sqrt{3}}{\pi}\frac{\Phi_0}{L_yL_\varphi}, \quad L_z \gg l_\varphi. \tag{5.144}$$

A full perturbation treatment, with evaluation of all the diagrams (and the variations in f_{mn}) suggests that the factor of $\sqrt{3} \to 1.2$ for the case of (5.143) [28]. As the sample dimensions change from a long quantum wire to a square two-dimensional region, the perturbation contribution from the first term of Eq. (5.141) no longer vanishes but instead begins to mix states with different m_y. This contribution is negative and therefore reduces the coefficient of the ΔB^2 term in the eigenvalue. Thus, higher dimensionality will reduce the coefficient and increase the value of the correlation magnetic field. This is expected to have less of a shape dependence, and the product $L_y l_\varphi \to l_\varphi^2$ for $L_z, L_y \gg L_\varphi$. We return to these arguments in the next section, where we examine the possible breakdown in the diffusive limit that has been adopted throughout this chapter.

5.6 Summary of universality

The conductance fluctuations that have been discussed here often are thought to have certain univeral properties – hence the name "universal conductance fluctuations." For example, it is clear in Fig. 5.9 that the amplitude of the fluctuations can easily be expressed as a simple quantity that is a function of the ratio of the wire length L to the coherence length l_φ. Moreover, the correlation field ΔB_c also is thought to depend on only the coherence length and perhaps the width of the wire in a narrow one-dimensional conductor. But how universal are these factors, and what is the nature of the deviations? In this section, we will examine just this point by reiterating some of the previous conclusions, and also by examining other data relevant to the question. We look first at the amplitude of the fluctuations and then at the dependence of these on the magnetic field, particularly at fields for which $\omega_c \tau \geq 1$.

5.6.1 The width dependence of the fluctuations

If we transition from a wide quasi-two-dimensional electron gas into a narrow quantum wire, we certainly expect to see some change in the nature of the fluctuations. However, once the wire limit is reached, for $W < l_\varphi$, will there continue to be a change in the nature of the dependence of the fluctuations on wire width? In fact, experimentally such a dependence is still found to be the case. Consider Fig. 5.26. Changing the bias voltage on the gates that are used to define a "split-gate" wire changes both the wire width and the carrier density in the wire. In this structure, the potential range over which the conductance fluctuates is larger than the thermal energy, so the equations should be modified (see Chapter 6). However, it is clear that the fluctuation amplitude is changing as the gate bias is varied. The fluctuations are thought to arise as the carrier density is varied by the gate bias [15]. However, the effect of the change in the gate width cannot be overlooked, and it is thought that the changes in the width will cause the fluctuations in the conductance. But why is the amplitude changing?

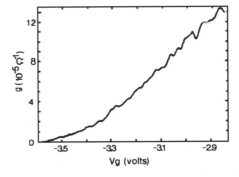

Fig. 5.26. Conductance fluctuations in the pinch-off characteristics of a split-gate wire. [After Thornton *et al.*, in "Granular Nanoelectronics," eds. D. K. Ferry *et al.* (Plenum Press, New York, 1991), pp. 165–179, by permission.]

We can consider the correlation function, as calculated in the last section. The amplitude of the fluctuation may be determined from (5.129) to leading order as

$$\delta g = \sqrt{F_\alpha} \sim \frac{e^2}{\pi \hbar} \frac{4}{\pi} \sqrt{\sum_m \frac{1}{\tilde{\lambda}_m^2}} \sim \frac{e^2}{\pi \hbar} \frac{4}{\pi} \sqrt{\sum_m \frac{1}{\left[m^2 + m_x^2 (L/W)^2\right]^2}}, \tag{5.145}$$

where we have used Eq. (5.133) in the absence of the correlation energy and set $L_x = W$, $L_z = L$. We can now assume that the length of the wire is sufficiently large that a very large number of longitudinal modes are included, and that the effective length is $L = l_\varphi$. The summation can be converted to an integral, and this leads to

$$\delta g \sim \frac{e^2}{\pi \hbar} \frac{2}{\sqrt{\pi}} \left(\frac{W}{l_\varphi}\right)^{3/2} \left(\sum_{m_x} \frac{1}{m_x^3}\right)^{1/2}. \tag{5.146}$$

In essence, the summation represents a summation over the number of contributing (transverse) modes. However, the transverse modal number m_x depends upon both the width and the carrier density. In the normal case, in which self-consistent calculations show that the potential profile is a flat bottom with parabolic sides [51], the number of modes remains nearly constant. (The mode separation is set by the parabolic walls, but the sheet density and Fermi level are set by the flat bottom, so the number of filled modes is nearly constant until the density begins to be depleted as the flat bottom disappears.) For near parabolic confinement, which is the usual case when the gate is depleting the density, the number of transverse modes in the wire is approximately

$$m_{x,\max} \sim \frac{W}{2} \sqrt{\frac{\pi \hbar^2 n}{2}}. \tag{5.147}$$

Will this make any difference? For a large number of modes, the summation is essentially $\zeta(3)$, where $\zeta(i)$ is the Riemann zeta function. If there is only a single mode, the summation is unity. For an infinite number of modes, the summation is the Riemann zeta function, and this is of order two. When we take the square root, it may be seen that the basic $W^{3/2}$ behavior remains in the problem. Consider, for example, the data in Fig. 5.27, which is for a split-gate wire in a GaAs/AlGaAs heterostructure [52]. The widths plotted are slightly different from those in the original paper (the current authors have carried out a slightly different evaluation of the width data), but the evidence is clear. The amplitude of the fluctuations has a strong W^3 dependence, both at zero magnetic field and at high magnetic field where $\omega_c \tau > 1$, and this dependence is not supported by the above theory. The flattening of the curve at the larger widths (more positive gate bias) is thought to occur for biases where the gate has actually lost control of the wire width. Other authors have suggested that such behavior could be due to reduction in the number of transverse modes [53], [54], but the above argument suggests that any such effect is really quite weak. The W dependence cannot be explained, even within the context of (5.146), and the number of transverse modes yields a term that remains quite constant as the gate bias is varied. The dependence of the fluctuations on wire width remains a questionable area where more data are required if the theory is to be universally accepted.

The above argument resides upon the assumptions that each mode of the wire acts nearly independent and that the summation over the modes in (5.129) corresponds to a summation over independent sources of fluctuation. This may not be the case in short wires in which the

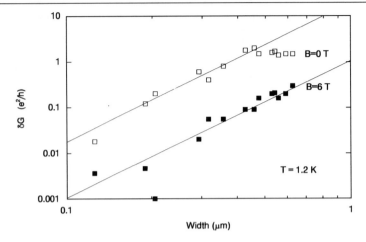

Fig. 5.27. Width dependence of the amplitude of the conductance fluctuations in a GaAs/AlGaAs split-gate wire. [After Ishibashi *et al.*, in "Science and Technology of Mesoscopic Structures," eds. S. Namba *et al.* (Springer-Verlag, Tokyo, 1992), by permission.]

wire diffusion propagator must be connected to quasi-two-dimensional regions. However, calculations by Chandrasekhar *et al.* [23], which follow those discussed above in connection with the weak localization, suggest that the major change is in the value of the preceeding constant, and that there is no major modification of the width dependence introduced into the amplitude of the fluctuations.

Other than this width dependence discussed here, the behavior of the amplitude of the fluctuations seems to be quite universal, depending solely on the ratio of the wire length to the coherence length. Although we have discussed this in terms of wires, quite the same behavior is expected in wide, quasi-two-dimensional structures.

5.6.2 Size variation of the correlation magnetic field

There have been few studies of the variations of the correlation magnetic field and/or the correlation energy in quantum wires. Yet we found in the above analysis that some width dependence should be expected. In Fig. 5.28, we show the variation of the correlation magnetic field (here called B_c, which is the ΔB_c used in the previous sections) [55]. There are several points of interest in these results. First, the correlation field in the 6-μm-long wire (the squares in the figure) are considerably larger than those of the 2-μm-long wire (the solid circles). The wires themselves are defined by the split-gate technique in a quasi-two-dimensional electron gas. Thus, the wire width can be readily varied by the applied gate bias. It should be noted that the coherence length inferred in these wires is thought to be of the order of 2.5 μm, if an analysis comparable to that of Eq. (5.31) is used to match the one-dimensional diffusion propagator to the two-dimensional contact regions. Thus, in some sense the wires are more properly thought of as quasi-ballistic wires rather than as true diffusive wires.

For the longer wire, it is probably reasonable to take the expected dependence of the correlation field on the width of the wire from Eq. (5.144). This may be restated as (in the

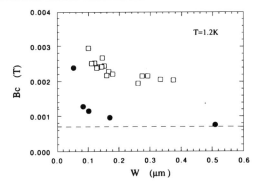

Fig. 5.28. The correlation magnetic field for split-gate-defined quantum wires. The open squares are for a 6 μm wire length, and the filled circles are for a 2 μm wire length. [After T. Onishi *et al.*, Physica B **184**, 351 (1993), by permission.]

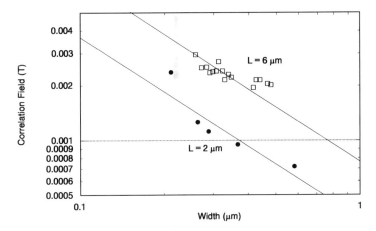

Fig. 5.29. A replot of the data for the variation of the correlation field. The upper set of data is for the 6 μm wire, and the lower data is for the 2 μm wire. The solid lines are a guide to the eye and have a variation as 1/W.

notation of the present section)

$$\Delta B_c = \frac{\sqrt{3}}{\pi} \frac{\Phi_0}{W l_\varphi}, \quad L \gg l_\varphi. \tag{5.148}$$

This suggests that the correlation field should vary as $1/W$, and this may be seen to be the case from Fig. 5.29, in which the data are replotted. The diagonal solid lines are a guide but have a slope that corresponds to $1/W$. It may be seen that this fits the data quite well. Actually, the magnitude is a fair fit as well. The prefactor in front of Eq. (5.148) is $\sqrt{3}/\pi = 0.55$, whereas the data have a value closer to 0.2–0.42 over the range of data shown in the figure for the longer sample (open squares in this data). This value is within a factor of two, but lower than the prefactor found from the theory.

The problem crops up with the smaller wire. If we try to use Eq. (5.143) for the case of a wire length small compared to the coherence length, then this equation leads us to expect a larger value of the correlation field. In fact, the data suggest a considerably smaller value of the field by almost a factor of two. The answer to this is likely to lie in the fact that the derivations used in arriving at these equations for the correlation field assume a long wire and neglect the coupling of the diffusion propagator to the quasi-two-dimensional electron gas. In fact, the amplitude and correlation field in the two-dimensional regime are expected to both be smaller than the equivalent values in the quantum wire (the replacement by the width W by an additional factor of l_φ is the simplest way to illustrate this change). Indeed, Chandrasekhar *et al.* [23] find a value for the coefficient mentioned above of about 0.42 in long wires, which is quite close to one end of the range seen in Fig. 5.29. In short wires, however, the equations are too complicated to evaluate other than by numerical simulation. Nevertheless, they find reductions in the correlation field by factors of two to three for local measurements (where the measuring path includes the current path). This is in keeping with the results discussed in this section as well. The moral seems to be that a proper comparison with experiment requires the use of quantum wires that are definitively longer than the coherence length so that the diffusion regime is well established.

5.6.3 Breakdown of universality in a magnetic field

In a high magnetic field, it has been speculated that the *universal* character of the UCF begins to break down. By "universal" we mean that both the amplitude of the UCF and the correlation magnetic field ΔB_c are described by a single phase-coherent property – the coherence length l_φ, the result found above. By this we mean that the amplitude scales with L/l_φ, and that ΔB_c is a simple function of l_φ through the effective area of the phase coherent "loop." Thus, the change in one should be accompanied by a corresponding change in the other. For high magnetic fields, this is no longer the case. It was first noticed by Timp *et al.* [56] that both the amplitude of the fluctuations and the correlation magnetic field ΔB_c measured in quantum wires increase at high magnetic fields when $\omega_c \tau > 1$, where ω_c is the cyclotron frequency ($= eB/m$). In a high-mobility quantum wire, they found that the correlation magnetic field increased from 10 mT at low values of the field to nearly 39 mT at 7 T. These measurements were for samples in which $L < l_\varphi$. They proposed that the fluctuations were due to Aharonov-Bohm effects and that the change in the correlation behavior was indicative of a change in the width of the distribution of electron trajectories across the wire as the magnetic field was varied.

Extensive measurements have subsequently been made by a variety of groups [8], [57]–[63] in longer samples, both for local and nonlocal situations. (Here, local means that the voltage measurements are made across a portion of the sample through which the bias current flows. The nonlocal measurements are made in a way such that the voltage measurements are made across a portion of the sample which is not in the current path.) A typical set of such data is shown in Fig. 5.30, which is due to the Nottingham group [62]. The data in Fig. 5.30a correspond to the local magnetoresistance $R_{xx} = R_{ag,ce}$, where the notation is that of Eq. (5.42) (see Chapter 3). At low values of the magnetic field, the UCF coexist with negative magnetoresistance due to the weak localization, although the latter effect is rather small and not easily distinguished in this trace. The latter allows a determination of $l_\varphi \simeq 0.3 \, \mu$m at 4.2 K. The next two panels show the value of the resistance fluctuation and a rectified voltage obtained from an a.c. current excitation. Finally, the nonlocal resistance $R_{ab,cd}$ is plotted in Fig. 5.30d. All of these measurements show a substantial increase in ΔB_c but there is no evidence of a corresponding change in the fluctuation amplitude; in fact,

In these samples, the scattering time is indeed determined by surface scattering, and therefore by the wire width. Thus, it is clear that, just as for the UCF above, the transition is set by the actual mobility in the wire, which arises from the surface scattering, and not by the bulk properties of the material. However, this clear behavior is not so evident in short wires where the phase coherence length includes a considerable fraction of the contacting region.

Now consider the equation for the correlation field expected for a quantum wire, as given in (5.148). First, one could assume that the motion of the current is contained uniformly across the width of the wire. Then, if the correlation magnetic field is to change in high magnetic fields, it must do so through a variation of the coherence length l_φ. Using the result (5.149), it is expected that l_φ will have two possible dependencies on the magnetic field. If the phase relaxation time is not varying with the field, then we expect $l_\varphi \sim B^{-1}$, which leads to $\Delta B_c \sim B$, and this behavior is seen in large parts of the data in the above figures. On the other hand, if the Landau levels (which anyway are not well formed) are being narrowed in the increasing magnetic field, then $\tau_\varphi \sim B^{1/2}$, and $l_\varphi \sim B^{-3/4}$, which leads to $\Delta B_c \sim B^{3/4}$. This is not very distinguishable from the linear behavior, and it is not clear if the various data are sufficiently good to separate such a field dependence. The variation in the phase coherence length, however, should be replicated by a change in the amplitude of the fluctuations. Since l_φ is decreasing in the magnetic field, we expect the effective length of the wire to be increasing with a consequent drop in the amplitude of the fluctuations. In fact, this behavior does not seem to be observed, which is why it is thought that the universal nature of the fluctuations no longer is valid.

The situation is actually more complicated than the cases discussed in the previous paragraph. At sufficiently high magnetic fields, the motion of the carriers changes from diffusive motion throughout the wire width to mainly edge-state transport. That is, the small size of the cyclotron orbit pushes the carriers into transport via skipping orbits along one side of the sample. This may still be viewed as diffusive transport if there remains significant diffusive surface scattering and impurity scattering. The Nottingham group has speculated that there are indeed two kinds of phase-coherent regions [62]. One of these is just the bulk region of the sample discussed in the last paragraph, while the other is the edge-state region. The change in the correlation magnetic field arises from the view that the effective area of the phase interference loops is clearly changing now, and the transport is dominated at high magnetic field by the latter phase-coherent region (i.e., transport changes from being defined over the former region to being defined over the latter region). However, the change in the correlation magnetic field is not due to a change in the phase coherence length, *but in the width of the current carrying area*. The extent of the edge-state region is inversely proportional to the magnetic field. In this view, the width of the wire W should be replaced by the cyclotron radius $r_c = k_F l_m^2 = \hbar k_F / eB$, and this gives an increase in the correlation magnetic field. In fact, if the coherence length is constant, one finds $\Delta B_c \sim 1/r_c l_\varphi \sim B$. This behavior has quite the same linear dependence on the magnetic field as that of the previous paragraph, but now the coherence length remains constant. Thus, *one would expect no change in the amplitude of the fluctuations for this case*. In this sense, there is no breakdown of the universal nature.

We need to reconsider the first case mentioned, that of Timp *et al.*, in which the amplitude actually *increased*. These measurements were thought to have been made in the quasi-ballistic regime, in which the coherence length is longer than the measurement region. The general feeling is that, even in quasi-ballistic wires, there is a change in behavior when $\omega_c \tau > 1$. Consider some more recent measurements for a quasi-ballistic wire. These

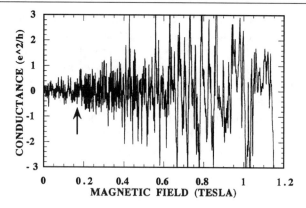

Fig. 5.33. Conductance fluctuations for a 1.25 μm wire after the Shubnikov–de Haas oscillations have been subtracted from the raw signal. The arrow marks the field at which the cyclotron orbit is calculated to be comparable to the wire width. [After Bird *et al.*, Phys. Rev. B **52**, 1793 (1995).]

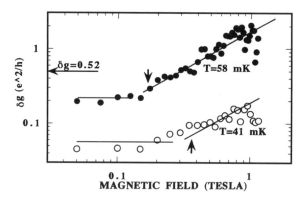

Fig. 5.34. Magnetic field dependence of the amplitude of the fluctuations for a 0.74 μm wire (open circles) and a 1.25 μm wire (solid circles). The arrows mark the points where the cyclotron radius is thought to be equal to the width. The solid lines are guides to the eye. [After Bird *et al.*, Phys. Rev. B **52**, 1793 (1995).]

measurements are shown in Fig. 5.33 for a wire of width 1.25 μm and length 30 μm [65]. In this case, the wire is defined by a wet chemical etch, so its width can not be varied. Nevertheless, the material is such that it is thought that the length is comparable to (or even less than) the coherence length, so that it is a quasi-ballistic wire. The arrow in the figure designates the value of magnetic field at which the cyclotron radius is equal to the wire width. It is clear that both the amplitude of the fluctuations and the correlation field are increasing with the magnitude of the magnetic field. The variation of the amplitude is shown in Fig. 5.34, and that of the correlation magnetic field is plotted in Fig. 5.35. Again, the arrows indicate the point where the cyclotron orbit is thought to equal the width of the wire. In these last two figures, the solid curve is a guide to the eye but represents a variation linearly proportional to the magnetic field. It is clear, though, that the variation of the amplitude of the fluctuations suggests a different interpretation than do the previous cases.

To consider this, let us return to the concept of the edge states and their width variation. In the skipping-orbit regime, the transport may not be diffusive in nature. Rather, it should

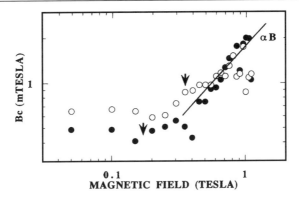

Fig. 5.35. Magnetic field dependence of the correlation magnetic field for the same wires as is Fig. 5.34. [After Bird *et al.*, Phys. Rev. B **52**, 1793 (1995), by permission.]

be more properly represented as quasi-ballistic, certainly in the shorter wires where $L < l_\varphi$. Here, the appropriate length for retention of phase coherence is no longer the coherence length l_φ but rather, as discussed in Chapter 1, the inelastic mean free path $l_{in} = v_F \tau_\varphi$. If this length is used in the results of the previous paragraphs, then we can conceive of the length of the phase-retaining trajectory as being on the order of $N\pi r_c = l_{in}$, where N is the number of "bounces" in the trajectory before losing phase coherence. The phase-coherent area for the fluctuation then can be considered to be $N\pi r_c^2/4$, and these two concepts can be combined to give

$$Wl_\varphi \rightarrow A = \frac{1}{4}r_c l_{in} = \frac{\hbar^2 k_F^2}{4m^*}\left(\frac{\tau_\varphi}{eB}\right). \tag{5.152}$$

This variation gives similar results as the last paragraph, but the coefficients have changed. If the phase relaxation time varies with the magnetic field, as discussed above, it is now possible to see variation as $\Delta B_c \sim B^{0.5}$. This is a new coefficient, not found previously. On the other hand, if the phase relaxation time is relatively constant, then we once again obtain $\Delta B_c \sim B$. It should be noted that in this quasi-ballistic regime, the correlation field is not expected to vary with the physical wire width, but it will vary with the density in the wire (the coefficient is the Fermi energy). Thus, magnetic depopulation can lead to values of the exponent even smaller than $1/2$. What about the amplitude of the fluctuation? In these quasi-ballistic wires, the important length changes from the coherence length at low values of the magnetic field to the inelastic mean free path at high values of the magnetic field. As pointed out in Chapter 1, these two lengths are different, and $l_\varphi = l_{in}\sqrt{\tau/2\tau_\varphi} < l_{in}$. If we replace the coherence length by the inelastic mean free path in the general scaling inherent in Fig. 5.9, then it is possible to draw the conclusion that the effective length of the wire is reduced and the fluctuation amplitude should correspondingly increase. The increase that is observed in Fig. 5.34 is about a factor of three to five. This seems to be on the order of magnitude that one would expect from the simple change in phase length discussed here, but this needs to be determined more effectively by experiment.

The variance in the experimental details that have been reported to date suggests that the understanding of this effect is not yet complete. Normally, universal conductance fluctuations are described by a single scaling length, the phase coherence length. If this were the case, the increase in the correlation magnetic field should be accompanied by a reduction in the amplitude of the fluctuations that would follow from a reduction in the coherence length l_φ, but this is not seen. Although it is tempting to argue that this is a failure of scaling, it should be noted that a simple interpretation in terms of the "effective" width of the wire changing by $W \to r_c$ yields a change in the correlation magnetic field without invoking a necessary change in the amplitude of the fluctuations, since it does not require a change in the coherence length. This view maintains universality in the behavior. The conclusion that can be drawn from the discussion of this last section is that currently not all is known about the nature of the conductance fluctuations in small structures. Particularly, in high magnetic fields there are a variety of variations of both the amplitude and the correlation magnetic field. There are suggestions that universal behavior is broken, but in most cases the results can be explained without this requirement. For sure, there is still further understanding that must be achieved in these very interesting structures. This is more than mere interest in an interesting physical phenomenon. As will be mentioned in the last chapter, and was discussed in the first chapter, these phenomena are expected to play a role in real devices in the ultra-submicron regime.

5.7 Fluctuations in quantum dots

Within the past seven to ten years, more effort has been expended in studying the ballistic transport in small constrained "quantum dot" regions. A great deal of the transport involving the Coulomb blockade was already treated in Chapter 4. However, the quantum dot has also become of interest for its ability to show a variety of quantum transport (and semiclassical transport) properties within a single device. In general, the quantum dot can be confined in a variety of ways, but the most common are the isolation by a pair of quantum point contacts (QPC) and the specific construction of confinement gates on nearly all sides of the dots. If the size of the dot is smaller than the coherence length, then the transport through the dot can become nearly ballistic. The most common aspect of this is the onset of "adiabatic" transport in a magnetic field, when the potentials near the QPCs are smoothly varying functions of positions. (In adiabatic transport, it is generally felt that there is no mixing between the various edge states moving through the quantum dot structure [66].) The QPCs themselves can act independently to restrict transport through them to a set of quantized subbands, or channels, leading to the conductance quantization already discussed in Chapter 3. When a high magnetic field is applied, the QPCs must be characterized by quantizing a set of magnetoelectric subbands as in quantum wires, and the edge states can be either transmitting or reflecting at the QPC. When the field is sufficiently high, the edge states dominate the transport through the dot, and the transport can be considered to be adiabatic. If the QPC is closed off, however, then one enters the tunneling or Coulomb blockade regime, and the magnetic field can affect the tunneling properties of the QPC and the resultant quantum dot properties.

At lower values of the magnetic field, the quantized channels within the QPC are not well described simply by the magnetic edge states, as the latter hybridize with the electrostatic confinement states in the QPC. At even lower values of the magnetic field, and when the QPC is relatively open, then nearly ballistic transport through the dot can occur. In these latter structures, it is possible to study the quantum transport of the dot that is closely associated with classical chaos in nearly closed cavities. The latter appear as aperiodic fluctuations in the conductance through the dot as a function of magnetic field or gate voltage, just as the universal conductance fluctuations of the previous sections. Thus, one double QPC used to define a quantum dot, or a group defining a series of quantum dots, can be biased with voltages and magnetic fields to cover a quite wide range of quantum transport regimes. And perhaps the most interesting is the latter, the chaotic aperiodic oscillations, for they may give insight into the still confusing world of quantum chaos. In this section, a number of the experimental investigations of these various regimes will be reviewed. No comprehensive treatment is attempted because the field is moving too rapidly for this. Rather, we try to merely illustrate the wealth of the investigations.

Some of the first experiments on quantum dots formed by two QPCs in series were those of Wharam *et al.* [67] and Kouwenhouven *et al.* [68]. In adiabatic transport through the dot, the conductance is completely determined by the QPC that is least transmitting, in terms of the number of modes allowed through the QPC. At the opposite extreme, when the dot is considerably wider (and larger) than the QPC, the electrons transmitted through one QPC can be scattered into all modes within the dot, so that the electron motion is randomized. In this limit the resistance of the overall structure is the sum of that of the two individual QPCs, rather than a phase-coherent single structure. The work of these investigators (and others in subsequent work) showed that a quantum dot could be ohmic (randomized scattering within the dot) at low magnetic fields and then become adiabatic at higher magnetic fields when the edge states were well formed. This transition is shown in Fig. 5.36 for a circular quantum dot of approximately 1.5 μm diameter. At low magnetic fields, the overall resistance is the sum of the two individual QPCs. Conductances G_A and G_B are the conductances through

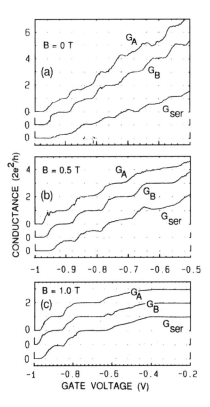

Fig. 5.36. Gradual change of the total conductance of the quantum dot. [After Kouwenhouven *et al.*, Phys. Rev. B **40**, 8083 (1989).]

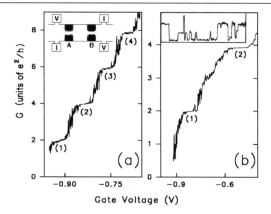

Fig. 5.37. Impurity induced fluctuations in a quantum dot. [After R. P. Taylor *et al.*, Phys. Rev. B **45**, 9149 (1992).]

the individual QPCs, while G_{ser} is the total conductance through the dot. The three curves in each panel are offset from one another for clarity. Already at 1.0 T, however, the overall resistance is the same as the individual QPCs. For the latter condition, the wide regions outside the quantum dot are characterized by three well-formed Landau levels, so that a small number of edge states are wending their way through the quantum dot.

As discussed in the earlier chapters, noise in a mesoscopic system can be observed due to the charging and discharging of a single impurity, as well as due to the universal conductance fluctuations. This has also been seen in connection with the QPCs in quantum dots [69]. At magnetic fields slightly below that necessary to enter the adiabatic regime, the fluctuations are seen to arise as one transitions from one electrostatic plateau in the QPC to the next, at low magnetic field. This is shown in Fig. 5.37. The numbers below the plateaus refer to the number of one-dimensional subbands and/or edge states occupied in the QPC at the magnetic field, which is zero in Fig. 5.37a and 1.8 T in Fig. 5.37b. In the inset of the latter panel is a temporal trace showing the characteristic impurity charging/discharging. Modeling of a QPC within one of the author's groups has recently shown that a single δ-function impurity located near the QPC can in fact introduce backscattering in a manner that leads to interferences in the transmission through the dot, yielding conductance curves like those shown in this latter figure [70].

Perhaps the clearest demonstration of the phase coherence of the adiabatic transport, however, is the observation of Aharonov-Bohm oscillations within the transport through the quantum dot [71]–[75]. In this sense, the oscillations are quite different from those normally encountered in a metallic (or semiconducting) ring, and which were discussed in earlier chapters. Here, the presence of these oscillations requires the establishment of well-formed edge states that can circle around the periphery of the quantum dot. In Fig. 5.38, the magnetoresistance is shown for a quantum dot that is approximately 1 μm square, with a quasi-two-dimensional carrier density in the heterostructure of 4.4×10^{11} cm^{-2} and a mobility of 3.5×10^5 cm^2/Vs. As can be seen from the notes in the figure, the measurements were made at quite low temperature, 10 mK. The dot itself is defined by the negative bias applied to the metallic gates surrounding the structure. The general plateau regions arise from the adiabatic edge-state transport and/or electrostatic channel definition within the QPCs that form the dot. In Fig. 5.39, an expanded portion of the magnetoresistance is shown with

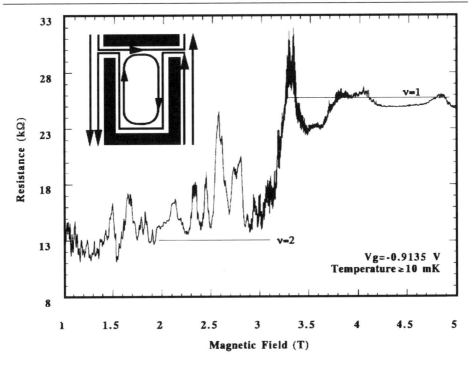

Fig. 5.38. Low-field magnetoresistance for the 1 μm dot. [After J. P. Bird *et al.*, Jpn. J. Appl. Phys. **33**, 2509 (1994).]

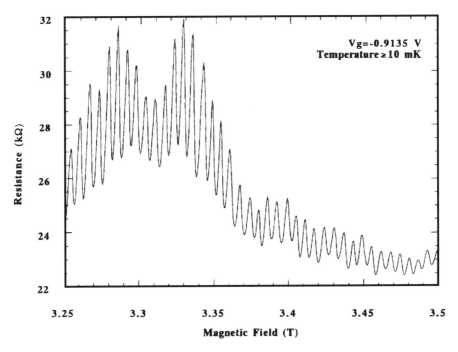

Fig. 5.39. Aharonov-Bohm oscillation in the magnetoresistance of the dot. [After J. P. Bird *et al.*, Jpn. J. Appl. Phys. **33**, 2509 (1994).]

clearly resolved Aharonov-Bohm oscillations in magnetic field. These oscillations are on the $\nu = 1$ plateau and have a period of oscillation of 6.3 ± 0.3 mT. This period corresponds to a square dot of side 0.81 μm, which means that the electrostatic confinement depletes the electron gas approximately 0.1 μm from each edge of the metallic gate. These oscillations were observable only at these low temperatures and were washed out already at 0.5 K, which is ascribed to thermal smearing of the discrete Aharonov-Bohm states. The actual period of oscillation had a small variation with absolute magnetic field, which is thought to be due to a magnetically induced change in the edge-state position [71], [76].

While it is possible that the small change in the period is actually related to the orbit size, as suggested, it also has been pointed out by Dharma-Wardana et al. [77] that the bare Aharonov-Bohm effect could be enhanced by the onset of the Coulomb blockade. This would become important particularly as one approached the tunneling regime in which the resistance rises above that of the $\nu = 1$ plateau. This is shown in Fig. 5.40 for a 2.0 μm dot. From Fig. 5.40a it is clear that when the gate is grounded, there are two edge states, corresponding to the lowest Landau level within the dot. When a sufficiently high negative bias is applied, however, the QPCs are pinched off, leading to the tunneling regime, and the resistance of the structure becomes quite high. The oscillation apparent in the tunneling regime is expanded in Fig. 5.40b for clarity. Using the single-electron properties discussed in Chapter 4, it is possible to calculate that the dot capacitance is 0.73 fF in order to account for the magnetocoulomb oscillations that are observed.

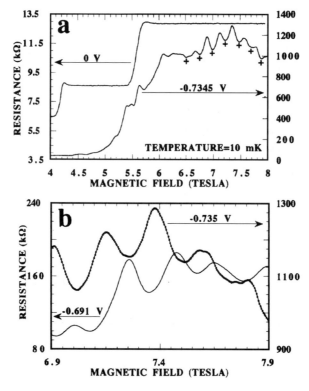

Fig. 5.40. (a) Magnetoresistance (expanded in (b)) of a 2 μm dot showing magnetocoulomb oscillations. [After J. P. Bird et al., Phys. Rev. B **50**, 14983 (1994).]

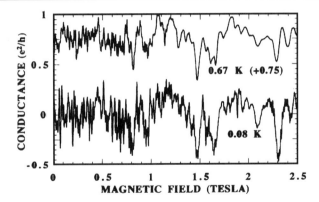

Fig. 5.41. Conductance fluctuations in a $1.0\,\mu$m square dot. The upper curve has been offset for clarity. [After J. P. Bird *et al.*, Phys. Rev. B **51**, 18037 (1995).]

Perhaps most unusual is that the quantum dots exhibit fluctuations and weak localization, in a manner similar to the diffusive wires, but for conditions in which the transport is clearly ballistic and not diffusive (the material is high mobility with few impurities present, which yields mean free paths considerably larger than the size of the quantum dots themselves). In fact, a number of observations of fluctuations in ballistic cavities [78]–[81], which were not explainable by the theory of the previous sections, have appeared. Typical fluctuaions are shown in Fig. 5.41 for a $1.0\,\mu$m square dot [82]. In both traces shown, a smoothed polynomial fit to the background magnetoresistance has been subtracted out of the signal. This average resistance was of the order of $16\,$kΩ in this case.

Because the particles can move throughout the interior of the quantum dot, it has been suggested that the fluctuations and weak localization are the result of chaotic transport within the dot. The suggestion that the observation of fluctuations in the conductance through a mesoscopic cavity could be used to probe the transition to quantum chaos probably is best credited to Jalabert *et al.* [83]. These authors demonstrated, through a series of theoretical calculations, that resistance fluctuations due to scattering from geometrical features should be observed in experimentally realizable mesoscopic quantum dots, provided that the transport through the dots was sufficiently ballistic in nature. These fluctuations arise from the complex scattering dynamics of geometric structures that can trap the electrons, the latter of which is quite natural in the quantum dot. These trapping times can be much longer than the ballistic transit time of a straight-through particle. The important point is that the characteristic correlation scale in magnetic field or Fermi energy (such as arise in the correlation function of the universal conductance fluctuations) for the fluctuations in the *quantum* transport regime can be predicted from knowledge of the equivalent irregular *classical* orbits in the same structure. In fact, it is possible to determine certain measures of the chaos in these quantum dots, since it is known that the particles escape from the dot with a nearly exponential decay function as [84]

$$N(t - t_0) = N(t_0)e^{-\gamma_{cl}(t-t_0)}, \tag{5.153}$$

where t_0 is the injection time of the particles. Here, $\gamma_{cl} = \lambda(1 - d)$ is the classical escape rate, λ is the Liapunov exponent for the chaotic "attractor" in the system, and d is the information dimension [85]. Moreover, one can define a swept path area in a magnetic field from the vector potential as

$$S = \frac{e}{\hbar B} \oint \mathbf{A} \cdot \mathbf{dl},$$ (5.154)

where B is the magnetic field. From this, the probability function for various swept areas in the chaotic quantum dot is found also to be exponentially distributed as [83]

$$N(S) \sim e^{-2\pi \alpha_{cl}|S|},$$ (5.155)

where α_{cl}^{-1} gives the rms area enclosed by typical trajectories in the structure. From this structure, it is then possible to write the actual correlation function for the fluctuations that arise in the quantum magnetotransport through the structure as

$$F(\Delta B) = \left| \int_{-\infty}^{\infty} N(S) e^{ie\Delta B S/\hbar} dS \right|^2 = \frac{F(0)}{\left[1 + \left(\frac{e\Delta B}{\alpha_{cl} h} \right)^2 \right]^2}.$$ (5.156)

The results obtained from a classical calculation were compared by Jalabert et al. [83] with the results obtained from a fully quantum-mechanical recursive Green's function calculation (such as those described in Chapter 3) and found to agree quite well with no free parameters.

The results of Jalabert et al. explained in an intuitive manner the presence of the weak localization (discussed further below) and the fluctuations. Subsequently, a number of investigators have examined a variety of quantum dots, with structures that should be regular and others that should be chaotic [86]–[88]. These authors have probed the transitions from regular to chaotic behavior and generally found agreement with the theory of the previous paragraph. The transition from regular to chaotic behavior has also been probed by varying the magnetic field [89], by distorting the shape [90], and by working in the tunneling regime [91]. More exotic behavior has been found in a tilted magnetic field, in which the field is not normal to the quasi-two-dimensional layer defining the electron gas of the quantum dot [92]. Of significance is the degree to which the results of the theory agree with the experiment. In Fig. 5.42, a typical correlation function is calculated from the classical chaotic behavior of a circular stadium, with input and output ports on opposite sides of the stadium. The ports create a set of side walls which convert the circular structure into a normal stadium with the two ends separated by the sidewalls. The two openings are 0.1 μm and the stadium

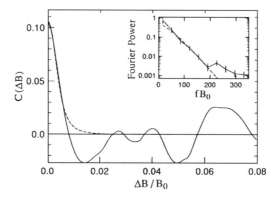

Fig. 5.42. Magnetic field correlation function for an open stadium (the curves are discussed in the text). [After Jalabert et al., Phys. Rev. Lett. **65**, 2442 (1990), by permission.]

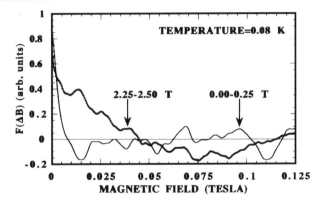

Fig. 5.43. Magnetic field correlation function for an open square dot. [After Bird *et al.*, Phys. Rev. B **51**, 18037 (1995).]

radius of curvature is $0.2\,\mu$m. The inset is the Fourier transform of the correlation function, and the dashed curve is the retransform of the exponential part of the Fourier transform. The half-width of the correlation function is about $\Delta B_c = 4$ mT (the normalization of the magnetic field in the figure is provided by $B_0 = mv_F/eW$, where W is the opening in the sidewall for input or output). In Fig. 5.43, experiments on a square $1.0\,\mu$m dot are shown [82]. Although the dots are of slightly different size, the agreement in both shape and the value of the correlation magnetic field is remarkable, particularly in the correlation function at low magnetic field (thin curve in the figure). The correlation magnetic field in the experiment is slightly less than 3 mT.

It is clear from Fig. 5.43 that the correlation magnetic field increases as the absolute value of the magnetic field increases, as in the case of universal conductance fluctuations in the previous section. While there is no good theory for the behavior of these fluctuations in the situation where edge states are beginning to form, it is clear that the fluctuations must still be related to interactions among the various edge states, particularly at the QPCs where they are forced to strongly interact (a point to which we return later). In fact, it may be rationalized that the arguments leading to (5.152) should be applicable, and Bird *et al.* have used a form of this latter equation to estimate the phase coherence time in the quantum dot [82], [93]. The variation of the correlation magnetic field, at two different temperatures, is shown in Fig. 5.44. The increase in the value of ΔB_c begins to set in when the cyclotron radius becomes smaller than one-half the dot diameter. The inset shows the skipping orbit picture that was used in the previous section to arrive at (5.152). It is the straight-line portions of the curve in the latter figure that are now used to estimate the phase coherence time τ_φ. The results of this are shown in Fig. 5.45 as a function of the temperature of the sample. Similar behavior of the phase coherence time has been obtained from a much more complicated determination of the temperature dependence of the classical area α_{cl} [94].

From Fig. 5.45, it is clear that the phase-breaking time exhibits two distinct regimes of temperature-dependent behavior. For temperatures above about 0.2 K, the phase-breaking time decays as $1/T$, a behavior expected for electron-electron-dominated phase-breaking processes (this will be discussed in the next chapter). Below this temperature, however, the phase-breaking time is essentially independent of the temperature. The transition between these two regimes is interpreted by the authors as a transition from mostly two-dimensional transport in the dot to mostly zero-dimensional transport, as defined by the density of states.

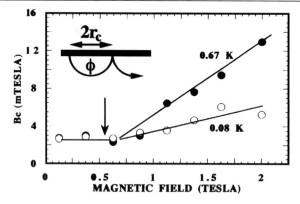

Fig. 5.44. Variation of the correlation magnetic field with the magnetic field. [After J. P. Bird *et al.*, Phys. Rev. B **51**, 18037 (1995).]

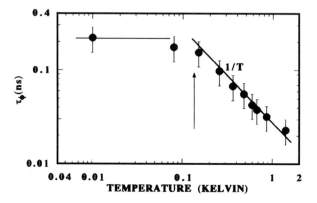

Fig. 5.45. The phase-coherence time deduced from the correlation magnetic field variation. [After J. P. Bird *et al.*, Surf. Sci., in press.]

At higher temperatures, the thermal smearing among the dot quantum levels makes the dot dominated by the two-dimensional electron gas properties. It is this two-dimensional density of states that leads to the $1/T$ behavior. At lower temperatures, however, the thermal broadening of the dot levels becomes smaller than the spacing Δ between the dot levels, and the system behaves as a fully quantized zero-dimensional system. For a dot of cross-sectional area A, the average level spacing can be estimated by $\Delta = h^2/2\pi m^* A$, and this may be estimated from the dot size found from the Aharonov-Bohm oscillations earlier to yield $\Delta/k_B = 130$ mK, which is in surprisingly good agreement with the break seen in the data. The saturation value of the phase coherence time is found to be 0.26 ns, as shown, in the 1.0 μm dot and 0.38 ns in a smaller 0.6 μm dot. It is likely that the actual value of the coherence time is closely associated with the escape time from the dot, in that the major phase-breaking interaction region may well be the QPCs themselves. More work is required to elucidate this point, however.

While the fluctuations are quite suggestive of chaotic transport, it is the weak localization that lends the strongest confirmation of this quite interesting behavior. Chang *et al.* [95] investigated two quantum dot structures, the first a stadium which is classically chaotic, and the second a circle which is classically regular (nonchaotic) at least in the closed geometry. It

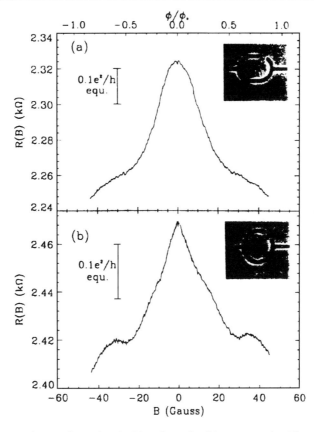

Fig. 5.46. Magnetoresistance for a chaotic (a) and regular (b) quantum dot. The curves have been averaged over 48 devices each. [After A. M. Chang *et al.*, Phys. Rev. Lett. **73**, 2111 (1994), by permission.]

was found, as shown in Fig. 5.46, that the "lineshape" of the weak localization was dramatically different for the two structures. In the chaotic geometry, the weak localization exhibits a Lorentzian-squared behavior consistent with Eq. (5.156), as is shown in Fig. 5.46a. On the other hand, for the regular geometry the weak-localization contribution to the resistance is a term that varies as $\Delta R_0 - \beta |B|$, that is it decays almost linearly with magnetic field. The possibility of such a lineshape was suggested by Baranger *et al.* [96] on the basis of a semiclassical argument that the regular orbits should have an area distribution function $N(S)$ that decayed as a power law rather than an exponential. It was argued that this would lead to the linear decay of the weak localization. They suggest that the rate of decay depends strongly upon the area S, and may even vanish. This rate of decay is found to be proportional to the square root of the escape rate γ_{cl} so that areas where this decay rate is small dominate the large-area behavior. Some indication of this is given in Fig. 5.42, where tailing away from the exponential decay of areas is evident. This leads to $N(S) \sim 1/S^2$, which gives a Fourier transform with the linear decay evidenced in the experiments. This suggests that, under the right conditions, the chaotic to linear change in the weak localization might be seen in a single quantum dot.

One can ask why the *square* quantum dots, for which experimental data have been

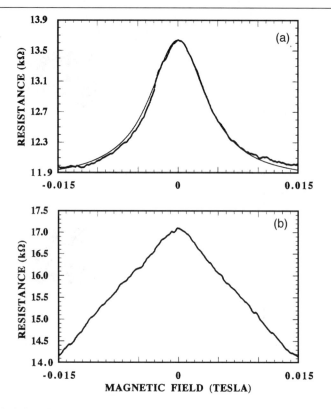

Fig. 5.47. Evolution of the weak localization peak in a 0.6 μm square dot as the QPCs are pinched off. [After J. P. Bird *et al.*, Phys. Rev. B, in press.]

presented in this section, should be chaotic. Classically, a *closed* square cavity is regular with stable orbits that do not transition to chaos. The reason that these dots show fluctuations and chaotic behavior lies in the openings through which particles and current must pass. This allows a wide distribution of entry angles to enter the dot, so that there are arbitrarily many orbits that sweep through essentially all the space within the dot. This allows phase interference between the different orbits and the onset of weak localization characteristic of chaotic behavior in the dot [96], [97]. In addition, of course, the experimentally defined square dots actually have rounding in the corners and a self-consistent potential that provides a number of reentrant diffusers, so that the dot is not exactly square. However, as the QPCs are pinched together, the angle of incoming trajectories is narrowed significantly and the dot can show a transition to regular behavior. Consider Fig. 5.47, in which data for a 0.6 μm dot are shown [98]. In Fig. 5.47a, the QPCs are relatively open with an overall dot resistance of 4 kΩ (the openings are thought to be about 0.1 μm). Here, the normal Lorentzian-squared lineshape is found, which is characteristic of chaotic transport within the dot. As the resistance is raised by closing the QPCs, however, the lineshape is seen to change. Figure 5.47b illustrates the lineshape for a dot resistance of >15 kΩ. No fit to this lineshape can be achieved with the Lorentzian-squared distribution. On the other hand, the decay is quite linear, and it is interpreted that a transition from chaotic to regular transport takes place within the dot at a total resistance near 15 kΩ. This linear behavior is less well pronounced in larger dots, which may be a result of bulk disorder [95].

This seems to make perfect sense, but not all is so clear. The problem is the dependence on the contact itself to transition between so-called chaotic and regular behavior in the closed dot. In fact, however, it has been observed that two quantum point contacts, *in the absence of any closed dot*, can exhibit a weak localization signal that has the lineshape expected for chaotic behavior – the Lorentzian lineshape [99], [100]. Clearly, these structures cannot be exhibiting chaotic behavior as there are no random closed orbits, only orbits reflected from one quantum point contact back to the other. We previously commented that such behavior could be seen in a single quantum point contact if impurities were present [70]. While this idea of the quantum point contact switching the behavior is satisfying, nearly all cases of the experiments are for only a few modes passing through the contact; this situation clearly establishes collimation of the input beam and should preclude the observation of chaotic behavior [101]. Indeed, closer examination of the correlation functions in the dots shows clear evidence of strong negative excursions and periodic oscillations (which are precluded by Eq. (5.156)), both of which are beginning to be evident in Figs. 5.42 and 5.43. These are evidence of regular, periodic structures in the quantum dots, which are presumed to be residual effects of the classical closed orbits; these are therefore quite stable in distinction from the assumptions of the chaos theory [102], [103]. These orbits remain quite stable even in the presence of dissipative, phase-breaking progresses [104]. Consequently, it seems that the presence (or absence) of chaos in these quantum dots is not as straightforward as first thought. Indeed, we may conclude that the general rule is that the presence of clear quantization within the dot structure is contrary to the continuum of states needed to support chaotic motion. We can only say that considerably more work must be done in this area before a final conclusion can be reached.

Bibliography

[1] B. L. Altshuler, A. G. Aronov, and D. E. Khmelnitskii, J. Phys. C **15**, 7367 (1982).

[2] D. K. Ferry, *Semiconductors* (Macmillan, New York, 1991).

[3] S. Chakravarty and A. Schmid, Phys. Rept. **140**, 193 (1986).

[4] C. W. J. Beenakker and H. van Houten, Phys. Rev. B **38**, 3232 (1988).

[5] L. D. Landau and E. M. Lifshitz, *The Classical Theory of Fluids* (Addison-Wesley, Reading, MA, 1962), p. 50.

[6] C. Umbach, C. van Haesondonck, R. B. Laibowitz, S. Washburn, and R. A. Webb, Phys. Rev. Lett. **56**, 386 (1986).

[7] C. de Graaf, J. Caro, and S. Radelaar, Phys. Rev. B **46**, 12814 (1992).

[8] J. A. Simmons, D. S. Tsui, and G. Weimann, Surf. Sci. **196**, 81 (1988).

[9] E. H. Sondheimer, Adv. Phys. **1**, 1 (1952).

[10] A. B. Pippard, *Magnetoresistance in Metals* (Cambridge Univ. Press, Cambridge, 1989).

[11] K. Fuchs, Proc. Cambridge Phil. Soc. **34**, 100 (1938).

[12] C. W. J. Beenakker and H. van Houten, in *Solid State Physics*, Vol. 44, eds. H. Ehrenreich and D. Turnbull (Academic Press, New York, 1991), pp. 1–228.

[13] T. J. Thornton, M. L. Roukes, A. Scherer, and B. P. van der Gaag, Phys. Rev. Lett. **63**, 2128 (1989).

[14] K. Forsvoll and L. Holwech, Phil. Mag. 9, 435 (1964).

[15] T. J. Thornton, M. L. Roukes, A. Scherer, and B. P. van der Gaag, in Granular Nanoelectronics, eds. D. K. Ferry, J. R. Barker, and C. Jacoboni (Plenum Press, New York, 1991), pp. 165–179.

[16] B. L. Altshuler, D. Khmelnitskii, A. I. Larkin, and P. A. Lee, Phys. Rev. B **22**, 5142 (1980).

[17] B. L. Altshuler and A. G. Aronov, Pis'ma Zh. Eksp. Teor. Fiz. **33**, 515 (1981) [JETP Lett. **33**, 499 (1981)].

[18] S. Hikami, A. I. Larkin, and Y. Nagaoka, Prog. Theor. Phys. **63**, 707 (1980).

[19] V. K. Dugaev and D. E. Khmelnitskii, Zh. Eksp. Teor. Fiz. 86, 1784 (1984) [Sov. Phys. JETP **59**, 1038 (1981)].

[20] A. Benoit, C. P. Umbach, R. B. Laibowitz, and R. A. Webb, Phys. Rev. Lett. **58**, 2343 (1987).

[21] W. J. Skocpol, P. M. Mankiewich, R. E. Howard, L. D. Jackel, D. M. Tennant, and A. D. Stone, Phys. Rev. Lett. **58**, 2347 (1987).

[22] A. Benoit, S. Washburn, C. P. Umbach, R. B. Laibowitz, and R. A. Webb, Phys. Rev. Lett. **57**, 1765 (1986).

[23] V. Chandrasekhar, D. E. Prober, and P. Santhanam, Phys. Rev. Lett. **61**, 2253 (1988); V. Chandrasekhar, P. Santhanam, and D. E. Prober, Phys. Rev. B **44**, 11203 (1991).

[24] B. L. Altshuler, Pis'ma Zh. Eksp. Teor. Fiz. **41**, 530 (1985) [JETP Lett. **41**, 648 (1985)].

[25] P. A. Lee and A. D. Stone, Phys. Rev. Lett. **55**, 1622 (1985).

[26] R. Landauer, IBM J. Res. Develop. **1**, 223 (1957).

[27] P. A. Lee, Physica **140A**, 169 (1986).

[28] P. A. Lee, A. D. Stone, and H. Fukuyama, Phys. Rev. B **35**, 1039 (1987).

[29] B. L. Altshuler and D. E. Khmelnitskii, Pis'ma Eksp. Teor. Fiz. **42**, 291 (1985) [JETP Lett. **42**, 359 (1985)].

[30] W. J. Skocpol, Physica Scripta **T19**, 95 (1987).

[31] S. Washburn, in *Mesoscopic Phenomena in Solids*, eds. B. L. Altshuler, P. A. Lee, and R. A. Webb (North-Holland, Amsterdam, 1991), pp. 1–36.

[32] M. Büttiker, Phys. Rev. Lett. **57**, 1761 (1986).

[33] H. Haucke, S. Washburn, A. D. Benoit, C. P. Umbach, and R. A. Webb, Phys. Rev. B **41**, 12454 (1990).

[34] R. P. Feynman and A. R. Hibbs, *Quantum Mechanics and Path Integrals* (McGraw-Hill, New York, 1965).

[35] J. W. Negele and H. Orland, *Quantum Many-Particle Systems* (Addison-Wesley, Redwood City, CA, 1988).

[36] A. L. Fetter and J. D. Walecka, *Quantum Theory of Many-Particle Systems* (McGraw-Hill, New York, 1971).

[37] G. D. Mahan, *Many-Particle Physics* (Plenum Press, New York, 1981).

[38] C. P. Enz, *A Course on Many-Body Theory Applied to Solid-State Physics* (World Scientific Press, Singapore, 1992).

[39] A. A. Abrikosov, L. P. Gor'kov, and I. Ye. Dzyaloshinskii, *Quantum Field Theoretical Methods in Statistical Physics* (Pergamon Press, Oxford, 1965).

[40] P. W. Anderson, Phys. Rev. **109**, 1492 (1958); Philos. Mag. B **52**, 505 (1985).

[41] J. S. Langer and T. Neal, Phys. Rev. Lett. **16**, 984 (1966).

[42] G. Bergmann, Phys. Repts. **107**, 3 (1984).

[43] S. Maekawa and H. Fukuyama, J. Phys. Soc. Jpn. **50**, 2516 (1981).

[44] P. G. Harper, Proc. Phys. Soc. (London) A **68**, 874 (1955).

[45] M. Ya. Azbel, Sov. Phys. JETP **17**, 665 (1963).

[46] M. Ya. Azbel, Sov. Phys. JETP **19**, 634 (1964).

[47] A. Rauh, G. H. Wannier, and G. Obermair, Phys. Stat. Sol. (b) **63**, 215 (1974).

[48] D. R. Hofstadter, Phys. Rev. B **14**, 2239 (1974).

[49] R. Fleischmann, T. Geisel, R. Ketzmerick, and G. Petschel Semicon. Sci. Technol. B **9**, 1902 (1994).

[50] J. Ma, R. A. Puechner, W. P. Liu, A. M. Kriman, G. N. Maracas, and D. K. Ferry, Surf. Sci. **229**, 341 (1991).

[51] S. E. Laux and F. Stern, Appl. Phys. Letters **49**, 91 (1986).

[52] K. Ishibashi, Y. Aoyagi, S. Namba, Y. Ochiai, and M. Kawabe, in *Science and Technology of Mesoscopic Structures*, eds. S. Namba, C. Hamaguchi, and T. Ando (Springer-Verlag, Tokyo, 1992), pp. 101–106.

[53] H. Tamura and T. Ando, Phys. Rev. B **44**, 1792 (1991).

[54] H. Higurashi, S. Iwabuchi, and Y. Nagaoka, Surf. Sci. **263**, 382 (1992).

[55] T. Onishi, Y. Ochiai, M. Kawabe, K. Ishibashi, J. P. Bird, Y. Aoyagi, and T. Sugano, Physica B **184**, 351 (1993).

[56] G. Timp, A. M. Chang, P. Mankiewich, R. Behringer, J. E. Cunningham, T. Y. Chang, and R. E. Howard, Phys. Rev. Lett. **59**, 732 (1987).

[57] C. J. B. Ford, T. J. Thornton, R. Newbury, M. Pepper, H. Ahmed, D. C. Peacock, D. A. Ritchie, J. E. F. Frost, and G. A. C. Jones, Appl. Phys. Lett. **54**, 21 (1989).

[58] R. P. Taylor, P. C. Main, L. Eaves, S. P. Beaumont, I. McIntyre, S. Thoms, and C. D. W. Wilkinson, J. Phys. Cond. Matter **1**, 10413 (1989).

[59] K. Ishibashi, S. K. Noh, Y. Aoyagi, S. Namba, M. Mizuno, Y. Ochiai, M. Kawabe, and K. Gamo, Surf. Sci. **228**, 286 (1990).

[60] A. A. Bykov, G. M. Gusev, Z. D. Kvon, A. V. Katkov, and V. B. Plyuchin, Superlatt. Microstruc. **10**, 287 (1990).

[61] J. P. Bird, A. D. C. Grassie, M. Lakrimi, K. M. Hutchings, P. Meeson, J. J. Harris, and C. T. Foxon, J. Phys. Cond. Matter **3**, 2897 (1991).

[62] C. V. Brown, A. K. Geim, T. J. Foster, C. J. G. M. Langerak, and P. C. Main, Phys. Rev. B **47**, 10935 (1993).

[63] A. K. Geim, P. C. Main, L. Eaves, and P. H. Beton, Superlatt. Microstruc. **13**, 11 (1993).

[64] S. Xiong and A. D. Stone, Phys. Rev. Lett. **68**, 3757 (1992).

[65] J. P. Bird, K. Ishibashi, Y. Ochiai, M. Lakrimi, A. D. C. Grassie, K. M. Hutchings, Y. Aoyagi, and T. Sugano, Phys. Rev. B **52**, 1793 (1995).

[66] I. Glazman, G. B. Lesovik, D. E. Khmelnitski, and R. I. Shekhter, JETP Lett. **48**, 238 (1988).

[67] D. A. Wharam, M. Pepper, H. Ahmed, J. E. F. Frost, D. G. Hasko, D. C. Peacock, D. A. Ritchie, and G. A. C. Jones, J. Phys. C **21**, L887 (1988).

[68] L. P. Kouwenhouven, B. J. van Wees, W. Kool, C. J. P. M. Harmans, A. A. M. Staring, and C. T. Foxon, Phys. Rev. B **40**, 8083 (1989).

[69] R. P. Taylor, S. Fortin, A. S. Sachrajda, J. A. Adams, M. Fallahi, M. Davies, P. T. Coleridge and P. Zawadzki, Phys. Rev. B **45**, 9149 (1992).

[70] A. Grincwagj, G. Edwards, and D. K. Ferry, Physica B **218**, 92 (1996); **227**, 54 (1996).

[71] B. J. van Wees, L. P. Kouwenhouven, C. J. P. M. Harmans, J. G. Williamson, C. E. Timmering, M. E. I. Broekaart, C. T. Foxon, and J. J. Harris, Phys. Rev. Lett. **62**, 2523 (1989).

[72] A. S. Sachrajda, R. P. Taylor, C. Dharma-Wardana, P. Zawadzki, J. A. Adams, and P. T. Coleridge, Phys. Rev. B **47**, 6811 (1993).

[73] P. J. Simpson, D. R. Mace, C. J. B. Ford, I. Zailer, M. Pepper, D. A. Ritchie, J. E. F. Frost, M. P. Grimshaw, and G. A. C. Jones, Appl. Phys. Lett. **63**, 3191 (1993).

[74] B. W. Alphanaar, A. A. M. Staring, H. van Houten, M. A. A. Mabesoone, O. J. A. Buyk, and C. T. Foxon, Phys. Rev. B **46**, 7236 (1992).

[75] J. P. Bird, K. Ishibashi, Y. Aoyagi, and T. Sugano, Jpn. J. Appl. Phys. **33**, 2509 (1994).

[76] J. P. Bird, K. Ishibashi, M. Stopa, Y. Aoyagi, and T. Sugano, Phys. Rev. B **50**, 14983 (1994).

[77] C. Dharma-Wardana, R. P. Taylor, and A. S. Sachrajda, Sol. State Commun. **84**, 631 (1992).

[78] M. L. Roukes, A. Scherer, S. J. Allen, Jr., H. G. Craighead, R. M. Ruthen, E. D. Beebe, and J. P. Harbison, Phys. Rev. Lett. **59**, 3011 (1987).

[79] G. Timp, H. U. Baranger, P. de Vegvar, J. E. Cunningham, R. E. Howard, R. Behringer, and P. M. Mankiewich, Phys. Rev. Lett. **60**, 2081 (1988).

[80] C. J. B. Ford, S. Washburn, M. Büttiker, C. M. Knoedler, and J. M. Honig, Phys. Rev. Lett. **62**, 2724 (1989).

[81] A. M. Chang, T. Y. Chang, and H. U. Baranger, Phys. Rev. Lett. **63**, 996 (1989).

[82] J. P. Bird, K. Ishibashi, D. K. Ferry, Y. Ochiai, Y. Aoyagi, and T. Sugano, Phys. Rev. B **51**, 18037 (1995).

[83] R. A. Jalabert, H. U. Baranger, and A. D. Stone, Phys. Rev. Lett. **65**, 2442 (1990).

[84] R. Blümel and U. Smilansky, Phys. Rev. Lett. **60**, 477 (1988).

[85] P. Gaspard and S. A. Rice, J. Chem. Phys. **90**, 2225 (1989); **90**, 2242 (1989); **90**, 2255 (1989).

[86] C. M. Marcus, A. J. Rimberg, R. M. Westervelt, P. F. Hopkins, and A. C. Gossard, Phys. Rev. Lett. **69**, 506 (1992).

[87] C. M. Marcus, R. M. Westervelt, P. F. Hopkins, and A. C. Gossard, Chaos **3**, 643 (1993).

[88] W. A. Lin, J. B. Delos, and R. V. Jensen, Chaos **3**, 655 (1993).

[89] Z.-L. Ji and K.-F. Berggren, Phys. Rev. B **52**, 1745 (1995).

[90] I. H. Chan, R. M. Clarke, C. M. Marcus, K. Campman, and A. C. Gossard, Phys Rev. Lett. **74**, 3876 (1995).

[91] C. M. Marcus, R. M. Westervelt, P. F. Hopkins, and A. C. Gossard, Surf. Science **305**, 480 (1994).

[92] G. S. Boebinger, G. Müller, H. Mathur, L. N. Pfieffer, and K. W. West, Surf. Science **361/362**, 742 (1996).

[93] J. P. Bird, K. Ishibashi, D. K. Ferry, R. Newbury, D. M. Olatana, Y. Ochiai, Y. Aoyagi, and T. Sugano, Surf. Science **361/362**, 730 (1996).

[94] R. M. Clarke, I. H. Chan, C. M. Markus, C. I. Duruöz, J. S. Harris Jr., K. Campman, and A. C. Gossard, Phys. Rev. B **52**, 2656 (1995).

[95] A. M. Chang, H. U. Baranger, L. N. Pfeiffer, and K. W. West, Phys. Rev. Lett. **73**, 2111 (1994).

[96] H. U. Baranger, R. A. Jalabert, and A. D. Stone, Phys. Rev. Lett. **70**, 3876 (1993).

[97] G. Edwards, A. Grincwajg, and D. K. Ferry, Physics B **227**, 14 (1996).

[98] J. P. Bird, D. M. Olatona, R. Newbury, R. P. Taylor, K. Ishibashi, Y. Ochiai, Y. Aoyagi, and T. Sugano, Phys. Rev. B **52**, 14336 (1995).

[99] R. P. Taylor, R. Newbury, R. B. Dunford, P. T. Coleridge, A. S. Sachrajda, and J. A. Adams, Phys. Rev. B **51**, 9801 (1995).

[100] J. A. Katine, M. A. Eriksson, R. M. Westervelt, K. L. Campman, and A. C. Gossard, Superlatt. Microstruc. **20**, 338 (1996).

[101] R. Akis, J. P. Bird, and D. K. Ferry, Phys. Rev. B. **54**, 17705 (1996).

[102] J. P. Bird, D. K. Ferry, R. Akis, Y. Ochiai, K. Ishibashi, Y. Aoyagi, and T. Sugano, Euro-phys. Lett. **35**, 529 (1996).

[103] J. P. Bird, D. K. Ferry, R. Akis, R. Newbury, R. P. Taylor, D. M. Olatona, Y. Ochiai, K. Ishibashi, Y. Aoyagi, and T. Sugano, Superlatt. Microstruc. **20**, 287 (1996).

[104] R. Akis, J. P. Bird, and D. K. Ferry, J. Phys. Condens. Matter **8**, L667 (1996).

6

Temperature decay of fluctuations

When the temperature is raised above absolute zero, the amplitudes of both the weak-localization, universal conductance fluctuations and the Aharonov-Bohm oscillations are reduced below the nominal value $e^2/\hbar$. In fact, the amplitude of nearly all quantum phase interference phenomena is likewise weakened. There is a variety of reasons for this. One reason, perhaps the simplest to understand, is that the coherence length is reduced, but this can arise as a consequence of either a reduction in the coherence time or a reduction in the diffuson coefficient. In fact, both of these effects occur. In Chapter 2, we discussed the temperature dependence of the mobility in high-mobility modulation-doped GaAs/AlGaAs heterostructures. The decay of the mobility couples to an equivalent decay in the diffuson constant (discussed in Chapters 2 and 5), $D = v_F^2 \tau/d$, where d is the dimensionality of the system, through both a small temperature dependence of the Fermi velocity and a much larger temperature dependence of the elastic scattering rate. The temperature dependence of the phase coherence time is less well understood but generally is thought to be limited by electron-electron scattering, particularly at low temperatures. At higher temperatures, of course, phonon scattering can introduce phase breaking.

Another interaction, though, is treated by the introduction of another characteristic length, the thermal diffuson length. The source for this lies in the thermal spreading of the energy levels or, more precisely, in thermal excitation and motion on the part of the carriers. At high temperatures, of course, the lattice interaction becomes important, and energy exchange with the phonon field will damp the phase coherence. This is introduced through the assumed thermal broadening of a particular energy level (any level of interest, that is). This broadening of the states leads to a broadening of the available range of states into which a nearly elastic collision can occur, and this leads to wavefunction mixing and phase information loss. Imagine two interfering electron paths from which an interference effect such as the Aharonov-Bohm effect can be observed. We let the enclosed magnetic field (or the path lengths) be such that the net phase difference is δ. At $T = 0$, all of the motion is carried out by carriers precisely at the Fermi energy. When the temperature is nonzero, however, the motion is carried out by carriers lying in an energy width of about $3.5k_BT$ (this is the full-width at half-maximum of the derivative of the Fermi-Dirac distribution), centered on the Fermi energy, where k_B is the Boltzmann constant. Thus, thermal fluctuations will excite carriers into energies near, but not equal to, the Fermi energy. The carriers with energy $E_F + k_BT$ will produce an additional phase shift $\delta\omega t$, where t is approximately the transit time along one of the interfering paths, where we assume that $\delta\omega \sim k_BT/\hbar$. This leads to a total phase difference that is roughly $\delta + k_BTt/\hbar$. In fact, the actual energies are distributed over a range of energies corresponding to the width mentioned above, so there will be a range of phases in the interfering electrons. The phase difference is always determined modulo 2π, so that when the extra phase factor is near unity (in magnitude), there will be a decorrelation of the phase due to the distribution of the actual phases of individual electrons, which means some phase interference is destroyed. Thus, one can think about the time over which this thermally induced phase destruction occurs as $t \sim \hbar/k_B T$. This time is a sort of thermal phase-breaking time, and so it can be used with the diffuson coefficient to define a thermal length $l_T = (D\hbar/k_B T)^{1/2}$ in analogy to the phase coherence length introduced in the previous chapters. Thus, l_T is the length over which dephasing of the electrons occurs due to the thermal excitations in the system. We call this length the *thermal diffuson length*. We note that the above process can be summarized by three steps: (1) the nonzero temperature means that a spread of energies, rather than a single well-defined energy, is involved in transport at the Fermi energy; (2) this spread in energies defines a dephasing time $\hbar/k_B T$, which describes interference among the ensemble of waves corresponding to the spread in energy; and (3) the normal phase coherence time τ_φ is replaced in the definition of the coherence length by this temperature-induced dephasing time. Thus, the thermal diffuson length describes dephasing introduced specifically by the spread in energy of contributing states at the Fermi surface.

The introduction of the thermal diffuson length means that mesoscopic devices are now characterized by a more complex set of lengths and the presence of an additional temperature-dependent quantity. We can also set a nominal temperature at which we do not need to worry about thermal effects. If $l_T > l_\varphi$, then the dominant dephasing properties are contained in the phase coherence length, and the thermal effects are minimal. On the other hand, if $l_T < l_\varphi$, then the primary dephasing interaction is due to thermal excitations in the system, and this becomes the important length defining the mesoscopic phenomena. Temperature dependence of mesoscopic phenomena are provided by both l_T and l_φ.

In this chapter we will examine how the temperature variations arise. First, in the next section, we examine the temperature dependence of the phase coherence length, as determined by measurements of weak-localization and universal conductance fluctuations. We also look at the few measurements of the decay with temperature of the phase coherence time itself. We then turn to how the temperature spread of the distribution function can introduce the dephasing. For this we will predominantly use zero-temperature Green's functions to describe the correlation function and carry out some averages over the energy spread to achieve the desired results. Finally, we turn to the decay of the phase coherence time itself through, for example, electron-electron scattering, and this entails a development of the temperature Green's functions.

6.1 Temperature decay of coherence

In this section, we want to begin to look at the decay of weak-localization and universal conductance fluctuations with temperature, both to illustrate the role of the thermal diffuson length l_T and to try to identify the mechanisms responsible for phase breaking in these mesoscopic systems. As mentioned, the basic resistance of the mesoscopic structure changes with temperature. The mobility of the electrons themselves is reduced as the temperature is increased, as discussed in Chapter 2. This carries over to nanostructures as well. In Fig. 6.1, we show the resistance that is measured in high-mobility GaAs/AlGaAs (mobility of 6.4×10^6 cm^2/Vs at 1.5 K) quantum wires as the temperature is varied [1]. Figure 6.1 describes the temperature dependence for wires of various widths and lengths. The resistances are estimated by using a sidewall depletion of about 0.2 μm in the calculation of total device

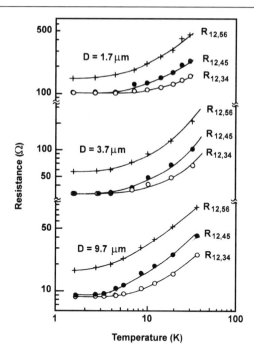

Fig. 6.1. Temperature dependence of high-mobility quantum wires. [After S. Tarucha *et al.*, IOP Conf. Series **127**, 127 (1992), by permission.]

width. (It should be pointed out that this sidewall depletion is only estimated, for thorough self-consistent treatments of the wire conductance with absolute determination of the wire width are completely lacking at this time.) At high temperature, all devices measured were in the diffusive limit, so the longitudinal resistance was proportional to the wire length. As the temperature was lowered, the length dependence of the transport became "quenched" and the transport itself was more quasi-ballistic. The width dependence clearly indicates the importance of the quasi-one-dimensional nature of the wires.

The solid lines in Fig. 6.1 are calculations based upon the multi-terminal Landauer-Büttiker formula, assuming a thermal scrambling of the various transfer coefficients $T_{ij,jk}$ for modes that differ in longitudinal and transverse quantum number. (Here, the multiple subscripts describe both the mode and terminal numbers, as discussed in Chapter 3.) This implies that the temperature effect is primarily a consequence of the thermal mixing of the modes, so one would assume that $l_T < L$ even though the wires are dominantly quasi-ballistic in nature. In these structures, the temperature dependence of the mobility within the wire itself is not as important to the overall resistance as the transmission coefficients for the various modes of the multi-mode structure. This is a clear example of temperature dependence due to the thermal spread in the energy at the Fermi surface, and the asymptotes tend toward a $T^{1/2}$ variation, characteristic of the thermal diffuson length.

6.1.1 Decay of the coherence length

As mentioned above, there are a variety of contributors to the decay of the coherence length itself with temperature. In many cases, however, these various mechanisms are not

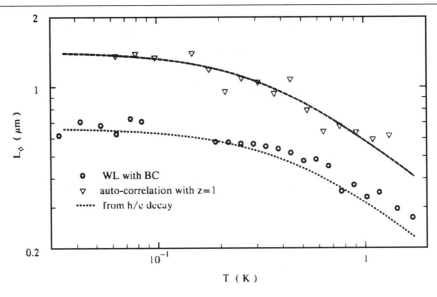

Fig. 6.2. Temperature variation of the coherence length, determined three different ways (discussed in the text). [After R. A. Webb *et al.*, in *Physics and Technology of Submicron Structures*, eds. H. Heinrich *et al.* (Springer-Verlag, Berlin, 1988) by permission.]

separated in an experimental measurement. Rather, only the change in the coherence length is measured by measuring, for example, the change in the behavior of the weak localization or the change in the amplitude of the fluctuations. In this section, we review some of the measurements of the temperature variation of the coherence length. It is important to remember, however, that this length is not itself a key factor, but that it summarizes a variety of temperature variations due to the diffuson "constant" and to the phase coherence time.

One of the first careful experiments concerning the temperature dependence tried to address an early discrepancy in the fit of various theories to the experiments [2]. In this set of experiments, metal loops were measured in a four-terminal configuration (the reader is referred to the discussion of the multi-terminal Landauer-Büttiker approach in Chapter 3), so that all of the various interference mechanisms were present. These authors then measured the voltage fluctuation under constant current conditions and determined the resistance fluctuations (and hence the conductance fluctuations). Figure 6.2 shows the results of measurements on Sb loops; several different methods of computing the coherence length are shown for comparison. The lower curve in the figure illustrates the determination of the coherence length l_φ from the weak localization. In addition, the correlation function for the fluctuations was determined. These latter experiments were carried out at high magnetic field so that the particle-particle contribution to the universal conductance fluctuation was absent. (As discussed in the previous chapter, the cooperon contribution from the particle-particle correlation function decays rapidly with the magnitude of the magnetic field.) The dotted curve corresponds to a determination of l_φ from the decay of the amplitude of the correlation function with temperature, where it is assumed that the fluctuation amplitude decays as $g = b \exp(-aL/l_\varphi)$, with the coefficient b found to be about 0.2 and $a = 1$. This particular functional form is not one that was discussed in the previous chapter. Nevertheless, the fit to the values obtained from the weak localization seems to be relatively good.

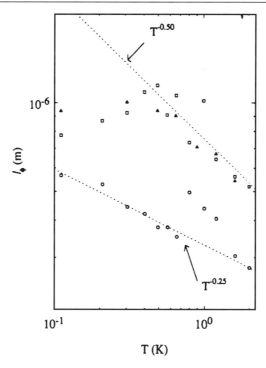

Fig. 6.3. Phase coherence lengths determined for a short wire from Eq. (5.30). The various data are discussed in the text. [After C. de Graaf *et al.*, Phys. Rev. B **12814** (1992), by permission.]

However, the value of the coherence length found from the half-width of the correlation function is somewhat different, and this is plotted as the triangles in the figure (the dashed line is merely a guide to the eye). This value yields results that are more than a factor of 2 larger and in essence require a different value of a. This deviation has also been found in MOSFETs [3] and in wires. The asymptote, at high temperatures, is a behavior slightly slower than T^{-1}.

We next turn to measurements made in a short wire, fabricated by the split-gate technique in a Si MOSFET structure. Here, the wire was fabricated to be rather short, so that the combination formula (5.30) for quasi-two-dimensional and quasi-one-dimensional structures was used to infer the coherence lengths for both the one-dimensional and two-dimensional regions [4]. The typical drawn length of the one-dimensional channel was 0.73 μm, although the actual length was longer due to fringing depletion around the metal edges of the gates. The magnetic field dependence of the weak localization was fit to Eq. (5.30), and values for $l_{\varphi,1D}$ and $l_{\varphi,2D}$ were determined from this fit. The results are plotted in Fig. 6.3. The squares and circles plots $l_{\varphi,2D}$ and $l_{\varphi,1D}$ obtained from this fit are shown as a function of the temperature of the sample. The full triangles plot $l_{\varphi,2D}$ is obtained from measurements in the homogeneous quasi-two-dimensional electron gas. The agreement of $l_{\varphi,2D}$ is certainly quite good between the two measurements, and the generally smaller value found for $l_{\varphi,1D}$ is consistent with the results discussed in the last chapter. It should be noted that the decays vary as $T^{-1/2}$ for the two-dimensional regions and as $T^{-1/4}$ for the one-dimensional region, shown by the dotted lines for reference, and these seem a reasonably good fit to the data. These decays seem to fit well to theoretical predictions for electron-electron scattering [5],

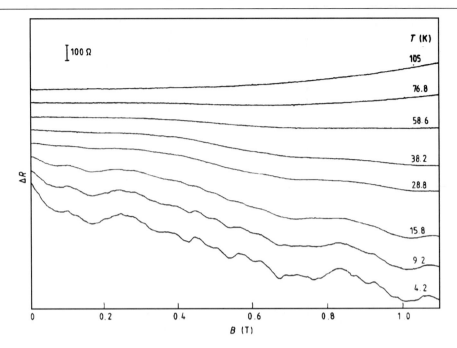

Fig. 6.4. Magnetoresistance for 0.32 μm wide etched wires in a high-mobility GaAs/AlGaAs structure. [After R. P. Taylor *et al.*, J. Phys. Cond. Matter **1**, 10413 (1989), by permission.]

[6] (which will be discussed later in this chapter):

$$l_{\varphi,1D} = \left[\frac{\pi W n_{2D}\hbar D}{a\sqrt{2}}l_T\right]^{1/2},\tag{6.1}$$

$$l_{\varphi,2D} = \left[\frac{4\pi n_{2D}\hbar D}{a}l_T^2\right]^{1/2}.\tag{6.2}$$

The parameter a has a theoretical value of unity at zero temperature and decreases with
the thermal length. The experimental factor for the ratio of $l_{\varphi,1D}/l_{\varphi,2D}$, from Fig. 6.3,
is 0.43 ± 0.04, while the theoretical value from the above equations is 0.29 at 1 K.
These latter authors consider that this fit is within the experimental error in determining the coherence lengths. Nevertheless, the dependence upon temperature and the fit to
these latter two equations suggest that the dominant temperature effects are the important introduction of the thermal diffuson length and the mixing of modes near the Fermi
surface.

Typical curves of resistance versus magnetic field, for high-mobility GaAs quantum
wires, are shown in Fig. 6.4 [7]. Here, the data are for an etched wire of width 0.32 μm. The
basic structure is a high-electron-mobility structure with a two-dimensional density of about
6×10^{11} cm^{-2} and a mobility of 4.7×10^5 cm^2/Vs. The sharp negative magnetoresistance
below 0.1 T is attributed to weak localization. The universal conductance fluctuations can
be quantified by the variance of the fluctuation amplitude, that is, by the correlation function
for the fluctuations. The effective dimensionality of the wire is really determined by the

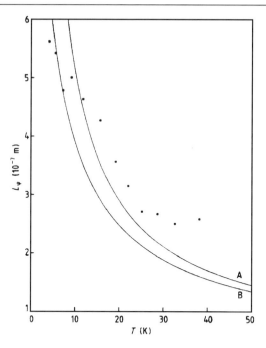

Fig. 6.5. The phase coherence length determined from the etched GaAs wires shown in Fig. 6.4. [After R. P. Taylor *et al.*, J. Phys. Cond. Matter **1**, 10413 (1989), by permission.]

relative magnitudes of l_φ and l_T. These give [8]

$$var(G) = \alpha \frac{e^2}{2\pi\hbar}\left(\frac{l_\varphi}{L}\right)^{3/2}, \quad l_\varphi \ll l_T, \tag{6.3}$$

and

$$var(G) = \beta \frac{e^2}{2\pi\hbar}\left(\frac{l_T^2 l_\varphi}{L^3}\right)^{1/2}, \quad l_T \ll l_\varphi. \tag{6.4}$$

Here, α and β are numerical coefficients with values near unity. Numerical simulations [9] suggest that $\alpha = \sqrt{6}$ and $\beta = \sqrt{4\pi/3}$. However, in many cases these two lengths are quite similar and one needs to interpolate between these two formulas, as

$$var(G) = \alpha \frac{e^2}{2\pi\hbar}\left(\frac{l_\varphi}{L}\right)^{3/2}\left[1 + \frac{9}{2\pi}\left(\frac{l_\varphi}{l_T}\right)^2\right]^{-1/2}. \tag{6.5}$$

Using this approach, the phase coherence length has been determined for these wires and is shown in Fig. 6.5.

At low temperatures, generally it is felt that the main phase-breaking mechanism is carrier-carrier scattering. This rate, for a one-dimensional channel, has been suggested to be [10], [11]

$$\frac{1}{\tau_\varphi} = \frac{\pi}{2}\frac{(k_B T)^2}{\hbar E_F}\ln\left(\frac{E_F}{k_B T}\right), \quad k_B T > \frac{\hbar}{\tau_c}, \tag{6.6}$$

$$\frac{1}{\tau_\varphi} = \left(\frac{k_B T}{D^{1/2}W\rho(E_F)\hbar^2}\right)^{2/3}, \quad k_B T < \frac{\hbar}{\tau_c}. \tag{6.7}$$

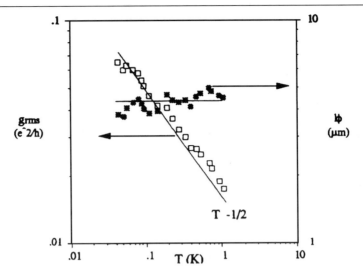

Fig. 6.6. Variance of the universal conductance fluctuations and the phase coherence length found in high-mobility wires at low magnetic field. [After J. P. Bird *et al.*, J. Phys. Cond. Matter **3**, 2897 (1991), by permission.]

Here, $\rho(E_F)$ is the density of states and τ_c is the momentum relaxation time (a weighted average of the scattering time as discussed in the last chapter). In Fig. 6.5, the lines are for a value of the coherence length using the phase-breaking time of Eq. (6.6) (curve A) and a phase-breaking time combining the effects of both equations (curve B). In this fit, the two-dimensional density of states was used to evaluate the phase-breaking time. The derivation of the proper phase-breaking times will be discussed later in this chapter.

Another set of data for a high-mobility quantum wire is shown in Fig. 6.6 [12]. Here, the width of the wires was in the range 0.5–1.2 μm, and the basic quasi-two-dimensional gas had a mobility as high as 10^6 cm^2/Vs. The wires were measured in the low-magnetic-field regime, and the universal conductance fluctuations decayed smoothly as $T^{-1/2}$. This behavior is consistent with a temperature-dependent phase coherence length that varies as Eq. (6.4), and this is also consistent with the data for the magnetoresistance. Surprisingly, these authors found that the phase coherence length saturated at a value smaller than the probe spacing in the four-terminal measurement scheme. Although not generally expected from the theory, this was felt to be consistent with a result that the correlation magnetic field was independent of the temperature (which supports a temperature-independent phase coherence length, contrary to the earlier assumption). This result suggests that there is likely to be a mobility-dependent length scale, on the order of several microns, which limits the extent of quantum diffuson processes. While these authors do not speculate on the cause of this quantity, the result is that phase-breaking processes may well be due to processes other than inelastic electron-electron scattering.

6.1.2 Decay of the coherence time

Most of the variations described in the previous section can be attributed primarily to the introduction of the thermal diffuson length l_T into the discussion of the effective lengths and coherence length. The transport parameters themselves are then not expected to show

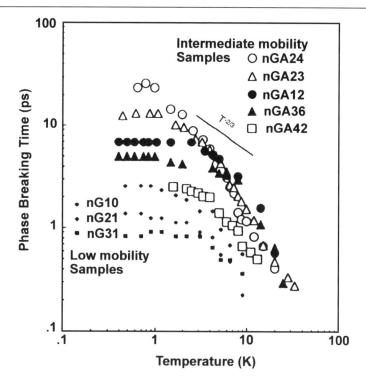

Fig. 6.7. Temperature dependence of the phase-breaking time for various samples. [After Ikoma *et al.*, IOP Conf. Series **127**, 157 (1992), by permission.]

much variation with temperature. As mentioned, the experiments are not usually analyzed to separate the variation of the transport parameters and, for example, the determination of the temperature dependence of the phase coherence time τ_φ.

However, one can separate out the phase coherence time and examine its temperature dependence in the proper circumstances. Indeed, the phase coherence time was measured directly in Fig. 5.45, and this remains one of the most significant direct measurements of the phase-breaking time. Here, we compare this with a variety of other measurements. In Fig. 6.7, the phase-breaking time, determined from a variety of GaAs/AlGaAs quantum wires, is shown as a function of temperature [13]. These samples include both modulation-doped single heterostructures and double heterostructures with doping in the GaAs channel itself. This allowed for sheet densities of 6.0–6.7×10^{11} cm^{-2} in the former case and of 0.34–1.3×10^{13} cm^{-2} in the latter case. The saturation of the phase-breaking time at low temperatures is interesting in that there is no acceptable theoretical basis for this. As seen from the figure, this saturation occurs in all samples; it has been reported by other authors as well [14]. The latter authors suggested that spin-orbit scattering might cause such a saturation, but this would introduce positive magnetoresistance not seen in these samples. Thus, it is thought that spin-orbit scattering does not play any role in these structures. Another possibility, which has not been thoroughly pursued, is surface-roughness scattering [15]. The diffusive nature of the surface scattering serves to mix modes of the quantum wire, and this scattering can introduce a localization length in the system, which in effect introduces a phase-breaking length [16]. Consequently, it is expected that such normally elastic scattering

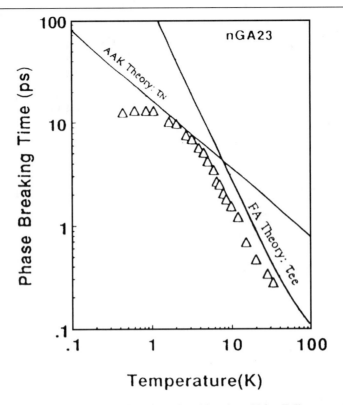

Fig. 6.8. The temperature dependence of the phase-breaking time. [After T. Ikoma *et al.*, IOP Conf. Ser. **127**, 157 (1992), by permission.]

processes can in fact introduce a phase-breaking process. At higher temperatures, the fall-off with temperature seems to be dominated by a $T^{-3/2}$ behavior, which is different from that expected for electron-electron scattering in a one-dimensional system.

In Fig. 6.8, measurements of the phase-breaking time for a single sample with relatively high mobility (21,000 at the lowest temperature) is shown. These measurements show that the phase coherence time decays rapidly for temperatures above about 5 K in these devices. The two solid curves are various theories for the breakup of phase by carrier-carrier processes and will be discussed later. We note, however, that the curve marked "AAK" decays roughly as $T^{-2/3}$, whereas the curve marked "FA" (these are authors' initials) decays roughly as $T^{-3/2}$, at least at the lower temperatures. These values correspond nicely to the temperature dependencies of Eqs. (6.7) and (6.6). Both of these theories can be thought of as forms of the electron-electron interaction, which will be discussed further below. The connection of the phase-breaking time to the phase coherence time depends on the dimensionality of the system and the temperature variation of the diffuson constant as well, so it is difficult to bring the various theories to bear on the actual temperature variation of the coherence length itself.

The general summary of these measurements, which are quite typical of those found in a variety of systems, is that the phase coherence length is relatively independent of temperature up to some critical temperature, and then decays for higher temperatures. The value of the critical temperature does not seem to be a universal quantity; rather it depends on the mobility (diffuson constant), the sample size, and the carrier density (the Fermi energy). The temperature dependence of the weak-localization and universal conductance fluctuations, in general, cannot be ascribed solely to the temperature variation of the phase coherence length, but must be studied for the effects of the thermal broadening and its introduction of the thermal length. In spite of early studies to the contrary, it appears that the temperature dependence of the coherence length obeys a power-law behavior, not an exponential behavior. This is consistent with expected thermal behavior in quantum diffuson, rather than simply an excitation argument. Moreover, it is generally thought that at these low temperatures the phase-breaking mechanism is electron-electron scattering. We turn to this in a later section. First, we discuss the general temperature dependence that is expected for the various parameters in the correlation functions developed in the last chapter. Nevertheless, it is important to note that the variety of experimental measurements provide a wide range of temperature variations, from which it is possible to conclude only that the experiments need further refinement, as does the theory. To draw any further conclusions would require a thorough investigation of the variations in the experimental conditions, and a subsequent evaluation of just what was really measured. That is, the implications concerning the temperature variation of the phase coherence length, or the phase coherence time, must be subject to de-convolution from the data that actually involve the temperature variations of a number of other possible parameters (density, surface-scattering rate, diffuson constant, etc.).

6.1.3 Summary

At this point, we should step back and review what the above experiments tell us. In the previous chapter, measurements of the phase coherence time in quantum dots led to a variation of the phase-breaking time as T^{-1}. In the experiments discussed in this section, a great variety of decay rates have been found. Webb *et al.* [2] find that l_φ decays slightly slower than T^{-1}, and the data suggest $T^{-2/3}$ in metal rings. De Graaf *et al.* [4], on the other hand, found that l_φ decays as $T^{-1/2}$ in two-dimensional structures and as $T^{-1/4}$ in one-dimensional structures of a Si MOSFET. In wires composed of high-mobility modulation-doped heterostructures, Taylor *et al.* [7] found that the phase-breaking length decay could be fit fairly well by the theory of [8], in which the phase-breaking rate decays slowly at $T^{-1/3}$. Conversion of these numbers to phase-breaking rates depends on knowledge of how the diffuson "constant" (or the elastic scattering rate) varies with temperature, and this is not usually given. Nevertheless, one may conclude that the phase-breaking rate decays slower with temperature than l_φ decays, if for no other reason than that it appears as a square-root factor in this last parameter. Direct measurements of the phase-breaking rate are fewer in number, as a consequence of the difficulty of the direct measurement, which is why the first data mentioned here are so important. For example, in earlier work, Bird *et al.* [12] found in long wires fabricated in high-mobility material that the only temperature variation arose from l_T and not from l_φ, mentioned above. Ikoma *et al.* [13] were able to separate out the phase-breaking rate and found that it decayed with temperature approximately as $T^{-3/2}$. This decay is much faster than can be inferred from any of the previous measurements but is much closer to the measurements mentioned in the last chapter. The problem is that if one is trying to infer the value of the phase-breaking time from measurements of conductance fluctuations, then this is complicated by the fact that the effective length of the sample, the diffuson constant for samples in the diffusive regime, the elastic mean-free path, and the thermal length are all varying with temperature, and all have an impact on the amplitude of the fluctuations and the correlation function. Therefore, before proceeding to a discussion of the electron-electron scattering process, which is usually inferred as the culprit for phase breaking (at low temperatures), we will review somewhat the role of temperature in the fluctuations.

6.2 The role of temperature on the fluctuations

In Chapter 5, we reviewed the correlation functions that arise from the universal conductance fluctuations at zero temperature. We want to begin to understand how the temperature variation will affect these results. For this, we basically follow the arguments of [8]. Early work in the theory of quantum transport in metals assumed that there were two characteristic lengths: the coherence length or inelastic diffuson length $l_\varphi = \sqrt{D\tau_\varphi}$, and the length L of the sample. For the cases in which $l_\varphi > L$, pure quantum transport was to be expected, whereas in the opposite case one should expect Boltzmann transport [17]. This suggests that in the former cases, the fluctuations should be governed by the zero-temperature theory worked out in the last chapter with the only temperature variation arising from l_φ. On the other hand, we expect that the UCF will exhibit a second temperature dependence once the temperature is sufficiently high that the latter situation is realized. This would lead us to accept the fact that the fluctuations should decrease with increasing temperature with one behavior in short samples and a different behavior in relatively long samples.

This is somewhat easier to understand in considering that any sample is composed of a group of small regions, each the size of the coherence length, and these will lead to an additive behavior of the fluctuation amplitude. That is, the sample can be thought of as being composed of a number of phase-coherent regions, through which a constant current is flowing. Then, in one dimension, the $var(R)$ is given by the summation of the voltage fluctuations across each region, as discussed in Section 5.2, since these voltage fluctuations are assumed to be independent of one another. Thus for a short sample, only a single phase-coherent region exists, and the voltage fluctuation is independent of length. However, in more than one dimension, there are a variety of parallel paths, so that $var(R)$ varies as L^0/L^{d-1}, where $d - 1$ is the number of parallel paths. Then, $var(R)/\langle R\rangle^2 \sim var(G)/\langle G\rangle^2 \sim L^{d-3}$, which leads to the $\delta G \sim L^{-2}$ behavior in one dimension discussed in connection with Fig. 5.9.

For a long sample, however, there are a number of phase-coherent regions that act independently of one another, and these cause the variance of each region to be added together in an almost classical manner so that $var(R)$ varies as L in one dimension. Since the resistances add linearly, we have $var(R)/\langle R\rangle^2 \sim 1/L$. In two dimensions (and in three dimensions), the classical conductance is essentially an additive function of the constituent conductances, and this result generalizes for the resistances. Then, $var(R)/\langle R\rangle^2 \sim var(G)/\langle G\rangle^2 \sim L^{-d}$, so that the $\delta G \sim L^{(d-4)/2}$, and the $\delta G \sim L^{-3/2}$ behavior in one dimension also discussed in connection with Fig. 5.9 arises.

In a conductor at elevated temperature, the term $var[G(l_\varphi)]$ will be modified by the spread in the Fermi distribution function, which introduces *energy averaging*. This means that several different modes, all with energies quite near the Fermi energy, will *mix* together. This will lead to a decay of the fluctuations with temperature. In general, however, this decay is only as a slow power law in T rather than an exponential as expected for excitation across a barrier or gap. The simple arguments that lead to the behavior above suggest that one can determine the length variation of the fluctuations once the latter are known within a subregion of size l_φ at a given temperature. The complication that arises in this argument is that there are *two* temperature-dependent length scales. The first of these is the coherence length itself, and the second is the thermal length $l_T = (\hbar D / k_B T)^{1/2}$. Both of these lengths enter the analysis of the variance of the conductivity. The major point to be shown below is that if $l_\varphi < l_T$, the $T = 0$ formulation is the proper approach. On the other hand, if $l_\varphi > l_T$, we have to address the proper role of the energy averaging that arises as a result of the spread in the Fermi function. In addition, one can conceive of an intermediate regime $l_\varphi > L > l_T$, where the temperature dependence arises only from the energy averaging and not from any change in the coherence length itself.

It is generally considered sufficient (and convenient) to calculate any changes in the fluctuations from the zero-temperature result through the integration

$$G(T) = -\int dE_k \frac{\partial f_k}{\partial E_k} g(E_k), \qquad (6.8)$$

which is the embodiment of a result found below in Eq. (6.161). In essence, $g(E_k)$ is the conductivity at energy E_k, and Eq. (6.8) is a restatement of the Boltzmann theory result that we must average this conductivity over the states that contribute to the conductivity. The derivative of the Fermi function f_k represents the weighting of each energy state. For example, at $T = 0$, the derivative is a delta function at the Fermi energy. Following this simple train of thought, the correlation function itself should be energy averaged in the simple process

$$F(\delta\mu, \Delta B, T) = \int dE_1 \int dE_2 \frac{\partial f(E_1, \mu)}{\partial E_1} \frac{\partial f(E_2, \mu + \delta\mu)}{\partial E_2}$$
$$\times \langle \delta g(E_1, B) \delta g(E_2, B + \Delta B) \rangle$$
$$= \int d(\Delta E) K(\Delta E, \delta\mu) F(|\Delta E|, \Delta B), \qquad (6.9)$$

where μ is the Fermi energy and

$$K(\Delta E, \delta\mu) = \int dE_1 \frac{\partial f(E_1, \mu)}{\partial E_1} \frac{\partial f(E_1 + \Delta E, \mu - \delta\mu)}{\partial E_1} \qquad (6.10)$$

is the convolution of the two Fermi functions. In principle, a more accurate approach is to recompute the correlation diagrams using the real-time Green's functions. It is reassuring that both methods seem to give the same result. Thus a simpler approach will be followed here.

Since the derivative of the Fermi function is large only for a region of order $k_B T$ on either side of the Fermi energy, the function $K(\Delta E)$ decays exponentially for $\Delta E > k_B T$ and is relatively constant for smaller values. To see this, let us rewrite (6.10) as

$$K(\Delta E, \delta\mu) = \beta e^\xi \int d\eta \left[\frac{e^\eta}{(1 + e^\eta)(1 + e^{\eta+\xi})} \right]^2, \qquad (6.11)$$

where $\beta = 1/k_B T$, $\eta = \beta(E_1 - \mu)$ and $\xi = \beta(\Delta E + \delta\mu)$. For small values of ξ, this may be rewritten as

$$K(\Delta E, \delta\mu) = \beta \frac{e^\xi}{(e^\xi - 1)^2} \int d\eta \left[\frac{1}{(1 + e^\eta)} - \frac{1}{(1 + e^{\eta+\xi})} \right]^2$$
$$\sim \beta \frac{\xi^2 e^\xi}{(e^\xi - 1)^2} \int d\eta \left[\frac{\partial f}{\partial \eta} \right]^2 \sim \beta, \quad \xi < 1, \qquad (6.12)$$

since the integral is of order unity. This function is relatively flat for values of $\xi < 1$; in fact, it has decreased to only about 85% of the peak value at $\xi = 1$. For large values of ξ, we can rearrange (6.11) by use of the identity

$$\frac{a}{a + e^x} \frac{e^x}{b + e^x} = \frac{a}{b - a} \frac{d}{dx}\left[\ln\left(\frac{a + e^x}{b + e^x}\right)\right] \tag{6.13}$$

to write

$$K(\Delta E, \delta\mu) = \beta \frac{e^\xi}{(e^\xi - 1)^2} \int d\eta \left[\frac{d}{d\eta}\ln\left(\frac{e^{-\xi} + e^\eta}{1 + e^\eta}\right)\right]^2$$

$$\sim \beta \frac{e^\xi}{(e^\xi - 1)^2} \int d\eta [f]^2 \sim \beta e^{-\xi}, \quad \xi \gg 1. \tag{6.14}$$

Thus, for most practical purposes, the kernel can be approximated as $K(\Delta E, \delta\mu) \sim \beta\Theta(\beta^{-1} - \Delta E - \delta\mu)$.

The zero-temperature correlation function has a dependence on the two energies at which the conductance is evaluated, which are different in Eq. (6.9). Thus, the energy decay of the correlation function enters into the averaging process even though the Fermi energies are the same for the two bubbles. This is where the temperature dependence enters into the magnetic field correlation function. If the decay of $K(\Delta E)$ is faster than the asymptotic decay of the zero-temperature correlation function, then the former term can be treated as a delta function in ΔE and the zero-temperature result is valid. On the other hand, if $K(\Delta E)$ decays more slowly than the zero-temperature function, which happens when $k_B T > E_c$, the averaging integral provides only a relatively sharp cutoff of the integral in energy at $k_B T$. In fact, we can evaluate the convolution for this simple behavior, assuming that the integral is cut off at the lower end at E_c. From Eq. (5.132), the correlation function basically varies with ΔE as $(\Delta E/\Delta E_c)^{-2}$, where ΔE_c is the "correlation length" or correlation energy. Then, in the asymptotic limit $\xi > 1$, Eq. (6.9) becomes

$$F(\Delta E, T) \sim e^{-\beta\Delta E_c}(\beta\Delta E_c), \tag{6.15}$$

which clearly illustrates that the power law dominates when $k_B T > E_c$, and that the exponential dominates in the opposite situation, where the zero-temperature result will be valid. We now consider the case for which the two bubbles are at the same energy ($\delta\mu = 0$), so that we are interested in the magnetic field–induced response.

The problem arises in the fact that the perturbative series for the correlation function in the diffusive systems, for example given by Eq. (5.130), is quite difficult to express in closed form due to the multiple contributions of comparable weight. Even when this is done, the results are expressable as derivatives of some special functions, which themselves are expressable only as power series. These are not very useful. Therefore, the simplest method of evaluating the temperature dependence is actually to evaluate the zero-temperature series, and then numerically integrate Eq. (6.9) at a given temperature. In lieu of this, we will (again following [8]) give only the asymptotic results in terms of the various parameters of the equations above. This will ignore prefactors of order unity, and so these results are really useful only in plotting the behavior of the correlation functions over a range of magnetic field or temperature to look at the trends rather than the absolute values. For this purpose, the asymptotic limiting forms will be obtained by setting $K(\Delta E) = \beta\theta(\beta^{-1} - \Delta E)$. It is also sufficient to consider just the terms arising from the diagram of Fig. 5.26a, since all the terms have comparable asymptotic behavior. For simplicity, a number of dimensionless variables will be utilized in the intermediate results. These are: $\eta = \Delta E / \Delta E_c$, $\delta = 1/\beta \Delta E_c$, and $\gamma = (L/\pi l_\varphi)^2$, and initially it will be assumed that all dimensions are the same.

6.2.1 Fluctuation amplitudes

In one dimension, the sums over the lateral modes are omitted from the expression for the correlation function (5.130)–(5.132), and the averaged value becomes on the order of

$$\bar{F}_{1D}(T) \sim \frac{1}{\delta} \sum_{m=1}^{\infty} \frac{1}{(m^2+\gamma)^2} \int_0^\delta d\eta \left[1 + \frac{\eta^2}{(m^2+\gamma)^2} \right]^{-2} . \tag{6.16}$$

The integral can be evaluated exactly, but the result is not particularly useful for examining the asymptotic behavior. Instead, it should be noted that the argument of the integral drops off rapidly for $\eta^2 > (m^2+\gamma)^2$. This suggests that one cuts off the integral at this point, and the difference will be whether or not this value is beyond the upper limit. If $(m^2+\gamma) > \delta$, then the integral is just δ. On the other hand, if $(m^2+\gamma) < \delta$, then the integral is just $(m^2+\gamma)$. Thus, we write the result as

$$\bar{F}_{1D}(T) \sim \frac{1}{\delta} \sum_{m=1}^{\infty} \frac{1}{(m^2+\gamma)^2} [(m^2+\gamma)\theta(\delta - m^2 - \gamma) + \delta\theta(m^2+\gamma-\delta)]$$

$$\sim \frac{1}{\delta} \left\{ \sum_{m=1}^{M} \frac{1}{(m^2+\gamma)} + \delta \sum_{m=M+1}^{\infty} \frac{1}{(m^2+\gamma)^2} \right\}, \tag{6.17}$$

where M is a cutoff where the theta functions transition. For $\delta \gg \gamma$, the first summation is just $\sim \gamma^{1/2}$ and the second is $\delta^{-1/2}$, so that [8]

$$\bar{F}_{1D}(T) \sim \frac{\sqrt{\gamma}}{\delta} + O(\delta^{-3/2}). \tag{6.18}$$

The fact that the first correction to the asymptotic behavior varies as $\sqrt{\gamma}/\delta \sim (l_T/l_\varphi)$ just points out that we are unlikely to be in any special strong limit, since the phase coherence length seldom exceeds the thermal length by more than an order of magnitude.

In two and three dimensions, we can follow the same procedure if we replace the sum over m^2 in Eq. (5.130) by one over either two or three sets of integers. If $\gamma \gg 1$, the transition in the theta functions can be set independent of the indices of the summations (in this case, $\gamma \gg M$ is actually required). Then, the double sums can be converted to integrals, and these are combined into two- or three-dimensional "spherical" integrals as

$$\bar{F}_d(T) \simeq \frac{1}{\delta} \left[\int_0^{\sqrt{\delta-\gamma}} \frac{m^{d-1}}{m^2+\gamma} dm + \delta \int_{\sqrt{\delta-\gamma}}^{\infty} \frac{m^{d-1}}{(m^2+\gamma)^2} dm \right]$$

$$\sim \frac{1}{\delta} \times \begin{Bmatrix} [\ln(\delta/\gamma)+1], & d=2 \\ \sqrt{\delta}, & d=3 \end{Bmatrix}. \tag{6.19}$$

It is now clear from these results that, up to a logarithmic correction in two dimensions, the phase coherence length has dropped out of the temperature dependence. We need to be reminded, however, that this assumed that $l_T < l_\varphi$.

Let us now consider the opposite extreme for which $l_T > l_\varphi$, but still for the high-temperature regime where $L \gg l_T > l_\varphi$. This is equivalent to $\gamma > \delta$. In this case, the zero-temperature correlation function varies as γ^{-2}, which is independent of η. In this regime, the zero-temperature correlation function can be removed from the integral, which itself just gives unity when combined with the prefactor of $1/\delta$. Hence, the results are those for the zero-temperature values. In one dimension, this was found to be $(l_\varphi/L)^{3/2}$. In general, the results found above can be used, and $rms[G(L)] = rms[G(l_\varphi)](l_\varphi/L)^{(4-d)/2}$.

We now recall that the rms value of the universal conductance fluctuations is proportional to the square root of the correlation function. We can summarize the high-temperative behavior where the lengths are small compared to the sample size L. This leads us to (here it is also assumed that $\tau_\varphi \sim T^{-p}$)

$$rms[G(T)] \sim \frac{l_T}{L}\left(\frac{l_\varphi}{L}\right)^{1/2} \sim T^{-\frac{1}{2}-\frac{p}{4}}, \quad d=1, \; l_T \ll l_\varphi < L, \tag{6.20}$$

$$rms[G(T)] \sim \left(\frac{l_T}{L}\right)^{(4-d)/2} \sim T^{(d-4)/2}, \quad d=2,3, \; l_T \ll l_\varphi < L, \tag{6.21}$$

$$rms[G(T)] \sim \left(\frac{l_\varphi}{L}\right)^{(4-d)/2} \sim T^{(d-4)/4}, \quad d=1,2,3, \; l_\varphi \ll l_T < L. \tag{6.22}$$

These results are consistent with the view that the behavior begins to become classical when the length scales become longer than l_φ. Hence, one can always write the results as

$$rms[G(T,L)] \sim rms[G(T,l_\varphi)]\left(\frac{l_\varphi}{L}\right)^{\frac{4-d}{2}}, \quad l_T \ll l_\varphi < L. \tag{6.23}$$

However, it must be pointed out that this is not equivalent to saying that one can divide the sample into small phase-coherent volumes of size l_T, since this particularly fails in the one-dimensional case given here. Care must be taken in low-dimensional systems in setting up arguments based on ensemble averaging phase-coherent regions.

6.2.2 Dimensional crossover

In the above discussion, it was assumed that all characteristic dimensions of the sample were the same; for example, $L_x = L_y = L_z = L$. This is seldom the case in real systems, and one must begin to consider the dimensions of actual samples. Thus, it is possible to have lengths large compared to the coherence (or thermal) length in one dimension but not in other dimensions. This was seen previously, where the effective dimensionality at zero temperature depended on the actual sample shape. The form that needs to be used for multiple dimensions, as found in the last chapter, is to replace m_x^2 of the last subsection with $(L_z/L_x)^2 m_x^2 = \alpha m_x^2$, and similarly for m_y^2 of the last subsection with $(L_z/L_y)^2 m_y^2 = \varsigma m_y^2$. To discuss the 1D-to-2D crossover, it is convenient to separate out the terms with $m_x^2 = m_y^2 = 0$, which then gives the proper 1D result. This allows us to write

$$\bar{F}_d(T) \simeq \bar{F}_{1D}(T) + \bar{F}_c(T), \tag{6.24}$$

where

$$\bar{F}_c(T) \sim \frac{1}{\delta}\int_0^\delta d\eta \sum_{m_x,m_y,m_z=1}^\infty \frac{\left(m_z^2 + \alpha m_x^2 + \varsigma m_y^2 + \gamma\right)^2}{\left[\left(m_z^2 + \alpha m_x^2 + \varsigma m_y^2 + \gamma\right)^2 + \eta^2\right]^2}. \tag{6.25}$$

To find the 2D-to-3D crossover, one would make a similar separation for the terms for which $m_y^2 = 0$, which would then produce the 2D result, with an additional term that creates the 3D correlation function.

Let us first consider the 2D-to-1D crossover, so we set $m_y^2 = 0$ in Eq. (6.25). This similarly omits the summation over the y variables. For this, we will convert the sums to integrals and rescale the variables with the new definitions $q_z = m_z$, $q_x = \sqrt{\alpha}m_x$. This raises the limit on the integral from unity to $\sqrt{\alpha}$ for the x-coordinate. The region of integration is not spherically symmetric, but the small region for $1 < q_z < \sqrt{\alpha}$ can be ignored for reasonably large γ. Then, the integral can be rewritten as

$$\bar{F}_c(T) \sim \frac{1}{\delta\sqrt{\alpha}} \int_0^\delta d\eta \int_{\sqrt{\alpha}}^\infty q\,dq \frac{(q^2 + \gamma)^2}{[(q^2 + \gamma)^2 + \eta^2]^2}. \qquad (6.26)$$

Except for the lower limit of the revised integral/summation, this is precisely the same integral as Eq. (6.16) above. Moreover, an equivalent procedure can be followed for the 2D-to-3D crossover, and the same integral is obtained if it is assumed that $\varsigma = 1$. Using the theta function separation discussed in the previous subsection, this becomes

$$\bar{F}_c(T) \sim \frac{1}{\delta\sqrt{\alpha}} \int_0^\delta d\eta \left\{ \int_{\sqrt{\alpha}}^{\sqrt{t-\gamma}} q^{d-1}\,dq \frac{1}{(q^2 + \gamma)} + \int_{\sqrt{t-\gamma}}^\infty q^{d-1}\,dq \frac{1}{(q^2 + \gamma)^2} \right\}. \qquad (6.27)$$

This assumes that $\delta > \gamma, \alpha$. In two dimensions, this gives a result essentially the same as previously with only the modification for the scale factor, as

$$\bar{F}_c(T) \sim \frac{1}{\delta\sqrt{\alpha}} \left[\ln\left(\frac{\delta}{\sqrt{\alpha} + \gamma} \right) + 1 \right] \sim \frac{1}{\delta\sqrt{\alpha}}, \quad d = 2. \qquad (6.28)$$

We can then write the amplitude of the fluctuations as

$$rms[G(T)] \sim \frac{l_T}{L_z} \left(\frac{l_\varphi}{L_z} + \frac{L_x}{L_z} \right)^{1/2}, \quad d = 2,\ l_T \ll l_\varphi < L_z. \qquad (6.29)$$

The crossover from 2D to 1D occurs when $L_x < l_\varphi$. Similarly, the crossover from 3D to 2D can be expressed by the result

$$rms[G(T)] \sim \left(\frac{l_T}{L_z} \right)^{1/2} \left(\frac{l_T}{L_z} + \frac{L_y}{L_z} \right)^{1/2}, \quad d = 3,\ l_T \ll l_\varphi < L_z, L_x, \qquad (6.30)$$

and crossover occurs when $L_y < l_T$. All of this is for $l_T \ll l_\varphi$. When the reverse case is true, the crossover always occurs when the sample size becomes smaller than the coherence length. These latter equations explain some of the unusual temperature behaviors found in the experiments discussed at the beginning of this chapter.

We note from this crossover behavior that a sample that is much wider than it is long can have a fluctuation amplitude much larger than e^2/h, and this effect has been seen in some samples [18]. It has been pointed out previously that this is precisely the ensemble behavior one would expect from putting L_x/l_φ parallel resistors together and then placing L_z/l_φ groups of these in series, when each resistor has a fluctuation amplitude of e^2/h [19].

6.2.3 Correlation ranges

With the results of the above scaling behavior, we are now in a position to estimate the temperature dependence of the correlation "lengths" in energy and magnetic field. In the last chapter, the correlation functions themselves were calculated, and it was assumed that the magnetic field did not alter the basic eigenvalues of the diffuson equation for the correlator. As long as this remains true, then the magnetic field can be treated as an additional factor, just as the inelastic length is introduced. Hence, the dimensionless ratio $b = \Delta B / B_c(0)$ gives an additive correction to the eigenvalues in the same way that the dimensionless ratio $\gamma = (L/\pi l_\varphi)^2$. Here, we note that the zero-temperature value $B_c(0)$ is given by Eq. (5.138) as $\sqrt{3}\Phi_0/L_y L_z$ when $L_y, L_z \ll l_\varphi$, and by Eq. (5.139) as $\sqrt{3}\Phi_0/\pi L_y l_\varphi$ when $L_y \ll l_\varphi \ll L_z$. The relationship for the similar nature of b and γ leads us to simply replace γ by $\gamma + b$ in (6.16), and proceed exactly as in the above sections. We then can ask just what value of b reduces the amplitude of the fluctuation by a factor of 2. In one dimension, reference to (6.18) tells us that this occurs about where $\gamma \sim b$ when $l_T < l_\varphi$. This same result is obtained in all dimensions when $l_T > l_\varphi$. Thus, in one dimension, the temperature dependence is always determined by the phase coherence length, and Eq. (5.139) holds so long as the dimensions of the sample satisfy the designated limits. Hence, the magnetic correlation function has a temperature dependence that arises solely from the temperature dependences of the inelastic scattering time (or phase relaxation time) and the diffuson constant. This result is independent of the relative sizes of l_T and l_φ.

If we have a quasi-two-dimensional sample in which $L_y \ll l_\varphi \ll L_z, L_x$ and we are interested in a case where the field is directed longitudinally along the sample length, then the sample is still governed by the same calculation of the eigenvalues for the diffuson correlator. Since the eigenstates in the direction parallel to the field are independent of the field, we can still use the approach of replacing γ by $\gamma + b$ to obtain the correlation magnetic field $B_c(T)$. Hence, for the longitudinal magnetoconductance,

$$B_c(T) \sim \frac{\Phi_0}{L_{\min} L_y}, \quad L_{\min} = \min(l_T, l_\varphi). \tag{6.31}$$

This completely describes the case where the area normal to the field is essentially one-dimensional.

Let us turn to the case where there is a large quasi-two-dimensional system with the magnetic field normal to the plane of the sample. The eigenvalues are the Landau levels themselves. When this summation is made in the above expressions for the correlation function, the zero-temperature function decays very slowly, with respect to any appreciable variation due to the inelastic cutoff γ. This means that the latter parameter plays no part in the decay, and the magnetic correlation length arises from the value $b \sim \delta$, independent of the thickness of the sample. This is true when $l_T < l_\varphi$. When the phase coherence length is the shortest appropriate length, then it dominates everything. These results suggest that the temperature-dependent correlation magnetic field is given by

$$B_c(T) \sim \frac{\Phi_0}{L_{\min}^2}. \tag{6.32}$$

For the energy correlation length, we already have determined the dependence of the convolution function $K(\delta\mu)$ in Eq. (6.11). The behavior of this relation tells us that μ_c can never be much larger than $k_B T$ because of the exponential decay of K. Thus, we really don't have to consider the case where $l_T > l_\varphi$, because it will give essentially the zero-temperature results. We need only to consider the opposite case, and this just modifies the limits on the η integration to the range $(\bar{\mu}, \delta + \bar{\mu})$, where $\bar{\mu} = \delta\mu/E_c$. Not surprisingly, similar results are obtained as for the magnetic correlation length; for example, in one dimension, $\bar{\mu} \sim \gamma$, and in two and three dimensions, $\bar{\mu} \sim \delta$. This leads to

$$\mu_c \sim \frac{\hbar}{\tau_\varphi}, \quad l_T < l_\varphi < L_z, \tag{6.33}$$

in one dimension, and

$$\mu_c \sim k_B T \tag{6.34}$$

for all other cases.

6.3 Electron-electron interaction effects

The study of the interacting electron gas has a long history. From the earliest days, there has been interest in computing the self-interactions that arise from the Coulomb force between individual pairs of electrons, as opposed to the interaction between the electrons and the impurities that was of primary interest in the last chapter. Early studies, using simple, straightforward perturbation theory, led to a singularity near the Fermi surface. That is, as the energy approached the Fermi energy from either above or below (in a degenerate system for which the Fermi energy lies within the conduction band), the interaction terms diverged; this divergence has been termed a Fermi-edge singularity. It was then realized that a proper calculation, taking into account the screening of any single pair-wise interaction by the remaining electrons (in a manner similar to carrying out the ladder-diagram summation in the last chapter, a calculation to which we return later in this section), removed all the singularities [20] except in one dimension, where the Fermi-edge singularity remains [21]. Following this early work, it was assumed that in general there would be no further complications arising from additional scattering by impurities [22]. However, as we have seen in the last chapter, this is certainly not the case in mesoscopic systems. Altshuler and Aronov [23] were the first to show that additional singularities arose from the presence of the impurity interaction, and that weak singularities would be seen in the tunneling density of states, particularly in Josephson junctions. In addition, it is now believed that the dominant phase-breaking interaction at low temperature in mesoscopic systems arises from the electron-electron interactions. The normal interaction gives rise to the Hartree or Hartree-Fock self-energy shifts and broadening from the self-energy. What is likely responsible for the phase-breaking is the redistribution of momentum and energy among the electrons, so that any one electron loses memory of its specific phase.

In this section, we want to examine the interaction effects, particularly as they are modified by the presence of the strong impurity scattering process for mesoscopic devices. However, these are temperature-dependent effects, so we cannot simply evaluate everything at the Fermi energy. We must begin to allow for a temperature broadening in the Fermi distribution, as discussed in the previous section, so that there is a range of energy in the vicinity of the Fermi energy that becomes involved. In order to treat this properly, we need to modify the Green's functions that we have been using, since they basically assume that $T = 0$. Our application of the temperature Green's functions will focus primarily on the electron-electron interaction and the manner in which this interaction leads to phase breaking. First, however, we review the overall properties of the electron-electron interaction, since there is considerable physics to be understood before calculations are addressed.

The simplest understanding of the electron-electron interaction can be found simply by a straightforward evaluation in terms of the Fermi golden rule. Such an evaluation will clearly point out the role played by various effects, but will keep the principle factors in sight. We may express the electron-electron scattering rate as

$$\frac{1}{\tau_{ee}} = \frac{2\pi}{\hbar}|V_q|^2\rho(E)n_e, \tag{6.35}$$

where V_q is the scattering matrix element arising from the screened Coulomb interaction, $\rho(E)$ is the density of final states, and n_e is the number of electrons with which the incident electron can scatter. Energy can be exchanged in this interaction, although the full two-particle interaction is energy-conserving. That is, the incident electron, with which we are concerned, can gain or lose energy to the scattering center, which in this case is another electron. Because energy can be exchanged, this interaction can lead to the loss of phase coherence on the part of the incident electron. The number of electrons with which the incident electron can scatter is given essentially by $\rho(E)\delta E$, where δE is the allowed range of interaction energy of the two electrons. In most cases we are interested in, $\delta E = k_B T$. If we insert this result into Eq. (6.35) we get

$$\frac{1}{\tau_{ee}} = \frac{2\pi}{\hbar}|V_q|^2\rho^2(E)\delta E. \tag{6.36}$$

In three dimensions, $\rho(E) \sim E^{1/2}$, and in nondegenerate semiconductors, δE can be taken as E as well. For an energy-independent matrix element, as in the case of long-range interactions (small q) that are fully screened, this leads to $1/\tau_{ee} \sim E^2$, and this behavior is found in the scattering rate for impact ionization in bulk semiconductors [24], which leads to the so-called soft threshold for ionization processes [25]. When the energy is averaged over a Maxwellian distribution, we find that the electron-electron scattering time, proportional in many cases to the phase-breaking time, varies as $\tau_{ee} \sim 1/T^2$. If the distribution is degenerate, then the density of states functions can be approximated by taking their values at the Fermi surface, and by using the fact that the important energy spread is given by the thermal spread of the Fermi-Dirac distribution, as in (6.36). Then, it is found that the scattering time varies as $\tau_{ee} \sim 1/T$.

In two dimensions, the density of states is constant, so the only variation in the scattering rate (other than that possible from the matrix element) arises from the energy spread involved in the interaction. Using $\delta E = k_B T$, one finds then that the scattering rate varies as $\tau_{ee} \sim 1/T$. This result arises whether the distribution is degenerate or nondegenerate. In one dimension, $\rho(E) \sim E^{-1/2}$, and this behavior is seen as the source of the Fermi-edge singularity. However, if we take the interaction energy range as simply the energy, we find that the scattering rate is independent of temperature. The same result is found in zero dimensions (such as in a fully quantized quantum dot), since the density of states is a delta function, and the number of scatterers is simply the number of electrons in the isolated energy level.

These give first-order approximations to the scattering rates in order to understand the source of the various temperature dependencies. Certainly, energy and temperature variations in the matrix element can modify these simple behaviors, and it must be pointed out that we have ignored the matrix element contributions. The calculations often will be quite complicated, so the basic understanding may well be lost in plodding through the equations. Thus, it is important to keep the simple results in mind throughout. Nevertheless, the above discussion assumes a variety of things: screened interaction, loss to electron-hole pairs, and so on. Before proceeding it is useful to examine what these concepts mean in the electron-electron interaction.

6.3.1 Electron energy loss in scattering

Electrons that will lose energy to the overall electron gas usually are considered to be in *excited states*. That is, for one reason or another the "incident" electron is in a state with an energy and momentum that place it above the Fermi energy (and Fermi momentum). This could as easily be the case for a hole lying below the Fermi energy, but here we will concentrate solely upon the excited electron because in this approach the results are most easily carried over to the nondegenerate case. In creating this excited state, the electron was excited from an initial state $\mathbf{k}_0$ lying below the Fermi energy to a state $\mathbf{k}$ lying above the Fermi energy. This leaves an empty state below the Fermi energy, which is the corresponding *hole* (or uncompensated electron at $-\mathbf{k}_0$). In order for this process to occur, one must provide a momentum $\mathbf{q} = \mathbf{k} - \mathbf{k}_0$ and energy $E(q) = (\hbar^2/2m^*)(k^2 - k_0^2)$ from an external source. In the previous chapter, electron and hole diagrams were treated in which the transferred energy and momentum were zero. This is a special case, and when the excitation comes from scattering from another electron, the momentum and energy transfer can be different from zero.

The energy and momentum of the excited electron, and the remaining hole, are not uniquely related to each other. (For a different view from that of low-temperature mesoscopic devices, one should review the arguments in cross-bandgap excitation of electron-hole pairs by hot carriers in impact ionization [24].) For a given momentum transfer $\hbar\mathbf{q}$, there is a range of allowed electron and hole momenta and energies. There are two cases to be considered. In the first case, $q < 2k_F$, only a fraction of the electrons lying within the Fermi sphere can participate in the interaction. That is because a significant fraction of the states with $\mathbf{k} = \mathbf{q} + \mathbf{k}_0$ are already occupied and therefore forbidden as possible final states. Thus, we set limits on k to be $k_F < k < k_F + q$. These limits must be carefully evaluated. When $k = k_F$, then k_0 can range from $-k_F$ to 0, and q correspondingly varies from 0 to $2k_F$. Throughout this variation, however, the excess energy of the excited particle $E_k = E(k) - E(k_F) = 0$. That is, the excited state lies right at the Fermi surface. When $k = k_F + q$, k_0 can range from $-k_F$ to k_F. Correspondingly, the transferred momentum q varies from 0 up to $2k_F$, as before. The excess energy of the excited particle now varies from $E_k = 0$ up to $8E(k_F)$. Thus, there is an entire range of energies and momenta possible for the interaction. This is why the two densities of states, mentioned in the introduction to this section, must be convolved with one another.

When $q > 2k_F$, the Pauli exclusion principle is no longer a determinant on the range of allowed energy and momentum. Rather, every possible k_0 is allowed for every value of k. In this case, the convolutions run over the full Fermi sphere. The possible final states that are allowed by the Pauli exclusion principle must lie at least an energy $(\hbar^2/2m^*)[(q - k_F)^2 - k_F^2]$ above the Fermi energy, and the maximum energy that can be transferred is, as previously, $(\hbar^2/2m^*)[(q + k_F)^2 - k_F^2]$. This leads to the limitations on the energy of the excited state as

$$q(q - 2k_F) < k^2 < q(q + 2k_F). \tag{6.37}$$

When $q > 2k_F$, we see that there is a minimum energy, and hence a minimum momentum, in the excited state that is produced by the interaction. Thus, there is a minimum amount of energy that must be transferred to the excited electron-hole pair. This energy must come from the interacting particle that gives up the energy $E(q)$ and momentum $\hbar\mathbf{q}$.

The previous arguments have been for degenerate semiconductors, and that generally is the assumption that we will make below. It is useful to review the arguments above for the case of a nondegenerate semiconductor, for which k_F is imaginary ($E_F < 0$, where the bottom of the conduction band is taken to be the zero of energy). Obviously, then, we need only the last case discussed, since the momentum transfer cannot be less than the Fermi momentum. The maximum energy that can be transferred is obviously $\hbar^2 k^2 / 2m^*$ for which $\mathbf{q} = \mathbf{k}$, but we should be quick to point out that the incident particle at $\mathbf{k}$ can actually gain energy and momentum from another electron, as well as lose it to the other electron. In this case, however, the roles of the two particles are interchanged. First, consider the case for which the electrons at $\mathbf{k}$ and $\mathbf{k}_0$ are on the same energy surface, so that $k^2 = k_0^2$. Consider first the simple case in which one electron gives up all of its energy to the other. If we set $\mathbf{q} = \mathbf{k}$, we are required to have the angle between $\mathbf{k}_0$ and $\mathbf{q}$ be $\pi/2$, which specifies that the initial two particles are propagating at right angles to one another. That is, if one particle is directed along the (001) axis, the other must lie in the plane normal to this axis. In a quasi-two-dimensional system, this is a very special case in which only one possible state is allowed. Now, consider the general case, in which the particle at $\mathbf{k}$ is scattered to $\mathbf{k} - \mathbf{q}$, while the particle at $\mathbf{k}_0$ is scattered to $\mathbf{k}_0 + \mathbf{q}$, which conserves momentum. Energy conservation then gives

$$q = \frac{k^2 - k_0^2}{k \cos \theta + k_0 \cos \theta_0}, \tag{6.38}$$

where θ is the angle between $\mathbf{k}$ and $\mathbf{q}$, and θ_0 is the angle between $\mathbf{k}_0$ and $\mathbf{q}$. This formula does not work for the simple case above, since setting $k = k_0$ when $\cos \theta = 1$ forces $\cos \theta_0 = -1$, and this gives $q = k + k_0$, which means that the two particles exchange momentum; this is not a real scattering event. Thus, Eq. (6.38) is good only off a common energy shell (i.e., $k \neq k_0$). Nevertheless, we can see that the maximum value of q for which the incident particle at $\mathbf{k}$ loses energy is $2k$. In fact, in order to satisfy energy conservation, the scattering momentum must lie in the range

$$-2k_0 \cos \theta_0 < q < 2k \cos \theta_0. \tag{6.39}$$

The ranges of allowed final states that arise from the energy and momentum transfer considerations define a spectrum of single particle excitations that can result from the electron-electron interaction. Outside of the range defined by (6.37), the electron-electron interaction is not effective, although we will see below that scattering from the *collective* modes of the electron gas is possible. While the discussion of this section has been concentrated on the excited electron-hole pair, we must remember that our primary interest lies in the properties of the particle that provides the excitation. This particle has been characterized here by the momentum $\hbar\mathbf{q}$ and energy $E(q)$. We now turn our attention to the Hamiltonian for the electron gas and discuss how we want to partition the various terms.

6.3.2 Screening and plasmons

The Coulomb interaction that enters into the scattering of one electron by the other electrons is a very long range interaction. Consequently, it is usually necessary to cut off the range of this interaction, as is done for impurity scattering as well. This screening is usually introduced *ad hoc*, but this is not necessary. In this section, we wish to show how this can be accomplished, while at the same time introduce the collective modes, the *plasmons*, into the discussion. The approach we follow is essentially adapted from that presented by Madelung [26].

The electron gas may be considered, for the case at present, as a uniform, homogeneous quantity. When an additional electron is introduced into the gas, the Coulomb interaction causes the other electrons to be slightly repelled from it, which leaves a positive charge around the new electron. This extra positive charge in essence *screens* the electron. However, the process by which this occurs is a dynamic process involving the movement of the entire electron gas through internal interactions. This process is such that the movement of the electrons away from the extra charge is usually too great, and so the electrons must move back again, leading to oscillations of the electron gas. These are the collective oscillations, which are the plasmons. So, all of these processes must be involved in the manner in which the screening of the simple Coulomb interaction arises. In this subsection we pursue a classical understanding of the effects, leaving the quantum-mechanical approach to later subsections, where we can deal with it properly in terms of Green's functions. Our starting point is the simple Hamiltonian

$$H = \sum_i \frac{p_i^2}{2m^*} + \frac{e^2}{8\pi\varepsilon_0} \sum_{i,j\neq i} \frac{1}{|\mathbf{r}_i - \mathbf{r}_j|}, \tag{6.40}$$

where the extra factor of two in the Coulomb term arises because of the double counting in summing over both subscripts completely. An *ad hoc* introduction of screening is often made by adding an exponential decay term within the second term of Eq. (6.40), such as $\exp(-\lambda|\mathbf{r}_i - \mathbf{r}_j|)$, where λ is an inverse screening length, such as the Debye length or the Fermi-Thomas length. However, this would eliminate the long-range part of the Coulomb interaction, which would have to be examined separately. Almost the same effect can be achieved by limiting the range of k in any Fourier transform of the Coulomb interaction. Consequently, with the Fourier transformation of the second term, defined by

$$\frac{e^2}{8\pi\varepsilon_0} \sum_{i,j\neq i} \frac{1}{|\mathbf{r}_i - \mathbf{r}_j|} = \frac{e^2}{2V_0\varepsilon_0} \sum_{i,j\neq i} \sum_{\mathbf{k}} \frac{1}{k^2} e^{i\mathbf{k}\cdot|\mathbf{r}_i - \mathbf{r}_j|}, \tag{6.41}$$

a separation will be introduced between long-range effects and short-range effects through the use of a cutoff wavevector $k_c = \lambda$. With this separation, Eq. (6.40) can be rewritten as

$$H = \sum_i \frac{p_i^2}{2m^*} + \frac{e^2}{2V_0\varepsilon_0} \sum_{i,j\neq i} \left(\sum_{k<\lambda} + \sum_{k>\lambda} \right) \frac{1}{k^2} e^{i\mathbf{k}\cdot|\mathbf{r}_i - \mathbf{r}_j|}. \tag{6.42}$$

Along with the screening of the individual electron by the rest of the electron gas, the collective oscillations act back upon the original charge through a self-consistent field produced by their own Coulomb potential. This may be described by the irrotational vector potential $\mathbf{A}(\mathbf{r}_i)$. This vector potential can be expressed through its own Fourier transform as

$$\mathbf{A}(\mathbf{r}_i) = \frac{1}{\sqrt{V_0\varepsilon_0}} \sum_{k>0} \frac{\mathbf{k}}{k} Q_{\mathbf{k}} e^{i\mathbf{k}\cdot\mathbf{r}_i}. \tag{6.43}$$

The unit vector inside the summation assures that the vector potential is irrotational. The requirement that the vector potential be real leads to $\mathbf{k}Q_{\mathbf{k}}^* = -\mathbf{k}Q_{-\mathbf{k}}$, so that $Q_{\mathbf{k}}^* = -Q_{-\mathbf{k}}$. The electric field that arises from the vector potential may be written as

$$\mathbf{E} = -\frac{\partial \mathbf{A}}{\partial t} = -\frac{1}{\sqrt{V_0\varepsilon_0}} \sum_{k>0} \frac{\mathbf{k}}{k} \frac{\partial Q_{\mathbf{k}}}{\partial t} e^{i\mathbf{k}\cdot\mathbf{r}_i} = \frac{1}{\sqrt{V_0\varepsilon_0}} \sum_{k>0} \frac{\mathbf{k}}{k} P_{\mathbf{k}}^* e^{i\mathbf{k}\cdot\mathbf{r}_i}, \tag{6.44}$$

where the momentum $P_{\mathbf{k}}^*$ conjugate to $Q_{\mathbf{k}}$ has been introduced in the last term. The $Q_{\mathbf{k}}$ and $P_{\mathbf{k}}$ can be taken to be the collective coordinates of the field that describes the collective motion of the electron gas as a whole. Thus, we introduce these coordinates into (6.42) in the following manner. First, the leading term which describes the kinetic energy of the electrons is modified by the replacement $\mathbf{p}_i \to \mathbf{p}_i + e\mathbf{A}(\mathbf{r}_i)$, so that the momentum responds to the collective self-consistent field of the electron gas as a whole. Second, the part of the summation in the second term, which represents the long-range part of the Coulomb interaction $(k < \lambda)$, is replaced by the energy of the self-consistent electric field interactions among the electrons. Thus, we may write (6.42) as

$$
H = \sum_i \frac{1}{2m^*}\left(\mathbf{p}_i + \frac{e}{\sqrt{V_0 \varepsilon_0}}\sum_{k>0}\frac{\mathbf{k}}{k}Q_{\mathbf{k}}e^{i\mathbf{k}\cdot\mathbf{r}_i}\right)^2 + \frac{e^2}{2V_0\varepsilon_0}\sum_{i,j\neq i}\sum_{k>\lambda}\frac{1}{k^2}e^{i\mathbf{k}\cdot|\mathbf{r}_i-\mathbf{r}_j|}
$$
$$
+ \frac{1}{2V_0}\sum_{k,k'<\lambda}P_{\mathbf{k}}P_{\mathbf{k}'}\frac{\mathbf{k}\cdot\mathbf{k}'}{kk'}\int e^{i(\mathbf{k}+\mathbf{k}')\cdot\mathbf{r}_i}\,dV. \tag{6.45}
$$

The integration in the last term produces a delta function, which allows us to write the last term as

$$
\frac{1}{2}\sum_{k<\lambda}P_{\mathbf{k}}^*P_{\mathbf{k}}. \tag{6.46}
$$

At this point, we depart from the classical approach and go over to the quantum-mechanical approach. This is achieved by replacing the various variables with noncommuting operators, which satisfy the commutation relationships

$$
[p_{i\nu}, r_{j\xi}] = -i\hbar\delta_{ij}\delta_{\nu\xi}, \tag{6.47}
$$
$$
[P_{\mathbf{k}}, Q_{\mathbf{k}'}] = i\hbar\delta_{\mathbf{k},\mathbf{k}'}, \tag{6.48}
$$

where ν and ξ refer to the x-, y-, and z-components of the position and momentum vectors. The first term in the brackets of Eq. (6.45) then contains the term

$$
\mathbf{p}_i e^{i\mathbf{k}\cdot\mathbf{r}} + e^{i\mathbf{k}\cdot\mathbf{r}}\mathbf{p}_i = 2e^{i\mathbf{k}\cdot\mathbf{r}}\mathbf{p}_i - \hbar\mathbf{k}e^{i\mathbf{k}\cdot\mathbf{r}}. \tag{6.49}
$$

Using this expression, (6.45) can be rearranged to yield

$$
H = \sum_i \frac{p_i^2}{2m^*} + \frac{e^2}{2V_0\varepsilon_0}\sum_{i,j\neq i}\sum_{k>\lambda}\frac{1}{k^2}e^{i\mathbf{k}\cdot|\mathbf{r}_i-\mathbf{r}_j|}
$$
$$
+ \frac{1}{2}\sum_{k<\lambda}\left(P_{\mathbf{k}}P_{\mathbf{k}}^* + \omega_p^2 Q_{\mathbf{k}}Q_{\mathbf{k}}^*\right)
$$
$$
+ \frac{e}{\sqrt{V_0\varepsilon_0}m^*}\sum_{k<\lambda}Q_{\mathbf{k}}e^{i\mathbf{k}\cdot\mathbf{r}_i}\frac{\mathbf{k}}{k}\cdot\sum_i(\mathbf{p}_i - \hbar\mathbf{k})
$$
$$
+ \frac{e^2}{2V_0 m^*\varepsilon_0}\sum_{k,k'<\lambda}[1 - \delta_{\mathbf{k},-\mathbf{k}'}]\frac{\mathbf{k}\cdot\mathbf{k}'}{kk'}\sum_i e^{i(\mathbf{k}+\mathbf{k}')\cdot\mathbf{r}_i}, \tag{6.50}
$$

where $\omega_p^2 = ne^2/m^*\varepsilon_0$ is the low-frequency plasma frequency (the plasmons have this characteristic frequency for frequencies well below the polar optical phonon frequency), and $n = N/V_0$ is the electron density.

In introducing the collective coordinates P_k and Q_k for the plasma oscillations, we have raised the number of degrees of freedom, so there must be some additional constraints connecting these two degrees of freedom. These extra constraints come from the Poisson equation, in which we require that $\nabla \cdot \mathbf{E} = -\rho/\varepsilon_0$. Expanding the charge density into Fourier coefficients, and using the above relationships for the electric field, the added constraints may be expressed as

$$P_\mathbf{k} - i\sqrt{\frac{e^2}{V_0\varepsilon_0 k^2}} \sum_j e^{i\mathbf{k}\cdot\mathbf{r}_j} = 0. \tag{6.51}$$

In the transition to quantum mechanics, this becomes an operator equation, and the constraint is applied by requiring that this operator, when applied to the wave equation, yields a zero result.

Now, let us turn to the meaning of Eq. (6.50). The first line corresponds to a gas of nearly free electrons interacting with one another through a *screened* interaction, where the screening is introduced by the wavevector cutoff. The second line of the equation describes the harmonic oscillator-like vibrations and energy levels of the collective plasma oscillations of the electron gas. The last two lines correspond to the interaction of the nearly free electrons with the collective modes of the electron gas. The third line is clearly the scattering of the nearly free electron through the emission or absorption of plasmons. This can be seen clearly by going over to the number representation for the electrons and plasmons, but this will not be pursued at this point. The last term is usually ignored by adopting what is called the *random phase approximation*, in which it is assumed that the phase in the exponential term varies so rapidly that no terms for which $\mathbf{k} \neq -\mathbf{k}'$ can really contribute to the energy. A careful expansion of the operators in this latter term in the number representation shows that it corresponds to the processes in which a momentum $\mathbf{k}+\mathbf{k}'$ is transferred to the emission or absorption of a *pair* of plasmons, or to the simultaneous emission and absorption of a plasmon. Use of the random phase approximation, in this context, implies ignoring terms in which more than a single electron interacts with more than a single plasmon.

What has been achieved by the introduction of the cutoff wavevector is that the Hamiltonian is now expressable in terms of the nearly free energy of the electrons and the plasmons separately, and two interaction terms in which the nearly free electrons interact with each other through a screened Coulomb interaction and in which an electron can emit or absorb plasmons. The crucial step was the introduction of the cutoff wavevector λ, which corresponds to the screening length. Since momentum is transferred to the collective modes, one must expand the plasmon energy to higher orders, and this leads to $E_{pl} \simeq \hbar\omega_p + \alpha\lambda^2$. Now, since pair excitations cannot excite or absorb plasmons, the plasmon interaction must come from the single particle spectrum, so we require that $E = \hbar^2 k^2/2m^* > E_{pl}$, which leads to a cutoff wavelength $\lambda \sim \omega_p/v_F \sim 0.7k_{FT}$, where k_{FT} is the Fermi-Thomas screening wavevector (hence the usual argument that the latter overestimates screening). In the nondegenerate semiconductor, the Fermi velocity is replaced by the thermal velocity, which leads to the Debye screening wavevector, a result found in careful studies of plasmon scattering in hot electron systems [27].

In most systems, however, the screening is not given by a simple cutoff wavevector but is described by a dynamic polarization function for the electron gas. In order to more fully treat this dynamic screening, and to properly compute the scattering self-energy for the electron-electron single-particle scattering, we will move to the Green's function treatment.

However, since the system is usually at a nonzero temperature, we must first introduce the appropriate Green's function formulation, to which we now turn.

6.3.3 Temperature Green's functions

At temperatures above absolute zero, there is an analogous Green's function to those that have been used in the previous chapter. However, this function is more complicated, and the calculation of the equilibrium properties and excitation spectrum also is more complicated. The first part, the determination of the equilibrium properties, is handled by the introduction of a *temperature Green's function*, often called the Matsubara Green's function, while the second step requires the computation of a time-dependent Green's function that describes the linear response of the system [28]. In principle, the definition of the temperature Green's function is made simply by taking account of the fact that there exist both a distribution function in the system and a number of both full and empty states. The principle change is that the average over the basis states represented by $\langle ..f(A)..\rangle$ goes into $Tr\{\hat{\rho}[..f(A)..]\}$, where $\hat{\rho} = \exp[\beta(\Omega - K)]$ is the system density matrix, $\beta = 1/k_B T$, Ω is the free energy (defined through the partition function), and $K = H - \mu N$ is the grand canonical ensemble, which includes the chemical potential μ (usually equal to the Fermi energy) and total number of electrons N. With this new Hamiltonian, we introduce the modified Heisenberg picture

$$A(\mathbf{x}\tau) = e^{K\tau/\hbar} A(\mathbf{x}) e^{-K\tau/\hbar}, \tag{6.52}$$

where a new complex variable τ has been introduced as a replacement for the time, but which can be analytically continued to it. That is, we have replaced the variable it with τ primarily to make the connection with the thermal distribution $e^{-\beta H}$ (although we shall use the grand canonical ensemble mentioned above), and we will have to be quite careful about various trajectories and paths that are taken, particularly for factors such as time ordering operators. This then leads to the single-particle Green's function

$$G_{rs}(\mathbf{x}\tau, \mathbf{x}'\tau') = -Tr\{\hat{\rho}T_\tau[\Psi_r(\mathbf{x}\tau)\Psi_s^\dagger(\mathbf{x}'\tau')]\}, \tag{6.53}$$

where r and s are spin indices (spin will now be important), T_τ is the "time ordering" operator in the imaginary τ domain, and the field operators are

$$\Psi_r(\mathbf{x}\tau) = e^{K\tau/\hbar}\Psi_r(\mathbf{x})e^{-K\tau/\hbar}, \quad \Psi_s^\dagger(\mathbf{x}\tau) = e^{K\tau/\hbar}\Psi_s^\dagger(\mathbf{x})e^{-K\tau/\hbar}. \tag{6.54}$$

These two field operators are not adjoints of one another as long as τ is real, but this property is recovered when the latter is analytically continued to it.

The temperature Green's function is useful because it will allow us to calculate the thermodynamic behavior of the system. If the Hamiltonian is time independent, then the Green's function depends only on the difference $\tau - \tau'$, as for the previous case. Similarly, a homogeneous system will lead to the Green's function being a function of $\mathbf{x} - \mathbf{x}'$. One useful property is shown by

$$\sum_s G_{ss}(\mathbf{x}\tau, \mathbf{x}\tau^+) = \sum_s Tr\{\hat{\rho}[\Psi_s^\dagger(\mathbf{x}\tau)\Psi_s(\mathbf{x}\tau)]\}$$

$$= e^{\beta\Omega}\sum_s Tr\{e^{-\beta K}e^{K\tau/\hbar}\Psi_s^\dagger(\mathbf{x})\Psi_s(\mathbf{x})e^{-K\tau/\hbar}\}$$

$$= e^{\beta\Omega}\sum_s Tr\{e^{-\beta K}\Psi_s^\dagger(\mathbf{x})\Psi_s(\mathbf{x})\} = \langle n(\mathbf{x})\rangle, \tag{6.55}$$

where we have used the anticommutation of the fermion field operators and the cyclic property of the trace. The total density is found by integrating this over all space. Here, it becomes obvious why the complex time variable τ has been introduced. It allows the temporal evolution operators to be cast into the same exponential form as the statistical density matrix itself, and allows directly for the commutation of these different functions of the Hamiltonian.

Let us consider a simple noninteracting system, in which the creation and annihilation operators $a_{\mathbf{k}\lambda}^{\dagger}$ and $a_{\mathbf{k}\lambda}$ create and destroy an electron in momentum state $\mathbf{k}$ with spin λ. Then, the field operators can be written as

$$\Psi(\mathbf{x}) = \frac{1}{\sqrt{V}} \sum_{\mathbf{k}\lambda} e^{i\mathbf{k}\cdot\mathbf{x}} \eta_{\lambda} a_{\mathbf{k}\lambda}, \tag{6.56}$$

$$\Psi^{\dagger}(\mathbf{x}) = \frac{1}{\sqrt{V}} \sum_{\mathbf{k}\lambda} e^{-i\mathbf{k}\cdot\mathbf{x}} \eta_{\lambda}^{\dagger} a_{\mathbf{k}\lambda}^{\dagger}, \tag{6.57}$$

where η_{λ} is a spin wavefunction and V is the volume (this usage differs from the previous chapter). The equation of motion for the creation and annihilation operators is easily found, for example, by using

$$\hbar \frac{\partial a_{\mathbf{k}\lambda}}{\partial \tau} = e^{K\tau/\hbar} \{K, a_{\mathbf{k}\lambda}\} e^{-K\tau/\hbar} = -(E_k - \mu) a_{\mathbf{k}\lambda} e^{-K\tau/\hbar}, \quad E_k = \frac{\hbar^2 k^2}{2m}, \tag{6.58}$$

or

$$a_{\mathbf{k}\lambda}(\tau) = a_{\mathbf{k}\lambda} e^{-(E_k - \mu)\tau/\hbar}, \quad a_{\mathbf{k}\lambda}^{\dagger}(\tau) = a_{\mathbf{k}\lambda}^{\dagger} e^{(E_k - \mu)\tau/\hbar}. \tag{6.59}$$

Then, the Green's function can be written as

$$\begin{aligned} G_{rs}(\mathbf{x}\tau, \mathbf{x}'\tau') &= -e^{\beta\Omega} Tr\left\{ e^{-\beta K} T_{\tau} \left[\Psi_s(\mathbf{x}\tau) \Psi_s^{\dagger}(\mathbf{x}'\tau') \right] \right\} \\ &= -\frac{1}{V} \sum_{\mathbf{k},\mathbf{k}'} \sum_{\lambda,\lambda'} e^{i\mathbf{k}\cdot\mathbf{x} - i\mathbf{k}'\cdot\mathbf{x}'} (\eta_{\lambda})_r (\eta_{\lambda'})_s \\ &\quad \times e^{-(E_k - \mu)\tau/\hbar + (E_{k'} - \mu)\tau'/\hbar} \langle a_{\mathbf{k}\lambda} a_{\mathbf{k}'\lambda'}^{\dagger} \rangle. \end{aligned} \tag{6.60}$$

The spin wavefunctions are orthonormal, so that $\lambda' = \lambda$. Similarly, the last average requires that the momenta states are the same, and $\langle a_{\mathbf{k}\lambda} a_{\mathbf{k}'\lambda'}^{\dagger} \rangle = 1 - \langle a_{\mathbf{k}'\lambda'}^{\dagger} a_{\mathbf{k}\lambda} \rangle = 1 - f_k$, where

$$f_k = \frac{1}{1 + e^{\beta(E_k - \mu)}} \tag{6.61}$$

is the Fermi-Dirac distribution function, with the chemical potential being exactly the Fermi energy in this case. Then,

$$G_{rs}^0(\mathbf{x}\tau, \mathbf{x}'\tau') = \begin{cases} -\frac{\delta_{rs}}{V} \sum_{\mathbf{k}} e^{i\mathbf{k}\cdot(\mathbf{x}-\mathbf{x}')} e^{-(E_k - \mu)(\tau - \tau')/\hbar} (1 - f_k), & \tau > \tau', \\ \frac{\delta_{rs}}{V} \sum_{\mathbf{k}} e^{i\mathbf{k}\cdot(\mathbf{x}-\mathbf{x}')} e^{-(E_k - \mu)(\tau - \tau')/\hbar} f_k, & \tau < \tau'. \end{cases} \tag{6.62}$$

One important point of these Green's functions is the periodicity in complex time of the functions themselves. One obvious reason for using the complex time version of the operators is that the Heisenberg temporal propagator is in a form that commutes with the density matrix itself. However, this practice creates some unusual behaviors. Consider the case in which, for convenience, we take $\tau = 0$, $\tau' > 0$, so that we can write Eq. (6.53) as (we take only the fermion case here)

$$
\begin{aligned}
G_{rs}(\mathbf{x}, 0, \mathbf{x}'\tau') &= e^{\beta\Omega} Tr\left\{e^{-\beta K}\Psi_s^\dagger(\mathbf{x}'\tau')\Psi_r(\mathbf{x}0)\right\} \\
&= e^{\beta\Omega} Tr\left\{\Psi_r(\mathbf{x}0)e^{-\beta K}\Psi_s^\dagger(\mathbf{x}'\tau')\right\} \\
&= e^{\beta\Omega} Tr\left\{\Psi_r(\mathbf{x}0)e^{-\beta K}\Psi_s^\dagger(\mathbf{x}'\tau')e^{-\beta K}e^{\beta K}\right\} \\
&= e^{\beta\Omega} Tr\left\{e^{-\beta K}\Psi_r(\mathbf{x}, \beta\hbar)\Psi_s^\dagger(\mathbf{x}'\tau')\right\} \\
&= -G_{rs}(\mathbf{x}, \beta\hbar, \mathbf{x}'\tau').
\end{aligned}
\tag{6.63}
$$

Here, because of the periodicity, we have to assume that $0 < \tau' < \beta\hbar$. Thus, we find that the temperature Green's function is *antiperiodic* in $\beta\hbar$. This can easily be shown to be true in the second time variable as well. (The boson version is periodic in $\beta\hbar$.) This result is very important and builds the properties of the density matrix into the results for the Green's function. The importance of this is that the perturbation series is now integrated only over the range $[0, \beta\hbar)$, instead of $[0, \infty)$. In the usual situation in which the Hamiltonian is independent of time, the Green's function depends only on the difference in time coordinates, and the time is readily shifted so that

$$
G_{rs}(\mathbf{x}, \mathbf{x}', \tau - \tau' < 0) = -G_{rs}(\mathbf{x}, \mathbf{x}', \tau - \tau' + \beta\hbar > 0).
\tag{6.64}
$$

For the noninteracting Green's function (6.62), this leads to the important result

$$
f_k e^{\beta(E_k - \mu)} = 1 - f_k,
\tag{6.65}
$$

which insures that the equilibrium distribution function is a Fermi-Dirac function.

With this periodicity in mind, we can now introduce the Fourier transform representation (in time) for the temperature Green's function. We note that both the boson and fermion temperature Green's functions are fully periodic in $2\beta\hbar$. We let $\tau'' = \tau - \tau'$, and define the Fourier transform from

$$
G_{rs}(\mathbf{x}, \mathbf{x}', \tau'') = \frac{1}{\hbar\beta} \sum_n e^{-i\omega_n\tau''} G_{rs}(\mathbf{x}, \mathbf{x}', \omega_n),
\tag{6.66}
$$

where, for the moment,

$$
\omega_n = \frac{n\pi}{\hbar\beta}.
\tag{6.67}
$$

We remark here that τ'' in this equation is restricted to lie within the principal periodic region (e.g., $-\hbar\beta < \tau'' < \hbar\beta$) since it is a complex quantity, and we must assure that the summation does not diverge because of this. Principally, this means that the real time is restricted to a finite range. This now leads to the transform itself,

$$
G_{rs}(\mathbf{x}, \mathbf{x}', \omega_n) = \frac{1}{2} \int_{-\hbar\beta}^{\hbar\beta} d\tau'' e^{i\omega_n\tau''} G_{rs}(\mathbf{x}, \mathbf{x}', \tau'').
\tag{6.68}
$$

It is convenient to separate this integral into two parts, the positive and negative time segments, as

$$
\begin{aligned}
G_{rs}(\mathbf{x}, \mathbf{x}', \omega_n) &= \frac{1}{2} \int_{-\hbar\beta}^{0} d\tau'' e^{i\omega_n \tau''} G_{rs}(\mathbf{x}, \mathbf{x}', \tau'') + \frac{1}{2} \int_{0}^{\hbar\beta} d\tau'' e^{i\omega_n \tau''} G_{rs}(\mathbf{x}, \mathbf{x}', \tau'') \\
&= -\frac{1}{2} \int_{-\hbar\beta}^{0} d\tau'' e^{i\omega_n \tau''} G_{rs}(\mathbf{x}, \mathbf{x}', \tau'' + \beta\hbar) \\
&\quad + \frac{1}{2} \int_{0}^{\hbar\beta} d\tau'' e^{i\omega_n \tau''} G_{rs}(\mathbf{x}, \mathbf{x}', \tau'') \\
&= \frac{1}{2}(1 - e^{-i\omega_n \beta\hbar}) \int_{0}^{\hbar\beta} d\tau'' e^{i\omega_n \tau''} G_{rs}(\mathbf{x}, \mathbf{x}', \tau'').
\end{aligned} \tag{6.69}
$$

The prefactor vanishes for n even. A similar approach shows that the prefactor will vanish for the bosonic form for n odd. For these cases, the prefactor is unity. This further implies that for fermions we use only the odd frequencies in (6.67), whereas for bosons we use only the even frequencies. Then, for fermions (6.68) reduces to

$$
G_{rs}(\mathbf{x}, \mathbf{x}', \omega_n) = \int_{0}^{\hbar\beta} d\tau'' e^{i\omega_n \tau''} G_{rs}(\mathbf{x}, \mathbf{x}', \tau''), \quad \omega_n = \frac{(2n+1)\pi}{\hbar\beta}. \tag{6.70}
$$

Since this Green's function is often called a *Matsubara Green's function*, the frequencies are referred to as the *Matsubara frequencies*.

To examine the nature of this function, let us expand the field operators in a set of basis functions $\{\phi_m\}$, so that in Dirac notation we can write (6.70) using (6.60) as

$$
\begin{aligned}
G_{rs}(i\omega_n) &= -e^{\beta\Omega} \sum_{m,m'} |\langle m|\Psi(\mathbf{x})|m'\rangle|^2 e^{-\beta E_m} \int_{0}^{\beta\hbar} d\tau'' e^{i\omega_n \tau''} e^{\tau''(E_m - E_{m'})/\hbar} \\
&= e^{\beta\Omega} \sum_{m,m'} |\langle m|\Psi(\mathbf{x})|m'\rangle|^2 \frac{e^{-\beta E_m} + e^{-\beta E_{m'}}}{i\omega_n + (E_m - E_{m'})/\hbar}.
\end{aligned} \tag{6.71}
$$

In particular, if the basis set is that for which the creation and annihilation operators are defined for the field operator, then the matrix elements are zero except for particular connections between the two coefficients; that is, $m' = m \pm 1$, with the sign determined by which of the two operators is used to define the matrix elements. However, there is no requirement that this basis set be the same as that used in the field operator definition. On the other hand, if we use momentum wavefunctions (the plane waves), then the difference in energies produces just the single momentum E_k, where $\mathbf{k}$ is the Fourier transform for the difference in position $\mathbf{x} - \mathbf{x}'$.

In direct analogy with the results of the previous chapter, it is now possible to write the retarded and advanced temperature Green's functions as

$$
G_{rs}^{r}(\mathbf{x}\tau, \mathbf{x}'\tau') = -\theta(\tau - \tau') Tr\{\hat{\rho}[\Psi_r(\mathbf{x}\tau), \Psi_s^{\dagger}(\mathbf{x}'\tau')]_{+}\}, \tag{6.72}
$$

$$
G_{rs}^{a}(\mathbf{x}\tau, \mathbf{x}'\tau') = \theta(\tau' - \tau) Tr\{\hat{\rho}[\Psi_r(\mathbf{x}\tau), \Psi_s^{\dagger}(\mathbf{x}'\tau')]_{+}\}. \tag{6.73}
$$

[Note that since we are using the curly braces to represent the quantity over which the trace is performed, we can no longer use these to indicate the anticommutator; therefore we will use the normal commutator with the subscript "+" to indicate that the anticommutator relationship is used for the fermions.] For now let us consider just the retarded function and take the Fourier transform of this quantity. First, we make the same basis function expansion as above, and this leads to

$$G_{rs}^r(\tau'') = -\theta(\tau'')e^{\beta\Omega} \sum_{m,m'} e^{-\beta E_m} \Big\{ |\langle m|\Psi(\mathbf{x})|m'\rangle|^2 e^{i\tau''(E_m - E_{m'})/\hbar}$$

$$+ |\langle m'|\Psi(\mathbf{x})|m\rangle|^2 e^{-i\tau''(E_m - E_{m'})/\hbar} \Big\}$$

$$= -\theta(\tau'')e^{\beta\Omega} \sum_{m,m'} |\langle m|\Psi(\mathbf{x})|m'\rangle|^2 e^{i\tau''(E_m - E_{m'})/\hbar}$$

$$\times (e^{-\beta E_m} + e^{-\beta E_{m'}}), \tag{6.74}$$

where we have interchanged the two dummy indices in the second term of the first line. It is important to note that the normal Heisenberg time variation has been used, and not the limited imaginary time form. We now Fourier-transform this quantity over the entire frequency range, so that

$$G_{rs}^r(\omega) = e^{\beta\Omega} \sum_{m,m'} |\langle m|\Psi(\mathbf{x})|m'\rangle|^2 \frac{(e^{-\beta E_m} + e^{-\beta E_{m'}})}{\omega + (E_m - E_{m'})/\hbar + i\eta}, \tag{6.75}$$

with the normal small quantity η required for convergence of the integration. Comparing this result with that of (6.71) leads to an informative result, that the Matsubara Green's function can be transformed into the retarded temperature Green's function through the limiting process

$$\lim_{i\omega \to \omega + i\eta} G_{rs}(i\omega_n) = G_{rs}^r(\omega). \tag{6.76}$$

This limiting process is referred to as analytic continuation [29]. A similar result, with the sign of the small quantity reversed, leads to the advanced temperature Green's function.

Another quantity of interest is the spectral density function $A(\mathbf{k}, \omega)$, which was defined in the previous chapter. This quantity is defined from the retarded or advanced Green's functions through

$$A(\mathbf{k}, \omega) = -2Im\big\{G_{rs}^r(\mathbf{k}, \omega)\big\} = 2Im\big\{G_{rs}^a(\mathbf{k}, \omega)\big\}. \tag{6.77}$$

From the form of the retarded function (6.75), the only imaginary part arises from the denominator of the last fraction. This term can be expanded as

$$\frac{1}{\omega + (E_m - E_{m'})/\hbar + i\eta} = P\frac{1}{\omega + (E_m - E_{m'})/\hbar} - i\pi\delta[\omega + (E_m - E_{m'})/\hbar]. \tag{6.78}$$

Using this in the definition of the retarded function, we arrive at

$$A(\mathbf{k}, \omega) = 2\pi e^{\beta\Omega} \sum_{m,m'} |\langle m|\Psi(\mathbf{x})|m'\rangle|^2 (e^{-\beta E_m} + e^{-\beta E_{m'}})$$

$$\times \delta[\omega + (E_m - E_{m'})/\hbar]$$

$$= 2\pi e^{\beta\Omega}(1 + e^{-\hbar\beta\omega}) \sum_{m,m'} |\langle m|\Psi(\mathbf{x})|m'\rangle|^2 e^{-\beta E_m}. \tag{6.79}$$

We note that the right side is positive definite, so the spectral density can be interpreted as a probability density function. This leads to the important result that

$$1 = \int \frac{d\omega}{2\pi} A(\mathbf{k}, \omega) = e^{\beta\Omega} \sum_{m,m'} |\langle m|\Psi(\mathbf{x})|m'\rangle|^2 (e^{-\beta E_m} + e^{-\beta E_{m'}})$$

$$= e^{\beta\Omega} \sum_m e^{-\beta E_m} \{\Psi(\mathbf{x})\Psi^\dagger(\mathbf{x}) + \Psi^\dagger(\mathbf{x})\Psi(\mathbf{x})\} = e^{\beta\Omega} Tr\{e^{-\beta K}\} = 1. \qquad (6.80)$$

In the noninteracting case, we expect the spectral density to be a delta function relating the energy to the momentum. The broadening of this function by the interaction eliminates the simple relationship between energy and momentum that exists in classical mechanics, and thus is the major introduction of quantum effects in this regard. Finally, we note that by comparing Eqs. (6.71), (6.75), and (6.79), we can write the Green's functions in terms of the spectral density as

$$G_{rs}(i\omega_n) = \int \frac{d\omega'}{2\pi} \frac{A(\mathbf{k}, \omega')}{i\omega_n - \omega'}, \qquad (6.81)$$

$$G_{rs}^r(\omega) = \int \frac{d\omega'}{2\pi} \frac{A(\mathbf{k}, \omega')}{\omega - \omega' + i\eta}. \qquad (6.82)$$

The connection between these two equations is an example of the process of analytic continuation. Here we take the complex frequency $i\omega_n$ and analytically continue it to the proper real frequency (with an imaginary convergence factor, as has been used in the previous chapters), $\omega + i\eta$.

6.3.4 One-particle density of states

In Chapter 2 we developed the concept of the density of states (per unit energy per unit volume). There, for example, it was shown that the density of states could be written as

$$\rho_d(\omega) = \frac{2}{\hbar} \int \frac{d^d\mathbf{k}}{(2\pi)^d} \delta(\omega - E_k/\hbar), \qquad (6.83)$$

where d is the dimensionality of the system, and the factor of $s = 2$ is for electrons where the opposite spins do not raise the spin degeneracy (or represents a summation over the diagonal spin indices of the Green's function). For three dimensions, one recovers the familiar

$$\rho_3(\hbar\omega) = \frac{1}{2\pi^2} \left(\frac{2m}{\hbar^2}\right)^{3/2} (\hbar\omega)^{1/2}. \qquad (6.84)$$

If we write this in the noninteracting Green's function form (with the latter described by plane-wave states and putting in the values of the creation and annihilation operators), this can be more easily related to the noninteracting Green's function as

$$\rho_d(\omega) = -\frac{s}{\pi\hbar} \int \frac{d^d\mathbf{k}}{(2\pi)^d} Im\{G_{rs}^r(\mathbf{k}, \omega)\}, \qquad (6.85)$$

and hence the density of states is related to the spectral density. This relationship can be expressed as

$$\rho_d(\omega) = \frac{s}{2\pi\hbar} \int \frac{d^d\mathbf{k}}{(2\pi)^d} A(\mathbf{k}, \omega). \qquad (6.86)$$

In general, the decay of excitations is much stronger in the disordered (large impurity scattering and diffusive transport) system than it is in the pure system. This is of concern, since most mesoscopic systems are treated as disordered systems because of the randomness introduced by the impurity scattering. One of the important properties, therefore, is the one-particle density of states. Now, by one-particle density of states we don't mean the quantity for a single electron, but rather the density of states that is appropriate for a single electron in the sea of other electrons and impurities. The temperature Green's function can be recognized through the reverse of the analytic continuation procedures, which was

$$\rho_d(\omega) = -\frac{s}{\pi \hbar} \int \frac{d^d \mathbf{k}}{(2\pi)^d} Im\{G_{rs}(\mathbf{k}, i\omega \to \omega + i\eta)\}, \tag{6.87}$$

and this may be related to the noninteracting Green's function and the self-energy by Dyson's equation as

$$\rho_d(i\omega_n) = -\frac{s}{\pi \hbar} \int \frac{d^d \mathbf{k}}{(2\pi)^d} Im\left\{ \frac{1}{\left[G_{rs}^0(\mathbf{k}, i\omega_n)\right]^{-1} - \Sigma(\mathbf{k}, i\omega_n)} \right\}, \tag{6.88}$$

where the self-energy includes those parts from impurities and the electron-electron interaction, which itself may be mediated by the impurity scattering in the diffusive limit. The result for the interacting Green's function is essentially the same as that of the last chapter for zero-temperature Green's functions, and arises from the properties of the linear perturbation expansion and resummation into Dyson's equation itself, and not from any particular property of any one type of Green's function.

Thus, the spectral density in the interacting system represents the entire density of states, including both the single particle properties and their modification that arises from the self-energy corrections arising from the presence of the interactions. In the next sections, we will examine the electron-electron interaction and its modification in the presence of strong impurity scattering. This will entail an approach that is quite different from the one usually found in high-mobility materials, where the electron-electron interaction is thought to be the dominant interaction process. Let us now turn to the self-energy for the electron-electron interaction.

6.3.5 The effective interaction potential

In a semiconductor, the interactions of the electrons can be with a variety of scattering centers, whether from impurities, from other electrons, or from phonons. Generally, one can write the diagrammatic expansion of the perturbation series as shown in Fig. 6.9. Only the impurity and the electron-electron interactions are indicated in this figure. It also is important to note that only the lowest-order terms have been retained. The first term in the electron-electron diagram series is valid generally only for small momentum exchange, and therefore it is sensitive to the bare interaction potential. There are higher-order terms, on the other hand, which tend to involve larger momentum exchanges, and one needs to consider the screening of the interaction potential by the other electrons. The Coulomb interaction between individual electrons is a long-range interaction, and never should it be taken into account only in first-order perturbation theory. Rather, one should use an effective interaction potential that takes into account the screening, as discussed above. This was the procedure used in the previous chapter, and it leads to the need to account for the potential and its self-screening, as described in Chapters 2 and 5. This interaction potential may be

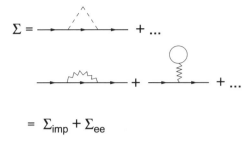

$$\Sigma = + \ldots$$

$$ = \Sigma_{imp} + \Sigma_{ee}$$

Fig. 6.9. Contributions to impurity (top) and electron-electron (bottom) self-energy.

found to be

$$U_{eff} = \frac{U}{1 - U\Pi_\omega},\tag{6.89}$$

where Π_ω is a *polarization* resummation, as discussed in connection with other diagrammatic expansions. However, as will be explained below, this polarization is slightly different in that the terms included in the summation are a different set of diagrams. We want to examine how this resummation of the terms occurs in this section. The interaction potential itself is the bare Coulomb interaction $e^2/4\pi\varepsilon r$, which leads to the Fourier transformations

$$U(\mathbf{q}) = \begin{cases} \dfrac{e^2}{\varepsilon q^2}, & d = 3, \\[2ex] \dfrac{e^2}{2\varepsilon q}, & d = 2, \\[2ex] \dfrac{e^2}{4\pi\varepsilon}\ln\left(1 + \dfrac{q_0^2}{q^2}\right), & d = 1, \end{cases}\tag{6.90}$$

where q_0 is an artificial cutoff in momentum for the Coulomb potential (which was discussed in a previous section) [30]. It is important to point out that the forms that appear in (6.90) are for true dimensionality d, since integration over the other directions in a quasi-d-dimensional system produces some modifications. These will not be dealt with here, but they can be quite important in some cases.

The summation represented by Eq. (6.89) is different than the earlier summations in some crucial details. Here, we are replacing the bare interaction line (the squiggly line in Fig. 6.9) with a summation of an infinite number of terms, which leads to a renormalized, dressed interaction. The set of diagrams that are most important in this interaction is shown in Fig. 6.10. Here, the polarization Π_ω is composed of the two Green's functions that form the ring in a summation of *ring* diagrams. (We emphasize that, as in all such cases, experience tells us just which set of diagrams should be resummed to give the dominant interactions.) That is, the polarization function that is important here is the closed pair of Green's functions, in which the expansion is a summation of such Green's function rings, connected at each end by an interaction line. Each ring corresponds to a density fluctuation, and the long-range correlations of these disparate density fluctuations are coupled through the carrier-carrier interaction into the effective interaction. The interaction lines at the input and output all carry momentum $\mathbf{q}$. However, the polarization is different here from that treated in the previous chapter in that the two Green's functions not only have different momentum $\mathbf{k}$ and $\mathbf{k} - \mathbf{q}$, but also different frequencies $\omega_n + \omega_1$ and ω_1. This is because it is a summation of

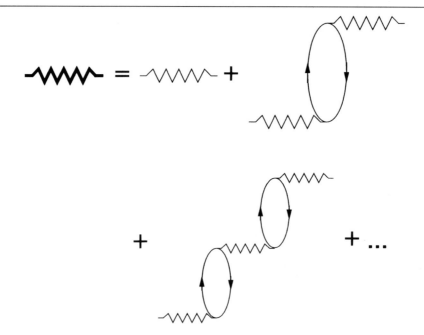

Fig. 6.10. The expansion of the interaction into ring diagrams.

ring diagrams rather than ladder diagrams. Thus, we denote this polarization Π_ω. However, it is also complicated by the fact that the two Green's functions are also representative of the diffusive nature of the transport in diffusive systems, so there can be a variety of corrections to the ring diagrams themselves. We deal with one such set of corrections in the next section. The expansion shown in Fig. 6.10 can be written as

$$U_{eff} = U + U\Pi_\omega U + U\Pi_\omega U\Pi_\omega U + \cdots = \frac{U}{1 - U\Pi_\omega}. \tag{6.91}$$

This is just the result of (6.89). In many cases, this is written as

$$U_{eff} = \frac{U_0}{\varepsilon(\omega)}, \tag{6.92}$$

where U_0 is the appropriate form from (6.90) with the dielectric constant removed. This leads to the general definition of the dielectric function as

$$\varepsilon(\omega) = \varepsilon[1 - U\Pi_\omega]. \tag{6.93}$$

In some cases, other contributions to the overall polarization, such as that due to the lattice in polar materials, can be added to the dielectric function.

The lowest-order polarization, composed of the ring diagrams, is still written as a summation over a pair of Green's functions, but now a summation over the new frequency ω_1 has been added, as well as over the momentum $\mathbf{k}$. The end result, of course, modifies the interaction potential, which is a function of $\mathbf{q}$, so that these new variables are integrated. It is this integration over interior variables in the pair of Green's functions that distinguishes this polarization, which is related to the density-density correlation function. We can now write the polarization as

$$\Pi_\omega(\mathbf{q}, \omega_n) = \frac{2}{\beta \hbar^2} \int \frac{d^d \mathbf{k}}{(2\pi)^d} \sum_{\omega_1} \left[\frac{1}{i\omega_1 - (E_k - \mu)/\hbar - i/2\tau} \right]$$

$$\times \left[\frac{1}{i\omega_1 + i\omega_n - (E_{k+q} - \mu)/\hbar + i/2\tau} \right], \quad (6.94)$$

and the role of the impurities has been included here through a lifetime in the Green's functions as a replacement for the normal factor η. The momentum integration is over a single set of spin states. A typical term in the frequency summation is of order $1/|\omega_1|^2$ for large values of the frequency, and the summation therefore converges absolutely [28]. Although one could do the contour integration, a more useful result is obtained by introducing a convergence factor (which must be handled carefully due to the complex frequencies) and rewriting the two sums as

$$\Pi_\omega(\mathbf{q}, \omega_n) = \frac{2}{\beta \hbar^2} \int \frac{d^d \mathbf{k}}{(2\pi)^d} \frac{1}{i\omega_n - (E_{k+q} - E_k)/\hbar - i/\tau} \sum_{\omega_1} e^{i\omega_1 \eta}$$

$$\times \left[\frac{1}{i\omega_1 - (E_k - \mu)/\hbar - i/2\tau} \right.$$

$$\left. - \frac{1}{i\omega_1 + i\omega_n - (E_{k+q} - \mu)/\hbar + i/2\tau} \right]. \quad (6.95)$$

At this point, we want to develop a useful identity that represents the frequency sums needed to evaluate (6.95). The summation that we would like to consider is the fermion sum

$$\sum_n e^{i\omega_n \eta} \frac{1}{i\omega_n - x}, \quad n \ odd. \quad (6.96)$$

This sum would diverge (we ignore the lifetime effects of τ in this argument) without the convergence factor, so η must remain positive until after the sum is evaluated (which implies directly that the separation done above is valid only when the convergence factor is included). The most direct approach is to use contour integration of a suitable function. The function $-\beta\hbar(e^{\beta\hbar z} + 1)^{-1}$ has simple poles at the values $z = i(2n + 1)\pi/\beta\hbar = i\omega_n$. Each pole has unit residue with the chosen normalization. Thus, we consider the contour integral

$$-\frac{\beta\hbar}{2\pi i} \int_C \frac{dz}{e^{\beta\hbar z} + 1} \frac{e^{\eta z}}{z - x}. \quad (6.97)$$

The choice now is the contour that we use. We pick a simple contour that encloses the imaginary axis (and includes the point $z = x$) in the positive sense (counterclockwise); that is, we use a circle of radius R with $R \to \infty$. The poles are then the set of Fermion Matsubara frequencies $z = i(2n+1)\pi/\beta\hbar = i\omega_n$ and the pole $z = x$. Now, as $R \to \infty$, the argument of the integral vanishes (for $\beta\hbar > \eta > 0$), so that the integral around the contour itself vanishes. Thus,

$$0 = \sum_n e^{i\omega_n \eta} \frac{1}{i\omega_n - x} - \frac{\beta\hbar \, e^{\eta x}}{e^{\beta\hbar x} + 1}. \quad (6.98)$$

After letting the convergence factor vanish, this gives [28], [30],

$$\sum_n e^{i\omega_n \eta} \frac{1}{i\omega_n - x} = \frac{\beta \hbar}{e^{\beta \hbar x} + 1} = \beta \hbar f(\hbar x). \tag{6.99}$$

The result for the frequency summation yields the Fermi-Dirac distribution (by construction). This result can now be used in Eq. (6.95) to give the polarization as

$$\Pi_\omega(\mathbf{q}, \omega_n) = -\frac{2}{\hbar} \int \frac{d^d\mathbf{k}}{(2\pi)^d} \frac{f_{k+q} - f_k}{i\omega_n - (E_{k+q} - E_k)/\hbar - i/\tau}. \tag{6.100}$$

Adding and subtracting a term $f_{k+q} f_k$ in the numerator allows us to rearrange this last equation as

$$\Pi_\omega(\mathbf{q}, \omega_n) = -\frac{2}{\hbar} \int \frac{d^d\mathbf{k}}{(2\pi)^d} \frac{f_{k+q}(1 - f_k) - f_k(1 - f_{k+q})}{i\omega_n - (E_{k+q} - E_k)/\hbar - i/\tau}. \tag{6.101}$$

Making the change of variables $\mathbf{k} + \mathbf{q} \to -\mathbf{k}$, and using the spherical symmetry of the energy bands, this equation can be rearranged into

$$\Pi_\omega(\mathbf{q}, \omega_n) = -\frac{4}{\hbar} \int \frac{d^d\mathbf{k}}{(2\pi)^d} f_{k+q}(1 - f_k) \frac{(E_{k+q} - E_k)/\hbar}{(\omega_n + 1/\tau)^2 + [(E_{k+q} - E_k)/\hbar]^2}. \tag{6.102}$$

There are a great variety of approximations that can be developed for the polarization. We discuss a few of these.

Static screening

In the case of static screening, we will essentially take the small-momentum, low-frequency limit of Eq. (6.100). In this approach, we ignore the frequency and the scattering rate as both being small. Then, the numerator and denominator are expanded as

$$f_{k+q} - f_k \simeq \mathbf{q} \cdot \frac{\partial f}{\partial \mathbf{k}} \simeq \left[\mathbf{q} \cdot \frac{\partial E}{\partial \mathbf{k}} \right] \frac{\partial f}{\partial E} \tag{6.103}$$

and

$$E_{k+q} - E_k \simeq \mathbf{q} \cdot \frac{\partial E}{\partial \mathbf{k}}. \tag{6.104}$$

For nondegenerate semiconductors we can also write

$$\frac{\partial f}{\partial E} \simeq -\beta f(E). \tag{6.105}$$

Now, using

$$2 \int \frac{d^d\mathbf{k}}{(2\pi)^d} f_k = n, \tag{6.106}$$

we can rewrite (6.100) as

$$\Pi_\omega(\mathbf{q}, 0) = -\frac{n}{k_B T}. \tag{6.107}$$

The dielectric function may then be written as

$$\varepsilon(0) = \varepsilon \left[1 + \frac{q_D^2}{q^2} \right], \quad q_D^2 = \frac{ne^2}{\varepsilon k_B T}, \tag{6.108}$$

where the latter quantity is the square of the reciprocal Debye screening length. This latter quantity provides exactly the cutoff behavior discussed in the above sections. For degenerate material, the average energy $k_B T$ is replaced by the Fermi energy E_F, and the quantity becomes the reciprocal of the squared Fermi-Thomas screening length.

Plasmon-pole approximation

Still another approximation occurs when we assume that the frequency is large compared to the energy exchange in the denominator of (6.100). This equation is first rewritten as

$$\Pi_\omega(\mathbf{q}, \omega_n) = -\frac{2}{\hbar} \int \frac{d^d\mathbf{k}}{(2\pi)^d} f_k \left[\frac{1}{i\omega_n - (E_k - E_{k-q})/\hbar - i/\tau} \right.$$
$$\left. - \frac{1}{i\omega_n - (E_{k+q} - E_k)/\hbar - i/\tau} \right], \qquad (6.109)$$

where a change of variables has been made in the first term. This can now be combined as

$$\Pi_\omega(\mathbf{q}, \omega_n) = -\frac{2}{\hbar} \int \frac{d^d\mathbf{k}}{(2\pi)^d} f_k$$
$$\times \frac{2E_k - E_{k+q} - E_{k-q}}{(i\omega_n - (E_k - E_{k-q})/\hbar - i/\tau)(i\omega_n - (E_{k+q} - E_k)/\hbar - i/\tau)}. \qquad (6.110)$$

The numerator of the fraction can be written as

$$2E_k - E_{k+q} - E_{k-q} \approx -\frac{\hbar^2 q^2}{m}, \qquad (6.111)$$

and the denominator is approximately $(i\omega_n)^2$. Using the summation (6.106), the dielectric function can now be written (after analytically continuing the frequency into the real domain)

$$\varepsilon(0) = \varepsilon \left[1 - \frac{\omega_p^2}{\omega^2} \right], \qquad (6.112)$$

where

$$\omega_p^2 = \frac{ne^2}{m\varepsilon} \qquad (6.113)$$

is the free-carrier plasma frequency.

In many cases, one wants to add the lattice contributions to the dielectric function, in which case the polarization in polar semiconductors may be added to (6.113) as [24]

$$\varepsilon(0) = \varepsilon \left[1 + \frac{\omega_{LO}^2 - \omega_{TO}^2}{\omega_{TO}^2 - \omega^2} - \frac{\omega_p^2}{\omega^2} \right], \qquad (6.114)$$

where ω_{TO} and ω_{LO} are the transverse and longitudinal optical phonon frequencies, respectively. When $\omega_p < \omega_{LO}$, there are two zeroes and two poles of the dielectric function. The poles are at zero frequency and at the transverse optical phonon frequency. The zeroes are at a down-shifted plasma frequency (evaluated at the low-frequency dielectric constant) and at the longitudinal phonon frequency. Here, the dielectric constant ε is inferred to be the high-frequency dielectric constant. In the opposite case, where $\omega_p > \omega_{LO}$, a zero at the transverse optical frequency cancels the pole at this frequency, so there remains only the poles at zero frequency and at the plasma frequency. This latter is the hybridized polar phonon-plasmon frequency, and corresponds to what Ridley [25] has called the *descreened* frequency.

In essence, the interaction between the single carrier and the background electron gas is evaluated at the zeroes of the dielectric function, which means that in this approximation the dominant scattering is from a single particle to either the longitudinal optical phonon or to a collective oscillation (the plasmons). In the simple approximation that has been used so far in this section, the only scattering is to the lattice or to the collective modes. This approximation has been termed the plasmon-pole approximation [31], [32], but it ignores the important single-particle scattering interactions. This is especially significant, since the single-particle interactions are thought to be dominant in phase-breaking processes. That is, in the last subsection for low frequencies (low energies), we predominantly found a screened interaction representing the scattering of an incident electron by the gas of nearly free electrons – single-particle scattering. Here, in the high-frequency (high-energy) situation, we predominantly find an interaction representing the scattering of an incident electron by the collective modes of the electron gas (and the lattice polar modes if that term is included). In the next subsection, we will probe between these two limits.

Dynamic screening

As a further consideration, we consider a number of factors. First, the product of the Fermi-Dirac functions for small $q \to 0$ give essentially the derivative of the function, which is equivalent to a delta function at the Fermi energy (at sufficiently low temperature). But, for this term the angular average vanishes. If we then expand the term in f_{k+q} and expand the energy differences, keeping only the lowest-order terms in q, the expansion of the energy around the momentum k yields a function of the angle between the two vectors, which is involved in the averaging process of the d-dimensional integration. (Alternatively, for nondegenerate materials, the filling factors can be ignored and the angular averages computed in the normal manner.) This leads to terms in Dq^2, as in the last chapter, except that here $D = v^2\tau/d$ and there is an extra factor in the numerator. This leads us to approximate the polarization as

$$\Pi_\omega(\mathbf{q}, \omega_n) = -\beta n \frac{Dq^2}{\tau(\omega_n + 1/\tau)^2 - Dq^2}. \tag{6.115}$$

We can see this as follows. We first utilize the polarization in the form (6.102). Then, the numerator is evaluated by (6.100), and the denominator is expanded into a power series, with both retaining terms up to second order in the $\cos\theta$ variation in order to avoid averaging to zero. Note that it is also important to expand the term f_{k+q}, since only the second term in this expansion leads to any nonzero terms. The series is then resummed with the above result. The resulting integral is performed normally, using the results of the last section, with the factor β being replaced by the inverse of the Fermi energy in strongly degenerate systems. The above result is exact in two dimensions but must be corrected by a factor of $2/d$ preceding each of the diffuson terms.

It is now possible to put together the effective interaction potential energy from Eqs. (6.89) and (6.90) and the last result. This gives

$$U_{eff} = \frac{e^2}{\varepsilon q^2 \left[1 - \frac{\chi_3 D}{\tau(\omega_n + 1/\tau)^2 - Dq^2}\right]}, \quad \chi_3 = \frac{ne^2\beta}{\varepsilon}, \quad d = 3, \qquad (6.116)$$

$$U_{eff} = \frac{e^2}{2\varepsilon q \left[1 - \frac{\chi_2 Dq}{\tau(\omega_n + 1/\tau)^2 - Dq^2}\right]}, \quad \chi_2 = \frac{n_s e^2\beta}{2\varepsilon}, \quad d = 2, \qquad (6.117)$$

$$U_{eff} = \frac{e^2}{4\pi\varepsilon \left[\left(\ln\left(1 + \frac{q_0^2}{q^2}\right)\right)^{-1} - \frac{\chi_1 Dq^2}{\tau(\omega_n + 1/\tau)^2 - Dq^2}\right]}, \quad \chi_1 = \frac{n_l e^2\beta}{4\pi\varepsilon}, \quad d = 1.$$

$$(6.118)$$

It is clear that the screening has a varying effectiveness as the dimensionality is reduced. In the limit of zero frequency and no scattering, we find that χ_3 is the square of the Debye screening wavevector (in nondegenerate material) and clearly agrees with the result we found above. Thus, the screening introduces a cutoff in the strength of the Coulomb interaction at small q, or large distances. A similar cutoff occurs in the two-dimensional case, so the screening also has a comparable effectiveness (the size of the cutoff wavevectors will be somewhat different), and this determines the range of the strengths between the two cases. Similarly, in one dimension, the small wavevector results are moderated by the cutoff of the screening effects. Thus, the screening remains strong in all dimensions and affects a cutoff at small wavevectors, or large distances, in the interaction itself.

In the limit of high frequency, however (and one must pass to the analytically continued limit for this), the plasmon-pole approximation is recovered, again in agreement with the previous subsection (the factor β is canceled by the additional factors in the integral arising from the remaining factor of D). It is important to note now that the plasma frequency in low-dimensional systems is not a constant value, as found in the previous subsection. In fact, in two dimensions, Eq. (6.117) leads to the result found in Chapter 2

$$\omega_{p2}^2 = \frac{n_s e^2}{2\varepsilon}q, \qquad (6.119)$$

which means that the plasma frequency actually vanishes at small wavenumber. Similarly, in one dimension we find

$$\omega_{p1}^2 = \frac{n_l e^2}{4\pi\varepsilon}q^2. \qquad (6.120)$$

It is the screening cutoff of the interaction that now allows us to assume that the electron-electron interaction is relatively independent of frequency and momentum in many cases. However, this assertion is true only for small values of $\omega_n + 1/\tau$. If the values of these latter terms are not small, then they become the dominant terms in the denominators of Eqs. (6.116)–(6.118). When these terms are not small, then the cancellation of the q^2 terms does not occur and the interaction is effectively *descreened*, as was discussed already in Chapter 2. To be sure, the strength of the interaction is modified somewhat, but the basic dependence on q at all distances returns to the interaction. We now turn to a more extensive momentum variation for the single-particle interactions (low frequency, hence low energy).

Momentum-dependent screening

Although the low-frequency approximation will be used here, the desire is to evaluate carefully the summation over the free carriers that is involved in the summation (6.109), which may be rewritten as

$$\varepsilon(q) = \varepsilon + \frac{e^2}{q^2} \int \frac{d^d\mathbf{k}}{(2\pi)^d} f_k \left[\frac{1}{E_{k+q} - E_k} - \frac{1}{E_k - E_{k-q}} \right]. \tag{6.121}$$

As in the previous sections, the energy denominators can be expanded as

$$E_{k\pm q} - E_k = \frac{\hbar^2 q^2}{2m} \pm \frac{\hbar^2 kq}{m} \cos\theta, \tag{6.122}$$

where the mass is, of course, the effective mass in the semiconductor of interest. The integration will be carried out for three dimensions but is readily extendible to lower dimensionality. The integration over the angles is straightforward, and this leads to

$$\varepsilon(q) = \varepsilon + \frac{me^2}{\hbar^2 q^3} \int_0^\infty f(k) \ln \left| \frac{k + 2q}{k - 2q} \right| k\,dk. \tag{6.123}$$

The form of the argument of the logarithm arises from having factored $\frac{\hbar^2 q}{2m}$ out of each term in the numerator and the denominator, so the magnitude sign is required to assure that the argument is positive definite. To proceed further, the following normalized variables are now introduced:

$$\xi^2 = \frac{\hbar^2 q^2}{8mk_B T}, \quad x^2 = \frac{\hbar^2 k^2}{2mk_B T}, \quad \mu = \frac{E_F}{k_B T}. \tag{6.124}$$

It may be noted that the temperature here is that of the distribution function and represents the electron temperature, not the lattice temperature, so that this formulation may well be used in nonequilibrium situations such as those of the next chapter. Although this has not been noted by any subscript on the temperature, it should not be confusing since this is the only temperature in the problem. By incorporating these normalizing factors into the expression (6.123), we obtain

$$\varepsilon(q) = \varepsilon \left[1 + \frac{q_D^2}{q^2} F(\xi, \mu) \right], \tag{6.125}$$

where

$$F(\xi, \mu) = \frac{1}{\sqrt{\pi}\xi F_{1/2}(\mu)} \int_0^\infty \frac{x\,dx}{1 + e^{x^2-\mu}} \ln \left| \frac{k + 2q}{k - 2q} \right|, \tag{6.126}$$

and $F_{1/2}(\mu)$ is a standard Fermi-Dirac integral

$$F_\eta(\mu) = \int_0^\infty \frac{x^{3/2+\eta}\,dx}{1 + e^{x^2-\mu}}. \tag{6.127}$$

In the case of a nondegenerate semiconductor (μ large and negative), $F(\xi, \mu)$ becomes Dawson's integral and is a tabulated function. In either case, however, best results are obtained from numerical simulation. In Fig. 6.11, the overall behavior of this function is plotted, and the topmost curve ($\mu = -30$) corresponds to the nondegenerate limit. The behavior is not very dramatic. As $q \to 0$ (i.e., $\xi \to 0$), $F(\xi, \mu) \to 1$ for nondegenerate material, and the usual Debye screening behavior is recovered. On the other hand, as $q \to \infty$, $F(\xi, \mu) \to 0$, and the screening is broken up completely. Thus, for high-momentum

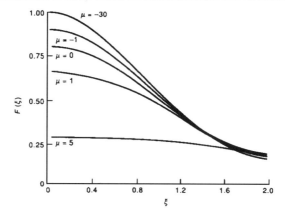

Fig. 6.11. Momentum dependence of the screening function.

transfer in the scattering process, the scattering potential is completely descreened $q \gg q_D$. The upshot of this is that the nonlinearity in Coulomb scattering appears once again as a scattering cross section that depends upon itself through the momentum transfer $\hbar q$ (and hence through the scattering angle θ). $F(\xi, \mu)$ has decreased to a value of one-half its maximum already at a value of $\xi = 1.07$, for which $q = 1.75q_T$, where $q_T = mv_T/\hbar$ is the thermal wavevector, corresponding to the thermal velocity of a carrier. For GaAs at room temperature, $q_T = 2.61 \times 10^6$ cm^{-1}. This value is only 3.5 times larger than the Debye wavevector at a carrier density of 10^{17} cm^{-3} and becomes smaller than the Debye screening wavevector at higher carrier densities. Although screening is not totally eliminated, it is greatly reduced, and this can lead to more effective single-particle scattering than expected.

On the other hand, as the carrier density is increased, the material becomes degenerate, and the Maxwellian approximation cannot be used. In the case of the Fermi-Dirac distribution above, the integrals are more complicated but still readily evaluated, as evident in Fig. 6.11. It is clear that the screening is strongly reduced at all wavevectors as the carrier density is increased, but the variation with the wavevector is also strongly reduced. The equivalent results for two dimensions have been evaluated by Ando *et al.* [33], and a figure equivalent to Fig. 6.11 appears in their review.

The self-energy

We would now like to calculate the self-energy for the single-particle electron-electron interaction to see how it yields the scattering dynamics. For this we will use the dynamic screening approximation (6.115) to the polarizability, and the effective interaction can then be written as

$$U_{eff} = \frac{e^2}{2\varepsilon q \left[1 - \frac{\chi_2 Dq}{\tau(\omega_n + 1/\tau)^2 - Dq^2} \right]} \tag{6.128}$$

for the two-dimensional case pursued here. For relatively low frequencies such as those involved in the single-particle scattering, this can be rearranged to yield (in lowest order)

$$U_{eff} = \frac{Dq^2 e^2}{2\varepsilon q [Dq^2 + \chi_2 Dq - \tau(\omega_n + 1/\tau)^2]}. \tag{6.129}$$

Since both the Green's function and the interaction strength are functions of frequency, and since these would be a product in real space, it is necessary to convolve them in frequency space, so the self-energy can be written as

$$\Sigma_{ee}^{r,a}(\mathbf{k}, i\omega_n) = -\sum_m \frac{1}{\beta\hbar^2} \int \frac{d^d\mathbf{q}}{(2\pi)^d} G^{r,a}(\mathbf{k} - \mathbf{q}, i\omega_n + i\omega_m)$$

$$\times \frac{Dq^2e^2}{2\varepsilon q[Dq^2 + \chi_2 Dq - \tau(\omega_m + 1/\tau)^2]}. \tag{6.130}$$

Fukuyama [34] argues that we are really interested in cases in which the momentum change is quite small, so that the energy E_k in the Green's function differs little from the Fermi energy. Fukuyama makes the additional observation that since we are interested in a small frequency change, the Green's functions can be approximated as

$$G^{r,a}(\mathbf{k} - \mathbf{q}, i\omega_n + i\omega_m) = \frac{1}{i\omega_n + i\omega_m - (E_{k-q} - \mu)/\hbar - i/2\tau} \sim i2\tau, \tag{6.131}$$

where τ is the broadening in the Green's function that arises from other scattering, such as from impurity scattering. With this approximation, the self-energy is momentum independent, since the last momentum variables will be integrated, as will the frequency dependence. However, this is not quite true if we take the zero of frequency at the Fermi energy for convenience; the range of the convolution integral is limited to the singular case for which one frequency is below the Fermi energy and the other is above the Fermi energy, or $\omega_n(\omega_n + \omega_m) < 0$. This limits the summation over frequency to those values for which $\omega_m < -\omega_n$. We can now rewrite the self-energy as

$$\Sigma_{ee}^{r,a}(\mathbf{k}, i\omega_n) = -\sum_{\omega_m < -\omega_n} \frac{2i\tau}{\beta\hbar^2} \int \frac{d^2\mathbf{q}}{(2\pi)^2} \frac{Dq^2e^2}{2\varepsilon q[Dq^2 + \chi_2 Dq - \tau(\omega_m + 1/\tau)^2]}. \tag{6.132}$$

It is clear that this formula has singularities. The self-energy is singular in two dimensions, in the limit of $q, \omega_n \to 0$, with a variation as $\ln(\omega_n) \sim \ln(T)$. Nevertheless, this integral can be rewritten as

$$\Sigma_{ee}^{r,a}(\mathbf{k}, i\omega_n) = -\sum_{\omega_m < -\omega_n} \frac{i\tau e^2}{2\varepsilon\pi\beta\hbar^2} \int_0^\infty \frac{q^2 dq}{[q^2 + \chi_2 q - \tau(\omega_m + 1/\tau)^2/D]}. \tag{6.133}$$

There is a general problem with which integration should be pursued first: that over q or that from the summation over ω_m. Either leads to the need to introduce a cutoff into the actual integration, either a cutoff on the largest value for q (which was discussed already in the leading sections of this chapter) or in the lower frequency limit. This is slightly easier in the case of the momentum, so we can rewrite the last equation in the leading terms as

$$\Sigma_{ee}^{r,a}(\mathbf{k}, i\omega_n) \sim -\sum_{\omega_m < -\omega_n} \frac{i\tau e^2}{2\varepsilon\pi\beta\hbar^2} \frac{\chi_2 q_{max}(q_{max} + \chi_2)v^2}{4(\omega_m + 1/\tau)^2}. \tag{6.134}$$

In this last form, a logarithmic term has been expanded under the assumption that the last fraction is small, and the diffuson constant has been expanded in terms of the velocity found earlier. The value of q_{max} is an upper cutoff on the momentum, and this was found earlier to be approximately $2k_F$. The summation over the frequencies can now be converted to an integral, and

$$\Sigma_{ee}^{r,a}(\mathbf{k}, i\omega_n) \simeq \frac{i\tau e^2}{2\varepsilon\pi\hbar} \frac{\chi_2 q_{max}(q_{max} + \chi_2)v^2}{4(\omega_n + 1/\tau)}. \tag{6.135}$$

The factor in the denominator is predominantly related to the Fermi energy, whereas the term in the velocity squared leads to a variation as T when the thermal averaging is carried out.

Chaplik [35] and Giuliani and Quinn [36] have carried out evaluations of the "lifetime" of an electron due to single-particle scattering from the Fermi sea. The latter is perhaps the most often cited for these calculations, and their result is

$$\frac{1}{\tau_{ee}} \sim \frac{E_F}{4\pi\hbar} \left(\frac{E_k - E_F}{E_F}\right)^2 \left[\ln\left(\frac{E_k - E_F}{E_F}\right) - \frac{1}{2} - \ln\left(\frac{2q_{FT}}{\hbar k_F}\right)\right]. \tag{6.136}$$

Here, the leading term can be rearranged as

$$\frac{E_k - E_F}{E_F} \approx \frac{p^2 - p_F^2}{p_F^2} = \frac{(p + p_F)(p - p_F)}{p_F^2} \sim \frac{2(p - p_F)}{p_F}. \tag{6.137}$$

Since the temperature in the strongly degenerate limit describes the fluctuations around the Fermi energy, the leading behavior in the electron-electron self-energy varies as $(p - p_F)^2 \sim mk_B T$, so the same linear behavior in temperature is found (Giuliani and Quinn, however, seem to have missed this point and claim a T^2 behavior), with the overall temperature behavior being $T \ln(T)$, a result found in some of the experiments cited at the start of this chapter. In Eq. (6.135), a similar behavior is found if we relate the velocity to the quasi-particle velocity [$v^2 \sim (p - p_F)^2/m^2$]: the cutoff to the Fermi momentum $q_{max}^2 \to E_F$, and $\hbar\omega_n \to E_F$, the same leading-order behavior is recovered. We will return to this discussion in the last section of this chapter in a more explicit discussion of the appropriate diagram terms, but this basic result (with some slight modification) will be recovered.

6.3.6 Electron-electron interactions in disordered systems – The self-energy

It is now clear that, as in the previous chapter, the nature of the temperature variations will be found to some extent in the self-energy, since it describes the important broadening of the single-particle density of states. This, in turn, relates to the mixing of phase-coherent states due to broadening of the Fermi distribution function. Thus, the task *in disordered systems* is to define the self-energy in the system in which both impurities and electron-electron interactions (and possibly others as well) exist. In treating the self-energy, it is necessary to decide how one wants to resum the various diagrams for the interactions. For example, in disordered systems it was assumed in the last chapter that the disorder-inducing impurity interactions are the dominant interactions, and the electron-electron interactions were ignored. In Chapter 2, however, where we dealt with screening of an interaction by the electrons, the electron-electron interaction was assumed to be the dominant interaction. In this section, we will continue to assume that the disorder-inducing impurity interactions are the dominant scattering process, and that the electron-electron interactions are a perturbation of this process. Now, clearly, the entire set of Feynman diagrams that relate to scattering by impurities and by the electron-electron Coulomb interaction can be separated into sets of terms which lead to a sum of two distinct self-energies. Thus, the impurity scattering processes are treated as previously, and we add a new self-energy term Σ_{ee}, which accounts for the interaction effects that are themselves mediated by the impurities. This was shown in Fig. 6.9, where the jagged line represents the electron-electron interaction and the dashed line represents the impurity (averaged) interaction. If these were the only diagrams in the perturbation expansion, life would indeed be quite easy. However, we must recall that, for the impurities alone, we wound up with a great many different types of diagrams. There was the simple summation of independent Coulomb scattering events which were impurity averaged and which led to the simple impurity self-energy

$$\Sigma_{imp}^{r,a}(i\omega_n) = \mp i \frac{\hbar}{2\tau(i\omega_n)}, \tag{6.138}$$

found in the last chapter in Eq. (5.77). In addition, however, the two-particle Green's function contribution to the polarization in the Kubo formula led to a set of ladder diagrams which led to the diffuson and a set of maximally crossed diagrams which led to the cooperon. These additional contributions were important for weak localization and for universal conductance fluctuations.

The problem with two perturbing species at hand is what to do with diagrams like Fig. 6.12, where the two perturbations interfere with one another. The question is really to which of the self-energies will we ascribe such terms (or should they be ignored), and then how they are to be included in the selected self-energy. The normal approach (normal in the sense of semiclassical transport with the Boltzmann equation) is to assume that the electron-electron interaction is the dominant scattering process. This usually results in some sort of assumption of a drifted Maxwellian (or Fermi-Dirac) distribution, with the interaction being subsequently ignored except for its role in screening other scattering processes. This screening arises just from the diagrams of the type in Fig. 6.12, although this is not often recognized. This leads to a philosophy of the following strategic approach for when the various perturbing interactions interfere with one another: (1) an assumption is made regarding the dominant perturbation interaction, and this self-energy is treated as if the other processes are absent; and (2) the other interactions are treated as if they are *screened* by the dominant perturbation through the introduction of *vertex corrections*. Normal screening by the electrons is such a vertex correction.

In disordered materials, where the disorder is induced by the heavy impurity scattering, the impurity scattering is the dominant perturbation. Thus, we will subsequently screen the electron-electron interaction by the impurity interaction, with the latter leading to a vertex correction. That is, we calculate the interaction effects not between free electrons (which would be simple plane-wave states) but between the diffusive electrons. This is handled by the means of a vertex correction for the Coulomb interaction between the diffusing electrons. The impurity self-energy is still assumed to be given by Eq. (5.77), but with the temperature dependence included. This temperature dependence does not occur within the self-energy but arises from the energy (frequency) dependence of $\tau(\omega)$, which is coupled to an energy dependence when the integration over frequency is performed in the

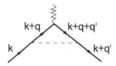

Fig. 6.12. Diagram with both impurity and electron-electron interaction.

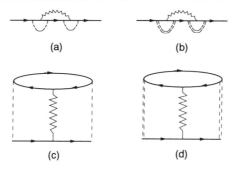

Fig. 6.13. The first-order Exchange interaction, (a) and (b), and Hartree interaction, (c) and (d), dressed by diffusons in (a) and (c), and by cooperons in (b) and (d).

calculation of the conductivity (and therefore of an average relaxation time). Thus, the sign on Eq. (6.138) is such that we no longer have to worry about the small quantity η (as we did in the last chapter). Here we are interested in investigating the self-energy term arising from the interaction between diffusing electrons in a sea of impurity scattering. We will calculate the phase-breaking time for this approach in the next section.

It was stated above that the electron-electron interaction would be subject to a vertex correction. Let us examine just what this means. Consider Fig. 6.13. In the figure we draw four possible electron-electron interactions within the sea of impurity scattering. Diagrams (a) and (b) refer to the exchange interaction, and diagrams (c) and (d) refer to the Hartree correction. Both pairs are first-order corrections to the energy of the "free" particles. In the diagrams, however, impurity interactions have been drawn around the points (the *vertices*) at which the interaction line connects to the Green's function lines. In diagrams (a) and (c), the single dashed line represents the diffuson interaction, whereas in diagrams (b) and (d) the double-dashed line refers to the cooperon contribution. But the diffuson and the cooperon are two-particle Green's function interactions. How do we describe these *dressings* of the vertex that are produced by the two-particle Green's function ladders (note that they actually connect with four Green's functions) that were treated in the last chapter? Consider the two-particle Green's function in Fig. 5.16 and the notation of momentum in Fig. 6.12. The two momenta $\mathbf{k}$ and $\mathbf{k}'$ on the right of (the left side of) Fig. 5.16 will become the momenta $\mathbf{k} + \mathbf{q}$ and $\mathbf{k} + \mathbf{q} + \mathbf{q}'$, respectively, in Fig. 6.12, where $\mathbf{q}'$ is the momentum transferred via the electron-electron interaction. Thus, the two Green's function lines on the right of the two-particle structure in Fig. 5.16 are pulled together to meet at the vertex of the electron-electron interaction. Now, $\mathbf{q}$ is the total momentum transferred to the impurities in the ladder diagram summation, so $\mathbf{k}''$ becomes the momentum $\mathbf{k}$ in Fig. 6.12. Similarly, the momentum $\mathbf{k}'''$ in Fig. 5.16 is associated with $\mathbf{k} + \mathbf{q}'$ in Fig. 6.12. Thus, the single impurity line in Fig. 6.12 becomes the full two-particle Green's function represented by the ladder diagrams that lead to the diffuson. Similarly, the set of impurity lines that cross one another from one side of Fig. 6.12 to the other lead to the set of maximally crossed diagrams, giving rise to the two-particle Green's function representation of the cooperon. The vertex corrections then are the resummations of the ladder diagrams and/or maximally crossed diagrams that represent the two-particle Green's function dressing of the vertex of the electron-electron interaction.

As a consequence, the vertex corrections lead to a Coulomb interaction not between two free electrons, but between two dressed particles described either by the diffuson or by the cooperon. These interactions are appropriate only for diffusive transport when the energy exchange between the carriers is quite small; that is, the diffusive interactions with the impurities involves very small energy exchange (it was assumed to be zero in the last chapter). The cooperon contribution to the vertex correction is given by the ladder summation of the (reversed) maximally crossed interactions $\Lambda = (1 - \Pi V_0)^{-1}$ of Eq. (5.102), or by reinserting the new frequency,

$$\Lambda_C(q, \omega_n) = \frac{1}{2\pi\hbar\rho(\mu)\tau^2} \frac{1}{Dq^2 + |\omega_n| + 1/\tau_\varphi}. \tag{6.139}$$

On the other hand, the diffuson contribution to the vertex correction is given by the resummation of the ladder diagrams themselves,

$$\Lambda_D(q, \omega_n) = \frac{1}{2\pi\hbar\rho(\mu)\tau^2} \frac{1}{Dq^2 + |\omega_n|}. \tag{6.140}$$

There is an important difference in these two equations, over and above the elimination of the phase-breaking time in the latter expression. For the diffuson, $q = k - k'$, where the latter two momenta are those of the Green's function ladder that gives rise to the polarization Π in the diffuson, while $q = k + k'$ in the cooperon. This difference is related to the conservation laws on the total number of particles. The diffusons represent the modification of the matrix elements by the dressed diffusing particles, while the cooperon relates directly to the phase interference properties of these diffusing electrons [34].

Altshuler and Aronov first studied the role of the interaction effects in a three-dimensional disordered material [23]. Later, Altshuler *et al.* extended this treatment to the two-dimensional situation [37]. In the first case, the authors considered only the diagram Fig. 6.13a, while in the latter work, they included this diagram plus the equivalent Hartree term from Fig. 6.13c. In these two cases, they found that the effect of these two terms led to a $\ln(T)$ dependence in the density of states, which is a noticeable renormalization of the energy structure around the Fermi energy at low temperatures. Fukuyama [38] later treated all four diagrams of Fig. 6.13 and found that they all contribute comparably in the lowest order of the interaction. Here, we want to illustrate the results but will treat only the first process, given by Fig. 6.13a. This diagram is redrawn in Fig. 6.14, where the curly line is the full electron-electron interaction, and the shaded triangles are the vertex correction, represented by the diffuson propagator.

By looking at the diagram in Fig. 6.14, it is clear that the self-energy for this term involves the product of one Green's function line with momentum $k - q$, and an interaction line at wavevector q. In the simplest case, this would lead to the product of just the interaction potential and the Green's function, integrated over the scattering wavevector q. However, as we have discussed, the simple interaction is modified by the vertex correction of the strong background of impurity scattering. In essence, we seek a correction that is one end of a ladder diagram in which there are two Green's functions at momentum k and $k - q$, which meet with the interaction line at momentum q. In the presence of strong impurity

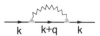

$$k \qquad k+q \qquad k$$

Fig. 6.14. The lowest-order Hartree diagram for the electron-electron self-energy.

scattering, one can consider that there may be an intermediate state connecting these lines, which represents a set of impurity scattering events organized so that the end momenta stay the same. This is just the set of ladder diagrams that we described above. Here, these modify the bare interaction potential in a way to allow for the weak interaction scattering in the "sea of impurity scattering." The ladder we need involves impurity-averaged scattering events between two Green's functions which always differ by the momentum **q**. This is just the ladder sum we used in the diffuson correction for weak localization. In short, we need the sum that appears to modify the bare impurity interaction in Eq. (6.140). Now, in this particular case, there are vertex corrections at each end of the interaction line, so the net interaction line carries a modification of the strength by $[\Lambda_D(q, \omega_n)/\Gamma_0]^2 = [2\pi\hbar\rho(\mu)\tau\Lambda_D(q, \omega_n)]^2$ (since we want only the summation and not the leading interaction term in the ladder diagram of the last chapter, which really introduces the number of impurities to the summation). The net interaction is now

$$\frac{U(\mathbf{q}, i\omega_n)}{[Dq^2 + |\omega_n|]^2\tau^2}. \tag{6.141}$$

Since both the Green's function and the interaction strength are functions of frequency, and since these would be a product in real space, it is necessary to convolve them in frequency space, and the self-energy can be written (just as in the last section) as

$$\Sigma_{ee}^{r,a}(\mathbf{k}, i\omega_n) = -\sum_m \frac{1}{\beta\hbar^2\tau^2} \int \frac{d^d\mathbf{q}}{(2\pi)^d} G^{r,a}(\mathbf{k} - \mathbf{q}, i\omega_n + i\omega_m) \frac{U(\mathbf{q}, i\omega_m)}{[Dq^2 + |\omega_m|]^2}. \tag{6.142}$$

Fukuyama [34] argues that we are really interested in cases in which the momentum change is quite small, so that the energy E_k differs little from the Fermi energy. Since we are interested in a small frequency change, Fukayama also observes that the Green's functions can be approximated as

$$G^{r,a}(\mathbf{k} - \mathbf{q}, i\omega_n + i\omega_m) = \frac{1}{i\omega_n + i\omega_m - (E_{k-q} - \mu)/\hbar - i/2\tau} \sim i2\tau. \tag{6.143}$$

With this approximation, the self-energy is momentum-independent, since the last momentum variables will be integrated, as will the frequency dependence. However, this is not quite true if we take the zero of frequency at the Fermi energy for convenience; the range of the convolution integral is limited to the singular case for which one frequency is below the Fermi energy and the other is above the Fermi energy, or $\omega_n(\omega_n + \omega_m) < 0$. This limits the summation over frequency to those values for which $\omega_m < -\omega_n$. We can now rewrite the self-energy as

$$\Sigma_{ee}^{r,a}(\mathbf{k}, i\omega_n) = -\sum_{\omega_m < -\omega_n} \frac{2i}{\beta\hbar^2\tau} \int \frac{d^d\mathbf{q}}{(2\pi)^d} \frac{U(\mathbf{q}, i\omega_m)}{[Dq^2 + |\omega_m|]^2}. \tag{6.144}$$

It is clear that this formula has singularities. Even if we ignore the variables in the interaction potential and treat the potential as constant, the self-energy is singular in two dimensions with a variation as $\ln(\omega_n) \sim \ln(T)$. The other diagrams in Fig. 6.13 have similar results and have to be taken into account on an equal footing.

If we take the interaction as constant within the germane region of momentum and frequency, $U(\mathbf{q}, i\omega_m) = U_0$, then the momentum integration can be carried out quickly. In two dimensions,

$$\Sigma_{ee}^{r,a}(\mathbf{k}, i\omega_n) = -\sum_{\omega_m < -\omega_n} \frac{iU_0}{\beta\hbar^2 D\tau} \frac{1}{2\pi} \int_0^\infty \frac{Dd(q^2)}{[Dq^2 + |\omega_m|]^2}$$

$$= -\frac{iU_0}{2\pi\beta\hbar^2 D\tau} \sum_{\omega_m < -\omega_n} \frac{1}{|\omega_m|}. \tag{6.145}$$

Thus, if we add this self-energy to the self-energy due to impurity scattering, or

$$\Sigma^{r,a}(\mathbf{k}, i\omega_n) = \Sigma_{imp}^{r,a}(\mathbf{k}, i\omega_n) + \Sigma_{ee}^{r,a}(\mathbf{k}, i\omega_n), \tag{6.146}$$

we find that the total lifetime is modified to be

$$\frac{1}{\tau} \rightarrow \frac{1}{\tau}[1 + \lambda g G_2(\omega_n, T)], \tag{6.147}$$

where $\lambda = \hbar/2\pi E_F\tau = \hbar/2\pi m D$ is a small dimensionless coupling constant, $g = U_0\rho_d(\mu)/\hbar = mU_0/\pi\hbar^3$ represents the strength of the interaction, and [34]

$$G_2(\omega_n, T) = \frac{2\pi}{\beta\hbar} \sum_{\omega_m < -\omega_n} \frac{1}{|\omega_m|} = \ln\left(\frac{\hbar\beta}{2\pi\tau}\right) - \psi\left(\frac{\hbar\beta\omega_n + 1}{2}\right), \tag{6.148}$$

where the latter function is the digamma function. In fact, each of the diagrams in Fig 6.13 contributes, and $g = g_a + g_b - 2(g_c + g_d)$, where the subscript refers to the particular diagram of the latter figure. The density of states was given above in Eq. (6.85). The imaginary part of the Green's function involves the scattering time in the numerator. The presence of the additional scattering (indeed, of any scattering at all) reduces this lifetime and reduces the density of states at the Fermi energy. Normally, one is familiar with broadening of the density of states at a subband edge due to the scattering. However, at the Fermi energy at low temperature, the strong interaction among the dense number of electrons excited just above the Fermi energy creates an interaction that lowers the density of states at this point. This term has a logarithmic divergence with reduction of the temperature, as given by the first term of the last equality in (6.148). The sign difference in the overall quantity g, which measures the strength of the interaction, arises from the fact that the Hartree and exchange terms have different signs, and the factor of two accounts for the spin summation in the Hartree terms.

6.4 Conductivity

At this point, it is useful to recompute the temperature dependence of the conductivity that arises merely from impurity scattering, without the complications of the electron-electron interaction. There is not much change in the self-energy given by Eq. (5.77), so the diagonal conductivity is still given by (5.78) to be (with the temperature Green's functions inserted in place of the zero-temperature functions)

$$\sigma = \frac{e^2\hbar^2}{m^2} \int \frac{d^d\mathbf{k}}{(2\pi)^d} \frac{k^2}{\beta\hbar^2}$$

$$\times \sum_{\omega_n} \frac{1}{i\omega_n - (E_k - \mu)/\hbar - i/2\tau} \frac{1}{i\omega_n - (E_k - \mu)/\hbar + i/2\tau}. \tag{6.149}$$

There are two aspects to this equation that lead to differences from (5.78). The first is that the frequency summation given here is actually the resolution of the delta function that appears in (5.78). The second is that the frequency sums are more difficult than those encountered in the previous section. This is because of the delta function incorporated into (5.77). This delta function is broadened at finite temperature, and some effect must occur to lead to this behavior. The actual complication arises from the presence of the self-energy terms (the scattering terms) in the Green's functions of Eq. (6.149). The presence of these energy-dependent self-energy terms means that the complex integration used in (6.96) becomes more complicated in that there may be branch cuts that arise from the presence of the self-energy. Consequently, the best method of attacking the problem is to go back to the actual polarization bubble itself and use the limiting process

$$\sigma = - \lim_{\omega \to 0} \left[\frac{Im(\Pi_r)}{\omega} \right]. \tag{6.150}$$

In this sense, we can write the lowest-order bubble that contributes to the conductivity as

$$\Pi(i\omega'_m) = \frac{2e^2\hbar^2}{m^2} \int \frac{d^d\mathbf{k}}{(2\pi)^d} \frac{k^2}{\beta\hbar^2} \sum_{\omega_n} G(\mathbf{k}, i\omega_n) G(\mathbf{k}, i\omega_n + i\omega'_m), \tag{6.151}$$

where we use the general Green's functions to simplify the notation, and in which the spin summation has been carried out. The procedure to be followed is essentially the same as that of (6.96) in that we define a contour integral

$$\frac{\beta\hbar}{2\pi i} \int_C \frac{dz}{e^{\beta\hbar z} + 1} g(z), \tag{6.152}$$

where $g(z)$ is "inspired" by the summation in (6.151). For simplicity, we will also use the reduced units $\xi = (E_k - \mu)/\hbar$. In this integration, however, the contributions from the simple pole of $g(z)$ that occurred in (6.96) must be replaced by the contributions from the two branch cuts at $z = \xi$ and $z = \xi - i\omega'_m$. The contour remains the basic circle of radius R in which the limit $R \to \infty$ is taken. This contour must be deformed, however, to create two line integrals, one above and one below each branch cut. Let us consider first the contribution from the branch cut at $z = \xi$. The contribution of this quantity to (6.152) is given by

$$\frac{\beta\hbar}{2\pi i} \left\{ \int_{-\infty}^{\infty} d\xi G(\mathbf{k}, \xi + i\delta) G(\mathbf{k}, \xi + i\omega'_m) \int_{\infty}^{-\infty} d\xi G(\mathbf{k}, \xi - i\delta) G(\mathbf{k}, \xi + i\omega'_m) \right\}, \tag{6.153}$$

where the order of the limits on the integrals is determined by the directions of the integrations. (The paths are both in the positive (counterclockwise) direction in closing the contour, which means that the path above the branch cut is in the direction of increasing energy, whereas the path below the branch cut is in the direction of decreasing energy.) Reversing the direction of integration in the second term allows us to rewrite this contribution as

$$\frac{\beta\hbar}{2\pi i} \int_{-\infty}^{\infty} d\xi [G(\mathbf{k}, \xi + i\delta) - G(\mathbf{k}, \xi - i\delta)] G(\mathbf{k}, \xi + i\omega'_m). \tag{6.154}$$

The first term in the square brackets is the retarded Green's function, and the second term is the advanced Green's function. The bracketed term contributes the spectral density. Contribution of the second branch cut, at $z = \xi - i\omega'_m$, can similarly be calculated (in this case, the second Green's function is expanded on either side of the branch cut). Thus, we can write the summation from (6.151) as

$$S = \sum_{\omega_n} G(\mathbf{k}, i\omega_n)G(\mathbf{k}, i\omega_n + i\omega'_m)$$

$$= -\frac{\beta\hbar}{2\pi} \int_{-\infty}^{\infty} d\xi\, A(\mathbf{k}, \xi)\big[G(\mathbf{k}, \xi + i\omega'_m) + G(\mathbf{k}, \xi - i\omega'_m)\big]f(\xi). \qquad (6.155)$$

To proceed further, it is time to take the analytic continuation of the last equation. This makes the change $i\omega'_m \to \omega + i\eta$, which again converts the two Green's functions in the square brackets into the retarded and advanced functions, respectively. Then, the imaginary parts of these two Green's functions (everything else is real) contribute just another factor of the spectral density. We make a change of variables in the second term, by shifting the axis of the ξ integration, and

$$Im(S) = -Im\left\{\frac{\beta\hbar}{2\pi} \int_{-\infty}^{\infty} d\xi\, A(\mathbf{k}, \xi)[G(\mathbf{k}, \xi + \omega + i\eta) + G(\mathbf{k}, \xi - \omega - i\eta)]f(\xi)\right\}$$

$$= \left\{\frac{\beta\hbar}{4\pi} \int_{-\infty}^{\infty} d\xi\, A(\mathbf{k}, \xi)[A(\mathbf{k}, \xi + \omega) - A(\mathbf{k}, \xi - \omega)]f(\xi)\right\}$$

$$= \left\{\frac{\beta\hbar}{4\pi} \int_{-\infty}^{\infty} d\xi\, A(\mathbf{k}, \xi)A(\mathbf{k}, \xi + \omega)[f(\xi) - f(\xi + \omega)]\right\}$$

$$\simeq -\frac{\beta\hbar\omega}{4\pi} \int_{-\infty}^{\infty} d\xi\, A(\mathbf{k}, \xi)A(\mathbf{k}, \xi + \omega)\frac{\partial f}{\partial \xi}. \qquad (6.156)$$

Using this in Eqs. (6.150) and (6.151) leads to the conductivity

$$\sigma = -\frac{e^2\hbar^2}{m^2} \int \frac{d^d\mathbf{k}}{(2\pi)^d} \frac{k^2}{2\pi\hbar} \int_{-\infty}^{\infty} d\xi\, A^2(\mathbf{k}, \xi)\frac{\partial f}{\partial \xi}. \qquad (6.157)$$

At low temperatures, the spectral density contributes to the delta function that occurred in (5.77). From (6.80), we know that the integral over the spectral density, with respect to energy, is unity (i.e., the density of states in momentum space is uniform). This follows from

$$\int_{-\infty}^{\infty} dE\, A(\mathbf{k}, E) = \frac{1}{\pi} \int_{-\infty}^{\infty} dE\, \frac{\Sigma_i}{E^2 + \Sigma_i^2} = 1. \qquad (6.158)$$

In a similar fashion,

$$\frac{1}{\pi} \int_{-\infty}^{\infty} dE\, A^2(\mathbf{k}, E) = \frac{1}{\pi} \int_{-\infty}^{\infty} dE\, \left(\frac{\Sigma_i}{E^2 + \Sigma_i^2}\right)^2 = \frac{1}{\Sigma_i} = \frac{2\tau}{\hbar}. \qquad (6.159)$$

To use this result, we assume that the scattering is weak so that the spectral density is sharply peaked, in which case we can bring the derivative of the distribution function out of the integral. Then, the conductivity can be written as

$$\sigma = -\frac{e^2\hbar^2}{m^2} \int \frac{d^d\mathbf{k}}{(2\pi)^d} k^2 \tau(E_k)\frac{\partial f(E_k)}{\partial E_k}$$

$$= -\frac{2e^2}{dm^2} \int dE_k \rho(E_k)\tau(E_k)E_k\frac{\partial f(E_k)}{\partial E_k}. \qquad (6.160)$$

This last result is essentially the classical result obtained with the Boltzmann equation in Chapter 2, which is reassuring, but it depends on a special interpretation of the result of using the frequency summations. We note that the angular integration factor normally found for impurity scattering, the factor $(1 - \cos\theta)$ which was shown in the last chapter to require the next higher-order terms, is missing here, since it must usually be put into the Boltzmann equation by hand. We can put the present result into the more usual form (see Chapter 2)

$$\sigma = \frac{ne^2\langle\tau\rangle}{m}, \quad \langle\tau\rangle = -\frac{2}{dn}\int \rho(E_k)\tau E_k\left(\frac{\partial}{\partial E_k}f_k\right)dE_k, \tag{6.161}$$

and we have used the fact that

$$n = \int \rho(E_k)n_k\, dE_k = \frac{2}{d}\int \frac{\partial(\rho(E_k)E_k)}{\partial E_k}f_k\, dE_k$$

$$= -\frac{2}{d}\int \rho(E_k)E_k\left(\frac{\partial}{\partial E_k}f_k\right)dE_k. \tag{6.162}$$

As previously, the first set of ladder diagrams results in the addition of the $(1 - \cos\theta)$ term and the retention of the angular averages that have been already done in (6.151). From the previous subsection it is also clear that the first correction that will arise from the interacting system is to modify the scattering time τ according to (6.147). Our main concerns are to the higher-order corrections represented by the diffuson and cooperon corrections.

It is important to note at this point that the main result of the integration contained in (6.161) is to take the effective zero-temperature conductivity and average it over an energy width of the order of $k_B T$ at the Fermi energy. This latter is given by the width of the derivative of the Fermi function. In the low-temperature limit, this derivative approaches a delta function at the Fermi energy, and this leads to the low-temperature conductivity. However, this simple averaging result confirms the approach that was utilized in the first sections of this chapter to compute the temperature dependencies discussed there.

The lowest-order diagram for the higher-order corrections to the conductivity is shown in Fig. 5.19, which is simply a bubble containing a major interaction to create the two-electron Green's function. The expansion of the terms that are to be considered here is shown in Fig. 6.15. Other diagrams that differ only in the directions of the electron lines are also important and should be considered, but it is the generic topology that is important [39]. When calculating the correction to the conductivity in this mixed interacting-impurity disordered system, it should be recalled that the dominant contributions will arise from the diagrams

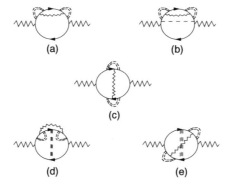

Fig. 6.15. The terms arising from the exchange interaction contribution to the conductivity.

containing the maximum number of diffuson poles per segment. For this to happen, the imaginary parts of the Green's functions meeting at each vertex should have opposite signs, which implies that the Matsubara frequencies also should have opposite signs. This places a constraint on the ranges of the summations over the frequencies in each expression. By these arguments, the first three of these are quite small, since they are nested interaction and diffuson lines which act more or less independently of one another. The most important diagrams are Fig. 6.15d,e, in which the two effects are nested and contain an additional pole besides the vertex corrections of the diffusons. Since the vertices have a vector nature, these latter two diagrams will vanish if the Green's functions to the right and left are taken in the small $\mathbf{q}$ limit. When the next-higher expansion terms are included for these Green's functions, an additional Dq^2 arises from the $\mathbf{q} \cdot \mathbf{v}$ terms, which actually cancels one of the diffuson poles, leaving only two such poles. Thus, only these two diagrams contribute to the corrections to the conductivity. These diagrams are those that arise from the lowest-order Exchange interaction, incorporating a diffuson. There are comparable terms for the cooperon and for the Hartree terms. The method will be illustrated with the diagram of Fig. 6.15d, for which the conductivity can be written as

$$
\delta\sigma = \frac{e^2\hbar^2}{2\pi\hbar\rho(\mu)m^2} \frac{1}{\beta\hbar^2} \sum_{\omega_n} \int \frac{d^d\mathbf{k}}{(2\pi)^d} G_r(\mathbf{k}, i\omega_n) G_a(\mathbf{k}, i\omega_n) \frac{1}{\beta\hbar^2}
$$

$$
\times \sum_{\omega_m} \int \frac{d^d\mathbf{q}}{(2\pi)^d} G_r(\mathbf{k}+\mathbf{q}, i\omega_n + i\omega_m) \frac{U(\mathbf{q}, i\omega_n)}{[Dq^2 + |\omega_n|]^3 \tau^3}
$$

$$
\times \int \frac{d^d\mathbf{k}'}{(2\pi)^d} \frac{kk'\cos\theta}{d} G_r(\mathbf{k}'+\mathbf{q}, i\omega_n + i\omega_m) G_a(\mathbf{k}', i\omega_n) G_r(\mathbf{k}', i\omega_n). \quad (6.163)
$$

The first two Green's functions arise from the two terms connected to the input current tail, and the last two Green's functions are those on the right side of the diagram and are connected to the output current tail. The other two Green's functions are in the top leg on either side of the diffuson connection spanning the top and bottom legs of the diagram. In the potential term at the center of the expression, two of the diffuson lines arise from the vertex correction, and the third is that spanning the bubble diagram. At this point, we begin to work with only the two-dimensional situation. The integration can be simplified by considering the terms involved in the first momentum summation, as [40]

$$
\mathbf{M}(\mathbf{q}) = \frac{\hbar}{m} \int \frac{d^2\mathbf{k}}{(2\pi)^2} kG_r(\mathbf{k}, i\omega_n) G_a(\mathbf{k}, i\omega_n) G_r(\mathbf{k}+\mathbf{q}, i\omega_n + i\omega_m)
$$

$$
\simeq \frac{\hbar}{m} \int \frac{d^2\mathbf{k}}{(2\pi)^2} kG_r(\mathbf{k}, i\omega_n) G_a(\mathbf{k}, i\omega_n) G_r^2(\mathbf{k}, i\omega_n + i\omega_m)\mathbf{v} \cdot \mathbf{q}
$$

$$
\simeq \frac{2\tau^3\mu}{\hbar} \mathbf{q}, \quad (6.164)
$$

where the leading term in the Taylor expansion for $G_r(\mathbf{k}+\mathbf{q}, i\omega_n + i\omega_m)$ has been ignored on the basis of the above arguments on size of the various terms (and the basic integral vanishes if the Green's functions are symmetric in the vector $\mathbf{k}$). The result (6.164) is obtained by converting the integration over momentum into an integration over energy. Then, it is noted that the electron-electron interaction basically requires an electron on one side of the Fermi energy to scatter from an electron on the other side of the Fermi energy.

Thus, as discussed above, if $\omega_n > 0$, we require $\omega_n + \omega_m < 0$. This means that only two of the Green's functions lead to poles inside the contour; it is further assumed that the Fermi energy μ is the dominant energy in the resulting approximations. The integrands for Fig. 6.15d,e are the same, except that the integrand for (d) is proportional to $\mathbf{M}(\mathbf{q}) \cdot \mathbf{M}(\mathbf{q})$ and that for (e) is proportional to $\mathbf{M}(\mathbf{q}) \cdot \mathbf{M}(-\mathbf{q}) = -\mathbf{M}(\mathbf{q}) \cdot \mathbf{M}(\mathbf{q})$, so that these two diagrams are of opposite sign. Care must be taken in writing down the actual integration. Here we follow the arguments of Altshuler *et al.* [40]. When $\omega_m < 0$, the contributions of these two diagrams cover the same region of integration, and the net result vanishes. On the other hand, when $\omega_m > 0$, the conditions for Fig. 6.15d,e are such that they add so long as the summation over ω_n is evaluated at ω_m. Then, with a factor of 2 for the two terms that contribute for positive frequency (they cancel for negative frequency), the conductivity can be written as

$$\sigma = \frac{2e^2\tau^2}{\hbar^2} \frac{1}{\beta\hbar^2} \sum_{\omega_n > 0} \int \frac{d^2\mathbf{q}}{(2\pi)^2} \frac{U(\mathbf{q}, i\omega_n)Dq^2}{[Dq^2 + \omega_n]^3}. \tag{6.165}$$

At this point, the fully screened Coulomb interaction is inserted from the general result of Eq. (6.117). This may be rewritten as

$$U_{eff} = \frac{e^2}{2\varepsilon q\left[1 + \frac{\chi_2 Dq}{(\omega_n) + Dq^2}\right]} = \frac{e^2}{2\varepsilon q} \frac{(\omega_n) + Dq^2}{(\omega_n) + Dq^2 + \chi_2 Dq}$$

$$\simeq \frac{e^2}{2\varepsilon q} \frac{(\omega_n) + Dq^2}{\chi_2 Dq}, \tag{6.166}$$

so that we can now write

$$\sigma = \frac{2e^2\tau^2}{\hbar^2} \frac{e^2}{2\varepsilon\chi_2} \frac{1}{\beta\hbar^2} \sum_{\omega_n > 0} \int \frac{d^d\mathbf{q}}{(2\pi)^d} \frac{1}{[Dq^2 + \omega_n]^2}. \tag{6.167}$$

There is another set of diagrams similar to Fig. 6.15d,e, with the interaction line in the lower Green's function. This changes the limits of integration, but they are symmetric so we can add the results. We now note that since the expressions are valid only for $\omega\tau < 1$, we need to cut off the integration in frequency, and

$$\sigma = \frac{e^2}{\pi\hbar} \frac{e^2}{2\varepsilon\chi_2} \frac{\tau^2}{\beta\hbar^3} \sum_{\omega_m > 0} \int^{1/\tau} \frac{q\,dq}{[Dq^2 + \omega_n]^2}. \tag{6.168}$$

These can now be integrated to yield

$$\sigma \simeq \frac{e^4\tau^2}{4\pi\varepsilon\chi_2\hbar^2 D} \ln\left(\frac{\tau}{\beta\hbar}\right). \tag{6.169}$$

There are other contributions to the conductivity from other diagrams, such as the Hartree terms. Generally, these are of the same order but do not change the basic temperature dependence by very much. The point here is to demonstrate how this temperature dependence can be obtained. Our major interest is how the interaction affects the fluctuations, and we turn to this in the next section by determining a form for the phase-breaking time. It is important to note that the result (6.169) is one form that has been obtained. Many different forms have been obtained, but the consistent equivalence lies in the logarithmic dependence of the temperature at low temperatures (for which this derivation is valid). At the lowest of temperatures, the additional terms here become important and can be quite important in evaluating the overall conductivity.

6.5 The phase-breaking time

There has always been some concern as to whether the electron lifetime, found above as the scattering rate for electron-electron interactions, was actually the appropriate quantity to use to estimate the phase-breaking time in mesoscopic systems. As a consequence, there have been other approaches to compute this quantity, and we review some of those approaches here.

In the discussion above, the self-energy for the interacting system was computed in the presence of disorder. This resulted in an effecive electron-electron scattering rate given by the combination of Eq. (6.145) and (6.148) as

$$\frac{1}{\tau_{ee}} \sim \frac{U_0}{2\pi^2 \hbar^2 D\tau} \ln\left(\frac{\hbar\beta}{2\pi\tau}\right), \tag{6.170}$$

where the digamma function term has been ignored as small compared to the logarithmic term, and τ is given by the impurity scattering. While this effectively defines the rate at which carriers are scattered by other carriers, it remains different in form from the experimental phase-breaking times (assumed to be due to intercarrier scattering) given in the early sections of this chapter, and from other calculations of the electron-electron scattering rate, such as

$$\Sigma_{ee}^{r,a}(\mathbf{k}, i\omega_n) \simeq \frac{i\tau e^2}{2\varepsilon\pi\hbar} \frac{\chi_2 q_{max}(q_{max} + \chi_2)v^2}{4(\omega_n + 1/\tau)}. \tag{6.171}$$

Now, it has to be pointed out that (6.170) is valid at very low temperatures, but (6.171) is valid over a much larger range of temperatures. Indeed, the latter tends to give the $T\ln(T)$ behavior found experimentally for the phase-breaking rate in recent studies of ballistic quantum dots. Nevertheless, the first of these two equations has the same basic behavior as Eq. (6.6), in that there is a logarithmic dependence on the temperature incorporated in the scattering time. We note that this form, which is from (6.145), will have the prefactor modified if the proper screened Coulomb interaction is averaged within the integrals that make up the computation of the self-energy. It should be noted, as mentioned, that this result does not give the proper phase-breaking time in any dimension $d < 3$. This is because it emphasizes the larger energy-exchange processes that can arise in the electron-electron interaction. In one and two dimensions, the important contribution to the lifetime comes from small energy-exchange collisions, so there is a significant difference between the phase-breaking time τ_φ and the lifetime defined in (6.170) [22]. For this reason, Eq. (6.171) is expected to be more appropriate. In any case, in this section we want to compute other forms for the phase-breaking time that arise from the interactions among the electrons due to their mutual Coulomb potential.

We recall from the discussion of Sec. 6.1 that there are two possible (suggested) theories leading to phase breaking from the interacting electrons. One of these assumes that the carriers interact directly and that this Coulomb interaction leads to a phase-breaking process, such as treated above. The other is a more indirect process involving thermal fluctuations of the electromagnetic environment. Both of these processes are found to have regions of validity if we believe the data of Fig. 6.8. In this section, these two approaches will be discussed in detail. We turn first to the more exotic approach, the latter one.

6.5.1 Interactions coupled to background fields

Altshuler *et al.* [41] have argued that the interaction among the electrons that arises from their Coulomb repulsion creates a self-consistent field that is equivalent to that of the thermal fluctuations of electromagnetic waves. The proper definition of the phase-breaking time is the equivalent time in which correlations among the electrons are broken up. The logical place to examine this is the diffuson equation for the cooperon, since the latter is most sensitive to the phase coherence of the diffusing electrons (at least in the disordered regime in which impurity scattering dominates all transport). Thus, the phase-breaking time τ_φ provides a real cutoff of the interference effect that leads to weak localization. Here we present the argument of these authors for the phase-breaking time.

The quantum correction to the conductivity that leads to weak localization is related to phase coherence between electronic states of wavevectors $\mathbf{k}$ and $-\mathbf{k}$, which are diffusing around a ring of impurity-scattering centers. This is characterized by the cooperon, which represents the particle-particle correlator. In essence, the conductivity is related to the correlation function through Eq. (5.1). The correlation function for the cooperon (and the diffuson for that matter) can be written in the form used in Eq. (5.11) (ignoring for the moment the vector potential) as

$$\left[\frac{\partial}{\partial \zeta} - D\nabla^2 + \frac{1}{\tau_\varphi^*}\right] C(\zeta, \zeta', \mathbf{r}, \mathbf{r}') = \frac{1}{\tau}\delta(\zeta - \zeta')\delta(\mathbf{r} - \mathbf{r}'), \tag{6.172}$$

where $\zeta = t - t'$ represents the difference time between the two bubbles. Now, as remarked in Chapter 5, the cooperon is sensitive to the *sum* of the magnetic fields at the two times, rather than the difference (as is the diffuson). Thus, in the presence of the magnetic field (from whatever external or internal self-consistent field that leads to a vector potential), this can be rewritten in the form

$$\left\{\frac{\partial}{\partial \zeta} - D\left[-i\nabla + \frac{e}{\hbar}\mathbf{A}(\mathbf{r}, t'' + \zeta/2) + \frac{e}{\hbar}\mathbf{A}(\mathbf{r}', t'' - \zeta/2)\right]^2\right.$$
$$\left. + \frac{1}{\tau_\varphi^*}\right\} C(\zeta, \zeta', \mathbf{r}, \mathbf{r}') = \frac{1}{\tau}\delta(\zeta - \zeta')\delta(\mathbf{r} - \mathbf{r}'), \tag{6.173}$$

where $t'' = (t + t')/2$ is the average time, since the two vector potential terms arise from the two different times in the problem. In these two equations, τ_φ^* is the phase-breaking time that arises from any other processes (such as phonons), but not from the small energy-exchange electron-electron interaction that we want to examine here. For example, it can arise from the interaction with phonons or from the large energy-exchange processes given by the lifetime for the electron-electron interaction obtained above.

At this point, Altshuler *et al.* [41] express the solution to (6.173) as a Feynman path integral and then average over the electromagnetic fluctuations. The term that arises in the action (the term in the argument of the exponential in the path integral) is the cross product of the momentum operator and the vector potential. This is equivalent to integrating out the time dependence in (6.173) and treating the cross term as a perturbation, which in second order leads to an addition term in the spatial differential equation as

$$\left(\frac{em}{\hbar}\right)^2 \int_{-\zeta}^{\zeta} dt \int_{-\zeta}^{\zeta} dt' v_\alpha(t) v_\beta(t') \langle A_\alpha(\mathbf{r}(t)) A_\beta(\mathbf{r}(t)) \rangle. \tag{6.174}$$

The vector potential correlation function can be written in terms of its Fourier transform as

$$\langle A_\alpha(\mathbf{r}(t)) A_\beta(\mathbf{r}(t)) \rangle = 2 \int \frac{d^d \mathbf{k}}{(2\pi)^d} \int \frac{d\omega}{2\pi} \left\{ \cos \left[\frac{\omega}{2}(t + t') \right] \right.$$
$$\left. + \cos \left[\frac{\omega}{2}(t - t') \right] \right\} e^{i\mathbf{k} \cdot (\mathbf{r} - \mathbf{r}')} \langle A_\alpha A_\beta \rangle_{\mathbf{k}, \omega}, \tag{6.175}$$

and the transformed correlator is given, in turn, by the fluctuation dissipation theorem. In three dimensions, in the classical limit, this is given by

$$\langle A_\alpha A_\beta \rangle_{\mathbf{k}, \omega} = -\frac{2 k_B T}{\varepsilon_s \omega^3}$$
$$\times Im \left[\left(\delta_{\alpha\beta} - \frac{k_\alpha k_\beta}{k^2} \right) \frac{\omega^2}{\omega^2 \varepsilon_t(\mathbf{k}, \omega) - c^2 k^2} + \frac{k_\alpha k_\beta}{k^2 \varepsilon_l(\mathbf{k}, \omega)} \right], \tag{6.176}$$

where ε_s is the high-frequency dielectric constant for the semiconductor, where the transverse and longitudinal dielectric functions are given by

$$\varepsilon_t(\mathbf{k}, \omega) = 1 + \frac{i\sigma}{\varepsilon_s \omega}, \quad \varepsilon_l(\mathbf{k}, \omega) = 1 + \frac{D\kappa}{-i\omega + Dk^2} \tag{6.177}$$

for small momentum and frequency, and where κ is the inverse screening vector. Using these results, and keeping only terms on the order of the imaginary part of the transverse dielectric function, Eq. (6.176) becomes

$$\langle A_\alpha A_\beta \rangle_{\mathbf{k}, \omega} = \frac{2 k_B T}{\sigma \varepsilon_s \omega^2} \left[\left(\delta_{\alpha\beta} - \frac{k_\alpha k_\beta}{k^2} \right) \frac{1}{1 + \delta^4 k^4} + \frac{k_\alpha k_\beta}{k^2} \right], \tag{6.178}$$

where $\delta^2 = c^2 \varepsilon_s / \omega\sigma$ is the classical skin depth. In the case of a two-dimensional film or a one-dimensional wire, both the conductivity and the skin depth have to be computed using the appropriate reduced-dimensionality density in the conductivity. For the wire, of course, the term in the parentheses vanishes, so only the latter term in the brackets is important. In fact, this is true in all dimensions for the longitudinal (or diagonal) fluctuations. In two or three dimensions, when the skin depth is such that $\delta^4 k^4 \gg 1$, the only term that contributes is the last term in the brackets. On the other hand, when $\delta^4 k^4 \ll 1$, the only term that contributes is $\delta_{\alpha\beta}$. In the usual case, we can assume that the skin depth is such that the former limit holds, and this is the same result as for the one-dimensional case.

The factor that becomes important in the calculation is the last term in the square brackets of (6.178), and we can use the following identity to simplify some of the calculations:

$$\int \frac{d^d \mathbf{k}}{(2\pi)^d} k_\alpha k_\beta v_\alpha(t) v_\beta(t') e^{i\mathbf{k} \cdot (\mathbf{r} - \mathbf{r}')} (\cdot) = \frac{\partial^2}{\partial t \partial t'} \int \frac{d^d \mathbf{k}}{(2\pi)^d} e^{i\mathbf{k} \cdot (\mathbf{r} - \mathbf{r}')} (\cdot). \tag{6.179}$$

Bringing these results together, we can now write the correction term as

$$
\left(\frac{em}{\hbar}\right)^2 \frac{k_B T}{\sigma \varepsilon_s} \int_{-\zeta}^{\zeta} dt \int_{-\zeta}^{\zeta} dt' \int \frac{d^d k}{(2\pi)^d k^2} e^{i\mathbf{k}\cdot(\mathbf{r}-\mathbf{r}')}
$$
$$
\times \int \frac{d\omega}{2\pi} \left\{ \cos\left[\frac{\omega}{2}(t+t')\right] - \cos\left[\frac{\omega}{2}(t-t')\right] \right\}, \tag{6.180}
$$

where the two partial derivatives with respect to the two times have already been performed. The integration over the frequency is simple but leads to restrictions that $t' = \pm t$, which impact the propagator exponential; the term can be rewritten (keeping only the real part) as

$$
\left(\frac{em}{\hbar}\right)^2 \frac{k_B T}{\sigma \varepsilon_s} \int_{-\zeta}^{\zeta} dt \int \frac{d^d k}{(2\pi)^d k^2} [1 - \cos\{\mathbf{k}\cdot[\mathbf{r}(t) - \mathbf{r}(-t)]\}]. \tag{6.181}
$$

One-dimensional case

In one dimension, the integration over the momentum can be treated by noting that the integrand is basically a delta function that will allow us carry out the evaluation at either of the two limits of the integral. Combining these with a factor of 2, we can now add the perturbing term to the diffuson equation as

$$
\left\{ \frac{\partial}{\partial \zeta} - D\frac{\partial^2}{\partial \rho^2} + \left(\frac{em}{\hbar}\right)^2 \frac{2k_B T}{\sigma_1 \varepsilon_s} |\rho| + \frac{1}{\tau_\varphi^*} \right\} C(\zeta, \rho, \rho') = \frac{1}{\tau}\delta(\zeta)\delta(\rho - \rho'), \tag{6.182}
$$

where $\rho = [\mathbf{r}(t) - \mathbf{r}(-t)]$. It should be noted that the differential equation has become that for Airy functions in the spatial coordinate. For the moment, we turn our attention to the initial value, which is the peak amplitude of the correlation function. This satisfies the equation

$$
\left\{ -D\frac{\partial^2}{\partial \rho^2} + \left(\frac{em}{\hbar}\right)^2 \frac{2k_B T}{\sigma_1 \varepsilon_s} |\rho| + \frac{1}{\tau_\varphi^*} \right\} C(0, \rho, \rho') = \frac{1}{\tau}\delta(\rho - \rho'). \tag{6.183}
$$

At this point, we will change variables to a dimensionless set, with $\rho = L_N x$, and

$$
\left\{ -\frac{\partial^2}{\partial x^2} + x + \left(\frac{L_N}{l_\varphi^*}\right)^2 \right\} C(0, \rho, \rho') = \frac{L_N}{D\tau}\delta(x - x'), \tag{6.184}
$$

where

$$
L_N = \left(D\frac{\sigma_1 \varepsilon_s \hbar^2}{2e^2 m^2 k_B T} \right)^{1/3} \equiv \sqrt{D\tau_N} \tag{6.185}
$$

is a characteristic length over which the correlation function decays. We now can connect the decay time τ_N with the phase-breaking time associated with the fluctuations of the vector potential, which are assumed to be due to the interaction between the electrons themselves. This leads to

$$
\tau_\varphi \equiv \tau_N = \left(\frac{\sigma_1 \varepsilon_s \hbar^2}{2e^2 m^2 k_B T \sqrt{D}} \right)^{2/3}. \tag{6.186}
$$

This is a simple power-law decay of the phase-breaking time and does not have the logarithmic singularity encountered earlier for the lifetime. This is just the result (6.7), within some constants.

Two-dimensional case

In two dimensions the key factor is the integral over the momentum,

$$\int \frac{kdkd\theta}{(2\pi)^2 k^2}[1 - \cos\{\mathbf{k}\cdot[\mathbf{r}(t) - \mathbf{r}(-t)]\}] \sim \frac{1}{2\pi}\ln(\rho/l_T), \qquad (6.187)$$

where we have arbitrarily cut off the logarithmic term at the thermal length. This result is good for $\rho > l_T$, and the integral is a small constant in the opposite regime. Thus we can add a slowly varying potential to the equation for the correlator, leading to

$$\left\{-D\frac{\partial^2}{\partial\rho^2} + U(\rho) + \frac{1}{\tau_\varphi^*}\right\}C(0, \rho, \rho') = \frac{1}{\tau}\delta(\rho - \rho'), \qquad (6.188)$$

where

$$U(\rho) \sim \left(\frac{em}{\hbar}\right)^2 \frac{k_B T}{\pi \sigma_2 \varepsilon_s}\ln(\rho/l_T). \qquad (6.189)$$

We again introduce reduced coordinates through $\rho = L_0 x$, and

$$L_0 = \left(D\frac{\pi \sigma_2 \varepsilon_s \hbar^2}{e^2 m^2 k_B T}\right)^{1/2} \equiv (D\tau_0)^{1/2}, \qquad (6.190)$$

so that

$$\tau_0 = \frac{\pi \sigma_2 \varepsilon_s \hbar^2}{e^2 m^2 k_B T}. \qquad (6.191)$$

The constant terms in the differential equation become

$$\left(\frac{L_0}{l_\varphi^*}\right)^2 + \ln\left(\frac{L_0}{l_T}\right), \qquad (6.192)$$

and the second term usually dominates the results. This suggests that the phase-breaking time is τ_0 when $L_0 < l_T$, and that

$$\frac{1}{\tau_N} = \frac{1}{\tau_0}\ln\left(\frac{L_0}{l_T}\right) \qquad (6.193)$$

in the opposite case. This result, when compared with the one-dimensional result above, suggests that the prefactor terms in τ_N in each case contain a temperature factor. This is clearly the case for the wire result quoted in (6.7). The additional logarithmic term contributes a temperature dependence only through the factor σ_2 in L_0, since the explicit terms in T and D cancel between the two lengths. More important, we note that the phase-breaking rate now has the same $T\ln T$ found in the discussions above, so this same result is obtained by a variety of approaches for the two-dimensional electrons.

Three-dimensional case

In three dimensions, the effective potential approach can also be used. Here, the potential falls off as $1/\rho$ when $\rho < l_T$, and is constant in the opposite regime. This leads to a small correction, and the conductivity is dominated primarily by the phase coherence length. This implies that τ_φ^*, which contains the contributions from the large energy–exchange collisions, is the proper phase coherence length, at least to leading order, and there is no major effect of small momentum–exchanging electron-electron interactions.

In summary, then, the interactions with the so-called background fluctuations of the electromagnetic spectrum (essentially just noise generated by density fluctuations arising from the carrier-carrier interactions) can lead to a phase-breaking process that is important in $d < 3$. The temperature dependence of this process is given basically by a power-law behavior of low order. This provides the behavior best described by (6.7) as one model of the phase-breaking process.

6.5.2 Modifications of the self-energy

The above discussion derives from the semiclassical equation of motion for the diffusons (and cooperons) developed in the last chapter for the dependence of the correlation function on the magnetic field. An alternative form for calculating the phase-breaking time is to go directly to the diagrammatic analysis followed in the previous section. This has been done by Fukuyama [42]. This argument corrects an earlier derivation by Fukuyama and Abrahams [43], which had found a $\ln(T)$ multiplier. (That is, they assert that the term depending on the logarithm of the temperature, which has arisen prominently in several derivations above, is incorrect, at least within their derivation.) The important point in these derivations is that it is not adequate to neglect the strong wavevector dependence of the self-energy, and hence of the lifetime τ_ε, that arises due to the electron-electron interaction. Of course this was done in one of the treatments above, but in the impurity-dominated process the interaction potential was taken outside of the integrals by assuming that there was some preferred momentum for the interaction. These authors argue, quite correctly, that it is the strong dependence of the interaction that leads to a correction for the small energy-exchange (large momentum) transfer interactions and to a more realistic phase-breaking time for the interaction self-energy. The self-energy was previously evaluated at quite small momentum exchange, which is the source of the problem, and the nature of screening in two dimensions leads to the corrections that need to be calculated here. In general, the self-energy can be written as in Eq. (6.142). However, there are a variety of extra diagrams that contribute to the electron-electron dephasing time. These are shown in Fig. 6.16. These will be evaluated for the case of a two-dimensional system.

The first diagram, Fig. 6.16a, contributes just the term described in (6.142), which can be rewritten once again as

$$\Sigma_{ee}^{(1)}(\mathbf{k}, i\omega_n) = -\sum_m \frac{1}{\beta\hbar^2\tau^2} \int \frac{d^d\mathbf{q}}{(2\pi)^d} G^{r,a}(\mathbf{k} - \mathbf{q}, i\omega_n + i\omega_m) \frac{U(\mathbf{q}, i\omega_m)}{[Dq^2 + |\omega_m|]^2}.$$

$$(6.194)$$

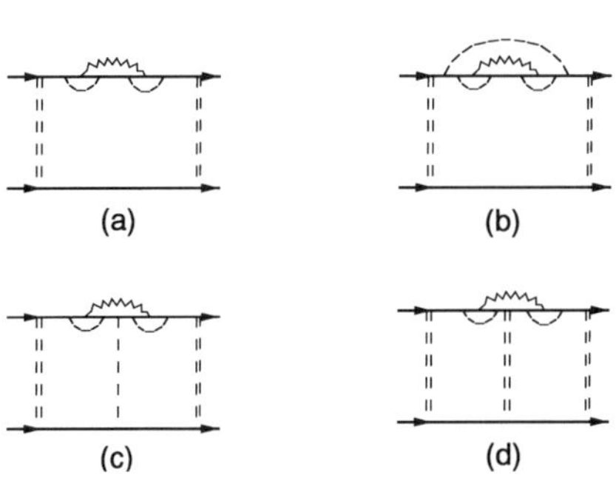

Fig. 6.16. The diagrams that contribute to the dephasing time for electron-electron scattering.

The diagram of Fig. 6.16b adds two more Green's functions and another diffuson loop, while that of Fig. 6.16c adds two more Green's functions (one on either leg of the two-particle Green's function) and a diffuson spanning the two Green's function legs. The terms in Fig. 6.16d eliminate the vertex corrections of the diffusons but add a pair of Green's functions and the cooperon spanning the two lines. These various terms can be combined by the techniques discussed above, and after carrying out the analytic continuation, the effective phase-breaking time can be obtained, as

$$
\frac{1}{\tau_\varepsilon} = \frac{2}{\hbar} Im\{\Sigma_{ee}(\mathbf{k}, i\omega_n)\}
$$

$$
= -\frac{2}{\pi} \int \frac{d^d\mathbf{q}}{(2\pi)^d} \int_{-\infty}^{\infty} dx\, Im\{U(\mathbf{q}, x)\} \left\{ \frac{f(x)}{Dq^2 - ix} + \frac{b(x)}{Dq^2 - ix + 1/\tau_\varphi} \right\},
$$

$$
\tag{6.195}
$$

where $b(x) = [e^x - 1]^{-1}$ is the Bose-Einstein distribution and $f(x)$ is the normal Fermi-Dirac distribution. The Bose distribution arises from the fact that the cooperon depends on two electrons propagating in opposite directions which are strongly coupled. This doubles the frequency of the Green's functions, making them depend on even integers instead of odd integers, which, in changing the Matsubara poles, converts the summation from one yielding a Fermi-Dirac distribution to one yielding a Bose-Einstein distribution. The second term in the curly brackets is the cooperon correction.

The dynamically screened Coulomb interaction, in two dimensions, is given by Eq. (6.117) and can be written as

$$
U(\mathbf{q}, x) = \frac{e^2}{2\varepsilon q} \frac{Dq^2 - ix}{D\kappa q + Dq^2 - ix} \sim \frac{e^2}{2\varepsilon q} \frac{Dq^2 - ix}{D\kappa q - ix}, \tag{6.196}
$$

where we have assumed that $\kappa > q$. We now also approximate this for wavevectors q such that $D\kappa q > x$. This essentially neglects the second term in the denominator of the second fraction, so that

$$
U(\mathbf{q}, x) \sim \frac{e^2}{2\varepsilon\kappa} \frac{Dq^2 - ix}{Dq^2}. \tag{6.197}
$$

The strong overall momentum dependence of the phase-breaking time arises from the screening wavevector, which suppresses the the divergence of the integration over the scattering momentum in (6.195). Now, these small momentum states do contribute to the conductance, and to the cooperon itself. Some compromise is required, and this is taken by cutting off the momentum integration at small momentum by $q > q_0 = (D\tau_\varphi)^{-1/2}$. For practical purposes the diffuson term can be replaced by the cooperon term with small error. Then, using

$$
f(x) + b(x) = \frac{1}{\sinh(\beta x)}, \tag{6.198}
$$

the phase-breaking rate can be written as

$$
\frac{1}{\tau_\varepsilon} = -\frac{1}{\pi^2} \int_{q_0}^{\infty} q\, dq \int_{-\infty}^{\infty} dx \left(\frac{-x}{Dq^2} \right) \frac{1}{\sinh(\beta x)} \frac{1}{Dq^2 - ix + 1/\tau_\varphi}
$$

$$
\sim \frac{1}{\pi^2 D} \int_0^{\infty} dx \frac{x}{\sinh(\beta x)} Re\left\{ \frac{\ln(-ix\tau_\varphi)}{-ix + 1/\tau_\varphi} \right\}
$$

$$
\sim \frac{2}{\pi D\beta} \ln(2\tau_\varphi/\hbar\beta) \sim \frac{4}{\pi m E_F \beta \tau} \ln(\pi m E_F \tau/2\hbar), \tag{6.199}
$$

where the last expression has been obtained by iteration once. Here it is clear that the temperature has been eliminated from the logarithmic term, so only the prefactor is temperature dependent. Again, the phase-breaking rate increases linearly with temperature, which agrees with (6.191), obtained for the interaction with the background fluctuations in the electromagnetic field. While the coefficients differ, this is a natural result of different approaches and not particularly a cause for great concern.

At this point, we may summarize that it is clear – from a number of derivations ranging from naive to very detailed – that the phase-breaking rate due to the electron-electron interaction in two dimensions increases linearly with the temperature. There may be an additional prefactor that varies as the logarithm of the temperature, but the presence of this term varies from one derivation to another. While absence of this latter term does not give the dependence found in (6.6) that is so evident in Fig. 6.8, it is more in keeping with the results in ballistic quantum dots discussed at the end of the last chapter. In discussing the presence of this additional factor, or the lack of it, there are two possibilities. First, the original derivation of Fukuyama and Abrahams [43] actually may have been correct, but this is disputed by Fukuyama himself. Second, it was remarked in the derivation of the previous subsection that another phase-breaking process, other than that of the small energy-exchanging carrier-carrier interactions, could be present and is represented by the term in τ_φ^*. The behavior observed in Fig. 6.8 may be due to a phase-breaking process which, although describable by (6.6), is not due to carrier-carrier interactions but to some other phase-breaking process – perhaps that of phonon scattering. At this point, the understanding of the theory is inadequate to determine this in any great detail. Both the detailed diagrammatic analysis of this section, and the electromagnetic treatment of the previous section, yield a carrier-carrier interaction–induced phase-breaking time $\tau \sim T^{-(d-1)}$. On the other hand, the self-energy that arises from the interactions among the electrons, which is described by the lifetime (6.170), does contain the logarithmic dependence, but the prefactor does not have the T^{-1} behavior of (6.6). Whether this becomes a more important term in the phase-breaking than initially expected or not, the more rapid temperature decay of the experiment and of (6.6) do not seem to be supported by an interpretation in terms solely of the interacting electron system.

Bibliography

[1] S. Tarucha, T. Saku, Y. Hirayama, J. R. Phillips, K. Tsubaki, and Y. Horikoshi, in *Quantum Effect Physics, Electronics, and Applications*, eds. K. Ishmail, T. Ikoma, and H. I. Smith, IOP Conf. Ser. **127**, 127 (1992).

[2] R. A. Webb, S. Washburn, H. J. Haucke, A. D. Benoit, C. P. Umbach, and F. P. Milliken, in *Physics and Technology of Submicron Structures*, eds. H. Heinrich, G. Bauer, and F. Kuchar (Springer-Verlag, Berlin, 1988), p. 98.

[3] W. J. Skocpol, P. M. Mankiewich, R. E. Howard, L. D. Jackel, D. M. Tennant, and A. D. Stone, Phys. Rev. Lett. **56**, 2865 (1986); **58**, 2347 (1987).

[4] C. de Graaf, J. Caro, and S. Radelaar, Phys. Rev. B **46**, 12814 (1992).

[5] E. Abrahams, P. W. Anderson, P. A. Lee, and T. V. Ramakrishnan, Phys. Rev. B **24**, 6783 (1981).

[6] R. G. Wheeler, K. K. Choi, A. Goel, R. Wisnieff, and D. E. Prober, Phys. Rev. Lett. **49**, 1674 (1982).

[7] R. P. Taylor, P. C. Main, L. Eaves, S. P. Beaumont, I. McIntyre, S. Thoms, and C. D. W. Wilkinson, J. Phys.: Cond. Matter **1**, 10413 (1989).

[8] P. A. Lee, A. D. Stone, and H. Fukuyama, Phys. Rev. B **35**, 1039 (1987).

[9] C. W. J. Beenakker and H. van Houten, Phys. Rev. B **37**, 6544 (1988).

[10] R. C. Dynes, Physica B **109 + 110**, 1857 (1982).

[11] K. K. Choi, D. C. Tsui, and K. Alavi, Phys. Rev. B **36**, 7751 (1987).

[12] J. P. Bird, A. D. C. Grassie, M. Lakrami, K. M. Hutchins, P. Meeson, J. J. Harris, and C. T. Foxon, J. Phys. Cond. Matter **3**, 2897 (1991).

[13] T. Ikoma, T. Odagiri, and K. Hirakawa, in *Quantum Effect Physics, Electronics, and Applications*, eds. K. Ishmail, T. Ikoma, and H. I. Smith, IOP Conf. Ser. **127**, 157 (1992).

[14] Y. K. Fukai, S. Yamada, and H. Nakano, Appl. Phys. Lett. **56**, 2133 (1990).

[15] J. H. Davies and J. A. Nixon, Phys. Rev. B **39**, 3423 (1989); J. A. Nixon, J. H. Davies, and H. U. Baranger, Phys. Rev. B **43**, 12638 (1991).

[16] Y. Takagaki and D. K. Ferry, J. Phys. Cond. Matter **4**, 10421 (1992).

[17] D. J. Thouless, J. Non-Cryst. Sol. **35/36**, 3 (1980).

[18] R. E. Howard, L. D. Jackel, P. M. Mankiewich, and W. J. Skocpol, Science **231**, 346 (1986).

[19] J. Imry, Europhys. Lett. **1**, 249 (1986).

[20] P. Nozières and D. Pines, *Theory of Quantum Liquids* (Benjamin, New York, 1966).

[21] S. Das Sarma, in *Quantum Transport in Ultrasmall Devices*, eds. D. K. Ferry, H. L. Grubin, C. Jacoboni, and A.-P. Jauho (Plenum Press, New York, 1995).

[22] See, for example, the discussion in P. A. Lee and T. V. Ramakrishnan, Rev. Mod. Phys. **57**, 287 (1985).

[23] B. L. Altshuler and A. G. Aronov, Sol. State Commun. **39**, 115 (1979); Zh. Eksp. Teor. Fiz. Pis'ma Red [JETP Letters **30**, 514 (1979)].

[24] D. K. Ferry, *Semiconductors* (Macmillan, New York, 1991).

[25] B. K. Ridley, *Quantum Processes in Semiconductors* (Oxford Univ. Press, Oxford, 1982).

[26] O. Madelung, *Introduction to Solid State Theory* (Springer-Verlag, Berlin, 1978) pp. 104–9.

[27] P. Lugli and D. K. Ferry, Appl. Phys. Lett. **46**, 594 (1985); IEEE Electron Dev. Lett. **6**, 25 (1985).

[28] A. L. Fetter and J. D. Walecka, *Quantum Theory of Many-Particle Systems* (McGraw-Hill, New York, 1971).

[29] G. D. Mahan, *Many-Particle Physics* (Plenum, New York, 1981).

[30] C. P. Enz, *A Course on Many-Body Theory Applied to Solid-State Physics* (World Scientific Press, Singapore, 1992).

[31] B. Vinter, Phys. Rev. B **13**, 4447 (1976).

[32] B. Vinter, Phys. Rev. B **15**, 3947 (1977).

[33] T. Ando, A. B. Fowler, and F. Stern, Rev. Mod. Phys. **54**, 437 (1982).

[34] H. Fukuyama, in *Electron-Electron Interactions in Disordered Systems*, eds. A. L. Efros and M. Pollak (North-Holland, Amsterdam, 1985).

[35] A. V. Chaplik, Sov. Phys. JETP **33**, 997 (1971).

[36] G. F. Giuliani and J. J. Quinn, Phys. Rev. B **26**, 4421 (1982).

[37] B. L. Altshuler, D. E. Khmelnitzkii, A. L. Larkin, and P. A. Lee, Phys. Rev. Lett. **44**, 1288 (1980).

[38] H. Fukuyama, J. Phys. Soc. Jpn. **49**, 644 (1980).

[39] B. L. Altshuler and A. G. Aronov, in *Electron-Electron Interactions in Disordered Systems*, eds. A. L. Efros and M. Pollak (North-Holland, Amsterdam, 1985).

[40] B. L. Altshuler, D. E. Khmelnitzkii, A. L. Larkin, and P. A. Lee, Phys. Rev. B **22**, 5142 (1980).

[41] B. L. Altshuler, A. G. Aronov, and D. E. Khmelnitsky, J. Phys. C **15**, 7367 (1982).

[42] H. Fukuyama, J. Phys. Soc. Jpn. **53**, 3299 (1984).

[43] H. Fukuyama and E. Abrahams, Phys. Rev. B **27**, 5976 (1983).

7

Nonequilibrium transport and nanodevices

The technological means now exists for approaching the fundamental limiting scales of solid-state electronics in which a single electron can, in principle, represent a single bit in an information flow through a device or circuit. The burgeoning field of single-electron tunneling (SET) effects, although currently operating at very low temperatures, has brought this consideration into the forefront. Indeed, the recent observations of SET effects in poly-Si structures *at room temperature* by Yano *et al.* [1] has grabbed the attention of the semiconductor industry. While there remains considerable debate over whether the latter observations are really single-electron effects, the resulting behavior has important implications to future semiconductor electronics, regardless of the final interpretation of the physics involved.

We pointed out in Chapter 1 that the semiconductor industry is following a linear scaling law that is expected to be fairly rigorous, at least into the first decade of the next century. This relationship will lead to devices with critical dimensions well below 0.1 μm. Research devices have been made with drawn gate lengths down to 20 nm in GaAs and 40 nm in Si MOSFETs. This suggests that such devices can be expected to appear in integrated circuits within a few decades (by 2020 if scaling rules at that time are to be believed). However, it is clear from a variety of considerations that the devices themselves may well not be the limitation on continued growth in device density within the integrated circuit chip. Factors such as resistance of metallization lines, time delays (and signal loss) in very long interconnects that must run across the entire chip or a significant part of the chip, and leakage currents within the devices may impact the continued scaling to a much larger degree than the physics of the devices themselves. Does this mean that we should not be concerned about the device physics and studies of quantum devices? Thankfully, no. Architecture is proceeding at a pace comparable to that of circuit scaling, and there is considerable effort in seeking ways in which the insertion of quantum devices within each cell of the integrated circuit design will provide for enhanced functionality. This means that we must seek the manner in which quantum effects in open device structures will carry over to these applications.

How do we couple the physics of the last several chapters to that needed for understanding the future ultrasmall microdevices expected for approaching generations of integrated circuits? Consider the implications of a small device that is fabricated as a GaAs MESFET. (The arguments are the same for MOSFETs or for HEMTs, so the approach with the MESFET will be adequate to illustrate several points.) A typical ultrasmall device can be considered with a gate length of 40 nm and a source-to-drain spacing of 100 nm. If the device is fabricated by growing a heavily doped, $n = 3 \times 10^{18}$ cm^{-3}, epitaxial layer 40 nm thick on a semi-insulating GaAs substrate, with a 80 nm width (twice the gate length), then

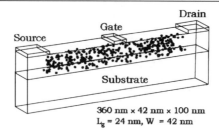

Fig. 7.1. Random distribution of impurities in an ultrasmall device.

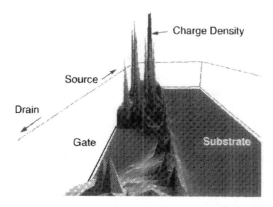

Fig. 7.2. The inhomogeneous carrier density found by self-consistent solutions to Poisson's equation for the impurity distribution of the previous figure.

there are (on average) only 384 impurity atoms under the gate, and fewer than 1000 in the entire device. These impurities are randomly located within the doped region, so there is a considerable variance in the actual number of dopant atoms under the gate (The rms variation under the gate is ±20 impurity atoms.) This leads to a considerable inhomogeneity in the actual electron concentration at any point in the device. Consider for example Fig. 7.1, in which we plot the actual dopant atoms (one possible distribution of the atoms) in an ultrasmall device with a slightly smaller gate length. It is clear from this figure that there are regions where the density of dopant atoms is above the average value, and other regions where the actual density is much less than the average value. In Fig. 7.2, the resulting electron density (under bias with current flowing) obtained by self-consistently solving Poisson's equation is plotted. Again, there are many peaks in the density, which correlate with the regions where the number of dopant atoms is high, and other regions where the density is low. The highest peaks occur in the region between the source and the gate, a region known to be important for both metering the actual current flow through the device and for impacting the series "source" resistance in the device. The peaks are broader and lower in the region between the gate and the drain due to enhanced carrier heating in this region, but they are still significant. A simple calculation suggests that each of the peaks in the figure corresponds to a small region about 10 nm on a side (in three dimensions) in which there are roughly 10 electrons. That is, the doped region throughout the device appears to be formed of a large number of quantum boxes (quantum dots in three dimensions). Although the current has been calculated in a semiclassical manner [2], it is clear that the real current in an actual device will be affected strongly by the potential barriers between the boxes, and that the actual potential shape (and value) will depend strongly on many-body effects within each quantum box. Will this transport be hopping transport from one box to another? Or will there be significant tunneling between the boxes? At this time, there are no answers to these questions. In quasi-two-dimensional systems such as HEMTs and MOSFETs, the carriers are localized at the interface (either between GaAs and AlGaAs or between Si and SiO_2). Nevertheless, the effects will continue to be there [3], except that the structures will now be essentially the quantum dot structures discussed in an earlier chapter.

The theoretical structures discussed so far in this book have been described in terms of an equilibrium system, either at zero temperature or in thermal equilibrium with the surroundings. It is possible to describe the thermodynamics of an *open, nonequilibrium* system through the use of nonequilibrium Green's functions, the so-called real-time Green's functions developed by Schwinger and colleagues [4], [5], by Kadanoff and Baym [6], and by Keldysh [7]. However, it has been pointed out that this interpretation is valid only when the entropy production (by dissipation) is relatively small and the system remains near to equilibrium [8], where, at least in lowest order, Liouville's equation remains valid as a equation of motion for the appropriate statistical ensemble representing the carrier transport under the applied fields. In *far-from-equilibrium* systems, however, there is usually strong dissipation, and the resulting statistical ensemble (even in steady state) is achieved as a balance between driving forces and dissipative forces; for example, it does not linearly evolve from the equilibrium state when the applied forces are "turned on." Enz [8] argues that there is no valid unitary operator (which describes the impact of perturbation theory) to describe the evolution through this symmetry-breaking transition to the dissipative steady state. As a consequence, there is no *general* formalism at this time that can describe the all-important far-from-equilibrium devices. On the other hand, the application of the real-time Green's functions to what are surely very strongly far-from-equilibrium systems – the excitation of electron-hole pairs in intense femtosecond laser pulses – has yielded results that suggest that their use in these systems is quite reliable for studying both the transition to the semiclassical Boltzmann theory and to explain experimentally observed details [9], [10]. As a consequence of these latter studies, as well as initial attempts to actually begin to model real devices with these Green's functions, the situation is believed much better than this pessimistic view would warrant. Nevertheless, it is essential that one move carefully with these functions to assure that the problematic approach is really meaningful.

In this chapter we will first review some of the experiments indicating nonequilibrium behavior in mesoscopic devices. We then turn to the development of the real-time Green's functions. We try to point out how they have been used for device simulations and studies. Finally, a brief discussion of alternate formulations of a statistical ensemble can be obtained from the real-time Green's functions.

7.1 Nonequilibrium transport in mesoscopic devices

Over the past few years, our understanding of the transport of electrons (and holes) through mesoscopic systems has increased significantly. Most of this transport work, described in the previous chapters, is based on a system either at (assumed) zero temperature or at least in thermal equilibrium with its surroundings (the RTD is a notable exception). Yet in most devices of interest, this will not be the case. In the past few years there have been studies of mesoscopic systems that have probed the nonequilibrium behavior of these systems, at least to the lowest order. Here we examine some of these effects.

7.1.1 Nonequilibrium effects in tunnel barriers

First, let us consider a structure in which a AlGaAs barrier is inserted between a heavily doped (n^+) GaAs layer and a lightly doped (n^-) GaAs layer. A second heavily doped (n^+) layer is placed at the other end of the lightly doped GaAs layer. This would normally be an $n^+n^-n^+$ structure, except the tunneling barrier is inserted at one end of the lightly doped region. It is this combination of structures that makes this example interesting. Hickmott *et al.* [11] were apparently the first to study such a structure, and they observed periodic oscillations in the applied voltage. These oscillations seemed to have a periodicity corresponding to the emission of a sequence of longitudinal (polar-) optical phonons. The basic structure is shown in Fig. 7.3 with an applied bias. The second derivative of the applied voltage as a function of the reverse bias current ("reverse" is defined here in the sense that the electrons flow through the barrier into the lightly doped region) is shown in Fig. 7.4.

An understanding of the principle at work arises from the consideration that the voltage drop is across three distinct regions [12], [13]: (i) the tunneling barrier, (ii) a fraction of the lightly doped (n^-) region that is depleted of carriers, and (iii) the lightly doped (n^-) region in which there are still a significant number of free carriers. Perhaps it is easiest to understand if we consider the constant current case instead of the constant potential case. Consider that there is a fixed number of electrons being injected through the tunnel barrier into the depleted portion of the lightly doped region. The voltage dropped across this region is denoted as V_D. When this voltage is less than the optical phonon energy ($eV_D < \hbar\omega_{LO}$), a relatively large drop must arise across this region, and the region is terminated by the portion that is still populated with carriers. The current is determined by the velocity of the carriers and by the residual doping ($j = n(x)ev(x)$, which requires the density to be a decreasing function of x since the velocity is an increasing function of x). When $eV_D > \hbar\omega_{LO}$, however, the electron

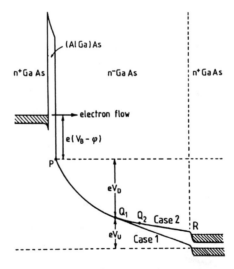

Fig. 7.3. The GaAs/AlGaAs/GaAs structure under a bias. [After Eaves *et al.*, in *Physics of Quantum Electron Devices*, ed. F. Capasso (Springer-Verlag, New York, 1990), by permission.]

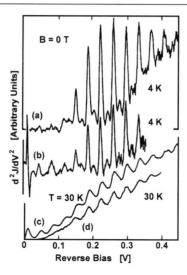

Fig. 7.4. The second derivative of the current with respect to voltage. [After Eaves *et al.*, in *Physics of Quantum Electron Devices*, ed. F. Capasso (Springer-Verlag, New York, 1990), by permission.]

can emit an optical phonon and drop to a lower kinetic energy state (hence lower velocity state). This requires that carriers move into the depleted region from the undepleted region to maintain the current constant. (More carriers at lower velocity are required to maintain the total current, which has been assumed constant.) This, in turn, tries to lower the total voltage drop that occurs across the device. In the end, the carrier movement actually tries to lower the voltage across the undepleted region as well, which leads to a feedback effect. In essence, these charge movements lead to the voltage swings required each time the voltage drop V_D passes the threshold for the emission of another optical phonon. The actual distribution of potential through the device must be found by solving Poisson's equation. The nonlinear feedback from the charge motion from the undepleted to the depleted region can lead to a kind of hysterisis in the current-voltage relationship, but it is this "switching" behavior in the actual charge distribution that leads to the observed oscillations.

The importance of the phonon emission is quite evident in resonant-tunneling diodes, where the carriers can be quite far from equilibrium in the quantum well. Normally one expects that the carriers can pass ballistically through the double barriers (and wells) without losing any energy. However, it is also possible in many cases for the carriers to emit a phonon. This can occur when the bias is above that necessary for the normal peak current, so that the particles entering the well lie more than $\hbar\omega_{LO}$ above an acceptable output level (one in which tunneling through the second barrier leaves them in the conduction band of the output side of the structure). This is enhanced significantly when the lower energy state is actually the bound state in the well, which leads to a potential distribution in which most of the potential drop is across the input barrier, while the bound state lies quite close to the Fermi level in the output barrier. Such a set of I-V curves is shown in Fig. 7.5 [13]. Here, the temperature is uniformly 4 K, and the resonant-tunneling diode is etched into a mesa shape of 100 μm diameter. The well is 11.7 nm wide, while the barriers are 5.6 nm thick. The principle resonant tunneling peak is evident at about 70 mV bias, and another peak (much smaller) appears at about 170 mV. This peak is much enhanced by the presence of a magnetic field oriented parallel to the current direction. Normally, an electron tunneling through a

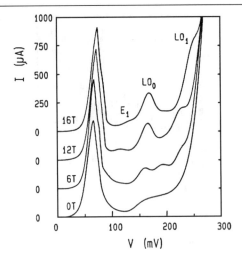

Fig. 7.5. The current-voltage characteristics for a resonant tunneling diode with a longitudinal magnetic field at 4 K. [After Eaves *et al.*, in *Physics of Quantum Electronic Devices*, ed. F. Capasso (Springer-Verlag, New York, 1990), by permission.]

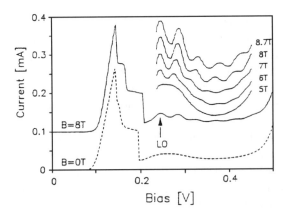

Fig. 7.6. The I-V characteristics at 1.8 K for a GaAs/AlGaAs resonant-tunneling diode for various values of the magnetic field. The peak marked LO is a phonon replica of the main peak. [After Yoo *et al.*, J. Soc. Sci. Technol. B **8**, 370 (1990).]

barrier conserves the transverse momentum. In the case of a phonon emission, however, this is no longer the case. The magnetic field helps to quantize the transverse momentum and highlight the phonon-assisted tunneling peak. The very weak peak marked E_1 in the figure is thought to be due to a nonresonant tunneling transition between the nth Landau level in the emitter and the $(n + 1)$st Landau level in the well. Such a transition is thought to be due to the effect of ionized impurities or interface roughness. The peaks marked LO_p correspond to the transition from the nth Landau level in the emitter to the $(n + p)$th Landau level in the well through the emission of a single optical phonon. The relative amplitudes of the main resonant tunneling peak and these other peaks provide qualitative indications of the contribution of various charge transport processes to the measured current.

The magnetic field is quite important in the observation of these phonon-assisted tunneling currents. In Fig. 7.6, we show another series of measurements for a GaAs/AlGaAs resonant-tunneling diode, this time at 1.8 K [14]. The I-V curves exhibit a main resonant peak at 144 meV and a first subsidiary peak, which has been marked as LO at 245 meV, in a magnetic field of 8 T. This subsidiary peak is attributed to tunneling with the assistance of a single phonon-emission process. In the absence of the magnetic field, there remains an indication of the process, but the peak is much less pronounced and spread in energy. This peak is enhanced by the reduction in phase space that arises from the application of a magnetic field. The inset curves show how the peaks develop as the magnetic field is increased. Although there are several peaks in the spectrum, only the first peak does not shift with magnetic field; all the other peaks move in magnetic field. As in [14], these latter peaks are thought to arise from transitions that involve changes in the Landau level index during the tunneling process, as well as include the emission of the optical phonon.

7.1.2 Ballistic transport in vertical and planar structures

When carriers are accelerated, or injected, into a region in which they are subject to an electric field, it takes a certain period of time before the transport can be characterized by a mobility, and/or a drift velocity. In this short time behavior (and also short spatial dimension behavior), the carriers move quasi-ballistically. That is, they move either with an initial injection velocity or under direct acceleration of the electric field. The time over which this behavior occurs is essentially the momentum relaxation time. In a high-mobility quasi-two-dimensional gas formed at the hetero-interface between GaAs and AlGaAs, where the mobility can be several million at low temperatures, this time can be of the order of many tens of picoseconds, which corresponds to an elastic mean free path of tens of microns (Table 1.1 gives values for a mobility of 10^5 cm^2/Vs, which can easily be exceeded by more than an order of magnitude). Many people have tried to construct devices that would utilize (and measure) this quasi-ballistic transport.

Hayes et al. [15] utilized an $n^+p^-np^-n^+$ structure (known as a planar-doped-barrier transistor – PDBT), and Yokoyama et al. [16] and Heiblum et al. [17] replaced the lightly doped p-regions with hetero-barriers. Both Yokoyama et al. [16] and Levi et al. [18] observed indications of ballistic transport. The most convincing measurements, which allow for spectroscopy of the injected hot carriers, were provided by Heiblum et al. [19]. These latter structures utilized a tunneling injector to provide a very narrow (on the order of 8 meV wide) spectrum of injected electrons [20]. These devices are known as Tunneling Hot Electron Transfer Amplifiers (THETAs) and have the ability to use a collector hetero-barrier to carry out energy spectral measurements of the injected carriers, as shown in Fig. 7.7. It may be noted from this picture that the base region forms a quantum well. The coherence of the injected beam of carriers can be estimated by the observation of oscillations in the transmission coefficient (from emitter to collector), which replicate the resonant levels in the well and the virtual resonances at energies above the barriers. This is shown in Fig. 7.8 for two different base widths and collector barrier heights (at the collection end of the collector in Fig. 7.7). Measuring the derivative of the emitter current with respect to the base-emitter voltage yields the transmission as a function of injection energy. With spectroscopic measurements of the injected beam, it is possible to identify the emission of LO phonons in the base region as an energy loss mechanism. However, this is not the major loss mechanism. Hollis et al. [21] have suggested that the dominant loss mechanism is due to the interaction of the injected electrons with the high density of cold electrons in the base region, and with plasmons (collective excitations) of this cold particle gas. Levi et al. [18] estimate that the mean-free

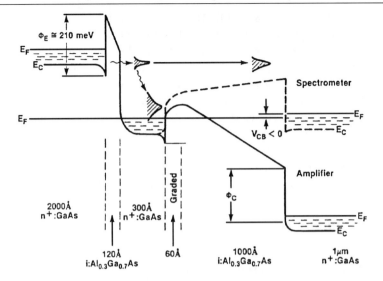

Fig. 7.7. Structure of a THETA device. [After Heiblum and Fischetti, in *Physics of Quantum Electron Devices*, ed. F. Capasso (Springer-Verlag, New York, 1990), by permission.]

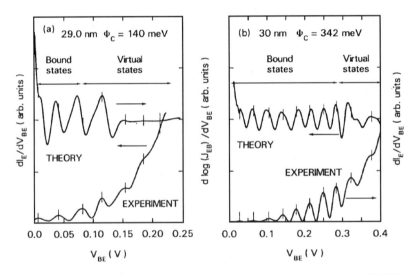

Fig. 7.8. Experimental and theoretical curves for the tunneling base current in a THETA device. [After Heiblum and Fischetti, in *Physics of Quantum Electron Devices*, ed. F. Capasso (Springer-Verlag, New York, 1990), by permission.]

path in their PDBT is only about 30 nm due to this carrier-carrier interaction. There is significant evidence that this carrier-carrier scattering is also important in THETA devices.

Very recent measurements have been carried out on structures in which a resonant-tunneling double barrier was inserted into the device at the collector edge of the gate [22]. The resonant level of this double barrier structure can then be used to more effectively scan through the spectrum of injected carriers and hopefully could produce a negative differential conductance as in other double barrier devices. However, no negative differential

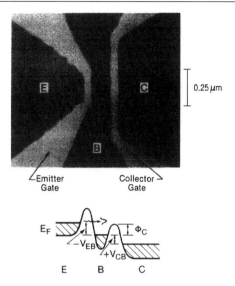

Fig. 7.9. SEM picture of a lateral THETA device and the schematic of the potential profile. [After Pavleski *et al.*, Appl. Phys. Lett. **55**, 1421 (1989), by permission.]

conductance was seen unless a high magnetic field was applied to the structures. The magnetic field constricts the phase space available for scattering and this greatly attenuates the carrier-carrier interaction. With a 30-nm-thick GaAs base, doped to $n = 10^{18}$ cm^{-3}, the carrier-carrier scattering limited mean-free path is estimated to be about 30 nm [20].

In lateral devices, which are constructed using surface gates on a quasi-two-dimensional heterostructure layer, the above loss to the high density of base electrons (or holes in some structures) can be avoided [23]. Such a lateral gate device is shown in Fig. 7.9. In these structures, further confinement of the electrons into a single, narrow lateral beam can be heightened by the use of side gates in the emitter region. Transfer efficiencies as high as $\alpha > 0.98$ can be achieved for injection energies below the optical phonon energy, and the elastic mean-free path is estimated to be greater than 0.48 μm [24], easily in keeping with the earlier discussion. This value for the mean-free path is comparable to that of the material itself (at low biases), so that it is conjectured that the value for the hot injected carriers is comparable to that of the cold background electrons, so long as the injection energy remains below the optical phonon threshold (all the measurements have been carried out at 4.2 K).

7.1.3 Thermopower in nanostructures

In general, when one passes a current through a semiconductor system, this electrical current is accompanied by a thermal current which constitutes the flow of energy through the conductor. In most cases, where the transport takes place in thermal equilibrium, this thermal current can safely be ignored. However, this is not always the case. The thermal current arises from the fact that the electrical current causes a temperature differential to exist in the semiconductor sample according to a relation in which the resulting voltage is given by

$$V = \frac{I}{G} + S\Delta T, \tag{7.1}$$

where S is known as the thermopower. Accompanying the temperature difference, one can define the thermal current as

$$Q = \Pi I - \kappa \Delta T, \tag{7.2}$$

where Π is known as the Peltier coefficient and κ is the total (electron plus lattice) thermal conductivity.

The earliest measurements of the nonequilibrium – generated thermopower S seem to have been those of Galloway *et al.* [25]. These authors actually measured the fluctuations in the conductance and related them to heating effects in the quantum structure. In any case of diffusion-limited transport, such as that most mesoscopic systems exhibit, the thermopower can be expressed in terms of the energy dependence of the conductance through

$$S = -\frac{\pi^2}{3} \frac{k_B^2 T}{eG} \frac{dG(E)}{dE}\bigg|_{E_F} \simeq -\frac{\pi^2}{3} \frac{k_B^2 T}{e\Delta E} \frac{\delta G}{G}, \tag{7.3}$$

where the derivative in the last form has been replaced by the conductance fluctuation and the average energy scale, with $\Delta E = \max(k_B T, \hbar D/l_{in}^2, \hbar D/L^2)$, and the terms have their normal meanings used throughout this text. In the nonequilibrium case, the appropriate energy scale is the first of these quantities, and

$$S = -\frac{\pi^2}{3} \frac{k_B}{e} \frac{\delta G}{G}. \tag{7.4}$$

It should be noted that the coefficient is composed primarily of known constants, and this formula is readily checked against experiment. The coefficient $(\pi^2 k_B/3e)$ takes the value of $284\,\mu\text{V/K}$. Thus, a measurement of the thermopower as a function of the relative amplitude of the fluctuations in conductance gives a measure of this value.

Galloway *et al.* [25] made measurements on samples such as those shown in Fig. 7.10, in which quantum wires in a bulk GaAs structure were measured. The wires had a physical width of $0.5\,\mu\text{m}$ and thickness of 50 nm. In these wires, it is thought that the thermal energy is larger than the other comparable energy-defining quantities listed under Eq. (7.3), at least

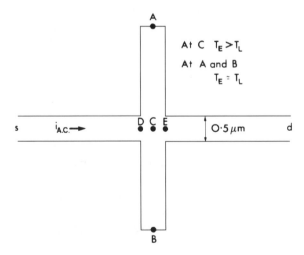

Fig. 7.10. Schematic diagram of the sample used for thermopower measurements. [After Galloway *et al.*, Surf. Sci. **229**, 326 (1990), by permission.]

for temperatures down to 0.1 K. The idea of the experiment is to heat the electrons by an applied current passed between the source s and drain d in the figure. While the electrons at the ends of the sidearm nominally will be at the lattice temperature, the carriers at point C will be at an elevated electron temperature T_E as a consequence of the heating by the joule current. The heating can be estimated by an energy balance equation, which semiclassically can be stated as [26]

$$\frac{\partial \langle E \rangle}{\partial t} = e\mathbf{J} \cdot \mathbf{F} - \frac{3k_B}{2\tau_\varepsilon}(T_E - T) \tag{7.5}$$

in three dimensions, where $\mathbf{F}$ is the applied electric field. In steady state, the electron temperature can be found as

$$T_E = T + \frac{2e\tau_\varepsilon \sigma}{3k_B}F^2. \tag{7.6}$$

We note that this heating is a heating of the electron system and not of the lattice as a whole. For a homogeneous conductor, the voltages V_{AC} and V_{CB} would cancel, but if the conductance fluctuations in these two sidearms are uncorrelated, then this will not be the case for the fluctuations. In these samples, it is felt that $l_\varphi \sim 0.25 \, \mu$m, while $l_{in} \sim 2.5 \, \mu$m, so that these voltages should be uncorrelated for the size of the sample used in the experiment. In Fig. 7.11, the voltage fluctuations measured at the three points A, B, and C (relative to point A) in the sample are shown as a function of the magnetic field. Here, the longitudinal field between points D and E is only 0.17 μV, and the scale on the fluctuations in the figure is 45 μV. The variation of the fluctuation voltage with the temperature differential $\Delta T = T_E - T$ is shown for a variety of lattice temperatures in Fig. 7.12. The linear dependence in this figure clearly confirms the expectations of Eq. (7.1) for a linear dependence of the fluctuating voltage on the temperature differential. The data give a slope of $220 \pm 20 \, \mu$V/K,

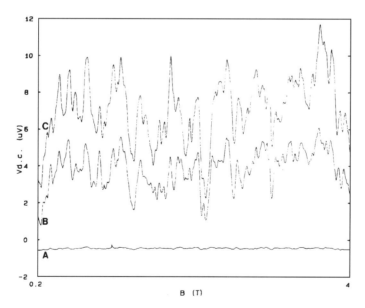

Fig. 7.11. The d.c. thermoelectric voltage fluctuations due to the electron temperature gradient. Curves B and C have been offset for clarity. [After Galloway *et al.*, Surf. Sci. **229**, 326 (1990), by permission.]

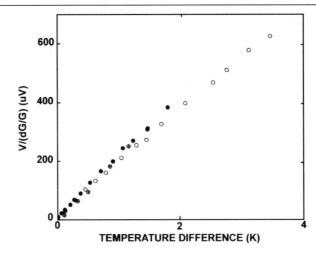

Fig. 7.12. The rms d.c. voltage fluctuation as a function of the temperature differential. The curves for open circles are at 1.08 K, those for the crosses at 2.0 K, and the solid circles at 4.0 K. [After Galloway *et al.*, Surf. Sci. **229**, 326 (1990), by permission.]

which is close to the theoretical value (within 20% in any case, considering that one needs to estimate the electron temperature from an indirect measurement, discussed below). These results seem to fit well to the theory of the thermopower in systems in which the electrons are heated by the conduction current.

In general, it is usually expected that carrier heating by the longitudinal current will produce voltages that are of odd order in the current, yet the above results are thought to be even in the current because they depend only on the temperature, which is quadratic in the current (or in the field as indicated above). Similar measurements have been shown by Molenkamp *et al.* [27] for the case in which the transverse voltage is coupled through point contacts. The sample is shown as an inset to Fig. 7.13, where the conductance through the transverse point contacts and the transverse voltage arising from the longitudinal heating current are shown. In this structure, the phase coherence length is small compared to the mean free path, so that the quantum-interference effects are not particularly important to the conductivity of the sample itself. The transport through the longitudinal channel is felt to be quasi-ballistic (e.g., not diffusive, but scattering is still considered to exist in order to create the enhanced electron temperature), and the two point contacts are not equivalent. If the two contacts were equivalent, then the total transverse voltage between the ends of the two sidearms would cancel, as discussed above. However, the inequality of the two point contacts breaks the inversion symmetry of the sample, which allows the observation of a transverse voltage (in the previous experiment above, only the fluctuations in the voltage could be observed), which is the difference in the thermopower-induced voltage across the two point contacts. The measurements are for a lattice temperature of 1.65 K, and the transverse voltage is plotted for a bias current of 5 mA and a fixed bias on one of the gates of -2 V. The curves in Fig. 7.13b are a calculation assuming an electron temperature of 4 K and a Fermi energy of 13 meV. This calculation assumes a quasi-equilibrium Fermi distribution for the carriers in the channel. The transverse voltage is even in the applied longitudinal current, and the large quantum oscillations occur as a result of the depopulation of the one-dimensional subbands in the point contacts. The actual voltage measured is proportional

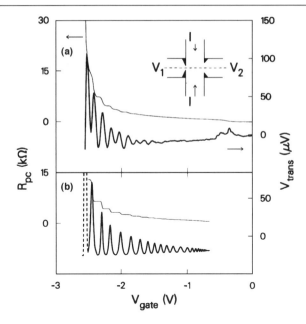

Fig. 7.13. Experimental traces of the conductance of the point contacts (thin curve) and the transverse voltage (thick curve). [After Molenkamp *et al.*, Phys. Rev. Lett. **65**, 1052 (1990), by permission.]

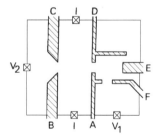

Fig. 7.14. Diagram of a 2-μm-wide channel coupled to a point contact (left) and a quantum dot (right). [After Molenkamp *et al.*, Semicond. Sci. Technol. **9**, 903 (1994), by permission.]

to the *difference* in the thermopower voltage of the two probes, as has been mentioned. Comparison of the peaks in the transverse voltage with the conductance through the point contacts confirms this behavior, in that the peaks are aligned (in voltage) with the steps in the conductance through the point contact. These authors have also measured the Peltier coefficient and the thermal conductivity of a point contact under nonequilibrium conditions [28] and some other thermoelectric effects [29].

The measurement of the thermopower can show other effects when it is used as a probe of adjacent quantum confined systems. In Fig. 7.14, the quantum wire is coupled to a quantum dot through one of the two transverse point contacts [30]. The active area of the dot is about 0.7×0.8 μm^2, which is adjacent to the 2-μm-wide, 20-μm-long channel. The dot area is tuned by the bias voltage applied to contact E. Heating is provided by a relatively large current flowing in the quantum wire, and the thermopower voltage is measured as the difference in voltages V_1 and V_2. In the previous measurement, the oscillations in the

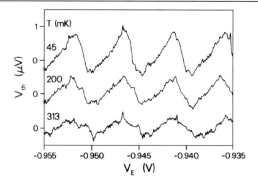

Fig. 7.15. Thermovoltage as a function of the tuning stub bias at various lattice temperatures. [After Molenkamp *et al.*, Semicond. Sci. Technol. **9**, 903 (1994), by permission.]

transverse voltage (and in the thermopower) arose from the change in the number of modes propagated through the point contact. Here, this change in transmission will arise from the single-electron tunneling to the states of the quantum dot. This behavior is shown in Fig. 7.15 for a variety of lattice temperatures. Clearly, the heating current of 18 nA, used in this figure, is sufficient to raise the electron temperature in the channel significantly above the lattice temperature. The oscillations in this case have a pronounced triangular shape. The period of the oscillations is the same as that of the conductance to the dot and arises from the periodic depopulation of the dot by a single electron as the tuning bias V_E is varied. The triangular shape is decreased as the lattice temperature is raised, presumably due to thermal smearing of the capacitive charging induced shape.

7.1.4 Measuring the hot electron temperature

In the previous section, the measurements of the thermopower were generally dependent on knowing (or estimating) the electron temperature in the channel. One part of the experiment, which was not discussed, is the subsidiary measurements necessary to ascertain an estimate of the carrier temperature T_E in the channel for a given value of the bias current. Measurements of this type have been demonstrated by, for example, Ikoma and associates [31]–[33].

It may be noted from (7.6) that simple measurements of voltage, current, and conductance are insufficient to determine the electron temperature in the channel. One still must ascertain the energy-relaxation time τ_ε that appears in this equation. Alternatively, one can probe the temperature itself and use Eq. (7.6) to determine the energy-relaxation time. This latter technique is based upon the measurements of the temperature dependence of the phase-breaking time τ_φ discussed in the last chapter. When carrier-carrier scattering dominates the phase-breaking time, then the dependence of τ_φ on carrier temperature is the same as its dependence on lattice temperature. By first measuring the lattice temperature dependence of the phase-breaking time, and then measuring the current (or field) dependence of the phase-breaking time, one can infer the electron temperature that corresponds to a given current. It should be pointed out, however, that such measurements tend to be *global* measurements, that is, measurements of the overall sample, whereas (7.6) is a *point* equation. Inhomogeneities in the field distribution, which are a natural consequence of the nonlinearity of the conductance with temperature, are a source of error in such measurements.

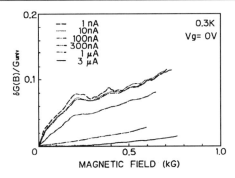

Fig. 7.16. Magnetoconductance spectra of a gated GaAs/AlGaAs quantum wire at 0.3 K for varying current levels. [After Ikoma *et al.*, in *Granular Nanoelectronics*, eds. D. K. Ferry *et al.* (Plenum Press, New York, 1991), by permission.]

For the experimental measurements, a quantum wire of 350 nm width and 10 μm length was formed in a GaAs/AlGaAs heterostructure by using focused ion beams to define the wire. In this approach, the ion beams produce damaged, low-mobility material adjacent to the defined wire. In Fig. 7.16, the magnetoconductance measured for a variety of bias currents is shown. These measurements are made at a lattice temperature of 0.3 K. As one can see, once the bias current exceeds about 0.3 μA, the conductance and the conductance fluctuations begin to decrease significantly. By fitting the conductance fluctuations (for magnetic fields above 0.2 T) with the theory of the previous chapters, one can determine the phase-breaking time. Here, one first determines the coherence length from the amplitude (and the correlation magnetic field) of the fluctuations, and the diffusion constant from the amplitude of the conductance itself. These two quantities then determine the phase-breaking time from $D\tau_\varphi = l_\varphi^2$. The values of the phase-breaking time determined in this manner are plotted in Fig. 7.17 for a variety of lattice temperatures. When the input power (I^2/G) exceeds 10^{-14} W, the phase-breaking time begins to decrease. In the previous chapter it was determined how the lattice temperature affected the phase-breaking time. Using such lattice temperature measurements of the phase-breaking time, one can now convert the values found in Fig. 7.17 to equivalent *electron* temperatures for each power level input to the device. In such a manner, it may be estimated that, for the case of $T = 3$ K and $P_e = 3.5 \times 10^{-13}$ W, the carrier temperature is 16 ± 5 K. This temperature corresponds to an energy relaxation time of 0.5 ns (determined by the temperature differential and input power through the energy balance equation), which is close to a predicted value for acoustic phonon relaxation in a GaAs/AlGaAs heterostructure [34]. This relaxation time for the energy is clearly three orders of magnitude larger than the phase-breaking time, which indicates that the phase-breaking mechanism is different than the energy relaxation mechanism. Presumably, carrier-carrier scattering is responsible for phase breaking, and phonon scattering is responsible for energy relaxation.

7.1.5 Hot carriers in quantum dots

Hot carriers are a common occurence in normal, large-scale semiconductor structures. Under the application of a large electric field, the electron temperature is raised, as has been discussed in the sections above. On the other hand, most studies of quantum structures

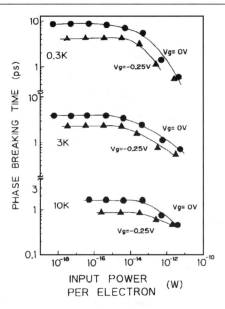

Fig. 7.17. The phase-breaking time as a function of the input power for various lattice temperatures. [After Ikoma *et al.*, in *Granular Nanoelectronics*, eds. D. K. Ferry *et al.* (Plenum Press, New York, 1991), by permission.]

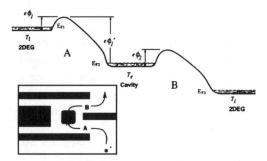

Fig. 7.18. Schematic of the energy-band profile for a lateral double-constriction quantum dot. The inset shows the configuration of the bias gates. [After Goodnick *et al.*, Phys. Rev. B **48**, 9150 (1993), by permission.]

have concentrated on the near-equilibrium properties. The quantum dot, which is a form of a lateral resonant-tunneling structure, can also exhibit extreme nonequilbrium behavior. Consider the quantum dot, as shown in the inset to Fig. 7.18. The top, bottom, and right-central gate lines define a pair of "point-contact" barriers, which are marked *A* and *B*. The central region forms the quantum dot. In essence, the dot region corresponds to the quantum well defined between a pair of tunneling barriers. The actual barrier properties can be modified by varying the voltages applied to the three defining gates. On the other hand, the large gate left of center in the figure corresponds to a "plunger" that tunes the bound states and population of the quantum dot itself. Under bias, the potential structure can appear

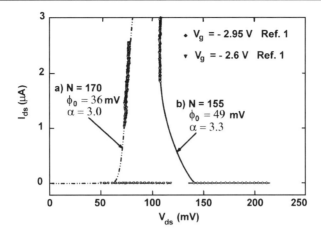

Fig. 7.19. Experimental and calculated current-voltage curves for two different bias configurations. [After Goodnick *et al.*, Phys. Rev. B **48**, 9150 (1993), by permission.]

as shown in Fig. 7.18, with the potential drop partially occuring across the first barrier and partially across the second barrier. When the region under the control gates is depleted, the current path must pass through the two barrier regions and through the quantum dot.

When the bias voltage is sufficiently high across the coupled structure, the current-voltage characteristics show a marked *S*-type negative differential conductance (this is called *S*-type because the lines through the various branches seem to form an *S*, and the current is multivalued for a given voltage) [35]. The transition from the low-current state to the high-current state is a strong function of the bias on the various gates in the structure. Typical current-voltage curves are shown in Fig. 7.19. In the figure are two different I-V curves. Each has a high-resistance line that is nearly flat at almost zero current (the extension to zero voltage is not shown for device "b"). Then, there is a high-current branch in which the voltage is almost constant, and for which negative differential resistance can be seen at lower current levels. As the voltage increases, the operating point follows the high-resistance branch until the switching voltage is reached, at which point operation jumps to the high-current branch. The controllable switching behavior in these structures can be explained by a balance approach in which energetic carriers injected through the emitter barrier will balance energy with the thermal bath of cool carriers residing in the quantum dot itself. Calculated current-voltage curves appear in Fig. 7.19. along with the data from the experiment [36].

Consider the potential barriers encountered by electrons as they traverse point *A* in Fig. 7.18. The barrier, as indicated in the figure, is created by depletion between the lower and right-central gates. This isolates the central region and allows the potential drop to occur between the dot and the outside regions. The separation between the Fermi level in the emitter region and the central dot region is given by

$$E_{F1} - E_{F2} = \frac{eV_{sd}}{2} + \frac{eQ}{C}, \tag{7.7}$$

where C is the charging capacitance of the dot region and $Q = e(N_0 - N)$ is the excess charge in the dot (N is the number of electrons in the dot, and N_0 is the equilibrium number).

Similarly, the difference in the Fermi energies of the dot and the collector region is given by

$$E_{F2} - E_{F3} = \frac{eV_{sd}}{2} - \frac{eQ}{C}, \tag{7.8}$$

and it has been assumed that the input and output barriers are identical.

While the Fermi energies determine the equilibrium states, the barrier heights themselves determine the kinetic rates. Under bias, the input barrier is reduced in the amount

$$\varphi_1 = \varphi_0 - \left(\frac{V_{sd}}{2} + \frac{Q}{C} \right) \Big/ \alpha, \tag{7.9}$$

and the factor α depends upon the abruptness of the barrier as discussed in Section 3.6.5. The output barrier is similarly lowered, but in the amount

$$\varphi_2 = \varphi_0 - \left(\frac{V_{sd}}{2} - \frac{Q}{C} \right) \Big/ \alpha, \tag{7.10}$$

For a perfectly abrupt barrier, α would be infinite (neglecting any image force lowering of the barriers). In the present structures, the barrier abruptness is limited by the fact that they are essentially depletion barriers, so that the barrier potential varies over a distance roughly comparable to the channel width W. Since the channel lengths are of the same order as the widths, we may estimate α as 3 as discussed in Chapter 3. An equally important parameter is the excess energy of the injected electrons (into the well), which is shown in Fig. 7.18 as $e\varphi_1'$. The latter is related to the applied bias by

$$\varphi_1' = \varphi_0 + \left(1 - \frac{1}{\alpha} \right) \left(\frac{V_{sd}}{2} + \frac{Q}{C} \right). \tag{7.11}$$

One assumption to be made is that the dimensions of the quantum dot and the outlying regions are sufficiently large that we can treat them as quasi-two-dimensional regions. A second, major assumption in determining the current-voltage relationship is that the injected electrons thermalize with the electrons already in the quantum dot region. This latter allows us to deal with the carriers in the dot through their Fermi energy and an equivalent electron temperature T_E. We can now write the current into the dot from the emitter as

$$I = eW v_{F1} \frac{m^* k_B T}{\pi \hbar^2} e^{-e\varphi_1/k_B T} \left[\exp\left(\frac{E_{F1}}{k_B T} \right) - \exp\left(\frac{E_{F2}}{k_B T_E} \right) \right], \tag{7.12}$$

where T is the lattice temperature. The bracketed term represents the difference in the Fermi factors for the currents into and out of the dot region. Once the current is known, and an equivalent amount of charge within the dot is determined along with the electron temperature, the input barrier can be determined from this equation. On the other hand, a reasonable approximation is to ignore the last term (large bias voltage) in the square brackets with respect to the first, and rewrite this as

$$I = eW v_{F1} \frac{m^* k_B T}{\pi \hbar^2} e^{-e\varphi_1/k_B T} \left[\exp\left(\frac{n_s \pi \hbar^2}{m^* k_B T} \right) - 1 \right], \tag{7.13}$$

where

$$n_s = \frac{m^* k_B T}{\pi \hbar^2} \ln\left[1 + \exp\left(\frac{E_F}{k_B T} \right) \right] \tag{7.14}$$

is the 2D density in a quasi-two-dimensional system as given in Chapter 2. (The Fermi energy is measured from the conduction band edge in this latter expression.) In a similar manner, the current flowing through the second constriction can be written as

$$
I = eW v_{F2} \frac{m^* k_B T_E}{\pi \hbar^2} e^{-e\varphi_2/k_B T_E} \left[\exp\left(\frac{E_{F2}}{k_B T_E}\right) - \exp\left(\frac{E_{F3}}{k_B T}\right) \right]
$$

$$
\sim eW v_{F2} \frac{m^* k_B T_E}{\pi \hbar^2} e^{-e\varphi_2/k_B T_E} \left[\exp\left(\frac{N\pi \hbar^2}{Am^* k_B T_E}\right) - 1 \right], \tag{7.15}
$$

and in the last expression the temperature corresponds approximately to that of a heated Maxwellian distribution. We have replaced the carrier density by the actual number of carriers N and the dot area A. The latter equation allows the determination of the barrier height of the second barrier under the same assumptions as mentioned above.

The energy flow into and out of the carriers in the quantum dot region can be defined by an energy balance equation, in keeping with that of Eq. (7.5). We can write this energy balance equation as

$$
I\left(\varphi_1' - \varphi_2 - \frac{k_B(T_E - T)}{e} \right) = N \left\langle \frac{dE}{dt} \right\rangle \Big|_{coll}, \tag{7.16}
$$

where the last term represents the dissipation of energy due to collisions. The first two terms in the parentheses are just the net potential drop across the quantum dot, and the third term represents the energy carried out by the carriers leaving through the second dot, which removes energy from the dot region. The energy loss rate of hot carriers in a GaAs quantum well (between two AlGaAs barriers) has been measured by Shah $et\ al.$ [37] using photoluminescence spectroscopy. Their data can be fit over a range of data by assuming both polar-optical phonon scattering and acoustic phonon scattering, described by

$$
\left\langle \frac{dE}{dt} \right\rangle \Big|_{coll} \sim \frac{\hbar \omega_{LO}}{\tau_{LO}} + \frac{k_B(T_E - T)}{\tau_{ac}}. \tag{7.17}
$$

For polar-optical scattering the nominal value of τ_{LO} is about 130 fs, while τ_{ac} is thought to be about 1.8 ns. However, it is likely that there are nonequilibrium polar-optical phonons as well [38], which would increase τ_{LO} to about 2 ps, which gives a better fit to the data of Shah $et\ al.$

The above equations are solved by first choosing a value of the current (since it is a multivalued function of the voltage), and then iterating the above equations to find a consistent set of values for the various parameters. The results are plotted against the data in Fig. 7.19, and the fit is quite remarkable. Two different values of the gate voltage are shown in the figure, and it can be seen from this that a small change in the gate bias produces a significant change in the barriers and the resulting current-voltage characteristic. A width $W = 200\,\text{nm}$ was used for the constrictions. The capacitance was not determined from the experiment, but a value of 10^{-16} F gives good agreement with the experimental data. Switching from the high-impedance state to the low-impedance state is found to occur essentially when the input barrier is pulled down to the emitter Fermi energy ($E_{F1} = e\varphi_1$). If there is no excess charge in the dot prior to the switching, then the switching voltage is just found to be

$$
V_s = 2\alpha \left(\varphi_0 - \frac{E_{F1}}{e} \right). \tag{7.18}
$$

As the gate bias is made more negative, the barrier heights are increased, and this leads to an increase of the switching voltage, as observed in the figure.

The observation of hot electron effects for carriers passing through quantum dots is quite important to the structures discussed in the introduction to this chapter. The dots experimentally studied here are much larger than those envisaged in ultra-small devices; the actual size enters into (7.18) only through the factor α. This represents the smoothness of the barrier potential and should be more or less the same regardless of the dot size. Consequently, one may expect to see significant switching and fluctuations in current transport within and between the dots that form, for example, in Fig. 7.2. In a device structure comparable to the latter figure, it is unlikely that the simple equations introduced here will be applicable, since the Fermi energies and temperatures will be varying throughout the structure. Approaches comparable to those of universal conductance fluctuations in Chapter 5 are thought to be necessary, since it will be necessary to determine long-range correlations throughout the structure. We turn to these functions in Section 7.2.

7.1.6 Breakdown of the Landauer–Büttiker formula

Near-equilibrium transport occurs when the energy associated with the bias voltage across the quantum structure is smaller than either the subband separation or the Fermi energy. Transport in short waveguides, in which the transport is in near equilibrium, usually satisfies the Landauer equation, with a conductance that is quantized in units of $2e^2/h$, where the quantization level corresponds to the number of occupied subbands. When the source-drain bias across the quantum waveguide (or point contact as the case may be) becomes comparable to the energy separation of the one-dimensional subbands, the number of subbands that are occupied for forward and reverse transmission differs, and nonlinear behavior can result. In Chapter 3, we showed that for relatively small source-drain bias across a quantum point contact, this nonlinear transport is still governed by the multichannel Landauer-Büttiker formula. The main experimental feature is an oscillatory behavior in the differential conductance versus source-drain voltage, which deviates from the near-equilibrium quantized conductance. This breakdown of the quantized conductance is due to the unequal population of the one-dimensional subbands as the Fermi energies on the left and right are moved up and down relative to one another. For larger source-drain bias, this oscillatory behavior damps out experimentally, whereas the Landauer-Büttiker formula (without explicit consideration of dissipative processes during transmission) still predicts structure in the differential conductance. As with resonant-tunneling diodes, the ideal picture of collisionless, coherent transport across a quantum point contact breaks down as the excess kinetic energy of the injected electrons is increased (relative to the Fermi energy in the absorbing contact) due to intercarrier and phonon scattering processes. Moreover, it has been argued that the high bias regime in QPCs will result in a situation in which the higher occupied subbands will act somewhat like current *filaments*, and this will lead to the possibility of negative differential resistance (in this case, multiple values of the current for a given voltage, in contrast to the NDR effect of resonant tunneling diodes discussed in Chapter 3) [39]. Some forms of this negative differential resistance have been observed in pinched quantum-dot devices (discussed in the previous section), but this is attributed to the nature of nonlinearity in the injection process over the first set of barriers (forming the quantum dot).

Waveguides that exhibit negative differential conductance have been formed by creating two overlapping gates on a uniform quasi-two-dimensional GaAs/AlGaAs structure, as

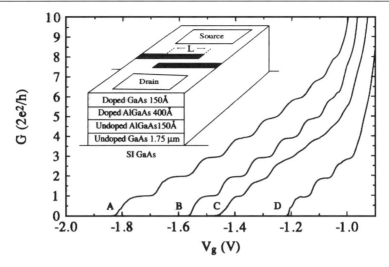

Fig. 7.20. The equilibrium conductance after light exposure of a zero overlap point contact, A, and the $L = 0.1\text{-}\mu\text{m}\,(B)$, $0.2\text{-}\mu\text{m}\,(C)$, and $0.5\text{-}\mu\text{m}\,(D)$ waveguides. The inset is a schematic of the material and the devices. [After Berven *et al.*, Phys. Rev. B **50**, 14630 (1994).]

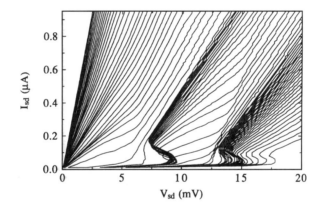

Fig. 7.21. The I-V characteristics of a 0.1-micron waveguide after illumination. The gate voltage increment is –20 mV. [After Berven *et al.*, Phys. Rev. B **50**, 14630 (1994).]

shown in the inset of Fig. 7.20. Here, the waveguide is formed in the region where the dots overlap, and this has been measured for overlap lengths of 0.1, 0.2, and 0.5 μm, which comprise relatively short wires. For low biases, the devices exhibit normal short wire quantized conductances. In the shortest wire, 0.1 μm, conductance plateaus up to $n = 5$ are easily measured. In the longer wires, the quantization is not as well pronounced due to the role the inhomogeneous confining potential plays, as discussed already in Chapter 3 [40], [41]. In Fig. 7.21, the I-V characteristics for large values of source-drain bias and various gate voltages are shown. At low bias, the I-V curves for various gate voltages are clustered around slopes corresponding to the quantized conductance. At higer bias, this particular sample exhibited two distinct regions of S-type negative differential resistance [42]. For the first, switching occurs for a bias of 6 meV, regardless of the applied potential on the two gates. The second switching occurs at a value of source-drain bias that was strongly

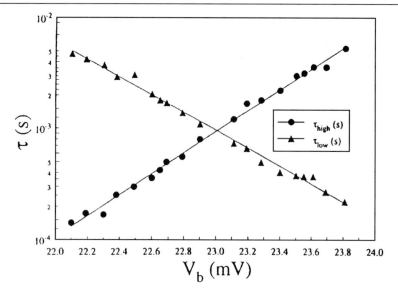

Fig. 7.22. The average times in the high- and low-current states as a function of source-drain bias for a fixed gate bias. [After Smith *et al.*, Physica B **227**, 197 (1997).]

dependent on the gate bias. The occurence of this SNDC behavior is found to be strongly correlated with the thermal history of the samples, and could be induced by *in situ* exposure to infra-red light.

Time dependent studies of the current-voltage characteristics reveal an even more surprising behavior. When biased in the region of instability, the current is observed to randomly switch between the off state and a current value along the d.c. I-V curve in the absence of SNDC [43]. This switching behavior has the characteristic behavior of random telegraph signals. However, the switching behavior of the present case is only observed in the region of instability. The average times in the low and high states follow one another such that their product is always constant over several orders of magnitude as shown in Fig. 7.22. Random telegraph signals originating from the independent fluctuations of an impurity do not generally exhibit this kind of lifetime behavior. Moreover, independent fluctuations of the quantum waveguide barrier should result in instabilities over the whole range of I-V characteristics rather than in localized bias regions.

Several transport mechanisms can be invoked to describe such an effect. One possible one is impact ionization of neutral donors. However, for this mechanism it would be expected that illumination should reduce the effect, but the opposite is observed – illumination enhances the negative differential resistance. Another possibility is the occurence of Coulomb blockade in the wire. The current expected from Coulomb blockade, however, is orders of magnitude below that seen in the experiments. In the previous section, negative diffential resistance in quantum dot structures was shown to require carrier confinement in the quantum dots. While the wire structures discussed here do not formally have quantum dots, the inhomogeneous potential expected from the impurities leads to multiple barriers and regions of trapped charge [44]. A further possible explanation is that the formation of such barriers in these wires, which are operated near pinchoff, creates barriers and puddles of trapped charge between these barriers, and subsequent hot electron effects lead to the negative differential resistance [42]. However, the correspondence of random telegraph switching with such heating behavior would have to provided by a dynamic rather than static model, as presented in the previous section [45].

7.2 Real-time Green's functions

Considerations of transport in semiconductors has traditionally been described by the application of the Boltzmann transport equation or by reduced forms of this, such as the drift-diffusion approximation [46]. Powerful Monte Carlo techniques have been developed to solve this equation for quite complicated scattering processes and in very far-from-equilibrium circumstances [47], [48].The general feature of semiclassical far-from-equilibrium transport is that the distribution function deviates significantly from the equilibrium Maxwellian form (or Fermi-Dirac form, in degenerate systems) [49]. The form of this distribution must be found from a balance between the driving forces (electric and magnetic fields, for example) and the dissipative scattering processes. In many cases, such as in excitation of electron-hole pairs by an intense laser pulse, the resulting far-from-equilibrium distribution cannot be evolved from any equilibrium distribution, so a gradual evolutionary process to the final steady-state distribution cannot be found. In fact, the formation of the dissipative steady state of the far-from-equilibrium distribution has been termed a type of phase transition [50].

In the case of quantum transport, we have utilized primarily the retarded and advanced Green's functions in the previous two chapters. These two Green's functions describe the evolution of the single-particle energy-momentum relationship into a many-body description in which the delta function $\delta(E - \hbar^2 k^2/2m^*)$, for a single parabolic band, evolves into the spectral density function $A(E, k)$. The latter is described by the imaginary parts of the retarded or advanced Green's functions. If we now move to the far-from-equilibrium dissipative steady state, we need additional functions that will describe the quantum distribution function that evolves to describe the balance between driving forces and scattering processes. These additional Green's functions, which are actually proper correlation functions, are provided by the nonequilibrium, or real-time, Green's functions developed over the past few decades [4]–[7].

Previously, we introduced the retarded and advanced functions through the definitions

$$G_r(\mathbf{r}, t; \mathbf{r}', t') = -i\Theta(t - t')\langle\{\Psi(\mathbf{r}, t), \Psi^\dagger(\mathbf{r}', t')\}\rangle, \tag{7.19}$$

$$G_a(\mathbf{r}, t; \mathbf{r}', t') = i\Theta(t' - t)\langle\{\Psi^\dagger(\mathbf{r}', t'), \Psi(\mathbf{r}, t)\}\rangle. \tag{7.20}$$

The need for the new functions arises from the property of the unitary operator that introduces the perturbing potential, which is used to describe the interaction of the carrier system with, for example, the dissipative mechanisms. We have defined this operator previously as

$$\exp\left[-\frac{i}{\hbar}\int_{t'}^{t} V(\eta)d\eta\right], \tag{7.21}$$

with the limits

$$t \to \infty, \quad t' \to -\infty. \tag{7.22}$$

These limits were acceptable because the system was assumed to be very near to equilibrium, so that the distribution function describing the carriers had the equilibrium form. Now, however, the first of these limits is unacceptable, since it assumes we know what happens in the infinite future. Since the distribution is not the equilibrium form, this limit can no longer

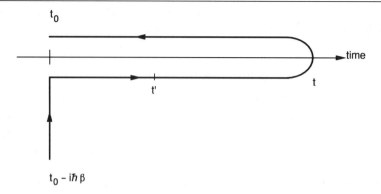

Fig. 7.23. The path of integration for the real-time Green's functions. The tail extending downward connects to the thermal equilibrium Green's functions, where appropriate.

be used. The second of the limits is problematical as well. To invoke such a limit means that the system evolution can be traced from the initial, equilibrium state. As remarked above, it is not clear that this procedure is valid either. At present, however, we will assume that such can be done, and we will return later to the treatment of failure of this process. We first address how to handle the upper limit of the integral in the unitary operator.

To avoid the need to proceed to $t \to \infty$ in the perturbation series, a new time path for the real-time functions was suggested by Blandin *et al.* [51]. (There may have been others who have suggested this approach, but this latter work seems to be the primary suggestion for the proper context.) This new contour is shown in Fig. 7.23. The evolution is assumed to begin with a thermal Green's function at $t' = t_0 - i\hbar\beta$, where β is the inverse temperature as introduced in the last chapter. It then evolves into the nonequilibrium, noninteracting Green's function at t_0. The contour then extends forward in time to the maximum of (t, t'), at which point it is returned *backward* in time to t_0. In many cases, the limit $t_0 \to -\infty$ is invoked when one is not interested in the initial condition (as in most far-from-equilibrium systems). The handling of the Green's functions when both time arguments are on either the upper branch or the lower branch is straightforward. On the other hand, when the two time arguments are on different branches of the time contour, the two new correlation functions must be defined [52]. These are the "less than" function

$$G^<(\mathbf{r}, t; \mathbf{r}', t') = i\langle\Psi^\dagger(\mathbf{r}', t')\Psi(\mathbf{r}, t)\rangle \tag{7.23}$$

and the "greater than" function

$$G^>(\mathbf{r}, t; \mathbf{r}', t') = -i\langle\Psi(\mathbf{r}, t)\Psi^\dagger(\mathbf{r}', t')\rangle. \tag{7.24}$$

In general, the four Green's functions defined here are all that are now needed, as the latter two will relate to the far-from-equilibrium distribution function, as we will see below. (This is true in lowest order, but higher-order multiparticle Green's functions may be needed also, as was the case in the previous chapters.) In practice, however, two further functions are useful. These two can be derived from the previous four, so they are merely combinations of the previous ones. The two new ones are the time-ordered Green's function

$$G_t(\mathbf{r}, t; \mathbf{r}', t') = i\langle T[\Psi^\dagger(\mathbf{r}', t')\Psi(\mathbf{r}, t)]\rangle$$
$$= \Theta(t - t')G^>(\mathbf{r}, t; \mathbf{r}', t') + \Theta(t' - t)G^<(\mathbf{r}, t; \mathbf{r}', t'), \tag{7.25}$$

where T is the time-ordering operator, and the antitime-ordered Green's function

$$G_{\bar{t}}(\mathbf{r}, t; \mathbf{r}', t') = \Theta(t' - t)G^>(\mathbf{r}, t; \mathbf{r}', t') + \Theta(t - t')G^<(\mathbf{r}, t; \mathbf{r}', t'). \tag{7.26}$$

These are obviously related to the previous Green's functions, and the various quantities can be related through

$$G_r = G_t - G^< = G^> - G_{\bar{t}} = \Theta(t - t')(G^> - G^<), \tag{7.27}$$

$$G_a = G_t - G^> = G^< - G_{\bar{t}} = -\Theta(t' - t)(G^> - G^<). \tag{7.28}$$

For systems that have been driven out of equilibrium, the ensemble average, indicated by the angular brackets, no longer signifies thermodynamic averaging or an average over the zero-temperature ensemble distribution function. Instead, the brackets indicate that some average needs to be performed over the available states of the far-from-equilibrium system, but in which these states are weighted by the far-from-equilibrium distribution function. As in semiclassical transport in which the equation of motion is the Boltzmann equation, finding this far-from-equilibrium distribution is usually the most difficult part of the nonequilibrium problem. The quantum transport problem has been solved for the nonequilibrium situation in only a very few cases, and usually under stringent appoximations.

7.2.1 Equations of motion for the Green's functions

It is obvious that solutions for the correlation functions will require that a set of equations of motion be developed for these quantities. With the added complications of four functions, however, these equations will be much more complicated than the simple Boltzmann equation, or the assumed forms used in previous chapters. Fortunately, only four of the six functions introduced above are independent in the general nonequilibrium situation.

Keldysh [7] introduced a general method of treating the set of Green's functions with a single matrix Green's function, and hence a single matrix equation of motion to be solved. This does not reduce the overall effort required, but it does simplify the equations in which these functions are described. We can illustrate this approach in the following manner, which will be done only for the functions defined on the two horizontal lines in Fig. 7.23. The approach can certainly be extended to the third, vertical line if necessary [53], but this is usually not required in the far-from-equilibrium situation. The general Green's function is defined in terms of two field operators. Each of these field operators can be on the upper line or the lower line of the overall contour in Fig. 7.23. Our matrix is thus a 2×2 matrix. The rows of the matrix are defined by the operator $\Psi(\mathbf{r}, t)$. That is, row one of the matrix corresponds when t is on the lower line, while row two is defined when t is on the upper part of the trajectory. Similarly, the operator $\Psi^\dagger(\mathbf{r}', t')$ defines the columns of the matrix. Column one is defined when t' is on the lower part of the trajectory, while column two is defined when t' is on the upper part of the trajectory. Thus, for example, we can write the 11 element by assuming contour ordering (time increases in the positive sense around the contour) as

$$G_{11}(\mathbf{r}, t; \mathbf{r}', t') = \Theta(t - t')G^>(\mathbf{r}, t; \mathbf{r}', t') + \Theta(t' - t)G^<(\mathbf{r}, t; \mathbf{r}', t') = G_t. \tag{7.29}$$

In a similar manner the 22 term yields the $G_{\bar{t}}$ function. The 12 term is achieved by beginning with the "normal ordering" of the operators (in which products arising from the use of Wick's theorem are ordered with the creation operators to the left). Then

$$G_{12} = i\langle T_C[\Psi^\dagger(\mathbf{r}', t')\Psi(\mathbf{r}, t)]\rangle = G^<. \tag{7.30}$$

Similarly, the 21 term becomes $G^>$. It should be remarked that the definition of rows and columns is not unique and that the transpose is used equally as often. Thus, we may write the contour ordered Green's function matrix as

$$\mathbf{G}_C = \begin{bmatrix} G_t & G^< \\ G^> & G_{\bar{t}} \end{bmatrix}. \tag{7.31}$$

It is useful at this point to introduce a coordinate transformation, often called a *rotation in Keldysh space* [7]. This rotation removes some of the degeneracy of the various elements, separating out the retarded and advanced functions as well as simplifying the matrix itself. The rotation is a spin-like rotation, in which the nonunitary matrices are

$$\mathbf{L} = \frac{1}{\sqrt{2}} \begin{bmatrix} 1 & -1 \\ 1 & 1 \end{bmatrix}, \quad \mathbf{L}^\dagger = \frac{1}{\sqrt{2}} \begin{bmatrix} 1 & 1 \\ -1 & 1 \end{bmatrix}, \quad \tau = \begin{bmatrix} 1 & 0 \\ 0 & -1 \end{bmatrix}. \tag{7.32}$$

These are used to modify the matrix Green's function into the Keldysh form $\mathbf{G}_K = \mathbf{L}\tau\mathbf{G}_C\mathbf{L}^\dagger$. The matrix Keldysh Green's function is then

$$\mathbf{G}_K = \begin{bmatrix} G_r & G_K \\ 0 & G_a \end{bmatrix}, \tag{7.33}$$

where $G_K = G^< + G^>$. (The reader is asked to pay particular attention to the fact that G_K is the matrix element representing the particular sum of the less-than and greater-than Green's functions, which is often called the Keldysh Green's function, and $\mathbf{G}_K$, which is in bold and is the actual matrix of Green's functions.)

We can now develop the equations of motion for the noninteracting forms of these Green's functions, that is, the equations that the functions will satisfy in the absence of any applied potentials and/or perturbing interactions. For this, we assume that the individual field operators are based on the wavefunctions that satisfy the basic Schrödinger equation. However, in the case of single-point potentials, such as those following from the Poisson equation, this potential can be included in the Schrödinger equation. So, in the case of only such potentials, this leads to

$$\left(i\hbar\frac{\partial}{\partial t} - H_0(\mathbf{r}) - V(\mathbf{r})\right)\mathbf{G}_{K0} = \hbar\mathbf{I}, \tag{7.34}$$

$$\left(-i\hbar\frac{\partial}{\partial t'} - H_0(\mathbf{r}') - V(\mathbf{r}')\right)\mathbf{G}_{K0} = \hbar\mathbf{I}, \tag{7.35}$$

where $\mathbf{I}$ is the identity matrix (unity on the diagonal and zero off the diagonal). The zero subscript has been added to indicate the noninteracting form of the Green's function.

Transport, as has been stated earlier, arises as a balance between the driving forces and the dissipative forces. To achieve a description of transport with the Green's functions, it is necessary to add some interaction terms to the Hamiltonian. As before, these will lead to self-energy terms Σ. With the new contour, it is possible to construct a Feynmann series expansion of the interaction terms with the new form of the contour-ordered unitary operator for the interaction. This procedure seems to continue to work in the case of the real-time Green's functions and has been pursued almost universally. The assumption is that the projection of the time axes back to the initial time allows the use of the pseudo-equilibrium to justify the use of an equivalent form of Wick's theorem. The various parts of the diagrams may then be regrouped into terms that represent the Green's function itself and terms that represent the interactions leading to the self-energy. The self-energy may also be expressed as a matrix, so it is possible to write the equation of motion for the full Green's function as

$$\left(i\hbar\frac{\partial}{\partial t} - H_0(\mathbf{r}) - V(\mathbf{r})\right)\mathbf{G}_K = \hbar\mathbf{I} + \mathbf{\Sigma}\mathbf{G}_K, \tag{7.36}$$

$$\left(-i\hbar\frac{\partial}{\partial t'} - H_0(\mathbf{r}') - V(\mathbf{r}')\right)\mathbf{G}_K = \hbar\mathbf{I} + \mathbf{G}_K\mathbf{\Sigma}, \tag{7.37}$$

where the self-energy matrix is now written in the Keldysh form as

$$\mathbf{\Sigma} = \begin{bmatrix} \Sigma_r & \Sigma_K \\ 0 & \Sigma_a \end{bmatrix}. \tag{7.38}$$

Generally, one would not expect the form of the self-energy to preserve the Keldysh form that is found here. In fact, if one applies the nonunitary transformations defined in and below (7.32) to the product functions (which are a convolution integral of the two Green's functions) that appear in the last terms of (7.36) and (7.37) before the Keldysh transform, the results differ greatly. However, it is important to note that care must be taken in correctly formulating the self-energy expressions in the contour-ordered Green's functions. The products that appear in (7.36) and (7.37) are integral products in the form

$$AB(1, 3) \rightarrow \int d2 A(1, 2)B(2, 3), \tag{7.39}$$

where the shorthand notation $1 = (\mathbf{r}_1, t_1)$ has been used, and the integration is over all internal variables. The proper way of writing these convolution integrals turns out to be that in which the Keldysh matrix Green's functions appear, as will be shown with the Langreth theorem.

7.2.2 The Langreth theorem

In truth, while we have written down the self-energy and the Green's function in Keldysh form in the last terms of (7.36) and (7.37), this result is not at all obvious. For example, when we begin with the easily understood Green's function matrix (7.31) and write the equations of motion, then the unit matrix on the right side of the equation is not appropriate. This is because the element in the 22 position should be $-\delta(t - t')\delta(\mathbf{r} - \mathbf{r}')$ [54]. The transform matrix τ converts this to a proper identity matrix and then the remaining two transformation matrices leave it unchanged. Thus, the transformations suggested above can easily be applied to the equation of motion for the noninteracting forms of the Green's functions.

For the product of the self-energy and the Green's function, however, the application of the nonunitary transformations in Keldysh space fail. The problem lies in defining just what the terms in the two matrices $\mathbf{\Sigma}\mathbf{G}$ and $\mathbf{G}\mathbf{\Sigma}$ should be. The solution to this problem, however, is to compute directly the matrix elements in the Keldysh form. The problem arises in the convolution form of the integration over the internal time variable in these structures. The factor we are interested in has its problems in the time integration, and not in the spatial integration. Consider the product

$$C(t, t') = \int A(t, t'')B(t'', t')dt''. \tag{7.40}$$

The approach that has become used is due to Langreth [55]. Consider, for example, the case for $C^<$. For this case, t is on the outgoing (bottom) leg of the contour in Fig. 7.23, while t' is on the incoming (top) leg of the contour. There are two possibilities. Langreth first deformed the contour by taking the contour at time t and pulling it back to t_0, and then returning it out to time t'. This creates two time loop "spikes," and the result depends on which leg the time t'' resides. We can split the integral into two parts, one for t'' on each of the two new paths. We call these subcontours C_1 and C_2, respectively. We can then write (7.40) as

$$C^<(t, t') = \int_t^{-\infty} A(t, t'')B(t'', t')dt'' + \int_{-\infty}^t A(t, t'')B(t'', t')dt''. \qquad (7.41)$$

Now, clearly in the first integral, which represents the C_1 contour, B is the less-than function, since its arguments always fold around the C_2 contour. The function A is the antitime-ordered function, but since $t > t''$, the theta function inherent in this last function gives us just the less-than function. In the second integral, A is the time-ordered function, and under the same limits it yields just the greater-than function. Thus, we can write these two integrals as

$$C^<(t, t') = \int_{-\infty}^t \Theta(t - t'')(A^> - A^<)B^< dt'' = \int_{-\infty}^t A_r(t, t'')B^<(t'', t')dt''. \qquad (7.42)$$

It is also possible to deform the contours not from t, but from t', which makes the A term always give the less-than function, and then the two paths contribute to yield the advanced function for B. These two possibilities must be combined to give the general result. Integrals such as this allow us to write the product formulas as

$$\begin{aligned} C^< = A_r B^< + A^< B_a, \quad C_r = A_r B_r, \\ C^> = A_r B^> + A^> B_a, \quad C_a = A_a B_a. \end{aligned} \qquad (7.43)$$

Similar arguments can be used to form triple products. The key point of interest here is that the use of the Langreth theorems tells us that the matrices that form the products ΣG and $G\Sigma$ should be written in the Keldysh form. Then, the proper matrix products are achieved in each case. One interesting point, however, is that the equation for the Keldysh Green's function G_K is actually two equations, which have been added for simplicity. In practice, however, the equations for the less-than and greater-than functions should be separated and treated as two disjoint equations of motion.

7.2.3 The Green-Kubo formula

The Kubo formula was developed in linear response to the applied fields, as the latter were represented by the vector potential. The use of the Kubo formula was a significant change from the normal treatment of the dominant streaming terms of the Boltzmann equation, or of the equivalent quantum transport equations, to the relaxation and/or scattering terms. With the Kubo formula, one concentrates on the relaxation processes through the correlation functions that describe the transport. Here, we talk about how the real-time Green's functions fit into the Kubo formula, with the combination termed the Green-Kubo formula. Now, if we note that the quantum-mechanical current is described by

$$j = \frac{e\hbar}{2im^*}\left[\Psi^\dagger(r)\frac{\partial\Psi(r)}{\partial r} - \frac{\partial\Psi^\dagger(r)}{\partial r}\Psi(r)\right], \qquad (7.44)$$

then it is not too difficult to develop the Green's function form of this quantity to use in the Kubo formula. To begin with, we note that this latter equation can be written in terms of the Green's functions as

$$\langle j(\mathbf{r}, t)\rangle = -\frac{e\hbar}{2m}\left(\frac{\partial}{\partial \mathbf{r}} - \frac{\partial}{\partial \mathbf{r}'}\right)G^<(\mathbf{r}, t; \mathbf{r}', t')|_{\mathbf{r}',t'\to\mathbf{r},t^+} + i\frac{ne^2}{m^*\omega}F(\mathbf{r}, t), \qquad (7.45)$$

where the last term represents the displacement current and will be ignored (in this term, F is the applied field and will also appear in the current-current correlation function itself). The actual value of the current found from the Kubo formula is given by the current-current correlation function, as

$$\langle \mathbf{j}(\mathbf{r}, t)\rangle = \frac{1}{\hbar}\int_0^t dt'\int d^3\mathbf{r}'\langle[j(\mathbf{r}, t'), j(\mathbf{r}', t - t')]\rangle \cdot \mathbf{A}(\mathbf{r}', t'). \qquad (7.46)$$

Here, $\mathbf{A}(\mathbf{r}, t)$ within the integral is the vector potential describing the driving fields. Forming the two currents with the aid of (7.44) and (7.45) and using a different choice of variable for the internal integration, we then can find the resultant current in terms of the real-time Green's functions as

$$\langle \mathbf{j}(\mathbf{r}, t)\rangle = -\frac{ie^2\hbar}{4m^{*2}}\lim_{\mathbf{r}',t'\to\mathbf{r},t^+}\left(\frac{\partial}{\partial \mathbf{r}} - \frac{\partial}{\partial \mathbf{r}'}\right)\cdot\int dt_s\int d^3\mathbf{s}\lim_{\mathbf{s}',t_s'\to\mathbf{s},t_s^+}\left(\frac{\partial}{\partial \mathbf{s}} - \frac{\partial}{\partial \mathbf{s}'}\right)$$
$$\times\left[G_r(\mathbf{r}, \mathbf{s}'; t, t_s')G^<(\mathbf{s}, \mathbf{r}'; t_s, t') + G^<(\mathbf{r}, \mathbf{s}'; t, t_s')G_a(\mathbf{s}, \mathbf{r}'; t_s, t')\right]\mathbf{A}(\mathbf{s}, t_s). \qquad (7.47)$$

In fact, as was discussed in Chapter 5, the Green's function product should actually be a two-particle Green's function involving four field operators. However, this has been expanded into the lowest-order Green's functions, as was done in that chapter. It should be remembered, though, that this is an approximation, and higher-order Green's function products may be needed to treat certain physical processes, as was also done in Chapter 5 to treat universal conductance fluctuations and weak localization. The result (7.47) can be Fourier-transformed to give the a.c. conductivity for a homogeneous system (required to take the spatial Fourier transforms), at least on the average scale of the response functions, giving (after a summation over spins)

$$\sigma(\mathbf{k}, \omega) = -\frac{e^2\hbar}{m^{*2}\omega}\int\frac{d^3\mathbf{k}'}{(2\pi)^3}\int\frac{d\omega'}{2\pi}\left(\mathbf{k}' + \frac{\mathbf{k}}{2}\right)\cdot\mathbf{k}$$
$$\times\left[G_r\left(\mathbf{k}' + \frac{\mathbf{k}}{2}; \omega'\right)G^<\left(\mathbf{k}' - \frac{\mathbf{k}}{2}; \omega' - \omega\right)\right.$$
$$\left. + G^<\left(\mathbf{k}' + \frac{\mathbf{k}}{2}; \omega'\right)G_a\left(\mathbf{k}' - \frac{\mathbf{k}}{2}; \omega' - \omega\right)\right]. \qquad (7.48)$$

After a few further changes of variables, the conductivity can be found as

$$\sigma(\mathbf{k}, \omega) = -\frac{e^2\hbar}{m^{*2}\omega}\int\frac{d^3\mathbf{k}_1}{(2\pi)^3}\int\frac{d\omega'}{2\pi}\mathbf{k}_1\cdot\mathbf{k}$$
$$\times[G_r(\mathbf{k}_1; \omega')G^<(\mathbf{k}_1 - \mathbf{k}; \omega' - \omega)$$
$$+ G^<(\mathbf{k}_1; \omega')G_a(\mathbf{k}_1 - \mathbf{k}; \omega' - \omega)]. \qquad (7.49)$$

This particular form differs somewhat from that used in Chapter 5, in that the real-time functions appear here. It should also be noted that the higher-order two-particle Green's functions have not been included, so this formulation is that of the lowest-order Green's functions, assuming that Wick's theorem is perfectly valid. We will remark about this again later. The form of Eq. (7.49) has not been utilized very much in transport calculations based on the real-time Green's functions. Nevertheless, it is important to note that the conductance here is an integral (actually, a double integral) over the current-current correlation function.

It is now useful to rearrange the terms into those more normally found in the equilibrium, and zero-temperature, forms of the Green's functions. For this, we make the ansatz

$$G^<(\mathbf{k}_1; \omega') = if(\omega')A(\mathbf{k}_1; \omega') = -f(\omega')[G_r(\mathbf{k}_1; \omega') - G_a(\mathbf{k}_1; \omega')]. \tag{7.50}$$

This ansatz is known to be correct in the zero-temperature formulation, in the thermal-equilibrium formulation, and in the Airy-function formulation to be discussed later. However, it is not known to be correct in all possible cases that may be found in the use of real-time Green's functions. There is no known proof of its correctness, and it is not in keeping with the various ansatze that have been proposed over the last decade or so for the decomposition of the less-than Green's function [56]. With this ansatz, however, we may rewrite (7.49) as

$$\sigma(\mathbf{k},\omega) = -\frac{e^2 \hbar}{m^{*2}\omega} \int \frac{d^3\mathbf{k}_1}{(2\pi)^3} \int \frac{d\omega'}{2\pi} \mathbf{k}_1 \cdot \mathbf{k}[G_a(\mathbf{k}_1; \omega')G_a(\mathbf{k}_1 - \mathbf{k}; \omega' - \omega)f(\omega')$$
$$- G_r(\mathbf{k}_1; \omega')G_r(\mathbf{k}_1 - \mathbf{k}; \omega' - \omega)f(\omega' - \omega)$$
$$- G_r(\mathbf{k}_1; \omega')G_a(\mathbf{k}_1 - \mathbf{k}; \omega' - \omega)\{f(\omega') - f(\omega' - \omega)\}]. \tag{7.51}$$

The first two products in the square brackets will cancel one another. This can be seen by changing the frequency variables as $\omega'' = \omega' - \omega$, and then using the fact that $\sigma(\mathbf{k},\omega) = \sigma^*(\mathbf{k}, -\omega)$ and $G_r = G_a^*$. Thus, we are left with only the last term in the square brackets. For low frequencies (and we will go immediately to the long-time limit of $\omega = 0$), the distribution function can be expanded about ω', so that

$$\sigma(\mathbf{k},\omega) = -\frac{e^2 \hbar}{m^{*2}\omega} \int \frac{d^3\mathbf{k}_1}{(2\pi)^3} \int \frac{d\omega'}{2\pi} \mathbf{k}_1 \cdot \mathbf{k} G_r(\mathbf{k}_1; \omega')G_a(\mathbf{k}_1 - \mathbf{k}; \omega' - \omega)\frac{\partial f(\omega')}{\partial \omega'}. \tag{7.52}$$

Finally, at low frequencies and for homogeneous material, we arrive at the form

$$\sigma(\mathbf{k}) = -\frac{e^2 \hbar}{m^{*2}} \int \frac{d^3\mathbf{k}_1}{(2\pi)^3} \int \frac{d\omega'}{2\pi} \mathbf{k}_1 \cdot \mathbf{k} |G_r(\mathbf{k}_1; \omega')|^2 \frac{\partial f(\omega')}{\partial \omega'}. \tag{7.53}$$

In the case of very low temperatures, one arrives back at the form used extensively in Chapter 5. That is, we replace the derivative of the distribution function as the negative of a delta function at the Fermi energy. The sum over the momentum counts the number of states that contribute to the conductivity and results in the density at the Fermi energy at low temperature. In mesoscopic systems, where only a single transverse state may contribute (in a quantum wire, for example), the Landauer formula can easily be recovered when one recognizes that $|G_r(\mathbf{k}_1; \omega')|^2$ represents the transmission of a particular mode. Even if there is no transverse variation, the integration over the longitudinal component of the wavevector will produce the difference in the Fermi energies at the two ends of the samples (as in, for example, the quantum Hall effect). The reduction of the less-than function has allowed a separation of the density-density correlation function from the current-current correlation function. This latter is represented by the polarization that appears as the magnitude squared of the retarded Green's function.

The approach (7.53) has been extensively utilized by the Purdue group to model mesoscopic systems with the equivalent Landauer formula for nonequilibrium Green's functions. For mesoscopic waveguides in the linear response regime, even with dissipation present, they have shown that this form can be extended to the use of a Wigner function (which will be discussed in a later section), which can then be used to define a local thermodynamic potential, and that reasonable results are obtained so long as these potentials are defined over a volume comparable in size to the thermal de Broglie wavelength $l_D = \sqrt{\pi\hbar^2/m^*k_BT}$ [57], [58]. They have been particularly successful in probing inelastic tunneling in resonant-tunneling diodes through the emission of optical phonons. In general, however, the expression (7.53) is an approximation, in that the magnitude term in the retarded Green's function is really a lowest-order representation of the actual polarization that appears in the conductivity bubble. This was addressed in Chapter 5 quite extensively, and the extended version becomes

$$\sigma = -\frac{e^2\hbar}{m^{*2}} \int \frac{d^3\mathbf{k}}{(2\pi)^3} \int \frac{d\omega'}{2\pi} \frac{\partial f(\omega')}{\partial\omega'} \Pi(\mathbf{k};\omega'), \qquad (7.54)$$

where the polarization satisfies the (reduced) Bethe-Salpeter equation

$$\Pi(\mathbf{k};\omega') = G_r(\mathbf{k};\omega')G_a(\mathbf{k};\omega')\bigg\{\mathbf{k}_1 \cdot \mathbf{k}\delta(\mathbf{k}_1 - \mathbf{k})$$

$$+ \int \frac{d^3\mathbf{k}_1}{(2\pi)^3} \frac{\mathbf{k}_1 \cdot \mathbf{k}}{k_1^2} T(\mathbf{k}_1 - \mathbf{k})\Pi(\mathbf{k}_1;\omega')\bigg\}. \qquad (7.55)$$

In this last expression, we have summed the conductivity over the Fourier variables in position to get the spatially averaged conductivity (equivalent to $\mathbf{r} = 0$, but with the origin at any point in the sample), and made a change of variables on the momentum variables for convenience. The polarization is the general product of the retarded and advanced functions that was dealt with in Chapter 5, at least on a formal basis. For isotropic scattering processes, the second term actually vanishes, and the earlier result is the formally exact result.

While these latter results appear to be quite simple in form, it is important to point out that the form of the distribution function that appears in the equations still must be determined by the balance between the driving forces and the dissipative forces: in essence, the less-than Green's function must be determined in the nonequilibrium system, just as the distribution function must be determined for the use of the Boltzmann treatment introduced in Chapter 2. In the next few sections, we consider some examples of the Green's function approach.

7.3 Transport in an inversion layer

Carriers at the interface, for example, between a silicon-dioxide layer and the bulk silicon, reside in a quasi-two-dimensional world as discussed in Chapter 2. Their motion normal to the interface is restricted by the high potential barrier of the oxide (more than 3 eV) and the depletion barrier within the bulk silicon. This potential well (between the oxide and the bulk) creates a quantization of the momentum in the motion normal to the interface. The carriers then reside in one of a series of minibands, with free motion only in the directions parallel to the interface. The general transport in this system has been studied by a great variety of approaches, and the area was extensively reviewed more than a decade ago [59]. In general, the motion parallel to the interface can be separated from that normal to the interface, with the latter quantized by the confinement potentials. In this normal direction, the subband minima are defined by the confinement energies, which result from the z-directed Schrödinger equation

$$H_{0\perp}\psi_n(z) = \varepsilon_n\psi_n(z). \tag{7.56}$$

The total energy is then composed of the plane wave energy in the transverse direction and the subband energy as

$$E_n = \frac{\hbar^2 k^2}{2m^*} + \varepsilon_n = \varepsilon_k + \varepsilon_n, \tag{7.57}$$

where $\mathbf{k}$ is now a two-dimensional momentum vector lying in the plane of the interface between the Si and the SiO_2. These quantum subbands that arise from the confining potentials have to be calculated self-consistently within the overall transport problem [60]. The self-energies that will be calculated later depend on the actual subband wavefunctions and energy levels. These self-energies, in turn, affect the actual population of the subbands and hence the self-consistent solution of the Schrödinger equation (7.56) and the Poisson equation for the exact confining potential within the semiconductor. Usually, the latter two equations are solved within the Hartree approximation, but a better approach is to include the role of the exchange-correlation corrections to the subband energies. This calculation itself is a formidable numerical challange if the confining potential is complicated in any way. Most often, some sort of approximation is made to this, but the results that are quoted below are obtained from a fully self-consistent solution, but without including the exchange-correlation corrections.

At low temperatures, the dominant scattering mechanisms in this structure are scattering by impurities, both within the bulk Si and in the oxide, and scattering off the potential fluctuations that arise from surface (interface) roughness. The latter is also important at higher temperatures (room temperature, for example) when the oxide field is high ($\simeq 10^6$ V/cm), but it is dominated by normal phonon scattering at lower oxide fields. Here, we will treat the low-temperature case.

7.3.1 Coulomb scattering

Scattering associated with Coulomb centers near the plane of the two-dimensional gas was introduced in Chapter 2. Here, we reiterate some pertinent points related to this scattering, since it will be introduced into the Green's function treatment. This scattering can be separated into contributions that arise from the ionized acceptors in the depletion layer, from trapped charge lying in the interface itself, and from charge in the oxide. These differ primarily by their possible sign and by the degree to which they also must be treated with an image charge, as has been developed by Stern and Howard [61]. The use of the Green's function approach allows an adequate treatment of the role of multiple scattering and dense numbers of impurities, which could create the diffusive transport regime dealt with in the preceding two chapters.

Scattering by the ionized acceptors (in an n-channel device) that create the depletion charge also must include the image of these charges that reflects across the dielectric discontinuity at the silicon-oxide interface. The square of the matrix element for the scattering from the single impurity located at depth z_i and lateral position (a two-dimensional position vector) $\mathbf{r}_i$ is given by

$$|\langle \mathbf{k}, n|U(\mathbf{q})|\mathbf{k}+\mathbf{q}, m\rangle|^2_{depl} = n_{depl}\left(\frac{e^2}{2\kappa q}\right)^2 A^2_{nm}(q)\int_0^\infty O^2_{nm}(q, z_i)dz_i \qquad (7.58)$$

for scattering between the n and m subbands, where $\kappa = (\varepsilon_{sc} + \varepsilon_{ox})/2$ is the average dielectric constant. Here, $\mathbf{q}$ is the Fourier-transformed momentum exchange associated with the scattering process. However, there is no momentum conservation in the direction normal to the interface, since the quantization removes the z-momentum as a good quantum number. Hence, the overlap integral between initial and final subbands does not prohibit inter-subband scattering; rather, this integral is just the form factor arising from the finite extent of the electron gas in the quantization direction as

$$A_{nm}(q) = \int_0^\infty \psi_n^*(z)e^{-qz}\psi_m(z)dz. \qquad (7.59)$$

The other factor in (7.58) describes the contribution of both the charge and its image. This latter expression is given by

$$O_{nm}(q, z_i) = \left(\frac{\varepsilon_{sc} + \varepsilon_{ox}}{2\varepsilon_{sc}}\right)e^{qz_i} + \left(\frac{\varepsilon_{sc} - \varepsilon_{ox}}{2\varepsilon_{sc}}\right)e^{-qz_i} + \left(\frac{\varepsilon_{sc} + \varepsilon_{ox}}{2\varepsilon_{sc}}\right)$$
$$\times \left[\frac{a_{nm}^{(+)}(q, z_i)}{A_{nm}(q)}e^{-qz_i} - \frac{a_{nm}^{(-)}(q, z_i)}{A_{nm}(q)}e^{qz_i}\right], \qquad (7.60)$$

where

$$a_{nm}^{(\pm)}(q, z_i) = \int_0^{z_i} \psi_n^*(z)e^{\pm qz}\psi_m(z)dz. \qquad (7.61)$$

Finally, it is important to note that n_{depl} is the sheet density of depletion charge, or depletion charge per unit area of the device.

At the interface between the silicon and the oxide, there are usually a significant number of defect states into which charge can be trapped. These states may arise from the disorder of the interface, from dangling bonds, or just from precipitated impurities. The effect of these charges is that of a sheet of scattering centers is located not exactly at the interface, but a small distance z_d (<0) into the oxide. This leads to a scattering strength of

$$|\langle \mathbf{k}, n|U(\mathbf{q})|\mathbf{k}+\mathbf{q}, m\rangle|^2_{it} = n_{it}\left(\frac{e^2}{2\kappa q}\right)^2 A^2_{nm}(q)e^{2qz_d}. \qquad (7.62)$$

Finally, by quite similar arguments, the scattering strength for a *uniform* distribution of charge within the oxide is given by

$$|\langle \mathbf{k}, n|U(\mathbf{q})|\mathbf{k}+\mathbf{q}, m\rangle|^2_{ox} = n_{ox}\left(\frac{e^2}{2\kappa q}\right)^2 A^2_{nm}(q)\frac{1 - e^{-2qd_{ox}}}{2q}, \qquad (7.63)$$

where d_{ox} is the thickness of the oxide layer itself. In contrast to the above, the oxide charge density n_{ox} is a three-dimensional charge density, which accounts for the extra factor of q in the denominator.

7.3.2 Surface-roughness scattering

Surface-roughness scattering is associated with disorder at the interface or edge of the quantum-confining structure, and the general semiclassical approach was discussed in

Chapter 2. The interface between the oxide and the semiconductor is not atomically smooth, and the variations in this interface lead to a random potential that can scatter the electrons. Early theories that were developed to treat surface-roughness scattering were based on the Boltzmann equation in which the surface appears as a boundary condition on the distribution function (see Sec. 5.1.3 for an equivalent approach in a quantum wire) [62]–[65]. The first proper quantum-mechanical treatment was apparently given by Prange and Nee [66]. Subsequently, the theory has followed essentially two different, but equivalent, paths.

In the first approach, the surface roughness is taken to be a variation in the confining potential, which appears as a boundary condition on the Hamiltonian itself. Since there is no simple perturbation theory to treat arbitrary changes in the boundary conditions, the problem of a complicated boundary condition is addressed by transforming the system through an appropriate coordinate transformation into a problem with a simpler (planar) surface. This transformation technique was proposed early [67], then used by Tesanovic *et al.* [68], and then later used by Trivedi and Ashcroft [69]. As a consequence of the transformation, additional terms appear in the Hamiltonian, and these can be treated by perturbation theory.

In the second approach, the effect of the surface roughness is buried in the potential and is incorporated by a Taylor series expansion of the potential. For example, the averaged potential (no roughness) is treated as a boundary condition on the Schrödinger equation, as before, and the variation in the boundary leads to the perturbing potential

$$\delta V = V(z_0 + \Delta(\mathbf{r})) - V(z_0), \tag{7.64}$$

where z_0 is the position of the averaged interface and $\Delta(\mathbf{r})$ is the z variation of the interface as a function of the *lateral* (two-dimensional) position $\mathbf{r}$. The lowest order of expansion of (7.64) gives [70]

$$\delta V \simeq \Delta(\mathbf{r}) \frac{\partial V}{\partial z}\bigg|_{z=z_0}. \tag{7.65}$$

Clearly, the second term on the right is an effective field describing the strength of the interface. It is tempting to take this field as the normal field produced by the gate and depletion charge, which will be some average over the actual field in the inversion layer; but it must be noted that the potential in (7.64) is the confining potential at the interface, and not the confining potential on the semiconductor side. The latter corresponds to the average electric field produced by the gate, whereas the former is the discontinuous potential produced by the conduction-band offset between the oxide and the semiconductor. The field of the latter can be quite large, approaching infinity for an abrupt interfacial band offset. Discussions about the proper choice of the field continue today, and clearly the experiments that relate the scattering to the inversion and depletion charge densities lend credence to the use of an average semiconductor field. This is where the two approaches come together, since a coordinate transformation within the quantum well of the z-coordinate changes the interface position variation to a variation in the confining potential in the semiconductor, and hence to a dependence on the average field within the semiconductor. Thus, it is possible to write the perturbing potential as

$$\delta V \simeq -e\Delta(\mathbf{r})F_{eff}, \tag{7.66}$$

and the problem now is to determine the effective field F_{eff}, a point we return to below. In Fig. 2.1, a high-resolution TEM lattice-plane image of the interface between Si and SiO_2

was shown. It is clear that the roughness is small but on the scale of one or two atomic layers, which suggests that Δ should be of the order of a few Ångstroms (less than a nm).

Computation of the scattering relies on the square of the matrix element, which upon Fourier-transforming leads to computation of the correlation function describing the auto-covariance of the random fluctuation $\Delta(\mathbf{r})$. Most early studies assumed that the correlation function was described by Gaussian correlation, or [71]−[80]

$$S_G(q) = \pi \Delta^2 \zeta^2 \exp\left(-\frac{q^2\zeta^2}{4}\right), \tag{7.67}$$

where Δ and ζ are the height and correlation length of the roughness, respectively. However, very careful measurements of the actual oxide-semiconductor interface made with high-resolution TEM lattice-plane imaging, with subsequent reconstructive modeling of the two-dimensional interface, suggests that the interface is not Gaussian correlated, but rather is much better described in terms of exponential correlation [81]. In Sect. 2.7.3, the differences in the autocorrelation function taken from Fig. 2.1 were discussed for these two models. It is clear that neither the central region nor the tails fit very well. The exponential fit to the correlation function is better in both the low-momentum central region and the high-momentum tails. This conclusion has been corroborated by more recent measurements using atomic-force microscopy [82]. In Fig. 7.24, we show results from the AFM measurements, where the Fourier transform of the autocorrelation function is plotted. For a spatial frequency above about 10^{-2} nm^{-1}, the slope is very close to the cubic behavior expected for exponential correlation, in which the variation is [82],[83]

$$S_E(q) = \frac{\pi \Delta^2 \zeta^2}{(1 + q^2\zeta^2/2)^{3/2}}. \tag{7.68}$$

Roughly speaking, this suggests that the interface is much better described as consisting of terraces that are a few nanometers in extent, with steps of one or two atomic layers at the edges. The parameters obtained from the AFM measurements suggest that Δ and ζ are 0.3 nm and 15 nm, respectively.

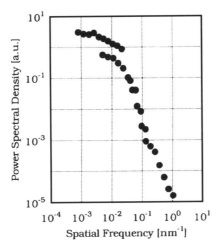

Fig. 7.24. The power spectral density (arbitrary units) of an AFM image of interface roughness in the oxide-Si interface. [After Yoshinobu *et al.*, Proc. Intern. Conf. Sol. State Dev. Mat., 1993, by permission.]

In general, then, the matrix element for scattering between subbands n and m for surface-roughness scattering is of the form

$$|\langle \mathbf{k}, n|U(\mathbf{q})|\mathbf{k} + \mathbf{q}, m\rangle|^2_{SR} = e^2 S(q) F_{eff}^2 A_{nm}^2(q), \qquad (7.69)$$

where the last factor has its previously defined meaning. In general, this potential interaction is unscreened at this point. Ando [84] and Saitoh [85] have pointed out that the potential can lead to variations in the local density, which means that there will be a screening of this potential. In general, the dielectric response enters through the replacement of the effective electric field by the charge density in the semiconductor (the depletion plus the inversion charge). We may write

$$F_{eff} = \frac{e}{\varepsilon(q)}(N_{depl} + \eta n_{inv}), \qquad (7.70)$$

where both densities are measured as sheet charges. There is some debate over the value for η, with values of 0.5 expected in simple cases [86], and $11/32$ quoted for a variational calculation of the energy levels and wavefunctions in a triangular potential [87]. With this result, the effective potential becomes

$$U_{eff}^2 = \frac{e^4 S^2(q) A_{nm}^2(q)(N_{depl} + \eta n_{inv})^2}{2\bar{\varepsilon}\left[1 + \frac{q_0}{q}F(q)\Pi(q)\right]^2}, \qquad (7.71)$$

where $\bar{\varepsilon}$ is the average dielectric constant between the oxide and the semiconductor, and the bracketed term in the denominator describes the screening due to the electron-electron interaction. This term was, in essence, derived for thermal Green's functions in Sect. 6.3.4 with the two-dimensional result being given by (6.95). We will derive this expression again in a later section. Here, it may be noted that q_0 is an effective screening wavevector, $F(q)$ is a slowly varying function that is unity at low densities and decreases with increasing Fermi energy, and $\Pi(q)$ is the polarization of the electron gas described to lowest order by the leading Green's function product $G_a G_r$ in (7.55).

7.3.3 The retarded function

It is still generally true that nearly all one-electron properties of the interacting electron system can be determined from the electron propagator itself, which is the interacting retarded Green's function. The remaining factor that must be known is the distribution function, which appears in (7.54). Here, we will concentrate on the retarded function, leaving the latter problem to the next section. In the interacting system, the retarded Green's function is related to the one-electron (noninteracting) retarded function by Dyson's equation. Thus, we need first to find the form for the noninteracting retarded Green's function, then to construct the self-energy function, and finally to formulate the interacting retarded Green's function.

The equation of motion for the noninteracting retarded function G_r^0 is found from Eq. (7.36) by setting both $V(\mathbf{r})$ and Σ to zero. Then we can rewrite this equation as

$$\left(i\hbar\frac{\partial}{\partial t} - H_0(\mathbf{r})\right)G_r^0(\mathbf{r}, \mathbf{r}', t - t') = \delta(\mathbf{r} - \mathbf{r}')\delta(t - t'), \qquad (7.72)$$

where equilibrium Hamiltonian $H_0(\mathbf{r})$ *includes* the confining potential of the quantum well at the interface. Because of the latter confinement, and because in general the system is separable, the noninteracting retarded Green's function will be expanded in a series over contributions from each of the subbands that form in the confinement potential. The three-dimensional vectors $(\mathbf{r}, \mathbf{k})$ will be written in terms of the two-dimensional vectors $(\mathbf{r}_p, \mathbf{k}_p)$ and the z-components (z, k_n), where k_n is defined by the confinement energy E_n of the nth subband. We can then write this function as

$$G_r^0(\mathbf{r}, \mathbf{r}', t - t') = \sum_n \psi_n^*(z')\psi_n(z)g_{r,n}^0(\mathbf{r}_p, \mathbf{r}_p', t - t'), \tag{7.73}$$

where the wavefunctions satisfy the completeness condition

$$\sum_n \psi_n^*(z')\psi_n(z) = \delta(z - z'). \tag{7.74}$$

Fourier-transforming with respect to the difference variables in the transverse direction (parallel to the interface), we find

$$G_r^0(\mathbf{k}_p, \omega) = \sum_n \psi_n^*(z')\psi_n(z)g_{r,n}^0(\mathbf{k}_p, \omega) = \sum_n \psi_n^*(z')\psi_n(z)\frac{\hbar}{\hbar\omega - E_n - E_{k_p} + i\hbar\eta}, \tag{7.75}$$

where η is a convergence factor, just as in (5.57). In fact, the noninteracting form here has the same generic form as the zero-temperature Green's functions, since the density of states per unit energy really doesn't change with temperature. Consequently, the unperturbed density of states function is still defined by

$$\rho_{0,2D}(\hbar\omega) = \frac{1}{\pi\hbar} \sum_n \int d^2\mathbf{k}_p a_{0,n}(\mathbf{k}_p, \omega) = \frac{g_v m^*}{\pi\hbar^2} \sum_n \Theta(\hbar\omega - E_n), \tag{7.76}$$

where g_v is the valley degeneracy (number of equivalent valleys in the conduction band), and where a factor of 2 has been included for spin degeneracy.

In a manner similar to the noninteracting Green's function, the interacting Green's function can be expanded in the subbands as

$$G_r(\mathbf{r}, \mathbf{r}', t - t') = \sum_n \psi_n^*(z')\psi_n(z)g_{r,n}(\mathbf{r}_p, \mathbf{r}_p', t - t'). \tag{7.77}$$

This diagonal approximation for the interacting retarded Green's function has been questioned, since a nondiagonal one might, for example, include a second summation over the various subband Green's function of different index within the first sum (and the two wavefunctions would then come from different subbands). However, the interaction between the various subbands is provided by the scattering processes themselves. We return to this point below. Thus, the appearance of diagonality in this equation is just an appearance; the result is fully nondiagonal once the self-energy is properly formulated. The subband retarded Green's function satisfies the Dyson equation, which appears after transformation as

$$g_{r,n}(\mathbf{k}_p, \omega) = \frac{\hbar}{\hbar\omega - E_n - E_{k_p} - \Sigma_{r,n}(\mathbf{k}_p, \omega)}. \tag{7.78}$$

The retarded self-energy that appears in this last equation may be calculated, by the same subband expansion, to be

$$\Sigma_{r,n}(\mathbf{k}_p, \omega) = \int dz \int dz' \psi_n^*(z)\Sigma_r(z, z'; \mathbf{k}_p, \omega)\psi_n(z'). \tag{7.79}$$

Since impurity and surface-roughness scattering are independent scattering mechanisms, both of which may be assumed to be weak, they can be treated separately. This is different

than the procedure used in the previous two chapters, where impurity scattering was sufficient to induce disorder (diffusive transport) so that impurity scattering and electron-electron
scattering had to be calculated by treatment of their interactions by multiple scattering. For
example, in Sect. 6.3.3, it was assumed that the electron-electron interaction was "screened"
by the impurity scattering–derived diffusons, which led to a vertex correction to the Coulomb
vertex. Here, it has been assumed that the electron-electron interaction is screened only by
itself, and in turn provides simple screening to the Coulomb and surface-roughness interactions. Both of the latter two interactions are assumed to be sufficiently weak that interaction
between these two scattering processes may be neglected. Hence, the self-energy can be
calculated separately for each and then added together. In this approach, the self-energy can
be written as

$$\Sigma_{r,n}(\mathbf{k}_p, \omega) = \sum_m \sum_{i=c,sr} \sum_{\mathbf{q}} |U_{i,nm}(\mathbf{q})|^2 g_{r,m}(\mathbf{k}_p - \mathbf{q}, \omega). \qquad (7.80)$$

The matrix elements for the scattering processes are given above, and there is a summation
over the two scattering processes. It is clear at this point that the interacting retarded Green's
function must be found self-consistently, not only because of the coupling among the Green's
functions themselves but also because the various subbands are explicitly linked within the
self-energy. The self-energy provides the renormalization of the subband energies as well
as the scattering rates themselves.

With this background, we can compare this approach with the nondiagonal one that often
appears in the literature. The interacting many-body propagator $G_{r,ij}(\mathbf{k}_p, \omega)$ is related to
the noninteracting free propagator $G_{r,ij}^0(\mathbf{k}_p, \omega)$ via the Dyson's equation [88]

$$[G_{r,ij}(\mathbf{k}_p, \omega)]^{-1} = [G_{r,ij}^0(\mathbf{k}_p, \omega)]^{-1} - \Sigma_{ij}(\mathbf{k}_p, \omega), \qquad (7.81)$$

where the self-energy is usually described in the noninteracting subband basis. Only in
rare and specialized models can one actually calculate the self-energy exactly. (Here it will
be done iteratively.) Since the noninteracting Green's function is exactly known and the
self-energy can be calculated, the interacting Green's function can be found by inverting
the matrix defined by (7.81). Moreover, the noninteracting Green's function is diagonal in
the noninteracting basis set, so that the last equation can be rewritten as

$$[G_{r,ij}(\mathbf{k}_p, \omega)]^{-1} = [G_{r,ii}^0(\mathbf{k}_p, \omega)\delta_{ij}]^{-1} - \Sigma_{ij}(\mathbf{k}_p, \omega). \qquad (7.82)$$

To reach the previous formulation, we sum over the j index in this last equation, and

$$[G_{r,i}(\mathbf{k}_p, \omega)]^{-1} = \sum_j [G_{r,ij}(\mathbf{k}_p, \omega)]^{-1} = [G_{r,ii}^0(\mathbf{k}_p, \omega)]^{-1} - \sum_j \Sigma_{ij}(\mathbf{k}_p, \omega). \qquad (7.83)$$

If we now define

$$\Sigma_i(\mathbf{k}_p, \omega) = \sum_j \Sigma_{ij}(\mathbf{k}_p, \omega), \qquad (7.84)$$

we can easily invert the equation and carry out the basis function operation defined in (7.77);
the result is just (7.78). Hence, the diagonal description used here is exact to the same order
as the matrix form of the Green's function that has appeared in the literature, which is
summarized in [88].

In Fig. 7.25, the density of states for the lowest subband that forms in the Si-SiO$_2$ interface
is shown for a combination of Coulomb and surface-roughness scattering. Here, Δ and ζ
have been taken to be 0.18 and 4 nm, respectively. It is clear that the sharp step in the density

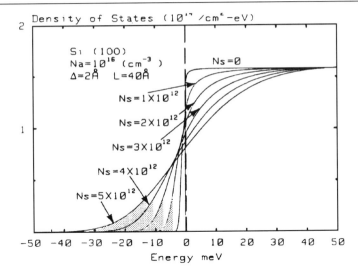

Fig. 7.25. The two-dimensional density of states in the lowest subband at the oxide-silicon interface. [After Goodnick *et al.*, Surf. Sci. **113**, 233 (1982).]

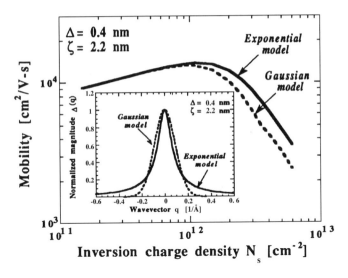

Fig. 7.26. The calculated zero-temperature mobility for the two models for surface roughness, with the same values of the scattering parameters.

of states is broadened considerably, and this effect increases as the density (and hence the scattering) increases. At high densities, where the effective field is quite large, the band tailing is actually dominated by the surface-roughness scattering.

The mobility at low temperature, where the distribution function can be taken to be a Fermi-Dirac function, is dependent on the model actually chosen for the correlation function of the surface roughness [89]. This can be seen in Fig. 7.26, where the mobility is adjusted to fit somewhat the mobility obtained in experiments by Kawaji *et al.* [90]. It is clear that the exponential model gives somewhat less-effective scattering, but this can be adjusted by changing the parameters in the model. In Fig. 7.27, theoretical fits to a variety of low-temperature data are shown [89]. The data here are the previously mentioned data as well as

Table 7.1. *Parameters used in the mobility fit.*

Parameter	Kawaji	Kobayashi	Goodnick	Units
Bulk Acceptor, N_a	0.9	2.8	1.0	10^{15} cm^{-3}
Depl. charge, N_{depl}	1.12	2.02	1.21	10^{11} cm^{-2}
Interface trap N_{it}	1.7	0.95	1.5	10^{11} cm^{-2}
Oxide charge N_{ox}	3.4		see text	10^{17} cm^{-3}
Delta, Δ	0.54	0.4	0.43	nm
Correl. length ζ	1.6	2.2	2.8	nm

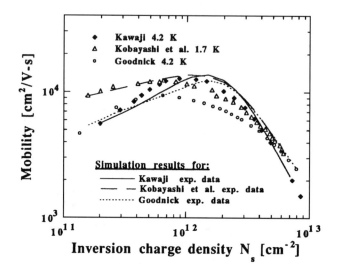

Fig. 7.27. The electron mobility as a function of inversion charge for several different sets of experimental data. The curves are discussed in the text.

that from Kobayashi *et al.* [91] and from Goodnick [92]. The latter data are from Hall-effect measurements and have been corrected for the Hall scattering factor. At this point, these results are fully equivalent to the use of the thermal Green's functions of the last chapter, since the introduction of the equilibrium Fermi-Dirac distribution really means that we are dealing with an equilibrium system. Nonequilibrium is obtained when we have to solve explicitly for the less-than Green's function in order to obtain the real distribution function. Exponential autocorrelation functions for the surface-roughness scattering have been used in the latter figure, and it can be seen that good agreement can be obtained by adjusting the two coupling constants in this scattering process. In the Goodnick data, the interface trapped charge, with a density of 4×10^{12} cm^{-2}, has been moved into the oxide layer by 6.5 nm, which represents a significant charging (due presumably to defects) in the oxide itself. The remainder of the fitting parameters for the various curves are tabulated in Table 7.1.

For all three devices shown in Fig. 7.27, the agreement between the theory and the experiment is quite good in both the low-density and the high-density regimes. The misfit between the experimental data and the theory in the mid-density regime occurs for all the samples and is an obvious neglect of some additional scattering process. One possible additional process is the electron-electron scattering, since these structures are characterized by a largely degenerate electron distribution in the quasi-two-dimensional gas.

7.3.4 The "less-than" function

We want now to turn our attention to the less-than function, which provides the quantum distribution function. The treatment here will be for a small value of the electric field, where the equations can be linearized in some sense. The high-electric-field case is left to the next section. Mahan [93] has given a full treatment of the linear-response approach to both the retarded and less-than Green's functions. In his equations of motion, which were developed for bulk transport, a gradient expansion in the self-energies was introduced to try to arrive at a closed set of equations. The approach here will be that of the multi-subbands present in an inversion layer, and the gradient expansion will not be used. Nevertheless, a closed Fredholm integral equation will be found for the scattering kernel in the distribution function (to be explained later), which is fully equivalent to that of Mahan, although the latter was done for phonon scattering.

We begin by writing down the equations of motion for the less-than Green's function from (7.36) and (7.37). These are

$$\left[i\hbar\frac{\partial}{\partial t} - H(\mathbf{r}) - e\mathbf{F}\cdot\mathbf{r} - V_c(\mathbf{r}) \right] G^<(\mathbf{r}, \mathbf{r}'; t, t') = \Sigma_r G^< + \Sigma^< G_a, \quad (7.85)$$

$$\left[-i\hbar\frac{\partial}{\partial t'} - H(\mathbf{r}') - e\mathbf{F}\cdot\mathbf{r}' - V_c(\mathbf{r}') \right] G^<(\mathbf{r}, \mathbf{r}'; t, t') = G_r \Sigma^< + G^< \Sigma_a, \quad (7.86)$$

where the product functions on the right sides are convolution integrals according to (7.39). We now make a change of variables to the so-called center-of-mass coordinates, through

$$\mathbf{R} = \tfrac{1}{2}(\mathbf{r} + \mathbf{r}'), \quad \mathbf{u} = \mathbf{r} - \mathbf{r}',$$
$$t_0 = \tfrac{1}{2}(t + t'), \quad \eta = t - t'. \quad (7.87)$$

Using these new coordinates, we can add and subtract the two equations (7.85) and (7.86) to arrive at the new equations

$$\left[i\hbar\frac{\partial}{\partial \eta} - \frac{\hbar^2}{2m}\nabla_\mathbf{u}^2 - \frac{\hbar^2}{8m}\nabla_\mathbf{R}^2 - e\mathbf{F}\cdot\mathbf{R} - V^+ \right] G^<(\mathbf{R}, \mathbf{u}; t_0, \eta)$$
$$= \frac{1}{2}[\Sigma_r G^< + \Sigma^< G_a + G_r \Sigma^< + G^< \Sigma_a], \quad (7.88)$$

$$\left[i\hbar\frac{\partial}{\partial t_0} - \frac{\hbar^2}{m}\nabla_\mathbf{u}\cdot\nabla_\mathbf{R} - e\mathbf{F}\cdot\mathbf{u} - V^- \right] G^<(\mathbf{R}, \mathbf{u}; t_0, \eta)$$
$$= [\Sigma_r G^< + \Sigma^< G_a - G_r \Sigma^< - G^< \Sigma_a], \quad (7.89)$$

where $V^\pm = V_c(\mathbf{R} + \mathbf{u}/2) \pm V_c(\mathbf{R} - \mathbf{u}/2)$. At this point, these two equations are Fourier-transformed on the difference variables. This yields

$$\left[\hbar\omega - E_\mathbf{k} - \frac{\hbar^2}{8m}\nabla_\mathbf{R}^2 - e\mathbf{F}\cdot\mathbf{R} - \frac{1}{2}(\Sigma_r + \Sigma_a) \right] G^<(\mathbf{R}, \mathbf{k}; t_0, \omega) = \frac{1}{2}(G_r + G_a)\Sigma^<, \quad (7.90)$$

$$\left[i\hbar\frac{\partial}{\partial t_0} - \frac{\hbar^2}{m}\mathbf{k}\cdot\nabla_\mathbf{R} - e\mathbf{F}\cdot\frac{\partial}{\partial\mathbf{k}} - (\Sigma_r - \Sigma_a) \right] G^<(\mathbf{R}, \mathbf{k}; t_0, \omega) = (G_r - G_a)\Sigma^<, \quad (7.91)$$

where the convolution integrals of functions have now become simple products in the Fourier space. The confinement potentials have been left out of these equations, but that is not correct. In order to treat the confinement potentials, we now separate the coordinates along the channel from those normal to the channel. The confinement potentials affect only the latter coordinates, and this is incorporated by expanding the Green's function into a subband series, as in (7.77). Then, the latter equations can be written in terms of the coordinates *along* the channel as

$$
\left[\hbar\omega - E_{\mathbf{k}_p} - \frac{\hbar^2}{8m}\nabla_{\mathbf{R}}^2 - e\mathbf{F} \cdot \mathbf{R} - \frac{1}{2}(\Sigma_{r,n} + \Sigma_{a,n}) \right] g_n^<(\mathbf{R}, \mathbf{k}; t_0, \omega) = \frac{1}{2}(g_{r,n} + g_{a,n})\Sigma_n^< ,
$$
(7.92)

$$
\left[i\hbar\frac{\partial}{\partial t_0} - \frac{\hbar^2}{m}\mathbf{k}_p \cdot \nabla_{\mathbf{R}} - e\mathbf{F} \cdot \frac{\partial}{\partial\mathbf{k}_p} - (\Sigma_{r,n} - \Sigma_{a,n}) \right] g_n^<(\mathbf{R}, \mathbf{k}; t_0, \omega) = (g_{r,n} - g_{a,n})\Sigma_n^< ,
$$
(7.93)

for the nth subband.

Our desire is to treat a homogeneous transport along the channel, for which we expect that the derivatives with respect to $\mathbf{R}$ and t_0 will vanish. However, care must be taken with this on account of the term in $e\mathbf{F} \cdot \mathbf{R}$ in (7.92). In essence, this term corresponds to a gauge shift as one moves along the channel and must be incorporated into the energy $\hbar\omega$ in order to be consistent [94]. In fact, Mahan suggests the change of variable $\hbar\omega + e\mathbf{F} \cdot \mathbf{R} \to \hbar\omega'$. In our case (of a Si inversion layer), the correction to the energy is quite small. The maximum size of the field term is of the order of $eF\lambda$. Now, in an inversion layer at 300 K, the mobility may be of the order of 500 cm^2/Vs, which leads to a mean free path $\lambda \sim 10$–15 nm, which leads to an energy correction for a field of 100 V/cm of 10–15 neV. The subband energies are typically tens of meV, so that this correction to the energy is quite small and may safely be ignored. A more important correction is the change in the gradient operation, in which

$$
\nabla_{\mathbf{R}} \to \nabla_{\mathbf{R}} + \frac{e\mathbf{F}}{\hbar}\frac{\partial}{\partial\omega} ,
$$
(7.94)

and the latter term will be quite important. Using these approximations and gauge changes, Eq. (7.93) can be written as

$$
\left[-e\mathbf{F} \cdot \mathbf{v}_p\frac{\partial}{\partial\omega} - e\mathbf{F} \cdot \frac{\partial}{\partial\mathbf{k}_p} - (\Sigma_{r,n} - \Sigma_{a,n}) \right] g_n^<(\mathbf{R}, \mathbf{k}; t_0, \omega) = (g_{r,n} - g_{a,n})\Sigma_n^< .
$$
(7.95)

This latter expression is often referred to as the *quantum Boltzmann equation*. This is the equation that needs to be solved in order to ascertain the subband less-than Green's function.

At this point, it is useful to introduce some further definitions, in order to ease the complexity of the following equations. For this purpose, we will write the retarded Green's functions and self-energies in terms of their real and imaginary parts, as

$$
g_{r,n} = (g_{a,n})^\dagger = W_n - i\frac{A_n}{2} , \quad \Sigma_{r,n} = (\Sigma_{a,n})^\dagger = S_n - i\Gamma_n ,
$$
(7.96)

where the Green's function has been defined using the accepted definition of the spectral density $A_n(\mathbf{k}_p, \omega)$. These functions were obtained in the previous section and can be carried over directly. We can also define the less-than self-energy in terms of the Green's function, using the extension of (7.80), as

$$
\Sigma_n^< = \sum_{i,m} \int \frac{d^2\mathbf{q}}{(2\pi)^2} W_{nm}^{(i)}(\mathbf{k} - \mathbf{q})g_m^<(\mathbf{q}, \omega) .
$$
(7.97)

An equivalent form for three-dimensional systems was introduced by Mahan [94]. Here, the summation over the index i sums over the various scattering processes that contribute to the self-energy. Finally, we introduce an assumed expansion for the less-than Green's function which is stimulated by its form in near equilibrium [94], but which does not really constrain the result to equilibrium, only to linear response in the electric fields. This form is

$$g_n^<(\mathbf{k}_p, \omega) = i A_n(\mathbf{k}_p, \omega) \left[f(\omega) - e\mathbf{F} \cdot \mathbf{v}_k \Lambda_n(\mathbf{k}_p, \omega) \frac{\partial f(\omega)}{\partial \omega} \right]. \tag{7.98}$$

The problem now is to find the kernel function $\Lambda_n(\mathbf{k}_p, \omega)$.

The task now is to introduce the expansion of the less-than Green's function into (7.95) along with the notational changes of (7.96). When this is done, the terms are of two types. Either they contain $f(\omega)$ or they contain the derivative of the distribution function with respect to ω. The former set of terms provides a differential equation, in $\mathbf{k}$ and ω, for the spectral density, but Mahan [93] has shown that it is basically the same form as the equilibrium one found in the previous section. That is, there are no corrections to this form that are linear in electric field. The second set of terms provides an integral equation for the kernel function:

$$\Lambda_n(\mathbf{k}_p, \omega) = \frac{1}{2\Gamma_n(\mathbf{k}_p, \omega)} \left[1 + \sum_{i,m} \int \frac{d^2\mathbf{q}}{(2\pi)^2} \frac{\mathbf{v}_{\mathbf{k}_p} \cdot \mathbf{v}_{\mathbf{q}}}{\mathbf{v}_{\mathbf{k}_p}^2} W_{nm}^{(i)}(\mathbf{k} - \mathbf{q}) \Lambda_m(\mathbf{q}, \omega) \right]. \tag{7.99}$$

This is the central result of this section. Once this integral equation is solved, the conductivity is easily found by the techniques introduced above as

$$\sigma_{2D} = -\frac{e^2}{\pi\hbar} \sum_n \int_0^\infty dE_{\mathbf{k}_p} \int \frac{d\omega}{2\pi} E_{\mathbf{k}_p} \Lambda_n(\mathbf{k}_p, \omega) A_n(\mathbf{k}_p, \omega) \frac{\partial f(\omega)}{\partial \omega}. \tag{7.100}$$

7.4 Considerations of mesoscopic devices

The previous section treated transport through a more-or-less uniform quasi-two-dimensional gas at the interface between the semiconductor and an insulator (or a wider-bandgap semiconductor such as that at the GaAs/AlGaAs interface). However, many of the mesoscopic devices of interest can not be described as existing in a uniform or homogeneous medium. Rather, devices such as the quantum dot (or lateral resonant tunneling diode at low bias and density of carriers) is strongly inhomogeneous, and the actual carrier densities within the dot region can be far from their equilibrium values, as indicated by the experiments discussed in the early sections of this chapter. In this section, we will begin to treat a somewhat different approach to the quantum mesoscopic device, one that involves in a sense a build-up from transfer Hamiltonians. This approach, applied to the quantum dot, follows the initial work of Meir *et al.* [95], in which conductance oscillations arising from the charging and discharging of the central dot region were analyzed with the Green's function technique. In the next paragraphs, we will largely follow the work of Jauho *et al.* [96], and at the end we will turn to a slightly different approach.

7.4.1 A model device

When considering a nonequilibrium tunneling structure, a Keldysh approach can be formulated as follows. In the remote past, the contacts at the right and left ends of the device and the central region (the quantum dot) may be assumed to be decoupled, and each region is in thermal equilibrium. In each of these three regions, the equilibrium distribution function is characterized by its respective Fermi energy (or chemical potential). These three chemical potentials do not have to correspond to one another, and the differences between them do not have to be small. One normally thinks of devices in which a single, uniform Fermi energy describes the entire device in the noncurrent-carrying equilibrium system. However, this is usually after some external connection *to each area of the device* has allowed charge redistribution to equilibriate the system overall. Here one can conceive of a device constructed in a manner in which the central dot region has not been allowed to equilibrate with the contacts through charge transfer; rather, the region continues to be isolated and cannot come to a common equilibrium. As one progresses from the distant past, the couplings between the different regions are established and treated as perturbations via the standard techniques described previously. It is important to note that these couplings do not have to be small, and therefore they must be treated to all orders (that is, the diagrams must be resummed properly to assure that renormalization occurs).

The time-dependent case can be treated similarly. In the distant past, the single-particle energies develop rigid time-dependent self-energy shifts which, in the case of the noninteracting contact reservoirs, translate into extra phase factors for the propagators. Although the perturbation theory has the same structure as in the stationary case, the calculations are made more difficult by the broken time-translational invariance.

The Hamiltonian for this system is broken into three pieces: $H = H_c + H_T + H_{cen}$, where H_c describes the contacts, H_T describes the coupling between the central region and the contacts via tunneling, and H_{cen} describes the central dot region. The contacts are assumed to broaden rapidly away from the dot barrier (the tunelling barrier, as described in Chapter 4) into metallic contacts. The electrons in these contact regions are assumed to be noninteracting except for some overall self-consistent potential. That is, any interactions among the carriers in these contacts have already been used to renormalize the electrons into quasi-particles that may be described as noninteracting particles. Applying a time-dependent bias (the potential difference between the right/drain contact and the left/source contact) means that the single-particle energies become time dependent: $E_{k\alpha}^0 \rightarrow E_{k\alpha}(t) = E_{k\alpha}^0 + \Delta_\alpha(t)$, where α is a label to denote a particular channel in the left or right contact region. The application of this bias does not change the occupation of the particular channel, however, and this occupation is determined by an equilibrium distribution function established in the distant past. Thus, we may write the contact Hamiltonian as [95]

$$H_c = \sum_{k, \alpha \in L, R} E_{k\alpha}(t) c_{k\alpha}^\dagger c_{k\alpha} , \tag{7.101}$$

where the $c_{k\alpha} (c_{k\alpha}^\dagger)$ are the appropriate fermion annihilation (creation) operators. The exact (in the distant past) Green's functions for the contacts in the uncoupled system are given by Eq. (7.23) as

$$g_{k\alpha}^<(t, t') = i\langle c_{k\alpha}^\dagger(t') c_{k\alpha}(t) \rangle = i f\left(E_{k\alpha}^0 \right) \exp\left[-\frac{i}{\hbar} \int_{t'}^{t} dt_1 E_{k\alpha}(t_1) \right], \tag{7.102}$$

where we have also used (7.50), the exponential function is the time version of the spectral density (a delta function in the noninteracting case), and

$$g_{k\alpha}^{r,a}(t, t') = \mp i\theta(\pm t \mp t') \langle \{ c_{k\alpha}(t), c_{k\alpha}^\dagger(t') \} \rangle$$

$$= \mp i\theta(\pm t \mp t') \exp\left[-\frac{i}{\hbar} \int_{t'}^{t} dt_1 E_{k\alpha}(t_1) \right]. \tag{7.103}$$

Here, the lowercase letters are used for the Green's functions to indicate the noninteracting forms of these functions.

The central, dot-region Hamiltonian depends on the geometry of the quantum region being simulated as well as upon the effects that are being investigated. The properties of interest here (and to the authors of [96]) are the relationship between the current and the local properties of the structure, such as the density of states, and Green's functions. Two forms for this Hamiltonian will be discussed. The first is to consider noninteracting time-dependent energy levels, for which the Hamiltonian may be written as

$$H_{cen} = \sum_m E_m(t)\mathbf{d}_m^\dagger \mathbf{d}_m , \tag{7.104}$$

where the operators $\mathbf{d}_m^\dagger(\mathbf{d}_m)$ create (destroy) an electron in state m. This model could be modified by the addition to this Hamiltonian of the Coulomb interaction $Un_\uparrow n_\downarrow$, which makes the model of the quantum dot an interacting model, but this will not be done here [95], [97]–[99]. Finally, the second model is that for a single state in the quantum dot, representing a resonant-tunneling level, with an electron-phonon interaction. This latter Hamiltonian becomes

$$H_{cen}^{e-ph} = E_0\mathbf{d}^\dagger\mathbf{d} + \mathbf{d}^\dagger\mathbf{d}\sum_{\mathbf{q}} M_{\mathbf{q}}\big[a_{\mathbf{q}}^\dagger + a_{-\mathbf{q}}\big]. \tag{7.105}$$

The second term in this Hamiltonian represents the interaction of an electron on the single site with the phonons. The full Hamiltonian must, of course, include the contribution from the free phonons, but this drops out of the interaction picture.

Finally, the coupling between the contacts and the central region can be expressed in terms of the hopping term

$$H_T = \sum_{k,\alpha\in L,R}\sum_n \big[V_{k\alpha,n}(t)\mathbf{c}_{k\alpha}^\dagger\mathbf{d}_n + h.c.\big]. \tag{7.106}$$

Here, the sets $\{\mathbf{d}_n^\dagger\}$ and $\{\mathbf{d}_n\}$ form a complete orthonormal set of single-electron creation and annihilation operators in the interacting region as expressed above. The hopping term describes the hopping of an electron from the dot to the contact (first term), and the hermitian conjugate operation, which is the hopping of an electron from one of the contacts into the dot region.

The current that flows into the device from the left contact (the cathode or source contact) is easily related to the change in the density in this contact due to transitions into the quantum dot region. Now, the current flowing from the contact into the dot corresponds to electrons flowing out of the dot, which leads to an increase in the density in the contact. This may be expressed as

$$J_L(t) = e\left\langle\frac{\partial N_L}{\partial t}\right\rangle = \frac{ie}{\hbar}\langle[H, N_L]\rangle, \tag{7.107}$$

where $N_L = \sum_{k,\alpha\in L,R}\mathbf{c}_{k\alpha}^\dagger\mathbf{c}_{k\alpha}$, H is given as above, and e is taken as a positive quantity. Since H_c and H_{cen} commute with the density in the left, one readily finds

$$J_L = \frac{ie}{\hbar}\sum_{k,\alpha\in L,R}\sum_n \big[V_{k\alpha,n}\langle\mathbf{c}_{k\alpha}^\dagger\mathbf{d}_n\rangle + V_{k\alpha,n}^*\langle\mathbf{d}_n^\dagger\mathbf{c}_{k\alpha}\rangle\big]. \tag{7.108}$$

To proceed, it is useful to define two correlation functions (Green's functions) that describe coupling between the dot and one of the contacts. These correspond to the less-than and greater-than Green's functions, and

$$G^<_{k\alpha.n}(t, t') = i\langle \mathbf{c}^\dagger_{k\alpha}(t')\mathbf{d}_n(t)\rangle, \tag{7.109}$$

$$G^>_{k\alpha.n}(t, t') = i\langle \mathbf{d}^\dagger_n(t')\mathbf{c}_{k\alpha}(t)\rangle. \tag{7.110}$$

If we now use the fact that these two Green's functions are related by $G^<_{k\alpha.n}(t, t') = -[G^>_{k\alpha.n}(t, t')]^*$, the right side of Eq. (7.108) can be rewritten as

$$J_L = \frac{2e}{\hbar} Re\left\{ \sum_{k,\alpha\in L,R} \sum_n V_{k\alpha,n}(t)G^<_{k\alpha,n}(t, t) \right\}. \tag{7.111}$$

It is now necessary to find an expression for this correlation function in order to evaluate the current. The approach to this is quite standard and is used for Green's functions as well as any other quantum functions. We develop an equation of motion for this correlation function fully in analogy to (7.107). This couples the correlation function to the Green's functions in the contact and in the central dot region with the coupling potential. This equation is then integrated. Since we are interested in the less-than Green's function in the interaction, the Langreth theorem is invoked for the Green's function product within the resulting equation of motion, and this can finally be written as [96]

$$G^<_{k\alpha.n}(t, t') = \sum_m \int dt'' V^*_{k\alpha,m}(t'')\left[G^r_{nm}(t, t'')g^<_{k\alpha}(t'', t')\right.$$

$$\left. + G^<_{nm}(t, t'')g^a_{k\alpha}(t'', t')\right], \tag{7.112}$$

where, for example, $G^t_{nm}(t, t') = -i\langle T\{\mathbf{d}^\dagger_m(t')\mathbf{d}_n(t)\}\rangle$. (The other Green's functions can be developed easily from the relations in a previous section. The lack of higher-order Green's functions is a direct result of the assumption of noninteracting contacts.) If we combine the above equations, the current can now be written as

$$J_L(t) = \frac{2e}{\hbar} Im\left\{ \sum_{k,\alpha\in L,R} \sum_{n,m} V_{k\alpha,n}(t) \int_{-\infty}^t dt'' V^*_{k\alpha,m}(t'') \exp\left[\frac{i}{\hbar}\int_{t''}^t d\eta E_{k\alpha}(\eta)\right] \right.$$

$$\left. \times \left[G^r_{nm}(t, t'')f_L(E_{k\alpha}) + G^<_{nm}(t, t'') \right] \right\}. \tag{7.113}$$

Finally, it is useful to recognize that the summation over k can be transformed into an integration over the energy, with the inclusion of a density of states, as

$$\sum_{k,\alpha\in L,R} \rightarrow \sum_{\alpha\in L,R} \int dE\rho_\alpha(E). \tag{7.114}$$

Now, using the separation in the energy expressed just above (7.101), we can define the *level-width functions*

$$\Gamma^L_{mn}(E, t', t) = 2\pi \sum_{\alpha\in L} \rho_\alpha(E)V_{k\alpha,n}(t)V^*_{k\alpha,m}(t') \exp\left[\frac{i}{\hbar}\int_{t''}^t d\eta \Delta_\alpha(\eta)\right]. \tag{7.115}$$

The interaction potential can be rewritten as $V_{k\alpha,n}(t) = V_{\alpha,n}(E,t)$, and we can finally write the current as

$$
J_L(t) = \frac{2e}{\hbar} Im \left\{ \sum_{n,m} \int_{-\infty}^{t} dt'' \int \frac{dE}{2\pi} e^{iE(t-t'')/\hbar} \Gamma_{mn}^L(E,t'',t) \right.
$$

$$
\left. \times \left[G_{nm}^r(t,t'') f_L(E) + G_{nm}^<(t,t'') \right] \right\}. \tag{7.116}
$$

This result is the main theorem from the work of Jauho *et al.* [96]. The current through the quantum dot structure is a function of local properties: the Green's functions in the quantum dot regime and the distribution function in the source contact. The first term in the last bracket is proportional to the occupation in the contact and is related to an *in-hopping* process. The last term is likewise related to an *out-hopping* process. An equivalent expression can be found for the current through the right contact. (In the time-independent case, Kirchoff's current law requires these two currents to be equal in magnitude with directions satisfying the overall conservation of current in the device.)

7.4.2 Proportional coupling in the leads

In the time-varying situation, we desire to develop a compact expression for the current through the leads. To do this, we will follow the lead of Jauho *et al.* [96] and make some assumptions on the line-width functions. The general assumption made by these authors is that the line-width functions defined by the coupling through the two ends of the dot may be related through

$$
\Gamma_{mn}^L(E,t',t) = \lambda \Gamma_{mn}^R(E,t',t). \tag{7.117}
$$

In general, this condition can be satisfied if the time-dependent level shifts in the two contacts are the same, that is, if $\Delta_\alpha^L(t) = \Delta_\alpha^R(t) = \Delta(t)$. This is not a restriction required in all cases, and the latter authors discuss other cases in which this restriction is removed. Here, however, we will retain this simplification, since the aim is merely to illustrate their approach to the treatment of mesoscopic devices in a far-from-equilibrium situation.

The manner in which the population in the central dot region can change is by the hopping of an electron out through either one of the two contacts, or by the opposite process of an electron hopping into the dot region from one of the leads. In this situation, the particle current through the contacts will differ for these contacts. Nevertheless, the continuity equation can be written for the central dot region as

$$
-e\frac{dN_{cen}(t)}{dt} = J_R(t) + J_L(t), \tag{7.118}
$$

where we recall that the currents are defined as flowing *into* the dot, and thus correspond to electrons moving out of the dot. Now, we can write the current as $J = xJ_L - (1-x)J_R$, which accounts for the fact that the particle current in the two leads may differ somewhat in the time-varying case. Equation (7.118) can be used to rewrite $J_L = xJ_L + (1-x)J_L$, and

$$
J_L(t) = xJ_L(t) - (1-x)\left[J_R(t) + e\frac{dN_{cen}(t)}{dt} \right]
$$

$$
= [xJ_L(t) - (1-x)J_R(t)] - (1-x)e\frac{dN_{cen}(t)}{dt}. \tag{7.119}
$$

We now introduce (7.116) and its equivalent for the right contact, so that

$$
J_L(t) = -(1-x)e\frac{dN_{cen}(t)}{dt} + \frac{2e}{\hbar}Im\left\{\sum_{n,m}\int_{-\infty}^{t}dt''\int\frac{dE}{2\pi}e^{iE(t-t'')/\hbar}\Gamma_{mn}^{R}(E,t'',t)\right.
$$
$$
\times\left[\lambda x G_{nm}^{r}(t,t'')f_L(E) - (1-x)G_{nm}^{r}(t,t'')f_R(E)\right.
$$
$$
\left.\left.+\lambda x G_{nm}^{<}(t,t'') - (1-x)G_{nm}^{<}(t,t'')\right]\right\}. \tag{7.120}
$$

The value of λ is now conveniently chosen so that $x = 1/(1+\lambda)$. This causes the expression on the last line of (7.120) to vanish, and this latter equation can be rewritten in the simpler form

$$
J_L(t) = -\frac{\lambda}{1+\lambda}e\frac{dN_{cen}(t)}{dt} + \frac{2e}{\hbar}\frac{\lambda}{1+\lambda}Im\left\{\sum_{n,m}\int_{-\infty}^{t}dt''\int\frac{dE}{2\pi}e^{iE(t-t'')/\hbar}\right.
$$
$$
\left.\times\Gamma_{mn}^{R}(E,t'',t)G_{nm}^{r}(t,t'')[f_L(E) - f_R(E)]\right\}. \tag{7.121}
$$

Jauho *et al.* [96] suggest that the time average of a function may be defined through

$$
\langle F(t)\rangle \equiv \lim_{T\to\infty}\int_{-T/2}^{T/2}dt\,F(t). \tag{7.122}
$$

Under d.c. bias, (7.122) is justifiable because it smooths the fluctuations that may arise in the device. In a perfectly periodic function, such as the exciting sinusoidal, time-varying potential envisioned in this model problem, this last equation also works just fine. The integral gives a T-independent result, and one can easily pass to the limit. In a nonlinear device with a time-varying excitation, however, this may not be the case. Nonlinear effects can cause a variety of frequency conversions, and the generation of subharmonics can arise. In particular, quadratic terms in the expansion of the nonlinear response will lead to rectification of the time-varying signal, which, under some circumstances, can lead to a slowly increasing d.c. bias on the device. The physics community usually adopts an assumption of linear response, for which the time-varying excitation is small in amplitude. The nonlinear mathematical community, however, has developed an extensive literature on *quasi-linear response*, for which the time-varying excitation is not required to be small. Rather, one looks for the overall, nonlinear response at the same frequency as the time-varying excitation. That is, one uses the averaging process

$$
\langle F(t)\rangle_\omega \equiv \int_{-\infty}^{\infty}dt\,F(t)\cos(\omega t), \tag{7.123}
$$

which synchronously detects the single-frequency component of the response of the device (similar techniques are, of course, used extensively by the experimental community). The approach that one wishes to use depends on just what response from the device is being studied, but care must be taken to assure that the definition is properly applied. Nevertheless, an averaging procedure will be followed, which will be indicated by the angular brackets.

The approach taken by the previous authors assumes that the tunneling couplings can be decomposed into energy- and time-dependent parts as $V_{k\alpha,n}(t,E) = u(t)V_{\alpha,n}(E_k)$. Thus, using the relationship between the two line-width functions – and this decomposition – Eq. (7.121) can now be rewritten as

$$\langle J_L(t)\rangle = \frac{2e}{\hbar} \int \frac{dE}{2\pi}[f_L(E) - f_R(E)]$$

$$\times Im\left\{\sum_{n,m} \frac{\Gamma_{mn}^R(E)\Gamma_{mn}^L(E)}{\Gamma_{mn}^R(E) + \Gamma_{mn}^L(E)} \langle u(t)A_{nm}(E,t)\rangle\right\}, \qquad (7.124)$$

where the first term is assumed to average to zero (no net change in the charge on the dot after each period of excitation), and

$$A_{nm}(E,t) = \int_{-\infty}^{t} dt'' e^{iE(t-t'')/\hbar} u(t'') G_{nm}^r(t,t'') \exp\left[\frac{i}{\hbar}\int_{t''}^{t} d\eta \Delta_\alpha(\eta)\right]. \qquad (7.125)$$

Because of the symmetry that now appears in the line-width function term, it is no longer necessary to distinguish between the left and right contacts, and the amplitude of the two currents are equal. The result (7.124) has the form of a Landauer formula, although the summation over the individual modes in each contact is still buried within the line-width functions. As a consequence, it is not a good representation of the multichannel Landauer formula at this point, although such a form can be achieved by some appropriate algebra.

7.4.3 A noninteracting resonant-level model

We now consider a model in which there is a single level in the quantum dot region. As above, we assume that the time dependence and the energy dependence (momentum dependence) of the tunneling coupling can be factorized. However, we assume that the time dependence of the two contacts may be slightly different: $V_{k\alpha,n}(t) = u_{L/R}(t)V_{\alpha,n}(E_k)$. In this approach, the real part of the self-energy, the level shift, will be ignored, and it is assumed that the line-widths are energy independent, so that $\sum_{\alpha \in L,R}\Gamma_\alpha \equiv \Gamma^{L,R}$. We also assume that only a single time-dependent function describes each of the contacts (in the energy and in the tunneling coefficient, etc.). Thus, we can write the retarded self-energy in terms of the the time-dependent functions and the line-widths as

$$\Sigma^r(t,t') = -i\sum_{\alpha \in L,R} u_\alpha^*(t)u_\alpha(t')\exp\left[\frac{i}{\hbar}\int_{t}^{t'} dt''\Delta_\alpha(t'')\right]$$

$$\times \int \frac{dE}{2\pi}e^{-iE(t-t')/\hbar}\theta(t-t')\Gamma_\alpha$$

$$= -\frac{i}{2}[\Gamma^L(t) + \Gamma^R(t)]\delta(t-t'). \qquad (7.126)$$

This result can be used in the Dyson equation for the interacting retarded Green's function (or more usefully in the equation of motion for this quantity) to give

$$G^{r,a}(t,t') = g^{r,a}(t,t')\exp\left\{\mp\frac{i}{2\hbar}\int_{t'}^{t} dt''[\Gamma^L(t'') + \Gamma^R(t'')]\right\}, \qquad (7.127)$$

where $g^{r,a}(t,t')$ is given by (7.103) without the subscripts $k\alpha,n$, since this Green's function describes the central dot state rather than the contact states. Within the spirit of these same approximations, we can write the less-than function for the central dot region as

$$G^<(t, t') = \frac{i}{\hbar} \int dt_1 \int dt_2 G^r(t, t_1) \Sigma^<(t_1, t_2) G^a(t_2, t')$$

$$= \frac{i}{\hbar} \int dt_1 \int dt_2 G^r(t, t_1) \left[\sum_{L,R} \int \frac{dE}{2\pi} e^{-iE(t_1-t_2)/\hbar} \right.$$

$$\left. \times f_{L,R}(E) \Gamma^{L,R}(E, t_1, t_2) \right] G^a(t_2, t'). \qquad (7.128)$$

This may be evaluated by the above approximations to give the current from (7.116) as [96]

$$J_L = \frac{e}{\hbar} \left[\Gamma^L(t) N(t) + \int \frac{dE}{2\pi} f_L(E) \int_{-\infty}^{t} dt' \Gamma^L(t', t) \right.$$

$$\left. \times Im\{ e^{iE(t-t')/\hbar} G^r(t, t') \} \right], \qquad (7.129)$$

where

$$N(t) = \sum_{L,R} \Gamma^{L,R} \int \frac{dE}{2\pi} f_{L,R}(E) |A_{L,R}(E, t)|^2 \qquad (7.130)$$

is the occupation of the central dot, and the notation

$$A_{L,R}(E, t) = \int dt' u_{R,L}(t') G^r(t, t') \exp\left[\frac{iE(t-t')}{\hbar} - \frac{i}{\hbar} \int_{t}^{t'} dt'' \Delta_{L,R}(t'') \right] \qquad (7.131)$$

has been introduced.

If we now assume that $u_R(t) = u_L(t) = u(t)$ and follow the same procedure as above for computing the "average" current, we find that this may again be written as a modification of (7.124):

$$\langle J \rangle = \frac{2e}{\hbar} \frac{\Gamma^L \Gamma^R}{\Gamma^L + \Gamma^R} \int \frac{dE}{2\pi} Im\{ f_L(E) \langle u(t) A_L(E, t) \rangle - f_R(E) \langle u(t) A_R(E, t) \rangle \}. \qquad (7.132)$$

These results can be extended almost trivially to the case of a phonon-assisted transition. For this, we recognize that the functions in the averaging are related to the spectral density. For this purpose, we return to Eq. (7.124), but with the approximations introduced here. This then leads to

$$J = \frac{ie}{\hbar} \frac{\Gamma^L \Gamma^R}{\Gamma^L + \Gamma^R} \int \frac{dE}{2\pi} [f_L(E) - f_R(E)] \int_{-\infty}^{\infty} dt e^{iEt/\hbar} [G^r(t) - G^a(t)]. \qquad (7.133)$$

The last bracketed term is recognized as the interacting spectral density. If one ignores the cold background of carriers in the system (the Fermi "sea"), then the Green's function can be calculated exactly as

$$G^r(t) = -i\theta(t) \exp\left[-\frac{it(E_0 - \Delta) - \Gamma t/2}{\hbar} - \Phi(t) \right], \qquad (7.134)$$

where

$$\Delta = \sum_{\mathbf{q}} \frac{M_{\mathbf{q}}^2}{\hbar \omega_{\mathbf{q}}} \qquad (7.135)$$

is the self-energy shift induced by the phonons, and

$$\Phi(t) = \sum_{\mathbf{q}} \frac{M_{\mathbf{q}}^2}{(\hbar\omega_{\mathbf{q}})^2} [N_q(1 - e^{i\omega_q t}) + (N_q + 1)(1 - e^{-i\omega_q t})]. \tag{7.136}$$

The straightforwardness of this result is felt by Jauho *et al.* [96] to clearly illustrate the power of the interacting current model developed here.

7.4.4 Another approach to the phonon-assisted tunneling

The second approach to be considered is one that is applied only to the steady-state situation of d.c. applied fields. First, the electron-phonon interaction will be treated in the self-consistent first Born approximation [100]. That is, only a single scattering process is included, but it is retained to all orders in the perturbation series. Second, the phonons are treated by the usual approximation that they consist of a bath of independent harmonic oscillators that interact with the electrons locally in position. This is not applicable to the polar mode of the optical phonons (which is a relatively long-range Coulombic interaction), but it does apply to the normal deformable-ion approximation used in the optical modes. Finally, the phonon modes themselves will be traced out of the problem, leaving only the electron coordinates and the electron-phonon interaction, as was done in the previous example. This approach will be applied to the large-scale equivalent of the previous sections – the resonant-tunneling diode. In this approach, the noninteracting electron Hamiltonian is described by the real-space terms

$$H_0 = \frac{(\mathbf{p} - e\mathbf{A})^2}{2m} + V(\mathbf{r}), \tag{7.137}$$

where $V(\mathbf{r})$ includes the linear potential drop, due to the applied bias, plus the conduction-band discontinuities that create the tunneling barriers. While the presence of the vector potential allows for the inclusion of a magnetic field, this term will be ignored here. Dephasing of the electrons is assumed to occur through their interaction with a bath of lattice vibrations, described by the creation and annihilation operators $\mathbf{a}_q^\dagger$ and $\mathbf{a}_q$, respectively, as in the above sections. The electrons interact with this bath through the term in (7.105) but are described here in the site representation as

$$H_{e-ph} = \sum_m U\delta(\mathbf{r} - \mathbf{r}_m)(\mathbf{a}_m^\dagger + \mathbf{a}_m). \tag{7.138}$$

If one assumes that a continuum of phonon modes exists, then the sum over the sites m becomes an integration over the modes: $\sum_m \rightarrow \int d\mathbf{r} \int d(\hbar\omega) J_0(\mathbf{r}; \hbar\omega)$. Here, J_0 is the density of oscillator modes and therefore describes the spectrum.

The self-energies for the electron-phonon interaction are evaluated in the self-consistent Born approximation, in which from the Langreth theorem one can write

$$\Sigma^{<,>}(\mathbf{r}, \mathbf{r}'; E) = \int dE' G^{<,>}(\mathbf{r}, \mathbf{r}'; E - E')D^{<,>}(\mathbf{r}, \mathbf{r}'; E'), \tag{7.139}$$

where the phonon Green's function is given by

$$D^{<,>}(\mathbf{r}, \mathbf{r}'; E) = \int d(t - t')e^{iE(t-t')/\hbar} \langle H_{e-ph}(\mathbf{r}, t)H_{e-ph}(\mathbf{r}', t')\rangle. \tag{7.140}$$

Rather than treat the full problem, it is assumed that the phonons scatter locally in space. In this sense, the retarded self-energy is written in terms of a simple energy shift and an imaginary part as

$$\Sigma^r(\mathbf{r}, \mathbf{r}'; E) = \left[\sigma(\mathbf{r}, E) - \frac{i}{\hbar 2\tau_\phi(\mathbf{r}, E)}\right]\delta(\mathbf{r} - \mathbf{r}'). \tag{7.141}$$

The retarded Green's function is then found from the equation of motion:

$$[E - H_0(\mathbf{r}) - \Sigma^R(\mathbf{r}, \mathbf{r}'; E)]G^r(\mathbf{r}, \mathbf{r}'; E) = \hbar\delta(\mathbf{r} - \mathbf{r}'). \tag{7.142}$$

Now, one can define the nonequilibrium occupation factor

$$f(\mathbf{r}, E) = \frac{n(\mathbf{r}, E)}{N_0(\mathbf{r}, E)}, \tag{7.143}$$

where

$$n(\mathbf{r}, E) = \frac{i}{2\pi}G^<(\mathbf{r}, \mathbf{r}; E), \quad p(\mathbf{r}, E) = -\frac{i}{2\pi}G^>(\mathbf{r}, \mathbf{r}; E),$$

$$N_0(\mathbf{r}, E) = n(\mathbf{r}, E) + p(\mathbf{r}, E) = -\frac{1}{\pi}G^r(\mathbf{r}, \mathbf{r}; E). \tag{7.144}$$

To proceed, it is necessary to define the less-than and greater-than self-energies. As above, these are assumed to be local functions, from which one may write

$$\Sigma^{<,>}(\mathbf{r}, \mathbf{r}'; E) = \mp i\frac{\hbar}{\tau_{p,n}(\mathbf{r}, E)}\delta(\mathbf{r} - \mathbf{r}'). \tag{7.145}$$

The Dyson's equation for the correlation functions may be written as

$$G^{<,>}(\mathbf{r}, \mathbf{r}'; E) = \int d\mathbf{r}_1 \int d\mathbf{r}_2 G^r(\mathbf{r}, \mathbf{r}_1; E)\Sigma^{<,>}(\mathbf{r}_1, \mathbf{r}_2; E)G^a(\mathbf{r}_2, \mathbf{r}'; E). \tag{7.146}$$

If we now introduce the local approximation for the self-energy, using Eq. (7.144), we can write the occupation factors as

$$f(\mathbf{r}, E) = \frac{1}{N_0(\mathbf{r}, E)}\frac{\hbar}{2\pi}\int d\mathbf{r}'\frac{|G^r(\mathbf{r}, \mathbf{r}'; E)|^2}{\tau_p(\mathbf{r}', E)}, \tag{7.147}$$

$$p(\mathbf{r}, E) = \frac{\hbar}{2\pi}\int d\mathbf{r}'\frac{|G^r(\mathbf{r}, \mathbf{r}'; E)|^2}{\tau_n(\mathbf{r}', E)}, \tag{7.148}$$

and

$$N_0(\mathbf{r}, E) = \frac{\hbar}{2\pi}\int d\mathbf{r}'\frac{|G^r(\mathbf{r}, \mathbf{r}'; E)|^2}{\tau_\phi(\mathbf{r}', E)}, \quad \frac{1}{\tau_\phi} = \frac{1}{\tau_p} + \frac{1}{\tau_n}. \tag{7.149}$$

In the last expression, the right side comes from combining the expressions like (7.148) for the particle and holes, and then using (7.141) and (7.144) to verify the relationship between the various scattering times. Thus, the last five equations form a consistent process once an expression is obtained for the two initial scattering times that appear in (7.145). This is found by reinserting these equations into the expression for the less-than Green's function (7.139). This leads finally to

$$\frac{1}{\tau_p(\mathbf{r}, E)} = \frac{\hbar}{2\pi}\int d(\hbar\omega)F(\mathbf{r}, \hbar\omega)N_0(\mathbf{r}, E + \hbar\omega)f(\mathbf{r}, E + \hbar\omega) \tag{7.150}$$

and

$$\frac{1}{\tau_n(\mathbf{r}, E)} = \frac{\hbar}{2\pi} \int d(\hbar\omega) F(\mathbf{r}, \hbar\omega) N_0(\mathbf{r}, E + \hbar\omega)[1 - f(\mathbf{r}, E + \hbar\omega)].$$ (7.151)

Finally, the phonon spectral function is given by

$$F(\mathbf{r}, \hbar\omega) = U^2 J_0(r, |\omega|) \times \begin{cases} N(\omega), & \omega > 0, \\ N(|\omega|) + 1, & \omega < 0, \end{cases}$$ (7.152)

where the distinction on the sign of the frequency is used to distinguish between absorption and emission of a phonon. Finally, one gets a closed integral equation for the distribution function

$$f(\mathbf{r}, E) = \frac{1}{N_0(\mathbf{r}, E)} \int d\mathbf{r}' |G^r(\mathbf{r}, \mathbf{r}'; E)|^2 \int dE' F(\mathbf{r}', E')$$

$$\times N_0(\mathbf{r}', E - E') f(\mathbf{r}', E - E')$$ (7.153)

by combining (7.147) and (7.150).

Two sets of boundary conditions are required in order to simulate an actual device. The first is on the retarded Green's function $G^r(\mathbf{r}, \mathbf{r}'; E)$. Lake and Datta [100] assume perfectly absorbing boundary conditions. The retarded Green's function itself is solved on a finite lattice and is then extended analytically to $\pm\infty$ [101]. Inelastic scattering is included throughout the boundary contacts, so the entire boundary regions act like parts of the contact. The second set of boundary conditions is for the distribution itself. For this, it is assumed that the distribution in the contacts is a proper Fermi-Dirac distribution with the appropriate chemical potential for that contact. Finally, by manipulating the above equations, one can arrive at an appropriate current density [102] (as a function of both energy and position):

$$J(\mathbf{r}, E) = \frac{e\hbar}{2\pi} \int d\mathbf{r}' \frac{|G^r(\mathbf{r}, \mathbf{r}'; E)|^2}{\tau_\phi(\mathbf{r}, E)} \left[\frac{f_0(E - \mu_{C1})}{\tau_\phi(\mathbf{r}', E)} - \frac{1}{\tau_p(\mathbf{r}', E)} \right],$$ (7.154)

which shows how the contact distribution propagates into the system. In Fig. 7.28, we illustrate some results of Lake and Datta for the energy-dependent current distribution throughout a resonant-tunneling diode, in which inelastic scattering is important. In the

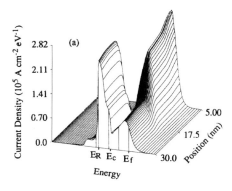

Fig. 7.28. The energy-dependent current-density for inelastic tunneling in an RTD. The parameters are discussed in the text. [After Lake and Datta, Phys. Rev. B **45**, 6670 (1992), by permission.]

figure, E_f is the Fermi energy in the emitter, E_c is the conduction band edge in the emitter, and E_r is the energy at the bottom of the resonance in the quantum well.

7.5 Nonequilibrium transport in high electric fields

In high electric fields, there can arise a significant interaction between the fields that drive the transport in the system and the dissipation within the system. The broadening that arises from the self-energy introduces the fact that there can be interference between the driving terms and the dissipative terms. This interference is of the same nature as the interference between the electron-electron interaction and the diffusive interaction with the impurities discussed in Chapter 6. Here, the interference between the field and the scattering processes is termed the *intra-collisional field effect* [103]. However, the effect was known earlier; it gives rise to field-induced effects in interband absorption that lead to the Franz-Keldysh effect [104]. Indeed, many of the properties of the Airy-function integrals that will be used here were worked out for a proper treatment of the Franz-Keldysh effect in interband absorption [105].

It has proven difficult to develop a tractable quantum transport approach, through the use of the real-time Green's functions, that incorporates both the collisional broadening that arises from the scattering processes and the intra-collisional field effect. Moreover, this task is further complicated by the need to deal with the length and time scales relevant to modern mesoscopic semiconductor devices. The general approach has followed that of the linear response used in previous chapters. Although the overall Green's function approach is rigorous in principle, most applications have been limited by the introduction of a center-of-mass transformation, in both space and time scales, and then by the use of a gradient expansion for slow variations around the center-of-mass coordinates. These two processes, especially the gradient expansion, tend to limit the results to low fields and prevent application to the far-from-equilibrium system. One of the first works to go beyond this was by Jauho and Wilkins [106], who treated high-field transport in a resonant level (the impurity level is in the conduction band) impurity scattering system. In this work, the self-energies for electron-phonon as well as impurity scattering were formulated; in the end, however, they were forced to invoke the gradient expansion in order to achieve a result for the kinetic equation. Nevertheless, the power of the Green's function approach was evident.

Another problem that becomes important is the actual collision duration – the actual time required, say, for the emission or absorption of a phonon. This problem has yet to be sorted out fully. There is little work that is really applicable to this, but one approach using the Green's functions is that of Lipavsky *et al.* [107]. In this latter work, it was shown that different definitions of the instantaneous approximation (i.e., single-time, which is a neglect of the ω dependence) for the self-energy will lead to different effects of the field on the collision, which emphasizes that proper renormalization of the Green's functions must be maintained. This collision duration can be quite important in fast processes [108].

The use of the center-of-mass transformations introduces some nonphysical variables into the description of transport. These, in fact, make the explicit assumption that the center-of-mass time $T = (t + t')/2$ has some physical significance (which it does in the limiting case $t = t'$). This isn't the case, usually, and nothing makes the point clearer than the need to modify the basic separation of the less-than Green's function into the product of the spectral density and the distribution function as done in Eq. (7.50). Approaches that have used the center-of-mass approach, with the gradient expansion, have been faced with the need to change the basic connection between $G^<$ and $f(\omega)$ found in (7.50) [109]. Consider the velocity autocorrelation function, which was the basis for the entire development of the diffusive transport of Chapter 5. This is a function of the time difference $t - t'$, where t' is the initial time for the correlation function. Except for the limiting case mentioned earlier, the center-of-mass time does not enter into any physical process of interest. Thus, an approach that avoids this need is preferable for investigating the far-from-equilibrium transport.

The transport in a high electric field is further complicated by the fact that the field actually breaks the symmetry along the field direction. In some sense, the momentum along the field can no longer be considered to be a "good quantum number," at least on the scale of the mean-free path. If the field is treated in the scalar potential gauge, in which $\mathbf{F} = -\nabla V$, it is the spatial translational symmetry that is broken. For this reason, many authors have tried to treat the field in a vector potential gauge, in which $\mathbf{F} = \partial \mathbf{A}/\partial t$, but this breaks the time-translational symmetry of the problem. These facts have greatly complicated the search for a high-field-transport solution for the Green's functions.

As a consequence of the many problems mentioned here, a different approach has been developed to treat high-field transport with the real-time Green's functions. This approach arises from the observations that the proper wavefunctions in a high electric field are the Airy functions [105]. Here, we transform the position variable along the field with a generalized Airy transform. This method is somewhat limited by the fact that it still assumes translational invariance in the transverse directions, through the use of the Fourier-transformed coordinates. The degree to which inhomogeneous systems can be treated by this approach is not yet known, although it is under investigation. The use of the Airy transformation will allow us to diagonalize the noninteracting Green's function in the presence of the field alone, and to achieve a simpler form of Dyson's equation [110].

The general approach is to utilize the Airy transformation of each portion of the coordinate system parallel to the electric field, which reduces the transverse coordinates effectively to a quasi-two-dimensional system. This approach differs, however, from the approach of the previous section in that the quantization in the z-direction has continuous eigenvalues rather than the discrete ones that arise in a true quantum well. The general Airy transform is defined by

$$F(\mathbf{k}_p, s) = \int d^2\mathbf{r}_p \int \frac{dz}{2\pi L} e^{i\mathbf{k}_p \cdot \mathbf{r}_p} Ai\left(\frac{z - s}{L}\right) f(\mathbf{r}_p, z), \qquad (7.155)$$

where $L = (\hbar^2/2m^* eF)^{1/3}$, $Ai(x)$ is an Airy function of the first kind, and s is a spatial variable (the transform variable) related to the zero crossings of the Airy function. In a triangular quantum well, with infinite confinement on one side and a linear potential on the other, the discrete wavefunctions are defined by the set of zero crossings s_n. Here, however, with the open system, s is continuous. For the Green's function, of course, there will be two Airy transformations for the two different positions. In this transform space, a function that is diagonal in momentum (only a single $\mathbf{k}_p$, which is assumed here) and in s variables is translationally invariant in the transverse plane, but not in the z-direction.

7.5.1 The retarded function

The retarded Green's function, in the presence of the electric field, still sastisfies Eqs. (7.36) and (7.37). In these equations, the electric field is introduced in the scalar potential gauge as eFz. In the absence of the self-energy, Eqs. (7.34) and (7.35) can be solved to give the solution for the noninteracting Green's function in the presence of the field, after Airy-transforming as

$$G^0_{r,F}(\mathbf{k}_p, s, s', t, t') = -i\Theta(t - t')e^{-iE_{\mathbf{k}_p,s}(t-t')/\hbar}\delta(s - s'), \tag{7.156}$$

where

$$E_{\mathbf{k}_p,s} = \frac{\hbar^2 k_p^2}{2m^*} + eFs \tag{7.157}$$

in parabolic bands. This form has the distinct advantage that the Fourier transformation of the difference variable in time leads to

$$G^0_{r,F}(\mathbf{k}_p, s, s', \omega) = \frac{\hbar\delta(s - s')}{\hbar\omega - E_{\mathbf{k}_p,s} + i\hbar\eta}, \tag{7.158}$$

which has the generic form expected for the retarded Green's function. There is a problem in using this form of the retarded Green's function because it assumes that the field was turned on in the infinite past (as indicated by the convergence factor η), but this ignores the transient part of the integration path used for the real-time functions. However, this is consistent with ignoring the "tail" segment connecting to the thermal equilibrium, which is consistent with seeking only the solution that is a balance between the driving force and the dissipative terms. Nevertheless, the approach here yields only the steady-state solution and cannot be used to evaluate the transient response of the carrier system.

The use of the above equilibrium form for the retarded noninteracting Green's function allows us to write the interacting Green's function as the integral solution of (7.36). In the Airy-transformed form, this equation then becomes

$$G_{r,F}(\mathbf{k}_p, s, s', \omega) = G^0_{r,F}(\mathbf{k}_p, s, \omega)\bigg\{\delta(s - s')$$

$$+ \int ds'' \Sigma_r(\mathbf{k}_p, s, s'', \omega)G_{r,F}(\mathbf{k}_p, s'', s', \omega)\bigg\}, \tag{7.159}$$

where the diagonality of the noninteracting function has been utilized to simplify the integral expression. Similarly, (7.37) can be transformed to yield

$$G_{r,F}(\mathbf{k}_p, s, s', \omega) = \bigg\{\delta(s - s') + \int ds'' G_{r,F}(\mathbf{k}_p, s, s'', \omega)$$

$$\times \Sigma_r(\mathbf{k}_p, s'', s', \omega)\bigg\}G^0_{r,F}(\mathbf{k}_p, s', \omega). \tag{7.160}$$

To proceed, it is now necessary to develop an expression for the self-energy that appears in the integrals on the right side of the last two equations. Here, we will treat the scattering due to optical phonons as the source for the self-energy. We have previously treated only scattering due to basically elastic processes, but in the presence of high fields, inelastic scattering due to the interaction of the carriers with the phonons is a dominant process. In treating this self-energy, a new consideration arises: the phonons are themselves characterized by a propagator describing their interaction. Normally, it is assumed that the phonons remain in thermal equilibrium (although much work in the laser excitation area has treated nonequilibrium phonons). Nevertheless, it is necessary to evaluate the paired propagators

$$\Sigma_r(\mathbf{r}, \mathbf{r}', t, t') = i[G(\mathbf{r}, \mathbf{r}', t, t')D(\mathbf{r}, \mathbf{r}', t, t')]_r. \tag{7.161}$$

It should be noted that the two Green's functions, $G(\mathbf{r}, \mathbf{r}', t, t')$ for the electron and $D(\mathbf{r}, \mathbf{r}', t, t')$ for the phonon, have the same set of spatial and temporal arguments. That is, they propagate in parallel. This means that they will create a convolution integral when Fourier-transformed into the momentum and energy domain. But which Green's functions do we take to get the retarded product? By the same arguments used in the Langreth theorem above, one can easily show that the terms needed are [52]

$$\Sigma_r(\mathbf{r}, \mathbf{r}', t, t') = i[G_r(\mathbf{r}, \mathbf{r}', t, t')D^>(\mathbf{r}, \mathbf{r}', t, t')$$

$$+ G^<(\mathbf{r}, \mathbf{r}', t, t')D_r(\mathbf{r}, \mathbf{r}', t, t')]. \tag{7.162}$$

The operator ordering in the last term is such that it vanishes as the density goes to zero. As a result, for nondegenerate semiconductors (which is the case most often, especially at high electric fields), it is possible to ignore the second term as being small in comparison with the first term. Moreover, in most semiconductors, the scattering by phonons is weak (there is very little multiple scattering) and the self-energy can be calculated in the Born approximation; essentially, the Green's function in the first term is replaced by the noninteracting form, which is the free, field-assisted propagator (but see [111]). Since the retarded Green's function is still solved self-consistently within Dyson's equation, collisional broadening still will appear in the final form, and the field's presence in the noninteracting Green's function introduces the interference between the field and the scattering process.

The phonon Green's function for nonpolar optical scattering can be expressed as (again, for the phonons in thermal equilibrium)

$$D_0^>(\mathbf{q}, \omega) = -i\pi \sum_{\pm} |M(\mathbf{q})|^2 \delta(\omega \pm \omega_0), \tag{7.163}$$

and it has been assumed that the phonons are nondispersive. The matrix element $M(\mathbf{q})$ also includes the phonon occupation factors – the Bose-Einstein distributions for the phonons, while the sum runs over emission and absorption processes. With these various approximations, the retarded self-energy can be written as

$$\Sigma_r(\mathbf{k}, \omega) = \pi \int \frac{d^3\mathbf{q}}{(2\pi)^3} \sum_{\pm} |M(\mathbf{q})|^2 G_{r,F}^0(\mathbf{k} \pm \mathbf{q}, \omega \pm \omega_0), \tag{7.164}$$

and it should be noted that this result *has not been Airy-transformed*. The momentum variable is $\mathbf{k} = \mathbf{k}_p + k_z \mathbf{a}_z$. To be useful, the momentum dependence in the field direction is transformed back to real space and then Airy-transformed. More properly, the product form is first generated in real space and then the result Airy- and Fourier-transformed using the delta function to break the convolution in Fourier space. In general, the matrix element for the nonpolar optical modes is not momentum dependent, so there is no complication in the matrix element in these various transformations. Finally, after carrying out the integrations over momentum, the self-energy is found to be [110]

$$\Sigma_r(\mathbf{k}_p, s, \omega) = \frac{m^*}{\hbar^2}\sqrt{\frac{m^*\xi}{2}} \sum_{\pm} |M(\mathbf{q})|^2 [M(s, \omega \pm \omega_0) + iN(s, \omega \pm \omega_0)], \tag{7.165}$$

where

$$M(s, \omega) = Ai'^2(\zeta) + uAi^2(\zeta), \tag{7.166}$$

$$N(s, \omega) = Ai'(\zeta)Bi'(\zeta) + \zeta Ai(\zeta)Bi(\zeta), \tag{7.167}$$

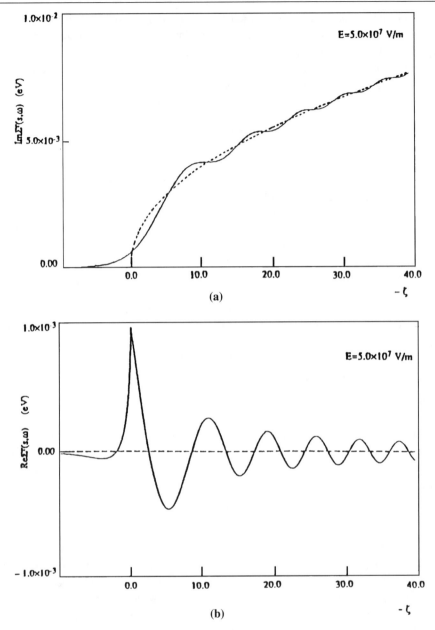

Fig. 7.29. The real (b) and imaginary (a) parts of the self-energy for nonpolar optical scattering in the Airy-transformed model.

and

$$\zeta = \frac{eFs - \hbar\omega}{\xi}, \qquad \xi^3 = \frac{3(e\hbar F)^2}{2m^*}. \tag{7.168}$$

The real and imaginary parts of the self-energy are plotted in Fig. 7.29. for parameters suitable for silicon. The oscillatory behavior of the real part of the self-energy is quite interesting and indicates that the interaction of the field and the scattering process is creating

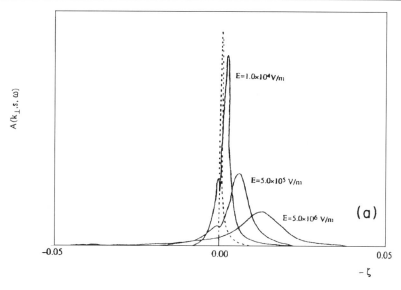

Fig. 7.30. The spectral density for three different values of the electric field. The transverse energy is taken to be the phonon energy in this case.

an equivalent quasi-two-dimensional behavior in the electron gas, which is reinforced by looking at the imaginary part. The step-like structure reinforces this interpretation.

Since the self-energy is diagonal in the Airy variable s, the integral in Dyson's equation collapses (due to an assumed delta function on the difference of the two arguments in the self-energy). There is no reason, then, not to also assume that the retarded Green's function is diagonal in the Airy transform variable. This leads to

$$G_{r,F}(\mathbf{k}_p, s, \omega) = \frac{\hbar}{\hbar\omega - E_{\mathbf{k}_p,s} - \Sigma_r(\mathbf{k}_p, s, \omega)}, \qquad (7.169)$$

which is in the noninteracting form and actually has the form of the zero-temperature Green's function as well. Here, however, the energy term contains the Airy-transformed scalar potential, which also appears in the self-energy. This result differs from that of the previous chapters. The spectral density is then twice the imaginary part of the retarded Green's function:

$$A(\mathbf{k}_p, s, \omega) = \frac{-2Im\{\Sigma_r(\mathbf{k}_p, s, \omega)\}}{\left[\hbar\omega - E_{\mathbf{k}_p,s} - Re\{\Sigma_r(\mathbf{k}_p, s, \omega)\}\right]^2 + [Im\{\Sigma_r(\mathbf{k}_p, s, \omega)\}]^2}, \qquad (7.170)$$

which, again, has the form as found in the equilibrium (or zero-temperature) Green's functions. In Fig. 7.30, the spectral density is plotted for three different values of the electric field, again for the parameters appropriate to silicon. The shift and distortion introduced by the electric field are clearly evident in the figure. Note, however, that the normalization is maintained. Any integration over the spectral density will therefore show only small effects arising from the field distortion and shift.

7.5.2 The less-than function

With the approximations in the above approach, it has been possible to get a good representation of the spectral density, which remains correct to all orders in the electric field since it is directly included without the need for any expansion in powers of the field (such as in the gradient expansion). Although the above approach is constrained to the case of weak scattering, higher orders of scattering would lead to the use of the self-consistent Born approximation in which $G_{r,F}$ itself is used in the self-energy, which leads to the need to interate the solutions. This, however, is only a complication in the solution procedure and does not invalidate the entire approach.

With the spectral function and the retarded Green's function now determined, it is possible to proceed to find the less-then function $G^<$. In the low-field case, it is often found that one needs to solve an integral equation for the latter function, especially in strong scattering where the scattering self-energy results from higher-order interactions treated by the Bethe-Salpeter equation, as in Chapter 5. In the weak-scattering limit here, the approach will be somewhat easier, but an integral equation is still expected for the less-than function, particularly because there is an integral over the product of the self-energy and the less-than function in Eqs. (7.36) and (7.37). This integral equation will actually be an integral equation for the distribution function, just as that which arises in Boltzmann transport [26].

The starting points, as in the case of the retarded equation, are the equations of motion for the Keldysh Green's function matrix. After introducing the Fourier and Airy transformations, two equations of motion result for the less-than function, as follows:

$$\left(\hbar\omega - E_{\mathbf{k}_p,s}\right)G^<(\mathbf{k}_p, s, s', \omega) = \int ds'' [\Sigma_r(\mathbf{k}_p, s, s'', \omega)G^<(\mathbf{k}_p, s'', s', \omega)$$

$$+ \Sigma^<(\mathbf{k}_p, s, s'', \omega)G_a(\mathbf{k}_p, s'', s', \omega)], \quad (7.171)$$

$$\left(\hbar\omega - E_{\mathbf{k}_p,s'}\right)G^<(\mathbf{k}_p, s, s', \omega) = \int ds'' [G_r(\mathbf{k}_p, s, s'', \omega)\Sigma^<(\mathbf{k}_p, s'', s', \omega)$$

$$+ G^<(\mathbf{k}_p, s, s'', \omega)\Sigma_a(\mathbf{k}_p, s'', s', \omega)]. \quad (7.172)$$

At this point, it has already been established that the retarded Green's function, the advanced Green's function (which is merely its complex conjugate), and the corresponding self-energies are functions of a single Airy variable. This eliminates the integration on the right sides of the two equations above. We can then formulate the sum and differences of these two equations, which in the present case yield the same resulting equation. In essence, this result means that these two equations are self-adjoint, which is an important advantage of the Airy transform approach. The resulting equation is then [112]

$$G^<(\mathbf{k}_p, s, s', \omega) = \frac{[G_r(\mathbf{k}_p, s, \omega) - G_a(\mathbf{k}_p, s', \omega)]\Sigma^<(\mathbf{k}_p, s, s', \omega)}{E_{\mathbf{k}_p,s} - E_{\mathbf{k}_p,s'} - \Sigma_a(\mathbf{k}_p, s', \omega) + \Sigma_r(\mathbf{k}_p, s, \omega)}$$

$$= \frac{1}{\hbar}G_r(\mathbf{k}_p, s, \omega)G_a(\mathbf{k}_p, s', \omega)\Sigma^<(\mathbf{k}_p, s, s', \omega). \quad (7.173)$$

Because the retarded and advanced Green's functions have imaginary parts that are sharply peaked, we expect the product of these two functions to be small unless the arguments correspond to one another. This leads us to assume that both the less-than self-energy and the less-than Green's function are diagonal in the Airy transform variable. With this approximation, it is then possible to recognize that the less-than Green's function satisfies the decomposition that has been used in deriving the Kubo formula above. That is, we can rewrite the last line of (7.173) as

$$G^<(\mathbf{k}_p, s, \omega) = i A(\mathbf{k}_p, s, \omega) f(\mathbf{k}_p, s, \omega), \tag{7.174}$$

since the first line of (7.173) contains the spectral function directly in the numerator. As a result, the quantum distribution function is obtained directly as the solution of

$$f(\mathbf{k}_p, s, \omega) = \frac{i \Sigma^<(\mathbf{k}_p, s, \omega)}{2 Im\{\Sigma_r(\mathbf{k}_p, s, \omega)\}}. \tag{7.175}$$

The less-than self-energy function can be developed as easily as the retarded self-energy for the case of the nonpolar optical phonons. As previously, it will be assumed that this function is diagonal in the Airy variable, so that we can write it (as in the previous section) as

$$\Sigma^<(\mathbf{k}_p, s, \omega) = \frac{1}{3^{1/6}L} \sum_\pm |M(\mathbf{q})|^2 \int d^2\mathbf{q}_p \int ds' Ai\left(\frac{s - s'}{3^{1/3}L}\right) G^<(\mathbf{k}_p - \mathbf{q}_p, s', \omega \pm \omega_0), \tag{7.176}$$

where the factor L has been defined previously. (The approach of the Langreth theorem allows us to write the parallel propagators as $\Sigma^< = D^< G^<$.) This last equation, of course, also depends upon the less-than Green's function. There is an important simplification that arises in the case of nonpolar optical phonon scattering. In this last scattering process, the scattering is *isotropic*, that is, it does not depend on the direction or magnitude of the scattering wavevector $\mathbf{q}$. This is not a general result, but it is one that arises particularly to the case of such isotropic scattering. Thus, a change of variables can be carried out in the integration over the scattering momentum, and the resulting integration removes all momentum dependence from the right side of the equation. Hence, for isotropic scattering the self-energy is independent of the momentum. A casual examination of (7.165) shows that the retarded function is independent of the momentum as well. Thus, both of the self-energies, and also the resulting distribution function, are independent of the momentum in the case of isotropic scattering. Thus, the resulting distribution function is dependent on only the energy ($\hbar\omega$) and the Airy variable. Thus, the integral equation for the distribution function can be written by combining (7.175) and (7.176) to give

$$f(s, \omega) = \frac{1}{2(3)^{1/6}L Im\{\Sigma_r(s, \omega)\}} \sum_\pm |M(\mathbf{q})|^2$$

$$\times \int ds' Ai\left(\frac{s - s'}{3^{1/3}L}\right) \rho_{s'}(s', \omega \pm \omega_0) f(s', \omega \pm \omega_0), \tag{7.177}$$

where

$$\rho_s(s, \omega) = \int d^2\mathbf{q}_p A(\mathbf{q}_p, s, \omega) \tag{7.178}$$

is the effective density of states at the final energy (in Airy transform space) of the scattering process. We note that the denominator of the prefactor plays the same role as the scattering function in the Boltzmann transport equation. In fact, the resulting formulation maintains much of the structure that is present in the path integral solution of the Boltzmann transport equation [26].

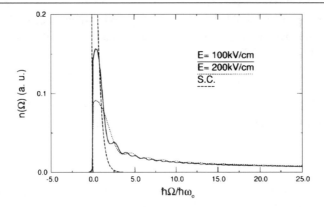

Fig. 7.31. The distribution function (or, more properly, the local density of particles) as a function of the electron energy. [After Bertoncini and Jauho, Phys. Rev. Lett. **68**, 2826 (1992), by permission.]

In Fig. 7.31, a distribution function appropriate to the above calculations for the retarded functions and the spectral density is shown (this is for silicon at room temperature) [113]. Although this solution has been found from an interative solution of (7.177), it should be remarked that it is possible to develop ensemble Monte Carlo solutions to the integral equation, as developed for the semiclassical Boltzmann equation.

7.5.3 Gauge-invariant formulations

It is not obvious that the present formulation retains the desired property of gauge invariance. In fact, such invariance is intrinsic to the understanding of just what the Airy-transformed equations represent. In treating the transport of carriers in a semiconductor, one has two different choices of the reference for energies and potentials. In one approach, it is possible to keep an external reference level as the zero of energy. Then as a particle is accelerated in an applied potential (by the electric field, which is the gradient of that potential), it actually moves ballistically, in that it maintains a constant energy. In consequence, the conversion of potential energy to kinetic energy must be accommodated by the conduction band edge's decreasing with position as the particle gains kinetic energy with position. Thus, the difference between the energy level of the particle (which remains constant in the ballistic motion) and the conduction band edge increases corresponding to the increase in kinetic energy of the particle due to the acceleration.

Normally, one follows a different representation in which the conduction band edge is treated as the absolute reference for the energies, and acceleration of the particle in an applied field corresponds to an increase in energy of the particle due to the acceleration. In this reference frame, the particle exhibits an increasing energy as it moves under the influence of the field. This is the reference frame that is normally utilized in essentially all treatments of classical transport as well as quantum transport. Nevertheless, the difference between this approach and the previous one is one of a gauge transformation. Gauge invariance asserts that the results that describe physical processes must be independent of the gauge used in the theoretical treatment. For the Airy transform approach, it is obvious that this should remain the case, a point that has been carefully proven by Bertoncini and Jauho [114],[115].

The use of the Airy transform approach is based on the former gauge, in which an external reference is maintained for the total energy. Then, the motion of the conduction band edge creates a system in which a linear potential is applied (actually this holds for either view), and this leads to the Airy function basis as the proper set of basis states for the solution of the transport problem. It is well known in quantum mechanics that the basis set used in solving quantum mechanical problems is not a unique property, and that any set that makes the problem easiest to solve is the "proper" choice. In the Airy transform approach, the energy gained from the field is represented by the term in the energy that varies as eFs. The Airy variable s can, in turn, be related to concepts such as effective turning points for the classical motion [116], although this connection must be pursued carefully. Nevertheless, the energy representation along the s-direction corresponds to that portion of the energy, so the reduced variables used above $(\hbar\omega - eFs)$ describe the transverse component of the energy. The total kinetic energy depends on just where the measurement is made, but it is clear that all energy components are properly included in the Airy transform approach, so that gauge invariance must follow.

Once the distribution function is found from the integral equation (7.177), then it can be transformed appropriately to a form (by retransforming with the Airy transform to a z-dependence, which is then included as the third component of a Fourier transform to momentum) that may now be used to compute the conductance from (7.53). For this, of course, the retarded and less-than Green's functions must be calculated within the same approach, as has been done in this section. We will pursue this argument of gauge invariance further here, following the basic approach of Bertoncini and Jauho [114].

To begin, we consider the product $g(\mathbf{r}, t; \mathbf{r}', t') = \Psi(\mathbf{r}, t)\Psi^\dagger(\mathbf{r}', t')$ of field operators. In terms of the "center-of-mass" coordinates from Eqs. (7.87), this product can be written as

$$g(\mathbf{R}, t_0; \mathbf{u}, \eta) = \Psi\left(\mathbf{R} + \frac{\mathbf{u}}{2}, t_0 + \frac{\eta}{2}\right)\Psi^\dagger\left(\mathbf{R} - \frac{\mathbf{u}}{2}, t_0 - \frac{\eta}{2}\right). \tag{7.179}$$

This may be Fourier-transformed in the difference coordinates to

$$g(\mathbf{R}, t_0; \mathbf{k}, \omega) = \int d^3\mathbf{r} \int dt\, e^{iw(\mathbf{R}, t_0; \mathbf{u}, \eta; \mathbf{k}, \omega)} g(\mathbf{R}, t_0; \mathbf{u}, \eta), \tag{7.180}$$

where

$$w(\mathbf{R}, t_0; \mathbf{u}, \eta; \mathbf{k}, \omega) = \int_{-1/2}^{1/2} d\xi \left\{ \eta\left[\omega + \frac{e}{\hbar}\varphi(\mathbf{R} + \xi\mathbf{u}, t_0 + \xi t)\right] \right.$$
$$\left. - \mathbf{r}\cdot\left[\mathbf{k} + \frac{e}{\hbar}\mathbf{A}(\mathbf{R} + \xi\mathbf{u}, t_0 + \xi t)\right] \right\}. \tag{7.181}$$

Here, φ and $\mathbf{A}$ are the scalar and vector potentials, respectively. This specific form of the Fourier transform contains explicit corrections to frequency and momentum that arise in a gauge field. We are familiar with the normal choice of a solely vector potential in the Peierl's substitution for treating a magnetic field. Even here, however, we are asssuming explicitly that the nature of the gauge fields in the Hamiltonian are such that the Hamiltonian does not contain spatially varying parameters whose influence does not produce contributions to the phase of the wavefunction that cannot be removed by a gauge transformation. The discussion of these latter quantities (which, for example, can lead to concepts such as the Berry phase) is beyond the scope of the present treatment, but it is known that in quasi-two-dimensional systems subject to both electric and magnetic fields, the invocation of periodicity in both dimensions leads to the need for phase corrections by nonanalytic potentials. However, there is some indication that such corrections may, in fact, be important in proper treatment of the quantum transport in far-from-equilibrium mesoscopic systems [117].

The assertion of gauge invariance is that $g(\mathbf{R}, T; \mathbf{k}, \omega)$ remains unchanged under a gauge transformation described through the function $\chi(\mathbf{r}, t)$ as follows:

$$\mathbf{A} \rightarrow \mathbf{A}'(\mathbf{r}, t) = \mathbf{A}(\mathbf{r}, t) + \nabla \chi(\mathbf{r}, t),$$

$$\varphi \rightarrow \varphi'(\mathbf{r}, t) = \varphi(\mathbf{r}, t) - \frac{1}{c} \frac{\partial \chi(\mathbf{r}, t)}{\partial t}, \qquad (7.182)$$

if φ and $\mathbf{A}$ are connected by the usual Lorentz gauge relationship, and if $\chi(\mathbf{r}, t)$ satisfies the source-free Helmholtz equation

$$\nabla^2 \chi(\mathbf{r}, t) - \frac{1}{c^2} \frac{\partial^2 \chi(\mathbf{r}, t)}{\partial t^2} = 0. \qquad (7.183)$$

The proof, which will not be given here, involves introducing the gauge field phase factor as $\Psi'(\mathbf{r}, t) = \exp[ie\chi(\mathbf{r}, t)/\hbar]\Psi(\mathbf{r}, t)$, which leads to an additional phase factor in the amount

$$\Delta\phi = -\frac{e}{\hbar} \int_{-1/2}^{1/2} d\xi \frac{d\chi(\mathbf{R} + \xi\mathbf{u}, t_0 + \xi t)}{dt}$$

$$= -\frac{e}{\hbar} \left[\chi\left(\mathbf{R} + \frac{\mathbf{u}}{2}, t_0 + \frac{\eta}{2}\right) - \chi\left(\mathbf{R} - \frac{\mathbf{u}}{2}, t_0 - \frac{\eta}{2}\right) \right], \qquad (7.184)$$

which is canceled by an equivalent term, with opposite sign, that arises from the exponential term of (7.180).

What we now want to do is to combine the gauge invariance argument with the Airy transform introduced earlier. To begin, we note that the frequency ω and the Airy transform variable s always occur in the form $\omega - eFs/\hbar$. Thus, the Airy transform variables and the energy are not independent variables. Therefore, we consider a function of the form

$$f(\mathbf{k}_p, \omega - eFs/\hbar, \omega - eFs'/\hbar). \qquad (7.185)$$

For this, we combine the gauge invariance argument with the general Airy transform as

$$\tilde{f}(\mathbf{k}_p, \mathbf{k}_z, \mathbf{k}_{z'}, \eta) = \int \frac{d\omega}{2\pi} e^{i(w + eFZ/\hbar)\eta} \int \frac{dz}{2\pi} \int \frac{dZ}{2\pi} e^{ik_z(Z+z/2)} e^{ik_{z'}(Z-z/2)}$$

$$\times \int\int \frac{ds\,ds'}{L^2} Ai\left(\frac{Z + \frac{z}{2} - s}{L}\right) Ai\left(\frac{Z - \frac{z}{2} - s'}{L}\right)$$

$$\times f(\mathbf{k}_p, \omega - eFs/\hbar, \omega - eFs'/\hbar). \qquad (7.186)$$

Here, the z-coordinates have been replaced by their center-of-mass values. If we now write

$$f(\mathbf{k}_p, \omega - eFs/\hbar, \omega - eFs'/\hbar) = \int\int dt\,dt'\, e^{i(\omega - eFs/\hbar)t} e^{-i(\omega - eFs'/\hbar)t'} f(\mathbf{k}_p, t, t') \qquad (7.187)$$

and use the integral representation of the Airy function, Eq. (7.186) reduces to

$$\tilde{f}(\mathbf{k}, \eta) = \frac{L\hbar}{eF} f\left(\mathbf{k}_p, \frac{\hbar k_z}{eF} + \frac{\eta}{2}, \frac{\hbar k_z}{eF} + \frac{\eta}{2}\right) \tilde{A}_F(k_z, \eta), \qquad (7.188)$$

where

$$\tilde{A}_F(k_z, \eta) = \frac{1}{\hbar} \exp\left[-i\frac{(eF)^2\eta^3}{24m^*\hbar} - i\frac{E_{k_p}\eta}{\hbar} \right] \tag{7.189}$$

is the equivalent gauge-invariant spectral density in the time domain.

In Eq. (7.188), the quantity $\hbar k_z/eF$ (and the equivalent in the z'-coordinate) can be indentified as the k_z-dependent contribution to the center-of-mass time, which moves in the coordinate of the traveling particle. Thus, in this formalism, the unperturbed, field-dependent spectral function can be factored out of all the functions that are of interest. This represents a formal analogy to the introduction of "reduced" functions, in which the spectral density and the distribution function, for example, can be separated in the less-than Green's function.

By the same procedure, a function $f(\omega - eFs/\hbar)$ can be put into gauge-invariant form. In this case, we may write

$$\tilde{f}(\mathbf{k}, \eta) = f(\mathbf{k}_p, \eta)\tilde{A}_F(k_z, \eta), \tag{7.190}$$

where

$$f(\mathbf{k}_p, \eta) = \int \frac{d\Omega}{2\pi} f(\Omega)e^{-i\Omega\eta}, \quad \Omega = \omega - \frac{eFs}{\hbar}. \tag{7.191}$$

Here, the k_z dependence is carried only by the unperturbed spectral density, although the latter also carries an explicit variation with the applied electric field. The result of this latter equation is interesting, for it clearly points out that the transports parallel and perpendicular to the electric field are clearly separable. This has been known for some time in classical transport but has not been so clear in quantum transport. The result (7.190) also points out that the simple product in real space becomes a convolution in the frequency space, as

$$\tilde{f}(\mathbf{k}, \omega) = \int d\Omega f(\mathbf{k}_p, \Omega)\tilde{A}_F(k_z, \omega - \Omega). \tag{7.192}$$

Using the Fourier transform of the free spectral density, this may be rewritten as

$$\tilde{f}(\mathbf{k}, \omega) = \int d\Omega f(\mathbf{k}_p, \Omega)\frac{1}{\xi}Ai\left(\frac{\hbar\Omega - (\hbar\omega - E_{k_z})}{\xi} \right), \tag{7.193}$$

where ξ is given in (7.168). This result simply states that, in order to transform a function defined in s, ω-space into a gauge-invariant form, we simply have to take its single Airy transform

$$f(s) = \int dz \frac{1}{L}Ai\left(\frac{z - s}{L} \right)f(z) \tag{7.194}$$

with L, z, and s replaced by ξ, $\hbar\Omega$, and $\hbar\omega - E_{k_z}$, respectively. Because of the nature of the results obtained in the above paragraphs, it is then easy to show that they are, in fact, gauge invariant, and that they may easily be transformed into a simple form exhibiting this property.

7.6 Screening with the Airy-transformed Green's functions

The formulation of a rigorous quantum mechanical theory for the role of the electric field in carrier screening is a long-standing and difficult problem. Indeed, at high electric fields, the system is in a far-from-equilibrium state, and the dielectric response must be found from the use of the nonequilibrium real-time Green's functions, which go beyond the equilibrium and/or semiclassical approaches. Nevertheless, the result is equivalent to the random-phase approximation result obtained in Chapter 2. Even in the Boltzmann transport picture in the semiclassical case, where dynamic screening is treated mainly through the Lindhard potential, a high electric field (and/or intense scattering) can cause descreening of the interaction [26], [118], usually in the case applied to the electron-electron interaction.

Within the nonequilibrium Green's function formalism, the problem of electron-phonon scattering is already quite involved, as we have seen in the previous section. Despite its importance, the problem of nonequilibrium dynamic screening, which at some level involves both the electrons and the phonons, has not yet been fully explored. Barker [118] discussed the field dependence of carrier screening in nonlinear transport theory, showing how a Lindhard-like dielectric function leads to descreening at high field, but did not derive an explicit expression for the effect. Barker and Lowe [119] developed a quantum theory of hot electron transport in inhomogeneous semiconductors based upon functional derivative methods, only to find that the field dependence of the carrier screening arises through the field dependence of the carrier distribution function and the energetics of the carrier-carrier scattering. More recently, Hu *et al.* [120], working within a gradient expansion formalism, extended the relaxation-time approximation to linear nonequilibrium screening and compared the result with that obtained from the semiclassical Boltzmann equation. This result, however, is essentially one that retains linear response. Recently, the Airy-transformed form of the real-time Green's function approach has been used to obtain the nonlinear, nonequilibrium screening and has shown how both the field and the scattering (collisional broadening) lead to descreening in the quantum dielectric response [121]. We follow this approach here. However, we work only with the polarization function itself, with the knowledge that the total screening function can be obtained using this polarization function in, for example, Eq. (6.82). (Screening with a thermal Green's function approach has been reviewed recently by Das Sarma [88].)

The polarization that leads to the screening function is generally obtained by the assumption that the advanced and retarded Green's functions are a response to the virtual absorption of a photon, propagation, and then the virtual emission of the same photon. In essence, this is a current-current response function, and the retarded polarization is just the product of Green's functions given in (7.47) and subsequent equations. Thus, the less-than polarization, which describes the electron response to a disturbing potential, may be written from the Langreth theorem as

$$\Pi^<(\mathbf{r}, \mathbf{r}', t, t') = -2i\hbar[G_r G^< + G^< G_a], \tag{7.195}$$

which is written schematically, and there are integrations over internal variables to be performed. However, these are simplified by Fourier-transforming on the transverse coordinates and Airy-transforming in the electric field direction, which leads to the transformed version

$$\Pi^<(\mathbf{q}_p, s, \omega) = -2i\hbar \int \frac{d^2\mathbf{k}_p}{(2\pi)^2} \int \frac{d\omega'}{2\pi} \int ds' \int ds'' W^2(s, s', s'') G^<(\mathbf{k}_p, s'', \omega')$$

$$\times [G_r(\mathbf{k}_p + \mathbf{q}_p, s', \omega' + \omega) + G_a(\mathbf{k}_p - \mathbf{q}_p, s', \omega' - \omega)], \tag{7.196}$$

where

$$W(s, s', s'') = \int dz Ai\left(\frac{z-s}{L}\right) Ai\left(\frac{z-s'}{L}\right) Ai\left(\frac{z-s''}{L}\right), \tag{7.197}$$

and L is the same normalized length used in the previous section.

One can simplify (7.196) further by taking into account the fact that the nonequilibrium Green's functions depend on both frequency and the Airy variable only through the combination $\hbar\omega - E_{\mathbf{k}_p,s} = \hbar\omega - E_{\mathbf{k}_p} - eFs = \hbar\Omega - E_{\mathbf{k}_p}$, where $\Omega = \omega - eFs/\hbar$ is a shifted frequency. Moreover, the function $W(s, s', s'')$ is sharply peaked around the point $s' \approx s''$, for a given value of s. Through a change of variable $\omega' \rightarrow \Omega + eFs/\hbar$, inside the frequency integral of (7.196), we can write the polarization function in a manner that depends only on the difference in Airy coordinates, and hence the integrated form is independent of any explicit variation on s. Therefore, the diagonal terms become

$$\Pi_r(\mathbf{q}_p, \omega) = -2i\hbar \int \frac{d^2\mathbf{k}_p}{(2\pi)^2} \int \frac{d\Omega}{2\pi} G^<(\mathbf{k}_p, \Omega)$$
$$\times [G_r(\mathbf{k}_p + \mathbf{q}_p, \Omega + \omega) + G_a(\mathbf{k}_p - \mathbf{q}_p, \Omega - \omega)]. \qquad (7.198)$$

This last equation is already in a gauge-invariant form and is a quite general expression for the nonequilibrium retarded polarization function. We can now introduce the general expressions for the retarded and advanced Green's functions, returning to the unshifted frequencies, and the result is quite reminiscent of the well-known Lindhard dielectric function [122]:

$$\Pi_r(\mathbf{q}_p, \omega) = 2i\hbar^2 \int \frac{d^2\mathbf{k}_p}{(2\pi)^2} \int \frac{d\omega'}{2\pi} G^<(\mathbf{k}_p, \omega') \left[\frac{1}{\hbar(\omega - \omega') + E_{\mathbf{k}_p - \mathbf{q}_p, s} + \Sigma_a(\omega' - \omega)} \right.$$
$$\left. - \frac{1}{\hbar(\omega + \omega') - E_{\mathbf{k}_p + \mathbf{q}_p, s} - \Sigma_a(\omega' + \omega)} \right], \qquad (7.199)$$

and, as before

$$E_{\mathbf{k}_p,s} = \frac{\hbar^2 k_p^2}{2m^*} - eFs. \qquad (7.200)$$

It has also been assumed that the dominant contribution to the self-energy is that of non-polar optical scattering, which is isotropic and leaves the self-energy independent of the momentum.

From (7.199), we see that the field dependence of the polarization function arises from both the correlation functions $G^<$ and the propagators $G_{r,a}$ through the self-energies $\Sigma_{r,a}$. This agrees with the concept put forward by Barker [118] and expected from the semiclassical approach [26]. The definition of the nonequilibrium Green's function $G_{r,a}$ is that they essentially describe the density response of the system. All the scattering processes are, in principle, introduced collectively in $G_{r,a}$ by the self-energy Σ: the renormalization of the electron energies due to the presence of the electric field and the scattering processes goes into the real part of the self-energy. On the other hand, the relaxation of the one-to-one correspondence between energy and momentum during the field-assisted scattering and the finite lifetime broadening of the spectral function are directly reflected in the imaginary part of the self-energy. The scattering of an electron by a nonpolar optical phonon was taken into account within the the lowest-order many-body perturbation theory, and the corresponding gauge-invariant electron self-energy could be evaluated in terms of the frequency and the strength of the field. The strength of the electric field appears directly in the argument of the self-energy through the frequency dependence. In this way, the so-called intra-collisional field effect is handled exactly in the evaluation of the interacting Green's functions.

The expression for the less-than Green's function was developed in the previous section, and this correlation function provides information about the distribution of electrons. As was found in the previous section, this Green's function may be easily decomposed into a product of the spectral density and the quantum distribution function, as expressed in (7.174). Using this decomposition, plus the properties of the retarded and advanced Green's functions, we can write (7.199) in the form

$$
\Pi_r(\mathbf{q}_p, \omega) = -2\hbar^2 \int \frac{d^2\mathbf{k}_p}{(2\pi)^2} \int \frac{d\omega'}{2\pi} f(\omega') A(\mathbf{k}_p, \omega')
$$

$$
\times [G_r(\mathbf{k}_p + \mathbf{q}_p, \omega' + \omega) + G_a(\mathbf{k}_p - \mathbf{q}_p, \omega' - \omega)]. \qquad (7.201)
$$

Because the form of the spectral density is known – and is itself an expansion in the retarded and advanced Green's functions – it is possible to carry out the integration over the momentum. There are four terms, each having the basic arctangent shape of the density-of-states function in two dimensions but with distortions introduced by the various energy-dependent terms. We do not need to actually carry out the integrations in order to understand the limiting behavior.

Let us first consider the limit $\omega \to 0$. In the large square brackets, the frequency argument becomes the same for both Green's functions, and the basic polarization becomes the simple expansion that is present in our earlier treatment of the Kubo formula. For large negative values of ω', the self-energy is zero and the spectral density vanishes, so there is no contribution to the integral from this region. The same is true for large positive values of ω', since the denominators begin to dominate the various Green's functions, while the self-energies vary only approximately as the square root of the frequency. Thus, the major contributions come from values of the argument of the self-energy in the vicinity of the range $(0, Re\{\Sigma_r\})$. The four terms that result may be paired into groups of two whose differences prescribe the basic strength of the integrand, and there is only a subtle noncancellation arising from the shifts in the momentum variables.

Moreover, it is clear that since the basic product of the Green's function and the spectral density varies approximately as $1/\Sigma^2$, the polarization is strongly reduced in the large scattering limit. Hence, the presence of strong scattering strongly damps the polarization and therefore the screening that arises. In addition, the electric field is primarily in the denominators of these two functions. Hence, very strong electric fields also lead to a strong reduction in the screening. Since the screening electrons have an ever smaller time in which to respond to an impurity, in the increasing electric field the field-induced descreening is related to the fall-off of the dielectric function at high frequency.

7.7 Other approaches to quantum transport in nonequilibrium systems

We have dealt so far with the nonequilibrium Green's functions, but this is not the only approach that has been applied to nanodevices. The Green's functions have been the primary choice for treatment of the transport in mesoscopic devices at low temperatures, so it has been natural to pursue them at the higher temperatures (and external driving fields) appropriate to the nonequilibrium situation. Indeed, one can return to the original kernel for the Green's function that was expressed in Chapter 5 as

$$
\Psi(\mathbf{r}, t) = \int d\mathbf{r}' \int dt' K(\mathbf{r}, t; \mathbf{r}', t') \Psi(\mathbf{r}', t'). \qquad (7.202)
$$

Indeed, the kernel in (7.202) describes the general propagation of any initial wavefunction at time t' to any other arbitrary time t (which is normally $>t'$, but not necessarily so). There are a number of ways in which to evaluate the kernel itself, one of which is by the method of Green's functions, the approach used so far. The most general case, even in nonequilibrium systems, is that in which the system is described in terms of a mixture of states. One can often separate the exponential time variation into the separate parts for the two times involved in (7.202), and then each time-varying basis is expanded in an arbitrary (but different) set of wavefunctions so that we have a properly mixed system. From this point, one may write the kernel as

$$K(\mathbf{r}, t; \mathbf{r}', t') = \sum_{n,m} c_{nm} \phi_n^*(\mathbf{r}, t) \phi_m(\mathbf{r}', t') = \Phi^*(\mathbf{r}, t) \Phi(\mathbf{r}', t'). \tag{7.203}$$

The equal-time version of this is termed the *density matrix* [123]

$$\rho(\mathbf{r}, \mathbf{r}', t) = \sum_{n,m} c_{nm} \phi_n^*(\mathbf{r}, t) \phi_m(\mathbf{r}', t) = \Phi^*(\mathbf{r}, t) \Phi(\mathbf{r}', t). \tag{7.204}$$

This function is in fact related to the Green's function, and it becomes the equal-time version of the less-than Green's function (because of the equal time in the arguments, much of the differentiation among the various Green's functions disappears). The diagonal terms in the density matrix yield the density at a point $\mathbf{r}$ at time t. On the other hand, the off-diagonal parts of the density matrix (for two different positions) describe the *spatial correlation* that arises in the quantum treatment of the nanodevice. Nevertheless, the assumption of equal-time arguments rules out any discussion of temporal correlation between the two states represented in the field operators.

The problem with the density matrix in many semiconductor problems is that it is defined only in real space, with the important quantum interference effects occuring between two separated points in space. Even so, it is a function of six variables (the two vector positions), plus the time (and/or the temperature), of course. In many cases, it would be convenient to describe things in terms of a phase-space function, whose six variables arise from a single (vector) position and a (vector) momentum. Such a function still depends upon six variables, plus the time (and/or the temperature). While this is not the normal case in quantum mechanics, it certainly can be arranged [124], [125]. To see how this may be achieved, we rewrite the density matrix in terms of a new set of coordinates, the center-of-mass and difference position coordinates. (We do not entertain such a redefinition in terms of the time coordinates, as has been discussed above.)

$$\mathbf{R} = \frac{1}{2}(\mathbf{r} + \mathbf{r}'), \quad \mathbf{s} = \mathbf{r}' - \mathbf{r}. \tag{7.205}$$

If we now Fourier-transform on the difference coordinates, the phase-space Wigner distribution function is obtained [124]

$$f_W(\mathbf{R}, \mathbf{k}, t) = \int d^3s \, \rho\left(\mathbf{R} + \frac{\mathbf{s}}{2}, \mathbf{R} - \frac{\mathbf{s}}{2}, t\right) e^{i\mathbf{k}\cdot\mathbf{s}}, \tag{7.206}$$

often called the Wigner-Weyl transform [126]–[128]. In this manner, the Wigner function represents a Fourier transform in the nonlocal coordinates, so that the momentum amplitudes correspond to the propagations that need to occur between two different points in space. In general, the Wigner function described in (7.206) is not positive definite. The latter is a consequence of the uncertainty relationship between position and momentum. If this equation were integrated either over all space (or momentum), the resulting momentum probability function (position probability function) would be positive definite. In the next two sections, we discuss briefly the modeling efforts that have been used for treating nanodevices with these two functions.

7.7.1 The density matrix

In general, the density matrix is best characterized in terms of the field operator form. The temporal equation of motion for the density matrix is then given by the Liouville equation as

$$i\hbar \frac{\partial \rho}{\partial t} = [H, \rho], \tag{7.207}$$

or

$$i\hbar \frac{\partial \rho}{\partial t} = \left[-\frac{\hbar^2}{2m} \left(\frac{\partial^2}{\partial \mathbf{r}^2} - \frac{\partial^2}{\partial \mathbf{r}'^2} \right) + V(\mathbf{r}) - V(\mathbf{r}') \right] \rho(\mathbf{r}, \mathbf{r}', t). \tag{7.208}$$

Sometimes, a higher-order operator algebra is used, since the Hamiltonian H is an operator in the Hilbert space defined by the basis states upon which the expansion for the density matrix is based. This higher-order operator algebra is a tensor operation that rotates the Hilbert space in a way in which the *superoperator* $\hat{H}$ can be defined through

$$i\hbar \frac{\partial \rho}{\partial t} = \hat{H}\rho. \tag{7.209}$$

This particularly simple superoperator is a *commutator-generating superoperator* [129], [130], but in general the superoperator is an operator in the space of operators that are allowed for the particular problem at hand. There is no problem incorporating a dissipative term in the Hamiltonian and treating it by perturbation theory. In fact, this is a quite viable method of treating irreversible transport [131], [132].

One problem with the Liouville equation is that it is an equation for the temporal evolution, *once the initial condition is known*. A mathematical proof exists to show that a sufficient way in which to find the initial condition, itself a pseudo-equilibrium solution (even in a far-from-equilibrium situation), is to actually use the adjoint equation to the Liouville equation. The latter is

$$-\frac{\partial \rho}{\partial \beta} = \left[-\frac{\hbar^2}{2m} \left(\frac{\partial^2}{\partial \mathbf{r}^2} + \frac{\partial^2}{\partial \mathbf{r}'^2} \right) + V(\mathbf{r}) + V(\mathbf{r}') \right] \rho(\mathbf{r}, \mathbf{r}', t). \tag{7.210}$$

The most reasonable derivation of this equation arises from using the anticommutator on the right side of (7.207), rather than the commutator, and passing to the limit of imaginary time $t \to -i\hbar\beta$, where β is the inverse energy of the temperature ($\beta = 1/k_B T$), as introduced for the thermal Green's functions.

In writing the density matrix in the above forms, it is assumed to be a single-particle quantity. In fact, the density matrix starts life as a many-body quantity [133]. In general, it must be projected onto a single-particle description in order to compare with the normal Boltzmann transport equation. A projection operator that achieves this, in effect, integrates over all coordinates of the particles above those of a single particle. Still, there remains a two-particle interaction that is basically due to the Coulomb interaction between particles. This correction is just the normal Hartree and exchange corrections to the single-particle

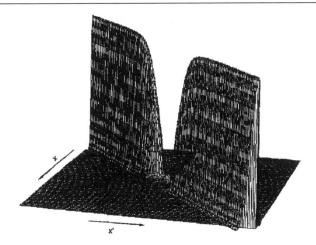

Fig. 7.32. The real part of the density matrix for a double barrier resonant tunneling structure. [From H. L. Grubin, private communication.]

energies. In the absence of these corrections, the equation of motion reduces to that for the single-particle density matrix. The latter will be used here.

The density matrix has been used to some extent in modeling and treating the transport through some mesoscopic nanostructures. In general, however, these approaches have not approached a detailed treatment of the dissipative processes that occur within the devices (for a review, see [46]). Rather, some approximations to simple elastic scattering have been made. In Fig. 7.32, we show the real part of the density matrix that arises from a simulation of a resonant tunneling structure in the absence of any applied bias. Here, one can see the small density in the potential well region between the two barriers. Frensley [134] has used the single-particle density matrix to study a double barrier resonant-tunneling diode, and also has looked at the case with more built-in barriers. The partial differential equation for the density matrix in Fig. 7.32 was solved using finite-difference techniques similar to those used to solve conventional semiconductor devices. For the simulations, the boundaries were treated as ohmic contacts, and it is found that the methods of solution are easier than the Wigner distribution function (see Section 7.7.2) but are more complicated to interpret. This latter work is the first step toward directly calculating the properties of real quantum devices, and it points out the importance of such simulations to gain insight into the operation of the devices. Nevertheless, incorporation of the ohmic boundary conditions greatly complicates the numerical simulation algorithm, and the system was subject to the growth of numerically unstable modes. This continues to be a problem in simulation of the density matrix in inhomogeneous systems. However, we point out below that within the Wigner formulation, there exists a methodology to overcome this problem.

Groshev [135] has also used the density matrix to simulate the Coulomb blockade (single-electron tunneling) regime of the resonant tunneling structure. In this case, he used a three-dimensionally configured structure in an attempt to identify lateral modes and fine structure in the tunneling current. However, he reduced the problem to a pseudo-hopping formalism and did not need to carefully study the spatial charge distribution self-consistently, nor did he carefully examine the effects of the boundary condition. Such approaches have also been used for the single-electron transistor, as mentioned in Chapter 4.

7.7.2 The Wigner distribution

The Wigner formalism offers many advantages for quantum modeling. First, it is a phase-space description, similar to the semiclassical Boltzmann equation. Moreover, scattering is a local process in real space, just as in the Boltzmann equation. Because of the phase-space nature of the distribution, and the conceptual similarity of the scattering processes to those of the Boltzmann equation, it is possible to rely on the correspondence principle to determine just where quantum corrections enter a problem. At the boundaries (the contacts), the phase-space description permits separation of the incoming and outgoing components of the distribution, as we will see below. This permits focusing on the nature of the contact in the problem, which is reduced to the more difficult problem of an open quantum system. Yet, there remain problems with the outgoing portions of the contacts, which will be discussed below. Another advantage of the Wigner function is that it is purely real, which simplifies some of the calculations and interpretations of the results. The Wigner equation of motion may be found by Fourier-transforming Eq. (7.208) as

$$\frac{\partial f_W(\mathbf{R}, \mathbf{k}, t)}{\partial t} + \mathbf{k} \cdot \frac{\partial f_W(\mathbf{R}, \mathbf{k}, t)}{\partial \mathbf{R}}$$

$$= \int \frac{d^3 k}{(2\pi)^3} W(\mathbf{R}, \mathbf{K}) f_W(\mathbf{R}, \mathbf{k} + \mathbf{K}, t) + \left(\frac{\partial f_W(\mathbf{R}, \mathbf{k}, t)}{\partial t} \right)_{coll.}, \qquad (7.211)$$

where

$$W(\mathbf{R}, \mathbf{K}) = \int d^3 s \sin(\mathbf{K} \cdot \mathbf{s}) \left[V\left(\mathbf{R} + \frac{\mathbf{s}}{2} \right) - V\left(\mathbf{R} - \frac{\mathbf{s}}{2} \right) \right]. \qquad (7.212)$$

By coupling this equation to Poisson's equation, we can obtain a fully self-consistent approach to modeling various devices. There have been a number of early attempts to model the transport of double barrier resonant-tunneling diodes with the Wigner function [136], [137].

A serious consideration of the Wigner formalism is the entry of nonlocal quantum-mechanical effects through the inherently nonlocal driving potential that appears in (7.212). In the limit of slow spatial variations, only the linear term coupling the first derivatives of the potential (the force) to the derivative of the Wigner distribution with respect to the momentum remains important, and Eq. (7.211) reduces to the Boltzmann equation. Thus, any quantum corrections must arise in the nature of the potential term of (7.212) and in the scattering processes that make up the collision integral (the last term in the first equation). In fact, if the potential has no spatial variation higher than quadratic in the difference variable, no quantum corrections arise, and the Wigner equation of motion will not reproduce the well-known quantization of the harmonic oscillator. That is because Eq. (7.211) is an equation for the time evolution of the distribution and not for the determination of the steady-state results. Thus, the initial condition for this equation must already include a great deal of the quantum information [138]. It is tempting to just neglect the time derivative term and solve for a steady state, but this is still known to fail for the harmonic oscillator [139]. In fact, the proper method of evaluating the equilibrium (steady-state) Wigner distribution is from the adjoint equation [138], which is the Fourier transform of (7.210). One further problem is that the correct boundary conditions presuppose knowledge of the state of the system both internally and at the boundaries, as the latter are a function of the former through the nonlocality of the internal potential. Thus, knowledge of the boundary conditions presupposes a full knowledge of the solution being sought, even without the equation of motion.

In order to include all orders of quantum corrections, one of two things can be done. The first is to extend the computational domain sufficiently far from the source of the quantum effects that the system is classical, so that a classical (or semiclassical) distribution can be used at the boundaries. It has been shown that quantum "corrections" heal over several thermal de Broglie wavelengths [140]. In a reasonable GaAs device at 300 K, this length is nearly 100 nm. In any case, one should make use of the adjoint equation, and this approach will be described here. If the potential approaches a constant, uniform value as $\mathbf{R} \to \pm\infty$ (where the device is supposed to localize near the coordinate origin), then the basis states at large distances can be taken to be plane waves in a pure momentum representation. In one dimension, we assume that $V(x) = V^-$ for $x < x^-$ and $V(x) = V^+$ for $x > x^+$, where $x^\pm$ describes the transition points beyond which the potential is uniform. For equilibrium situations, in which $V^- = V^+$, this gives scattering states incident from the left, with $k > 0$,

$$\psi_k(x) = \frac{1}{\sqrt{2\pi}}[e^{ikx} + r(k)e^{-ikx}] \tag{7.213}$$

for $x < x^-$, and

$$\psi_k(x) = \frac{1}{\sqrt{2\pi}}t(k)e^{ikx} \tag{7.214}$$

for $x > x^+$, where $t(k)$ is the transmission coefficient and $r(k)$ is the reflection coefficient. In a similar manner, states incident from the right are defined by

$$\psi_k(x) = \frac{1}{\sqrt{2\pi}}[e^{-ikx} + r(k)e^{ikx}] \tag{7.215}$$

for $x > x^+$, and

$$\psi_k(x) = \frac{1}{\sqrt{2\pi}}t(k)e^{-ikx} \tag{7.216}$$

for $x < x^-$. (It should be noted that the device is bilateral in equilibrium.) The density matrix is then defined in the semiclassical equilibrium regions of the boundaries as

$$\rho(x, x') = \frac{1}{Z}\left[\sum_n \psi_n(x)\psi_n^*(x') + \int dk\psi_k(x)\psi_k^*(x')f(E_k)\right], \tag{7.217}$$

where Z is the partition function, the sum over n is over bound states that may be localized at some point, E_k is the energy of a scattering state, and $f(E)$ is the Fermi-Dirac distribution function. From an unnormalized basis function, and using translation matrices, an unnormalized state may be computed on the entire domain. These states are normalized by applying scattering theory, in which wavefunctions in the presence of a scatterer (the built-in potential) are compared to those in a reference space. These are related through the Lippmann-Schwinger equation. A useful consequence of this equation is that the scattering states satisfy precisely the same orthonormality relations as the unperturbed states [140], [141]. Each state contributes to the density matrix according to the thermal distribution function $f(E)$. The partition function is found by considering the limit of large x, for which normalization may be invoked, as

$$Z^{-1} = 2\beta\sqrt{\pi}e^{\beta V}\lim_{x \to -\infty}\rho(x, x). \tag{7.218}$$

An algorithm is thus available for computing the equilibrium Wigner function, since the resulting density matrix can be easily Fourier-transformed. Such a result for a self-consistent

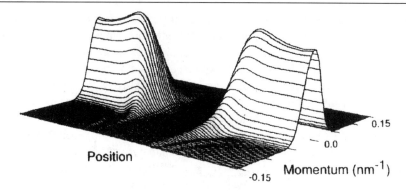

Fig. 7.33. The equilibrium Wigner function for a 5 nm well between 5 nm barriers in GaAs/AlGaAs.

solution to the double barrier resonant-tunneling diode is shown in Fig. 7.33. [142]. The Wigner distribution is characterized by a thermal distribution far from the barriers. The oscillations near the barriers are a result of the quantum repulsion from the barriers, which causes a depletion of carriers in this region, although this is not a depletion in the normal sense. This depletion arises from a momentum-dependent quantum repulsion which, in a sense, is complementary to barrier penetration by the wave function. It should be remarked, however, that the device simulation maintains overall charge neutrality for the structure.

Boundary conditions

Simulation of a real device includes some model of the interface between the interior simulation region and the supposed boundary/contact layer. At the very minimum, the external circuit consists of a battery that fixes the potential across the device and wires that carry the current from the battery to the device. These circuit parameters are usually included in device simulations as boundary conditions. In many models, the contact serves as an infinite reservoir of thermally distributed carriers [143]. This reservoir maintains a fixed distribution at the contact where the particles enter the simulation domain, and impacts the simulation of such open quantum structures [144]. Conditions at the contact/boundary must be consistent with physical reality. If a current is flowing through the device, current continuity requires that an identical current be flowing in the external circuit. This implies that the external circuit, and thus the contact regime, must be characterized by some distribution that reflects current flow, such as a shifted (in momentum space) Fermi-Dirac distribution.

In the flow field, there are left and right boundaries at $x = -L/2$ and at $x = L/2$. Mathematical constraints permit us to specify only one boundary, however; the linear term is only first-order in the spatial derivative. The model may be thought of as a coupled set of systems, one for the positive-momentum half-space and one for the negative-momentum half-space. Each of these momentum half-spaces has a boundary, or contact, from which electrons enter the slice, and they then leave the slice from the opposite end of the device. The entering contact is "constrained" by the current. Just inside the device, a current exists. Because an identical current must be present in the contact, and because the distribution within the contact is assumed to be a near equilibrium–shifted thermal distribution, the amount of the shift and therefore the distribution function within the contact are known. The distribution within the contact is then matched to the corresponding momentum half-space of the interior model region. This procedure must be carried out self-consistently, since the current within the device is a function of the boundary conditions, which in turn are functions of the current within the device. Thus, the boundary distribution is adjusted self-consistently to provide the necessary transfer of carriers into the device on the source end of each momentum half-space.

A second important property of the contacts is that they must effectively remove the equivalent amount of current (or carriers). In the ideal contact, the carriers may be thought of as being perfectly extracted with no backscattering. In real devices, this contact condition may not actually be present, and backscattering can arise from such events as phonon scattering through a large angle at the contact boundary or carrier-carrier scattering from the large density of carriers in the contact. Nevertheless, most attempts in device modeling assume that ideal extraction takes place at the outgoing contact. The nonlocal potential creates a problem because there is usually a weak discontinuity of the potential at the contact, which leads to an artificial reflection of carriers from the boundary region (an effect not known in classical device modeling) [145]. This result deviates from the idea of a perfectly absorbing contact, but it is well-known in wave propagation. The solution to this effect lies in coupling the adjacent incoming distribution to the outgoing distribution in a manner that cancels these artificial reflections, particularly those from the fastest particles, which cause the most trouble. This is done during the time-stepping procedure of the solution loop. If we rewrite (7.211) as

$$\frac{\partial f_W(x, k, t)}{\partial x} = a(k)\frac{\partial f_W(x, k, t)}{\partial t} + \int \frac{dK}{2\pi} M(x, K) f_W(x, k + K, t), \qquad (7.219)$$

where $a(k) = -1/v(k)$ and $M(x, K) = W(x, K)/v(k)$, the boundary condition can be found by a suitable operator normalization procedure [146]. This leads to the boundary condition at $x = -L/2$, where $k > 0$ represents the momentum half-space of interest for incoming waves, to be expressed as the temporal adjustment

$$\frac{\partial f_W(-L/2, k, t)}{\partial t} = -\int \frac{dK}{2\pi} \frac{M(-L/2, K)}{a(k) - a(K)} f_W(-L/2, k + K, t). \qquad (7.220)$$

Similarly, the boundary condition for the incoming particles at $x = L/2$, where the half-space of interest is $k < 0$, is given by the temporal adjustment

$$\frac{\partial f_W(L/2, k, t)}{\partial t} = -\int \frac{dK}{2\pi} \frac{M(L/2, K)}{a(k) - a(K)} f_W(L/2, k + K, t). \qquad (7.221)$$

These two equations constitute an absorbing boundary condition for the Wigner equation of motion that removes artificial reflections from the outgoing boundary. The reflected waves are absorbed at least to second order.

If we are to accurately model heterostructures, then there are also interior boundaries to consider, such as may arise at the interface between two materials, where there are differences in the effective mass. If the bands are parabolic, this does not introduce much of a problem, as it can be handled in a simplified manner by renormalizing the barrier potential (since in the scattering state basis the discontinuity in the momentum wavevector is defined by the product of the mass and the energy). However, this may not insure current continuity across the interface, which could introduce a source of error. The proper interface characteristics will take this into account [147].

Transport in waveguides and single barriers

The techniques of the above boundary conditions and iteration of (7.211) for a simple waveguide with a single tunneling barrier has been simulated both for the propagation of Gaussian pulses [136], [137] and for the steady state density in the the waveguide [148]. In the latter, the variation of the conductance of a quantum waveguide with a potential barrier was studied. In the Landauer approach (see Chapter 3), the intensity transmission probability of the waveguide modes is usually computed from the single-particle Schrödinger equation, assuming phase-coherent transport through the entire waveguide (normally phase-breaking interactions limit the experimental observation to short waveguides, or point contacts). On the other hand, the Wigner function approach allows the computation of a particle distribution, excited at one end of the the structure, to be determined throughout the waveguide. Moreover, in the Wigner function approach, one can also study the temporal behavior of the conductance as the barrier is switched.

In Fig. 7.34, the structure of the potential is illustrated. The waveguide is assumed to be 56 nm in length, with a barrier located in the center. The amplitude of this barrier is easily varied. The Wigner function itself is decomposed into a summation of a set of subband Wigner functions, one for each lateral quantum number of the waveguide. The boundary conditions for the steady state were set simply by matching to overall Fermi-Dirac distributions in the contacts. Scattering is treated in a relaxation time approximation (discussed below). In Fig. 7.35, the total Wigner distribution is shown for two different biases of the barrier, one for enhanced density and one for depleted density.

In these simulations, quantized conductance is observed when the applied potential (between the two ends of the waveguide) is small and the barrier is set to zero. When the voltage between the ends is raised, the conductance becomes nonlinear. In Fig. 7.36, we show this effect for a simulation at 300 K (with the barrier potential set to zero) for various waveguide widths [149]. When the applied potential is small, the current flowing in the waveguide is determined by a balance between the Fermi-Dirac functions at the two ends of the sample, in keeping with the Landauer formula. However, when the applied potential is made larger than the Fermi energy in the contacts (here about 10 meV), the current flowing through the waveguide is determined primarily by a single Fermi-Dirac function at the cathode. In essence, the current is metered into the waveguide from the cathode contact in

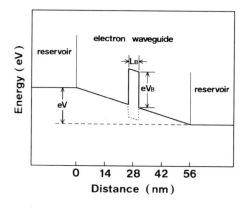

Fig. 7.34. Simple potential profile for solution of the Wigner distribution with a potential barrier. [After H. Tsuchiya et al., Jpn. J. Appl. Phys. **30**, 3853 (1991), by permission.]

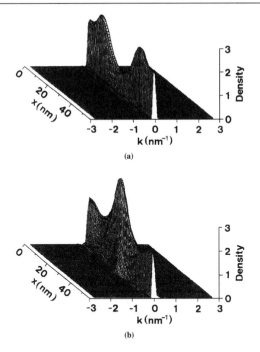

Fig. 7.35. Waveguide Wigner distributions for barrier heights of (a) 20 meV and (b) –20 meV. [After Tsuchiya *et al.*, Jpn. J. Appl. Phys. **30**, 3853 (1991), by permission.]

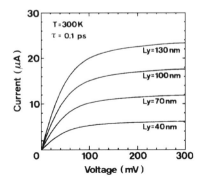

Fig. 7.36. Calculated current-voltage curves for an electron waveguide at room temperature. [After Tsuchiya *et al.*, IEEE Trans. Electron Dev. **39**, 2465 (1992), by permission.]

a manner resembling thermionic emission. This does not occur right at a potential equal to the Fermi energy due to the presence of the scattering, and the level of the current is set by a balance between the applied potential and the relaxation time τ. The current increases as the relaxation time is made longer.

Double barrier structures

We now turn to the simulation of a double barrier resonant-tunneling diode. For this purpose, we assume a GaAs/AlGaAs structure with two 5 nm barriers 0.3 eV high and a 5 nm quantum

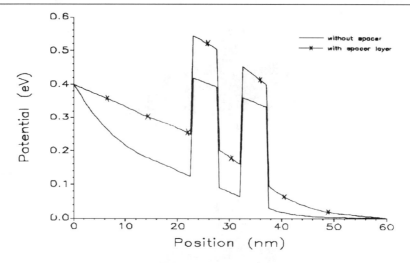

Fig. 7.37. Comparison of the potential drop across the RTD, with and without the spacer layers.

well [142]. A similar structure has been simulated more recently by Tsuchiya *et al.* [150]. The regions outside the barriers are assumed to be doped to a level of 10^{18} cm^{-3}. In many cases, a thin spacer layer (an undoped layer) is situated adjacent to the barriers. In the absence of the spacer layer, a bias will often deplete the region between the cathode and the first barrier. This depletion occurs for two reasons: (1) a quantum well is formed in this region and the presence of the quantized level leads to a lower carrier density, and (2) there usually is a Landauer contact resistance which creates a significant potential drop at the cathode interface [151], an effect well known in classical nonlinear semiconductor devices [152]. The effect of the spacer layer is to provide a resistance matching layer that overrides the cathode boundary resistance. It also tends to create an accumulation layer near the first barrier, which helps to create a reservoir of carriers for transmission through the barriers. The variations in the actual potential distribution through the RTD are shown in Fig. 7.37, where the effect of the barrier layer is illustrated. Here the undoped spacer layers are also 5 nm thick.

The steady state I-V characteristics of the device are calculated by applying an incremental (negative) bias potential to the cathode contact, and then by solving (7.211) to steady state by a time iteration procedure. The self-consistent potential is solved each time step from the Poisson's equation. A relaxation time approximation is used for the scattering/dissipative processes in this approach. The I-V characteristics of the structure, with the spacer layers, is shown in Fig. 7.38. The presence of the spacer layers increases the peak-to-valley ratio by concentrating the potential drop in the barrier region itself, and they provide a lower overall resistance by reducing the cathode potential drop.

Intrinsic bistability is obvious in the negative differential conductance regime in Fig. 7.38. This is thought to arise from charge storage in the quantum well as discussed in Chapter 3. During the upsweep of the voltage, the quantum-well resonant state sweeps through the full conduction band states so that as the minimum of the current is approached, the well is full of carriers (which remain there through a sequential tunneling process, as opposed to a complete resonant tunneling process). On the downsweep of the voltage, however, the well is basically empty as it comes into alignment with the full conduction band states. The

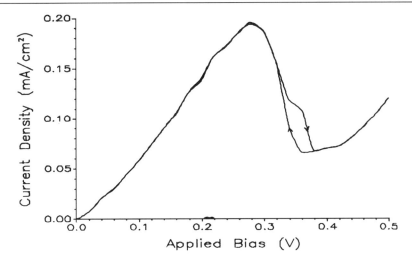

Fig. 7.38. The I-V curves of the RTD with spacer layers.

difference in charge in the quantum well, between the upsweep and the downsweep, changes the self-consistent potential (and the current) to reflect this bistability. This interpretation can be supported just by simple circuit arguments and from the presence of the negative differential conductance [153].

Finally, we remark on the use of the relaxation time approximation. This is certainly an over-simplification of the scattering processes (as can be seen by reference to the Green's function treatment of the RTD in a previous section), and it would have been better to utilize at least an energy-dependent relaxation time. The literature contains many allusions to an inaccuracy of the relaxation time approximation itself, in that it is often asserted that it is not charge conserving. In fact, *it should not be charge conserving on the local basis*, but it should be so on a global basis since charge is actually moved around in most semiconductor devices. Indeed, if we required that

$$\int \frac{d^3k}{(2\pi)^3}\left[\frac{f(\mathbf{R},\mathbf{k}) - f_0(\mathbf{R},\mathbf{k})}{\tau(\mathbf{k})}\right] = 0, \tag{7.222}$$

there would not even be a drift-diffusion model in classical transport. We must recall that the latter is found by rearranging the Boltzmann transport equation with the relaxation time approximation in steady state to obtain

$$f(\mathbf{R},\mathbf{k}) = f_0(\mathbf{R},\mathbf{k}) - \tau(\mathbf{k})\left[\frac{\hbar\mathbf{k}}{m^*}\cdot\nabla f(\mathbf{R},\mathbf{k}) - (e\nabla V(\mathbf{R}))\cdot\frac{\partial f(\mathbf{R},\mathbf{k})}{\hbar\partial\mathbf{k}}\right]. \tag{7.223}$$

If Eq. (7.222) were valid, then there would be no current flow because the bracketed term would have to average to zero as well. Yet the bracketed term contains all the driving forces for both drift and diffusion. Hence, the equality in (7.222) is not correct, and (7.223) illustrates that the current, obtained by taking the first moment of this latter equation with respect to momentum (one usually approximates the distributions on the right side by their equilibrium values), arises from a proper balance between the driving forces (contained in the square brackets) and the dissipation (represented by τ). If we take the first moment, we find that

$$\int \frac{d^3\mathbf{k}}{(2\pi)^3} \frac{\hbar \mathbf{k}}{m^*} [f(\mathbf{R}, \mathbf{k}) - f_0(\mathbf{R}, \mathbf{k})] = \mathbf{J} = e\nabla(Dn) + ne\mu\mathbf{E}, \qquad (7.224)$$

as expected. However, it can be shown equally as easily that the equality in (7.222) is valid only in those cases where there is no gradient in the density and no electric field present, hence in homogeneous thermal equilibrium systems.

Bibliography

[1] K. Yano, IEEE Trans. Electron Dev. **41**, 1628 (1994).

[2] J.-R. Zhou and D. K. Ferry, IEEE Trans. Electron Dev. **40**, 421 (1993).

[3] H. S. Wong and Y. Taur, *Proc. Intern. Electron Dev. Mtg.* (IEEE Press, New York, 1993), p. 705.

[4] P. C. Martin and J. Schwinger, Phys. Rev. **115**, 1342 (1959).

[5] J. Schwinger, J. Math. Phys. **2**, 407 (1961).

[6] L. P. Kadanoff and G. Baym, *Quantum Statistical Mechanics* (Benjamin, New York, 1962).

[7] L. V. Keldysh, Sov. Phys.-JETP **20**, 1018 (1965).

[8] C. P. Enz, *A Course on Many-Body Theory Applied to Solid-State Physics* (World Scientific Press, Singapore, 1992), p. 76.

[9] H. Haug, in *Quantum Transport in Ultrasmall Devices*, eds. D. K. Ferry, H. L. Grubin, and C. Jacoboni (Plenum Press, New York, in press).

[10] T. Kuhn and F. Rossi, Phys. Rev. B **46**, 7496 (1992).

[11] T. W. Hickmott, P. M. Solomon, F. F. Fang, F. Stern, R. Fischer, H. Morkoç, Phys. Rev. Lett. **52**, 2053 (1984).

[12] L. Eaves, P. S. S. Guimaraes, B. R. Snell, D. C. Taylor, and K. E. Singer, Phys. Rev. Lett. **53**, 262 (1985).

[13] L. Eaves, F. W. Sheard, and G. A. Toombs, in *Physics of Quantum Electron Devices*, ed. F. Capasso (Springer-Verlag, New York, 1990), pp. 107–146.

[14] H. Yoo, S. M. Goodnick, J. R. Arthur, and M. A. Reed, J. Vac. Sci. Technol. B **8**, 370 (1990).

[15] J. R. Hayes, A. F. J. Levi, and W. Wiegmann, Electron. Lett. **20**, 851 (1984); Phys. Rev. Lett. **54**, 1570 (1985).

[16] N. Yokoyama, K. Imamura, T. Ohshima, N. Nishi, S. Muto, K. Kondo, and S. Hiyamizu, *Intern. Electron Dev. Mtg. Proc.* (IEEE Press, New York, 1984), p. 532.

[17] M. Heiblum, D. C. Thomas, C. M. Knoedler, and M. I. Nathan, Appl. Phys. Lett. **47**, 1105 (1985).

[18] A. F. J. Levi, J. R. Hayes, P. M. Platzman, and W. Wiegmann, Phys. Rev. Lett. **55**, 2071 (1985).

[19] M. Heiblum, M. I. Nathan, D. C. Thomas, and C. M. Knoedler, Phys. Rev. Lett. **55**, 2200 (1985).

[20] M. Heiblum and M. V. Fischetti, in *Physics of Quantum Electronic Devices*, ed. F. Capasso (Springer-Verlag, Berlin, 1990), pp. 271–318.

[21] M. A. Hollis, S. C. Palmateer, L. F. Eastman, N. V. Dandekar, and P. M. Smith, Electron. Dev. Lett. **4**, 440 (1983).

[22] V. H. Y. Lam, M. W. Dellow, S. J. Bending, M. Elliott, D. J. Arent, and P. Guéret, Semicond. Sci. Technol. **10**, 110 (1995).

[23] A. Pavleski, M. Heiblum, C. P. Umbach, C. M. Knoedler, A. N. Broers, and R. H. Koch, Phys. Rev. Lett. **62**, 1776 (1989).

[24] A. Pavleski, C. P. Umbach, and M. Heiblum, Appl. Phys. Lett. **55**, 1421 (1989).

[25] T. Galloway, B. L. Gallagher, P. Beton, J. P. Oxley, M. Carter, S. P. Beaumont, S. Thoms, and C. D. W. Wilkinson, Surf. Sci. **229**, 326 (1990).

[26] D. K. Ferry, *Semiconductors* (Macmillan, New York, 1991).

[27] L. W. Molenkamp, H. van Houten, C. W. J. Beenakker, R. Eppenga, and C. T. Foxon, Phys. Rev. Letters **65**, 1052 (1990).

[28] L. W. Molenkamp, Th. Gravier, H. van Houten, O. J. A. Buijk, M. A. A. Mabesoone, and C. T. Foxon, Phys. Rev. Letters **68**, 3765 (1992).

[29] L. W. Molenkamp and M. J. M. de Jong, Sol.-State Electron. **37**, 551 (1994).

[30] L. W. Molenkamp, A. A. M. Staring, B. W. Alphanaar, H. van Houten, and C. W. J. Beenakker, Semicond. Sci. Technol. **9**, 903 (1994).

[31] T. Ikoma, K. Hirakawa, T. Hiramoto, and T. Odagiri, Sol.-State Electron. **32**, 1793 (1989).

[32] H. Hirakawa, T. Odagiri, T. Hiramoto, and T. Ikoma, in *Proc. Symp. on New Phenomena in Mesoscopic Structures*, eds. S. Namba and C. Hamaguchi, December 1989.

[33] T. Ikoma and T. Hiramoto, in *Granular Nanoelectronics*, eds. D. K. Ferry, J. R. Barker, and C. Jacoboni (Plenum Press, New York, 1991), pp. 255–276.

[34] K. Hirakawa and H. Sakaki, Appl. Phys. Lett. **49**, 889 (1989).

[35] J. C. Wu, M. N. Wybourne, C. Berven, S. M. Goodnick, and D. D. Smith, Appl. Phys. Lett. **61**, 1 (1992).

[36] S. M. Goodnick, J. C. Wu, M. N. Wybourne, and D. D. Smith, Phys. Rev. B **48**, 9150 (1993).

[37] J. Shah, A. Pinczuk, A. C. Gossard, and W. Wiegmann, Phys. Rev. Lett. **54**, 2045 (1985).

[38] W. Pötz and P. Kocevar, in *Hot Carriers in Semiconductors: Physics and Applications* (Academic Press, Boston, 1992), pp. 87–120.

[39] R. J. Brown, M. J. Kellly, M. Pepper, H. Ahmed, D. G. Hasko, D. C. Peacock, J. E. F. Frost, D. A. Ritchie, and G. A. C. Jones, J. Phys. Cond. Matter **1**, 6285 (1989).

[40] G. Timp, R. Behringer, S. Sampere, J. E. Cunningham, and R. E. Howard, in *Nanostructure Physics and Fabrication*, eds. M. A. Reed and W. P. Kirk (Academic Press, Boston, 1989), p. 31.

[41] Y. Takagaki and D. K. Ferry, J. Phys. Cond. Matter **4**, 10421 (1992).

[42] C. Berven, M. N. Wybourne, A. Ecker, and S. M. Goodnick, Phys. Rev. B **50**, 14630 (1994).

[43] J. C. Smith, C. Berven, S. M. Goodnick, and M. N. Wybourne, Physica B **227**, 197 (1997).

[44] J. A. Nixon, J. H. Davies, and H. U. Baranger, Phys. Rev. B **43**, 12638 (1991).

[45] A. Wacker, Phys. Rev. B **49**, 16785 (1994).

[46] D. K. Ferry and H. L. Grubin, in *Solid State Physics*, vol. 49 eds. H. Ehrenreich and H. Turnbull (Academic Press, Boston, 1995) pp. 283–448.

[47] C. Jacoboni and L. Reggiani, Rev. Mod. Phys. **55**, 645 (1983).

[48] D. K. Ferry, M.-J. Kann, A. M. Kriman, and R. P. Joshi, Comp. Phys. Commun. **67**, 119 (1991).

[49] D. K. Ferry, in *Handbook of Semiconductors*, Vol. 1, 2nd ed., ed. P. T. Landsberg (North Holland, Amsterdam, 1992), pp. 1039–78.

[50] I. Prigogine, Acad. Roy. Belg., Bull. Classe Sci. **31**, 600 (1945).

[51] A. Blandin, A. Nourtier, and D. W. Hone, J. Physique **37**, 369 (1976).

[52] J. Rammer and H. Smith, Rev. Mod. Phys. **58**, 323 (1986).

[53] M. Wagner, Phys. Rev. B **44**, 6104 (1991); **45**, 11595 (1992).

[54] G. D. Mahan, in *Quantum Transport in Semiconductors*, eds. D. K. Ferry and C. Jacoboni (Plenum Press, New York, 1992), pp. 101–140.

[55] D. C. Langreth, in *Linear and Nonlinear Electron Transport in Solids*, eds. J. T. Devreese and E. van Doren (Plenum Press, New York, 1976).

[56] P. Lipavský, V. Špička, and B. Velický, Phys. Rev. B **34**, 3020 (1986).

[57] M. J. McLennan, Y. Lee, and S. Datta, Phys. Rev. B **43**, 13846 (1991).

[58] Y. Lee, M. J. McLennan, G. Klimeck, R. K. Lake, and S. Datta, Superlatt. Microstruc. **11**, 137 (1992).

[59] T. Ando, A. B. Fowler, and F. Stern, Rev. Mod. Phys. **54**, 437 (1982).

[60] F. Stern and S. Das Sarma, Phys. Rev. B **30**, 840 (1984).

[61] F. Stern and W. E. Howard, Phys. Rev. **163**, 816 (1967).

[62] J. J. Thomson, Proc. Cambridge Phil. Soc. **11**, 1120 (1901).

[63] K. Fuchs, Proc. Cambridge Phil. Soc. **34**, 100 (1938).

[64] E. H. Sondheimer, Adv. Phys. **1**, 1 (1952).

[65] J. M. Ziman, *Electrons and Phonons* (Clarendon Press, Oxford, 1960).

[66] E. Prange and T. Nee, Phys. Rev. **168**, 779 (1968).

[67] J. Mertsching and H. J. Fishbeck, Phys. Stat. Sol. **41**, 45 (1970).

[68] Z. Tešanovic, M. Jaric, and S. Maekawa, Phys. Rev. Lett. **57**, 2760 (1986).

[69] N. Trivedi and N. W. Ashcroft, Phys. Rev. B **38**, 12298 (1988).

[70] R. F. Greene, in *Molecular Processes on Solid Surfaces*, eds. E. Draugle, R. D. Gretz, and R. J. Jaffee (McGraw-Hill, New York, 1969), p. 239.

[71] A. V. Chaplik and M. V. Entin, Sov. Phys.–JETP **28**, 514 (1969).

[72] M. Entin, Sov. Phys.–Sol. State **11**, 7 (1969).

[73] S. M. Goodnick, R. G. Gann, D. K. Ferry, and C. W. Wilmsen, Surf. Sci. **113**, 233 (1982).

[74] A. Gold, Sol. State Commun. **60**, 531 (1986).

[75] A. Gold, Phys. Rev. B **35**, 723 (1987).

[76] A. Gold, Phys. Rev. B **38**, 10798 (1989).

[77] A. Gold, Sol. State Commun. **70**, 371 (1989).

[78] G. Fishman and D. Calecki, Phys. Rev. Lett. **62**, 1302 (1989).

[79] H. Sakaki, T. Noda, K. Hirakawa, M. Tanaka, and T. Matsusue, Appl. Phys. Lett. **51**, 1934 (1987).

[80] Y. C. Cheng, in *Proc. 3rd Conf. Sol. State Dev., Tokyo, 1971* [Suppl. to J. Jpn. Soc. Appl. Phys. **1**, 173 (1972)].

[81] S. M. Goodnick, D. K. Ferry, C. W. Wilmsen, Z. Lilliental, D. Fathy, and O. L. Krivanek, Phys. Rev. B **32**, 8171 (1985).

[82] T. Yoshinobu, A. Iwamoto, and H. Iwasaki, in *Proc. 3rd Intern. Conf. Sol. State Devices and Mater.*, Makuhari, 1993.

[83] R. M. Feenstra, Phys. Rev. Lett. **72**, 2749 (1994).

[84] T. Ando, J. Phys. Soc. Jpn. **43**, 1616 (1977).

[85] M. Saitoh, J. Phys. Soc. Jpn. **2**, 201 (1977).

[86] A. G. Sabnis and J. T. Clemens, *Proc. Intern. Electron Dev. Mtg.* (IEEE Press, New York, 1979), p. 18.

[87] F. Fang and W. E. Howard, Phys. Rev. Lett. **16**, 797 (1966).

[88] S. Das Sarma, in *Quantum Transport in Ultrasmall Devices*, eds. D. K. Ferry, H. L. Grubin, A.-P. Jauho, and C. Jacoboni (Plenum Press, New York, in press).

[89] D. Vasileska, P. Bordone, T. Eldridge, and D. K. Ferry, J. Vac. Sci. Technol. B **13**, 1841 (1995).

[90] S. Kawaji and J. Wakabayashi, Surf. Sci. **58**, 238 (1976); Y. Kawaguchi, T. Suzuki, and S. Kawaji, Sol. State Commun. **36**, 257 (1980).

[91] Kobayashi, cited in Y. Matsumoto and Y. Uemura, Jpn. J. Appl. Phys. Suppl. **2**, 367 (1974).

[92] S. Goodnick, Ph.D. Thesis, Colorado State University, 1983, unpublished.

[93] G. D. Mahan, Phys. Repts. **110**, 321 (1984).

[94] G. D. Mahan, J. Phys. F **13**, L257 (1983); **14**, 941 (1984).

[95] Y. Meir, N. S. Wingreen, and P. A. Lee, Phys. Rev. Lett. **66**, 3048 (1991).

[96] A. P. Jauho, N. S. Wingreen, and Y. Meir, Phys. Rev. B **50**, 5528 (1994).

[97] Y. Meir and N. S. Wingreen, Phys. Rev. Lett. **68**, 2512 (1992).

[98] Y. Meir, N. S. Wingreen, and P. A. Lee, Phys. Rev. Lett. **70**, 2601 (1993).

[99] P. W. Anderson, Phys. Rev. **124**, 41 (1961).

[100] R. Lake and S. Datta, Phys. Rev. B **45**, 6670 (1992).

[101] M. J. McLennan, Y. Lee, and S. Datta, Phys. Rev. B **43**, 13846 (1991).

[102] S. Datta, J. Phys. Cond. Matter **2**, 8023 (1990).

[103] J. R. Barker, J. Phys. C **6**, 2663 (1973).

[104] W. Franz, Z. Naturforsch.**13a**, 484 (1958); L. V. Keldysh, Zh. Eksperim. i Teor. Phys. **34**, 1138 (1958) [Sov. Phys. JETP **7**, 788 (1958)].

[105] D. E. Aspnes, Phys. Rev. **147**, 554 (1966).

[106] A. P. Jauho and J. W. Wilkins, Phys. Rev. B **29**, 1919 (1984).

[107] P. Lipavský, F. S. Khan, A. Kalvová, and J. W. Wilkins, Phys. Rev. B **43**, 6650 (1991).

[108] D. K. Ferry, A. M. Kriman, H. Hida, and S. Yamaguchi, Phys. Rev. Lett. **67**, 633 (1991).

[109] P. Lipavský, F. S. Khan, and J. W. Wilkins, Phys. Rev. B **43**, 6665 (1991).

[110] R. Bertoncini, A. M. Kriman, and D. K. Ferry, Phys. Rev. B **44**, 3655 (1991).

[111] J. R. Barker, J. Physique **C7**, 245 (1981).

[112] R. Bertoncini, A. M. Kriman, and D. K. Ferry, Phys. Rev. B **41**, 1390 (1990).

[113] R. Bertoncini and A.-P. Jauho, Phys. Rev. Lett. **68**, 2826 (1992).

[114] R. Bertoncini and A.-P. Jauho, Phys. Rev. B **44**, 3655 (1991).

[115] R. Bertoncini and A.-P. Jauho, Semicond. Sci. Technol. **7**, B33 (1992).

[116] L. Reggiani, P. Poli, L. Rota, R. Bertoncini, and D. K. Ferry, Phys. Stat. Sol. (b) **168**, K69 (1991).

[117] J. R. Barker, in *Quantum Transport in Submicron Devices*, eds. D. K. Ferry, H. L. Grubin, C. Jacoboni, and A.-P. Jauho (Plenum Press, New York, 1995).

[118] J. R. Barker, Sol. State Commun. **32**, 1013 (1979).

[119] J. R. Barker and D. Lowe, J. Physique **C7**, 293 (1981).

[120] Y.-K. Hu, S. K. Sarker, and J. W. Wilkins, Phys. Rev. B **39**, 8468 (1989).

[121] K.-S. Yi, A. M. Kriman, and D. K. Ferry, Semicond. Sci. Technol. **7**, B316 (1992).

[122] J. Lindhard, K. Dan. Vidensk. Selsk. Mat. Fys. Medd. **28**, 8 (1954).

[123] U. Fano, Rev. Mod. Phys. **29**, 74 (1957).

[124] E. Wigner, Phys. Rev. **40**, 749 (1932).

[125] G. A. Baker, Phys. Rev. **109**, 2198 (1958).

[126] J. E. Moyal, Proc. Cambridge Phil. Soc. **45**, 99 (1949).

[127] T. B. Smith, J. Phys. A **11**, 2179 (1978).

[128] A. Janussis, A. Streklas, and K. Vlachos, Physica **107A**, 587 (1981).

[129] R. Zwanzig, J. Chem. Phys. **33**, 1338 (1960).

[130] H. Mori, Prog. Theor. Phys. **34**, 399 (1966).

[131] W. Kohn and J. M. Luttinger, Phys. Rev. **108**, 590 (1957).

[132] P. N. Argyres, in *Lectures in Theoretical Physics*, Vol. 8-A (Univ. Colorado Press, Boulder, 1966), p.183.

[133] R. P. Feynman, *Statistical Physics, A Set of Lectures* (Benjamin/Cummings, Reading, MA, 1972).

[134] W. R. Frensley, J. Vac. Sci. Technol. B **3**, 1261 (1985).

[135] A. Groshev, Phys. Rev. B **42**, 5895 (1990).

[136] U. Ravaioli, M. A. Osman, W. Pötz, N. C. Kluksdahl, and D. K. Ferry, Physica **134B**, 36 (1985).

[137] N. C. Kluksdahl, W. Pötz, U. Ravaioli, and D. K. Ferry, Superlatt. Microstruc. **3**, 41 (1987).

[138] P. Carruthers and F. Zachariesen, Rev. Mod. Phys. **20**, 245 (1983).

[139] H. Steinruck, Z. Ang. Math. Phys. **42**, 471 (1991).

[140] A. M. Kriman, N. C. Kluksdahl, and D. K. Ferry, Phys. Rev. B **36**, 5953 (1987).

[141] A. Szafer, A. M. Kriman, A. D. Stone, and D. K. Ferry, unpublished.

[142] N. C. Kluksdahl, A. M. Kriman, D. K. Ferry, and C. Ringhofer, Phys. Rev. B **39**, 7720 (1989).

[143] R. Landauer, Z. Phys. B **68**, 217 (1987).

[144] W. R. Frensley, Rev. Mod. Phys. **62**, 745 (1990).

[145] B. Engquist and A. Majda, Math. Comput. **31**, 629 (1977).

[146] C. Ringhofer, D. K. Ferry, and N. C. Kluksdahl, Trans. Theory Stat. Phys. **18**, 331 (1989).

[147] H. Tsuchiya, M. Ogawa, and T. Miyoshi, IEEE Trans. Electron Dev. **38**, 1246 (1991).

[148] H. Tsuchiya, M. Ogawa, and T. Miyoshi, Jpn. J. Appl. Phys. **30**, 3853 (1991).

[149] H. Tsuchiya, M. Ogawa, and T. Miyoshi, IEEE Trans. Electron Dev. **39**, 2465 (1992).

[150] H. Tsuchiya, M. Ogawa, and T. Miyoshi, Jpn. J. Appl. Phys. **31**, 745 (1992).

[151] R. Landauer, J. Phys. Cond. Matter **1**, 8099 (1989).

[152] M. P. Shaw, H. L. Grubin, and P. Solomon, *The Gunn-Hilsum Effect* (Academic Press, New York, 1979).

[153] F. W. Sheard and G. W. Toombs, Appl. Phys. Lett. **52**, 1228 (1988).

Index

.

PHYSICS LIBRARY

1-MONTH